HANDBOOK OF NEUROCHEMISTRY

VOLUME V

METABOLIC TURNOVER
IN THE NERVOUS SYSTEM

PART B

HANDBOOK OF NEUROCHEMISTRY

Edited by Abel Lajtha

HANDBOOK OF NEUROCHEMISTRY

Edited by Abel Lajtha

New York State Research Institute
for Neurochemistry and Drug Addiction
Ward's Island
New York, New York

VOLUME V

METABOLIC TURNOVER IN THE NERVOUS SYSTEM

PART B

ℙ SPRINGER SCIENCE+BUSINESS MEDIA, LLC

Library of Congress Catalog Card Number 68-28097

ISBN 978-1-4615-7171-1   ISBN 978-1-4615-7169-8 (eBook)
DOI 10.1007/978-1-4615-7169-8

© 1971 Springer Science+Business Media New York
Originally published by Plenum Press, New York in 1971

Softcover reprint of the hardcover 1st edition 1971

Contributors to this volume:

William A. Brodsky — Departments of Physiology and Biophysics, Mount Sinai Medical and Graduate Schools of the City University of New York, New York (pages 645 and 683)

R. R. Fritz — Division of Molecular Biology, Department of Pediatrics, The University of Texas Medical Branch, Galveston, Texas (page 439)

Robert L. Herrmann — Department of Biochemistry, Boston University School of Medicine, Boston, Massachusetts (page 481)

A. Lajtha — New York State Research Institute for Neurochemistry and Drug Addiction, Ward's Island, New York (pages 49 and 551)

N. Marks — New York State Research Institute for Neurochemistry and Drug Addiction, Ward's Island, New York (pages 49 and 551)

A. V. Palladin — Institute of Biochemistry of the Ukrainian Academy of Science, Kiev, U.S.S.R. (page 489)

N. M. Poljakova — Institute of Biochemistry of the Ukrainian Academy of Science, Kiev, U.S.S.R. (page 489)

D. A. Rappoport — Division of Molecular Biology, Department of Pediatrics, The University of Texas Medical Branch, Galveston, Texas (page 439)

Isaac Schenkein — Irvington House Institute, Department of Medicine, New York University School of Medicine, New York, New York (page 503)

Irving L. Schwartz — Departments of Physiology and Biophysics, Mount Sinai Medical and Graduate Schools of the City University of New York, and The Medical Research Center, Brookhaven National Laboratory, Upton, New York (pages 645 and 683)

W. N. Scott   The Departments of Physiology, Biophysics, and Ophthalmology, Mount Sinai Medical and Graduate Schools of the City University of New York, New York; The Institute for Medical Research and Studies, New York, New York; and The Medical Research Center, Brookhaven National Laboratory, Upton, New York (page 683)

Adil E. Shamoo   Departments of Physiology and Biophysics, Mount Sinai Medical and Graduate Schools of the City University of New York (page 645)

J. R. Smythies   Department of Psychiatry, University of Edinburgh, and Neuroscience Program, University of Alabama (page 631)

Louis Sokoloff   Section on Developmental Neurochemistry, Laboratory of Cerebral Metabolism, National Institute of Mental Health, United States Department of Health, Education and Welfare, Public Health Service, Bethesda, Maryland (page 525)

H. R. Wyssbrod   The Departments of Physiology, Biophysics, and Ophthalmology, Mount Sinai Medical and Graduate Schools of the City University of New York, New York; The Institute for Medical Research and Studies, New York, New York; and The Medical Research Center, Brookhaven National Laboratory, Upton, New York (page 683)

S. Yamagami   Division of Molecular Biology, Department of Pediatrics, The University of Texas Medical Branch, Galveston, Texas (page 439)

PREFACE

Volume V deals with the problems of turnover in the nervous system. "Turnover" is defined in different ways, and the term is used in different contexts. It is used rather broadly in the present volume, and intentionally so. The turnover of macromolecules is only one aspect; here "turnover" indicates the simultaneous and coordinated formation and breakdown of macromolecular species. The complexities of cerebral protein turnover are shown in a separate chapter dealing with the synthesis of proteins, in another on breakdown, and in still another on the relationship of these two (showing how the two halves of turnover are controlled). The fact that most likely the two halves of protein turnover, synthesis and breakdown, are separated spatially and the mechanisms involved are different further emphasizes the complexity of macromolecular turnover.

"Turnover" is used in a different context when the turnover of a cycle is discussed; but here again a number of complex metabolic reactions have to be interrelated and controlled; some such cycles are discussed briefly in this volume, additional cycles have been discussed with metabolism, and some cycles still await elucidation or discovery.

A different type of reaction which can also be interpreted as "turnover" is the exchange of a compound in the brain for another identical or similar one from outside this tissue, with no prior metabolic transformation of the compound to be replaced. Replacement via exchange involves processes of diffusion and transport; these are also treated in the present volume, along with a fairly detailed theoretical background for them, since they have special importance in the nervous system, which represents perhaps the highest complexity of membranes and is perhaps the richest in membrane structures among all the organs.

Other subjects discussed in Volume V are processes that often are not strictly considered part of the processes of turnover, such as hormonal factors affecting metabolism, growth, and development.

Some years ago the question could be asked whether there is turnover of the major components of the nervous system, especially because this organ functions as storage of memory. Since it is a depository for information of a permanent nature, it would not be unreasonable to assume that most of its structures are permanent. More recent studies, some of which are reviewed in the present volume, have shown, however, that not only does metabolic activity of all kinds proceed with great rapidity and intensity in the nervous system, but the turnover of most compounds occurs at fairly high rates, so that one can assume that only a very small portion of this organ is permanent.

What portion of this activity—turnover, metabolism, exchange, etc.—is specific for the brain and not present in other organs is a difficult question to answer at the present time, although there are a number of compounds and perhaps specific metabolic reactions that are present only in nervous structures and no other elements in the organism. Certainly most components and most reactions present in the brain are also present in other organs; this perhaps shows the unity of life processes. A somewhat more pertinent question, and one that is easier to answer, is whether the processes that one can study in nervous tissues are the ones responsible for the function of the brain. The more we learn about the nervous system, the more we find that its functions, its control processes, and its malfunctions can be understood in terms of biology and that the reactions that are described in the present volume have a great deal to do with the requirements and functions of the nervous system. As perhaps also in other areas, the more we know, the better we understand, the more remains to be found out and understood. This, however, should not—and, I am sure, does not—in any way discourage present and future workers in this field. What is really shown here is that even formidable problems are not unapproachable and their solution will not escape us forever. The rapid growth of investigations in this area clearly emphasizes that the hope of understanding and the importance of acquiring knowledge in this area are recognized increasingly by many.

Abel Lajtha

New York, New York
July 1971

CONTENTS OF PART B

Chapter 18

Chapter 19

Chapter 20

Chapter 21

CONTENTS OF PART A

Chapter 9

The Tricarboxylic Acid Cycle 283
by Sze-Chuh Cheng

Chapter 10

Metabolism and Function in Nerve Fibers...................... 317
by P. Greengard and J. M. Ritchie

Chapter 13

DEVELOPMENT

D. A. Rappoport, R. R. Fritz, and S. Yamagami*

Division of Molecular Biology
Department of Pediatrics
The University of Texas Medical Branch
Galveston, Texas

I. INTRODUCTION

In spite of the complexity of the brain structure and the growth patterns of its diverse cells and their processes, each cell is the product of the sequential activation and repression of genomes which uniquely determine its protein components.[1–5] In general, cell growth can be defined as that phase during which there is a net accrual of proteins. At maturation, when growth has ceased, cells no longer accrue proteins, although turnover of protein continues.

Recent studies[1,3] on the embryogenesis of various organisms have partially clarified the biochemical events that occur during cell proliferation, differentiation, and enlargement, and have shown that these events are the result of variations in the kinds and amounts of protein biosynthesized under temporal genetic control. In this report, an attempt will be made to describe briefly the changes in the biochemistry of the central nervous system during embryogenesis, fetal, and postnatal growth.

Spector[6] and Himwich[7] recently reviewed the biochemical aspects of the development of the central nervous system, while some of the morphological changes during development of the central nervous system were reviewed by Skoglund[8] and Wechsler.[9]

The growth patterns as seen in the emergence of brain function vary among species, as noted in a study of the various stages of brain growth.[7] Some investigators[10–12] have found very little muscular coordination and no rhythmic electrical activity in the brain of newborn rats. On the other hand, the newborn guinea pig is self-sufficient at birth and responds to a variety of stimuli, as evidenced by complex rhythmic electrical activity

* Present address: Department of Neuropsychiatry, Osaka City University Medical School, Osaka, Japan.

in the brain.[12,13] The development of the newborn human infant is somewhat intermediate between that of the newborn rat and the guinea pig.[7]

McIlwain[14] categorizes the development of the brain, based on the morphology, function, and chemical composition, in four stages as follows:

Stage I — Cell Proliferation and Differentiation. At the end of this first period the number of brain cells almost reaches adult values. Functionally, no nerve impulses are transmitted in the brain during this period, which lasts until the birth of the rat or approximately three-quarters of gestation in the guinea pig and man.

Stage II—Cell Growth and Differentiation. In this period, there is an increase in size of the brain cells and an outgrowth of axons in dendritic connections. In the rat, this stage extends to 10 days after birth and in man, it lasts until birth.

Stage III—Rapid Myelination. In this phase, the overall growth of the brain decreases but there is very rapid formation of myelin. This represents the period from the twelfth day to the twenty-fifth day after birth in the rat, from 46 days gestation to birth in the guinea pig, and from birth to 120 days of age in man.

Stage IV—Maturation. During this stage, myelination is completed and the brain attains its full morphological and functional maturity.

II. CELL PROLIFERATION AND DIFFERENTIATION

During development, proliferation, differentiation, and enlargement occur in a temporal sequence in which the homologous primordial cells originating from the zygote evolve successively into the blastula, gastrula, and finally into the highly differentiated fetal organism which ultimately matures postnatally to a functional organism. During the blastular and gastrular stages, the formation of the essential biopolymers during cellular proliferation go through a cyclic activation and suppression in an orderly sequence as depicted in Fig. 1. Although cell mitosis occurs with decreasing frequency in the brain as the organism approaches maturation, the sequence of biochemical events in mitosis remains the same. Atlas and Bond[15] have noted that the biochemical changes associated with gene activation and repression during mitosis require an interval of only 1 hr in the total cell replicating interval of 11 hr. The daughter cells then enter the G_1 stage in which RNA and protein syntheses are active.[16] In this stage, where neither DNA nor histones are synthesized, the cell is preparing for reinitiation of DNA synthesis.[16] In entering the S stage ($5\frac{1}{2}$ hr duration), DNA and histone syntheses are initiated in the nucleus and in the cytoplasm respectively.[17] The histones, which are part of the chromatin material, are synthesized by the microsomal fraction in the cytoplasm and must be transferred to the nucleus.[18] Recent studies[19–21] show that the synthesis of DNA occurs stepwise; DNA of 8–10S is formed first, followed by an intermediate size of 24–30S, and finally the characteristic DNA found in the chromatin

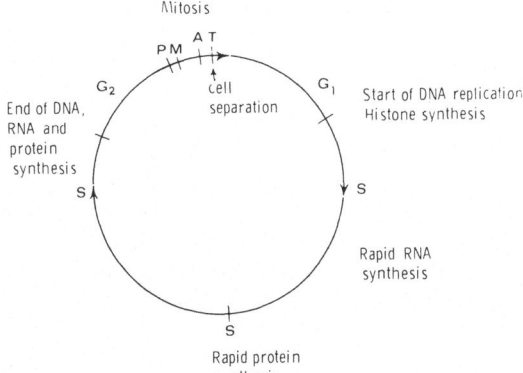

Fig. 1. Schematic representation of the cell replication cycle and the corresponding changes in biopolymer synthesis. (From Stubblefield.[197])

is formed. The smaller DNA units are ligated to form larger molecules by a separate process.[21]

Studies by Gallwitz and Mueller[17] have shown that histone synthesis in HeLa cells *in vivo* is dependent on the synthesis of DNA; however, the mode of dependency is unknown. Gurley and Hardin[22] suggest that some histone turnover does occur when DNA synthesis is blocked by thymidine in synchronized cultures of Chinese hamster cells. All five histone fractions are produced simultaneously.[22]

The synthesis of RNA in the G_1 and S phases includes the formation of rRNA, tRNA, and mRNA. Among the RNA's in the microsomes from synchronized HeLa cells at mid-S stage, Gallwitz and Mueller (unpublished data) found an 8S RNA which they suggest may be the mRNA for histones.[21] Kedes and Gross[23] have identified three species of RNA's synthesized in the nuclei of the cleaving sea urchin embryo cells which sediment between 9 and 10S and have been tentatively identified as mRNA's for thee classes of histones. It is to be noted that the latter authors[23] found histone synthesis in the nucleus of sea urchin eggs, while Gallwitz and Mueller[17] and Mueller[21] report histone synthesis only in the cytoplasm of HeLa cells.

In contrast to the histones, the nonhistone proteins in the chromatin are continuously synthesized through the S phase, independent of DNA synthesis. It was also found that the rate of labeled amino acid incorporation into nonhistone proteins increases during the S phase. Separation of nonhistone proteins on polyacrylamide gels has revealed many bands (at least 60). The nature of these proteins, formed during active and blocked synthesis of DNA, was strikingly similar.[21]

Active, continuous synthesis of RNA and protein through the latter portion of the S stage is required for subsequent cell division during the G_2 stage.[24] In the G_2 stage, some protein but no DNA synthesis occurs preliminary to mitosis.[25] What initiates the synthesis of DNA in the beginning of the S phase and what factors stop the synthesis at the end of the S phase are unknown, but it is assumed that both events are under gene control.

The enzymes which may play a primary role in the control of DNA synthesis are thymidylate and thymidine kinases since thymidine triphosphate is essential for this synthesis.[26,27] It is conceivable that the formation of the thymidine kinase in the G_1 phase as an essential protein is one of the steps prerequisite to the initiation of the S phase. Similarly, the disappearance of the thymidine and thymidylate kinases can explain the cessation of DNA replication at the beginning of the G_2 phase. In this manner, the appearance and disappearance of thymidine and thymidylate kinase activities can be visualized as a means of mitotic regulation through the periodic activation of the gene which transcribes the mRNA for this enzyme.[28-30]

A. Early Differentiation

The formation of three germinal layers, the ectoderm, mesoderm, and endoderm during the late gastrular stage of the embryo precedes the initiation of neural primordia. The ectodermal layer thickens and indents slightly to form a neural plate which subsequently folds to a neural groove. Later the edges of the groove converge and fuse to form the neural tube (Fig. 2). Experimental embryologists have reported that certain components within the embryonic cells are capable of inducing formation of neural elements (forebrain, hindbrain) from the ectoderm, and trunk elements (muscle, gut epithelia) from the mesoderm.[31] These inducing factors or evocators have been isolated by Tiedemann[31] from chick embryos, and identified as microsomal bound proteins. When chick embryos were subjected to phenol extraction, the insoluble protein found at the interphase between phenol and water induced neuralization of the ectoderm to form forebrain and hindbrain in amphibian embryo. By addition of methanol to the phenol phase, a protein precipitate was obtained which induced formation of tail and trunk structures in amphibian embryos. The latter, a mesodermal inducer, was purified and found to be a protein with a molecular weight of 25,000. However, a neural inducing factor has not yet been purified. Yamada[32,33] found that these

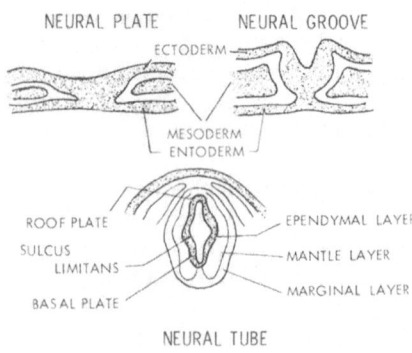

Fig. 2. Schematic representation of changes in embryonic rudiments as the neural plate evolves to the neural tube.

inducers are present not only in amphibian and chick embryos, but also in adult mammalian tissues such as marrow and liver.[34,35]

Initiation of differentiation of the neuroepithelial cells into neuroblasts (primordial neurons) and spongioblasts (primordial glial cells) occurs at the closure of the neural groove.[36–38] Using tritiated thymidine, Langman[36] and Altman[38] showed that neuroblasts do not incorporate thymidine, indicating that these cells do not synthesize DNA and cannot further divide, a recognized characteristic of differentiated neurons.

III. CHANGES IN DNA

In the gestational and postnatal periods of brain growth, an increase in DNA, which is an indication of cell proliferation, is observed in the rat up to 10 days of age.[39] The rate of DNA synthesis is maximal during gestation.[40] The enzyme responsible for DNA synthesis, designated as DNA-dependent DNA polymerase or DNA nucleotidyltransferase, has been studied in the chick embryo by Margolis[40] and in the rat brain by Bharucha and Murthy.[41] Margolis[40] found that the DNA polymerase activity was high in the 12-day chick embryo, decreased on the fourteenth day, increased on the fifteenth day, and then showed a progressive decrease up to hatching. He found the greatest activity in the cerebellar regions of the fetal brain and approximately one-third of this activity in the cerebral areas. After hatching, the cerebellar DNA polymerase remained at a relatively constant but low activity up to the seventh day, while the cerebral enzyme continued to decrease. The results also indicate that the DNA content increased continuously in the cerebellum and decreased in the cerebrum and optic lobes. After hatching, there was an apparent decrease in the DNA in all tissues, primarily because of an increase in the size of the brain. This, however, indicates that the DNA actually remains constant, while tissue mass increases by cell enlargement.[40]

A study of the DNA polymerase in rat brain during postnatal growth[41] showed that there were three types of DNA polymerases in rat brain; one in the soluble nuclear extract, another in the insoluble nuclear fraction, and a third in the soluble cytoplasm of the brain cells. This study showed the highest polymerase activity at birth, which subsequently decreased continuously to the sixth week and remained at this low level through the twelfth week postnatally.[41] Sung[42] studied the in vitro formation of DNA using slices and cell-free extracts from 6-day-old rat brain and found that the rate of DNA synthesis was 20 times higher in the cerebellum than in the cortex and 10 times higher than in the olfactory bulbs. Using slices of cerebellum from rats at different ages, he found that maximum brain DNA formation was maintained to the sixth day and then decreased rapidly to 10% of the maximum value at 18 days.[42] Changes in DNA during brain development reported in various species are summarized in a previous volume of this series by Rappoport, Fritz, and Myers.[43]

IV. CHANGES IN RNA

The RNA in the transcription of genetic information (mRNA) as components of ribosomes (rRNA), and as an intermediate in the activation of amino acids (acyl tRNA) is responsible for the active synthesis of new proteins during fetal and postnatal development.[44] Although new protein synthesis ceases at maturation, RNA synthesis continues at a diminished rate and a rapid turnover persists throughout the life span of the organism.[45–48] In the brain of the fetal rat and in the early postnatal period, the formation of new glial elements and the extension of axons, as well as the proliferation of dendritic processes, reflect the active and continuous formation of RNA as well as other cell constituents.[47,49] The composition and rate of formation of RNA's in glia and neurons in adult brain are different.[50] The incorporation of bases into glial RNA is twice as great as in the neurons isolated from rabbit hypoglossal cells.[50] Most of the RNA is synthesized in two compartments within the cell nuclei. The $17S$, $28S$, and a $4S$ rRNA are synthesized in the nucleoli from a $45S$ RNA precursor.[47–49,51,52] The chromatin is the second compartment in which at least three types of RNA's are formed, mRNA and tRNA which ultimately are extruded into the cytoplasm, and an RNA with a DNA-like base sequence found in the chromatin.[53,54] This indicates that RNA polymerase is present in nucleoli, where it catalyzes formation of rRNA, and is also associated with chromatin, where it is responsible for the synthesis of both mRNA and chromosomal RNA.[55] Barondes[56] and Furusawa and Rappoport[57] found that the activity of RNA polymerase in the rat was maximal at birth, decreased up to the age of 30 days, and remained at the minimal level in the adult brain (Fig. 3).

The enzyme responsible for the synthesis of RNA is known to exist in the soluble nucleoplasm, nucleoli, and in the chromatin aggregate. Recent studies on DNA-dependent RNA polymerase from rat liver indicate the existence of at least two distinct RNA polymerases; one localized in the nucleoli and another in the nucleoplasm.[55] Roeder and Rutter[55] separated the two types of rat liver RNA polymerases from the solubilized extracts of

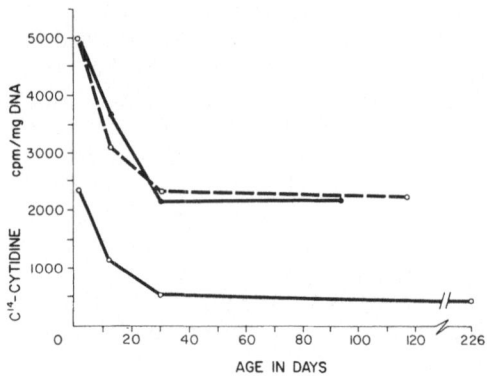

Fig. 3. Changes in RNA polymerase activity in the growing rat brain determined by CTP-2-^{14}C incorporation into RNA in isolated rat brain nuclei. The above curves represent three different experiments.

sonicated rat liver nuclei. Because of the compartmentalization, these findings suggest that the nucleolar RNA polymerase, which is responsible for the formation of rRNA's, can be regulated independently from other RNA polymerases in the nucleoplasm and in the chromatin. This may account for the independent formation of large quantities of rRNA which are acquired during the growth of a tissue.[55] These findings are not unique in rat liver nuclei. Roeder and Rutter[55] also showed the presence of three distinct RNA polymerases in the sea urchin gastrula. The nucleolar RNA polymerase was found to increase relative to another type of RNA polymerase during development of the sea urchin from the gastrular to the blastular stages. These findings demonstrate that there are multiple forms of DNA-dependent RNA polymerase in eukaryotic organisms.[55]

The precursors for RNA synthesis, nucleoside triphosphates, were shown by Mandel and Edel-Harth[58] to vary somewhat during growth of the rat brain as shown in the table below:

Nucleoside Triphosphates in Growing Rat Brain
(μmoles/100 g wet weight)

	1	2	7	14	21	28	35	Ad
ENTP	287	210	221	255	278	281	287	242
ATP	190	146	165	184	210	216	223	200
GTP	35	24	30	32	36	34	34	24
UTP	50	33	30	34	29	29	28	18
CTP	13	7	5.5	56	2.4	1.3	3.1	—

It can be seen that ATP constitutes 70% of the total nucleotide triphosphates during growth of the brain. Itoh and Quastel[59] recently reported that the level of ATP in both the infant and adult rat brain is one of the factors in the control of RNA synthesis.

Injection of orotic acid-6-^{14}C into the cisterna magna of rats at 16 and 90 days of age, and subsequent extraction of nuclear and microsomal RNA at different intervals, showed that nuclear RNA synthesis in the 16-day rat brain rapidly reaches a plateau in 3 hr and remains at approximately this level for the next 17 hr (Fig. 4) after injection.[60] During this interval, micro-

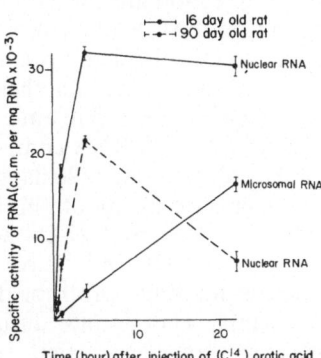

Fig. 4. Time course of incorporation of orotic acid-6-^{14}C injected into the brains of rats 16 and 90 days of age. At the indicated time intervals postinjection, the nuclear and microsomal RNA's were extracted and the specific activity of the newly formed RNA's determined.

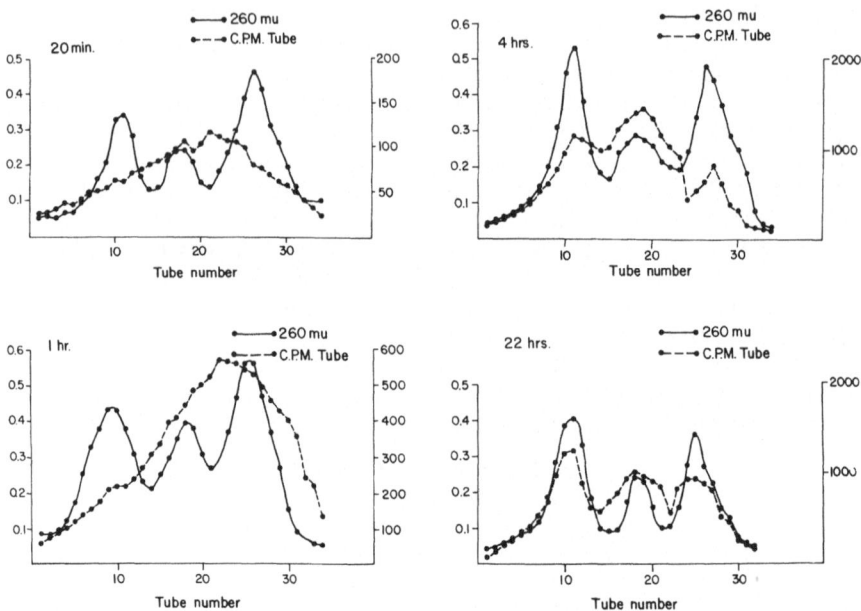

Fig. 5. Changes in the sucrose gradient profiles of newly formed brain nuclear RNA's from injected orotic acid-6-^{14}C into 16-day rats. The rats were sacrificed at 20 min, 1, 4, and 22 hr postinjection and the brain nuclear RNA was isolated and subjected to sucrose gradient centrifugation. The RNA fractions were isolated and the radioactivity and OD determined. The $S_{20,w}$ was estimated for each RNA fraction from its position in the sucrose gradient.

somal RNA in the 16-day rat brain incorporates orotic acid-^{14}C at a much slower rate (Fig. 4). Brain nuclei from 90-day rats rapidly incorporate the orotic acid within the first 3 hr after injection, but this is followed by a progressive decrease in orotic acid incorporation during the next 17 hr (Fig. 4). Nuclear RNA profiles made at various time intervals following orotic acid-6-^{14}C injection into the cisterna magna of both 16 and 90-day rats showed a change in the rate of formation of various RNA's during the first hour compared to the fourth and twenty-second hours (Fig. 5). Thus, in the 16-day rat brain, in the first 20 min, there was the appearance of radioactive RNA (14–18S) which increased in amount the first hour. At 4 and 22 hr after injection, 28S, 18S, and 9S RNA components were clearly visible, as shown in Fig. 5. In the older rat brain (90 days), a 16S peak appeared in 20 min after injection and increased in amount in the first hour in a manner similar to the 16-day rat brain. At 4 and 22 hr the 4S, 18S, and 28S RNA's appeared (Fig. 6). Egyházi and Hydén[51] have shown, however, that in the lateral vestibular nucleus in rabbits, orotic acid-^{3}H was incorporated into a 16S nuclear RNA in 15 min. They found, during the next 45 min, a change in synthesis of heterogeneous RNA's that resulted in a maximal size of 30S. After 3 hr, they

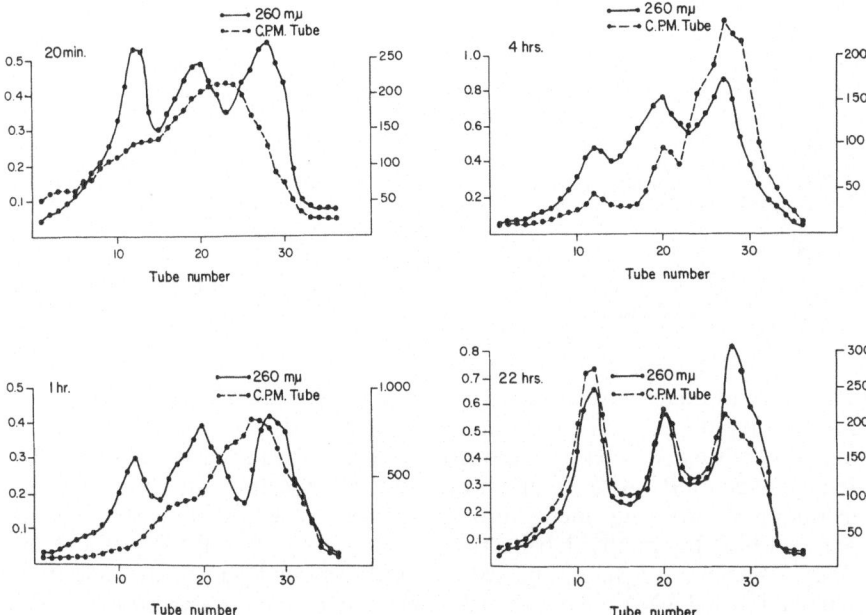

Fig. 6. Changes in the sucrose gradient profiles of newly formed brain nuclear RNA's from injected orotic acid-6-^{14}C into 90-day rats. The rats were sacrificed at 20 min, 1, 4, and 22 hr postinjection, and the brain nuclear RNA was isolated and subjected to sucrose gradient centrifugation. The RNA fractions were isolated and the radioactivity and OD determined. The $S_{20,w}$ was estimated for each RNA fraction from its position in the sucrose gradient.

noted a stable profile of 18 and 28S RNA. At 30 min, cytoplasmic RNA (microsomal RNA) had a gradient profile similar to the 15-min profile of nuclear RNA. Maximal radioactivity in cytoplasmic RNA appeared in the 10–12S region.[51]

In order to estimate the amount of "mRNA-like" activity present in various nuclear RNA fractions in young (10 day) and adult (90 day) rat brains, nuclear RNA was isolated and subjected to sucrose gradient fractionation (Fig. 7). Four fractions in each group were isolated (Table I). Each nuclear RNA fraction was added to ribosomal preparations from 16- and 90-day rat brains, respectively, and L-phenylalanine-1-^{14}C incorporation into protein was measured (Table I). It can be seen that fractions 3 and 4 (Table I) enhance the incorporation of phenylalanine into ribosomal protein to a greater extent than fractions 1 and 2, in both 16- and 90-day ribosomal preparations. However, the nuclear RNA fractions from the 16-day rat brain had a somewhat greater influence than the same fractions from 90-day rat brains (Table I).

Dutton et al.[61] investigated the labeling pattern of nuclear RNA in rat cerebral cortex (75–100 g rats) and found that, within the first hour,

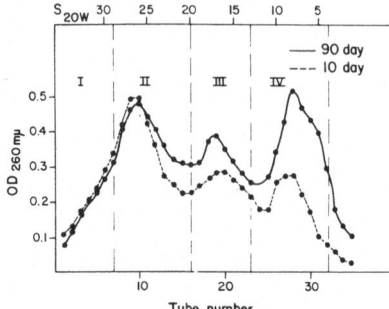

Fig. 7. Sucrose gradient profiles of nuclear RNA's extracted from 10- and 90-day rat brains and separated by centrifugation on a sucrose gradient. The $S_{20,w}$ values were calculated from the position of the RNA bands in the sucrose gradient.

maximum radioactivity appeared in the fraction between the 28 and 50S, and shortly thereafter in the fraction greater than 50S. However, the 5S fraction showed some early radioactivity. They also found labeled 18S and 28S ribosomal RNA 2 hr after intracerebral injection of ^{14}C-methyl-L-methionine, indicating methylation of rRNA's in contrast to the first 40 min in which the methyl label was observed only in the tRNA or the 4–6S RNA's. Stevenin et al.[62] injected ^{32}P intracisternally into rats and then isolated labeled RNA greater than 30S which had the base sequence of DNA from both the nuclei and microsomes.

TABLE I

Enhanced Incorporation of L-Phenylalanine-1-^{14}C by Rat Brain Nuclear RNA into Brain Ribosomal Protein

Added RNA	S_{20}	Ribosome and pH 5 enzyme (cpm/mg protein)	
		16-day	90-day
A. Nuclear RNA fractions from 10-day rats			
None added		2664	1512
Fraction 1	30	2936 (10)[a]	2124 (41)
Fraction 2	20–29	3628 (36)	2358 (56)
Fraction 3	13–19	3988 (50)	2474 (130)
Fraction 4	4–12	4368 (64)	3713 (146)
B. Nuclear RNA fractions from 90-day rats			
None added		2083	1057
Fraction 1	30	1892 (−10)	962 (−9)
Fraction 2	20–29	2244 (8)	1723 (63)
Fraction 3	13–19	2913 (40)	2099 (92)
Fraction 4	4–12	3021 (45)	2199 (101)

[a] Percentage change based on controls.

V. CHANGES IN PROTEIN

There is a higher rate of amino acid incorporation in young (1–20 days of age) rat brain than in adult (60 days or older) and this observation reflects both quantitative and qualitative changes in the components of the protein synthesizing system such as mRNA, pH 5 fraction (synthetases, transferases, tRNA's), and polysomes. These observations were made from in vitro studies using brain microsomes,[63,64] ribosomes,[47,65] isolated mitochondria,[66,67] tissue slices,[68,69] whose cell suspensions,[70] as well as from in vivo experiments with injected radioactive precursors[60,71] at various stages of postnatal development.

Since the structure of an organ is the result of specific genome expression, certain proteins should be unique to the organ. In this category, at least three groups of proteins have been observed to originate in nerve tissue: Moore's S-100 protein,[72] myelin protein consisting of proteolipids and encephalitogen, and brain antigens. Moore and Perez[72] reported the isolation of an acidic protein, found only in brain, containing 25–30% of glutamic and aspartic acids and a large amount of phenylalanine, and the complete absence of tryptophan. According to Hydén and McEwen,[73–75] the protein is mainly found in the glia. Levine[76] reported that the protein first appears in low concentrations in the 5-day chick embryo and then in increasing amounts up to 10–15 days. Moore and Perez[72] found that in the rat, this protein was absent at birth and appeared in increasing amounts 16 days postnatally, at the time when the rat opened its eyes. The rate of formation of S-100 protein was maximal between the sixteenth and twenty-third day.[72] The appearance of this protein in the brain preceded myelination. This protein is present in 30 vertebrate species including teleost, reptilian, avian, and mammalian organisms, and it is widely distributed in the brain.[72]

The proteolipids make up an appreciable portion of the white matter of the brain but are present in small amounts in the gray matter. The appearance of proteolipids correlates with the formation of myelin and this in turn correlates with the increase in brain cerebrosides. Mokrasch[77] and Klee and Sokoloff[78] state that the proteolipids are synthesized in vitro in the mitochondria-rich fraction. Klee and Sokoloff[79] found that the amino acid composition of the proteolipid fraction from neonatal mitochondria differed in composition from the proteolipids isolated from the adult mitochondria.

The discovery and isolation of a basic protein from myelin by Kies et al.[80] stemmed from the initial finding that injection of myelin proteins into guinea pigs induced experimental allergic encephalomyelitis. This protein (or a monomer of a protein complex) has recently been purified and its amino acid composition determined.[81] Immunofluorescent techniques have established the presence of this protein in myelin and this implies that it first appears at the initiation of myelination in the fetal or newborn animal brain.[82] This encephalitic protein has a range of molecular weights between 10,000 and 40,000 and recent studies have shown that this is probably due to partial proteolytic hydrolysis during isolation.[83] It contains a high concentration of lysine, arginine, glutamic and aspartic acids, but lacks cysteine. Immunological studies showed that these proteins were organ specific but were species nonspecific.[84]

Recently, Dutton and Barondes[85] investigated the incorporation of L-leucine-1-^{14}C into neuronal microtubular proteins in the developing mouse and found that the turnover of microtubular protein in the 5-day mouse brain was at approximately 4 days. They found that this protein represents 40% of the total soluble fraction in the young mouse brain and 15–20% in adult brain.

Wenger[86] and Friedman and Wenger[87] made studies of the brain proteins in the developing chick embryo and compared them to that of the adult. They found that antisera to adult chick brain homogenates reacted with proteins from brain homogenates of 5½-day chick embryos and the presence of the adult type protein increased in the developing embryo, reaching maximal levels on the twelfth day.[86,87] Similar results were obtained by McCallion and Langman[88] who prepared soluble extracts of chick embryos. These authors also noted that extracts from 12-day chick embryos are inhibitory to neuronal differentiation when added to earlier embryos.

Bondy and Perry,[89] using rabbit brain microsome preparations, found that the incorporation of DL-valine-1-^{14}C was twice as high in the fetal brain as in adult brain. They noted that the distribution of radioactive amino acid in the soluble protein fraction was similarly isolated from both *in vitro* and *in vivo* systems and they concluded that this pattern of incorporation was not unique to a particular stage of development.

Murthy and Rappoport[90] reported that microsomal preparations from neonatal rat brains incorporate L-leucine-1-^{14}C at a higher rate than those from adult brains, similar to the findings of Bondy and Perry.[89]

Gelber et al.[91] found that DL-leucine-1-^{14}C incorporation into mixed microsome-mitochondria from immature rat brain was three to four times more rapid than into such preparation from adult brain. When they replaced the mitochondria in this mixture from adult brain with mitochondria from immature brain or by an ATP-generating system, the rate of L-leucine-^{14}C incorporation was elevated to that observed with the microsome-mito-chondria preparation in mature brain. These observations, and those of Itoh and Quastel,[59] indicate that protein synthesis in dependent, among other things, on the generation of ATP and that there is a reduction in ATP generation via oxidative phosphorylation in the adult brain in contrast to the very young brain.

Comparing the protein synthesis in white and gray matter of the adult rat brain, Suzuki et al.[92] found that the white matter was less active than the gray and that this synthesis was greater in white and gray matter in the 12-day rat brain. They also noted that microsomes from young human brains were more active in protein synthesis than microsomes from adults.

Adams and Lim[93] made a comparison of L-valine-^{14}C incorporation by cell-free systems from young and adult rat brains on the basis of the

Michaelis–Menten kinetics. They found that microsomes from young rat brains had a greater capacity for amino acid incorporation than those from the adult. They attributed this difference to a supernatant factor based on the observation that microsomes from young animals, combined with adult soluble fractions, incorporated valine at a rate identical to that of the adult preparations.[93] It is very likely that the lower content of the amino acyl tRNA synthetases in the adult supernatant fractions were in part responsible for the decrease in valine incorporation into the microsomes from young rat brains to the levels of the adult preparations; this is similar to the results with leucine reported by Yamagami et al.[47] in their studies with rat ribosome systems.

Recently Johnson and Belytschko[94] noted that isolated brain microsomes from mice showed a progressive decrease in protein synthesis within the first 10 days after birth. Their findings indicate that none of the essential intermediate components of the protein synthesizing system were rate-limiting, such as microsome or polysome, mRNA, nor the tRNA binding, or the synthetase activities. They suggest that there may be a change in the nature of ribosomal particles.

The apparent decrease in amino acyl RNA synthetase activities may not be due to the actual decrease in the activities of these enzymes but could be attributed to the enzymatic activity responsible for the attachment of the —CCA group on the tRNA's and/or a change in activity of the enzyme which removes this group from the tRNA's. Johnson (personal communication) indicates that there is an 80% decrease in tRNA's with attached —CCA groups when protein synthesis in mouse brain decreases from very young to adult levels. This has not been tested in the rat and thus the active form of tRNA (tRNA—CCA) may also be decreased in the rat brain between the ages of 35–50 days, accounting for the 80% drop in the observed protein synthesis (Fig. 8).

In studies carried out by Rappoport et al.[47,54,95] on the changes in protein synthesis during development of the rat brain using an isolated ribosomal system, they found that there was a high rate of ribosomal protein synthesis in the newborn rat brain up to the age of 35 days and subsequently,

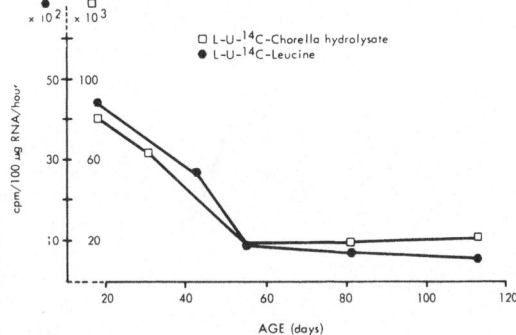

Fig. 8. Incorporation of L-leucine-U-^{14}C and L-amino acid-U-^{14}C mixtures (^{14}C-Chlorella protein hydrolysate) into the soluble RNA (tRNA) of the brain pH 5 fraction isolated from various age groups.

this synthesis decreased to a minimal level in the 50-day rat and remained at this level in the adult. Isolated ribosomes from rat brains at various ages were incubated with the pH 5 fraction prepared from a single age group (30 day) and the degree of L-(U-^{14}C) leucine incorporation into ribosomal protein was markedly reduced when adult brain pH 5 fractions were used. In another set of experiments, ribosomes from a single age group (18-day rats) were incubated with pH 5 fractions from various age groups and the results showed that the ribosomes from the older rats incorporated leucine to a smaller extent than the ribosomes from the very young rat brains. The decrement in protein synthesis, due to some change in pH 5 fraction, was greater than that observed with ribosomes from older animals. Further analysis of the components in pH 5 fractions revealed that the amino acyl RNA synthetase activity was maximal at birth but decreased progressively, reaching a minimal level in the 55-day rat brain and remaining at this level through adulthood (Fig. 8). The activity of the amino acyl transferase enzymes in the pH 5 fraction from rat brains in all the age groups tested was found not to be rate-limiting in respect to leucine uptake by the ribosome.

The conclusion from the above experiments was that in the translation phase of protein synthesis, the decrease in ribosomal protein synthesis in the growing rat brain was mainly due to a decrease in amino acyl RNA synthetases activity and in part due to some changes in the ribosomes (or polysomes).

Additional studies by Yamagami, Fritz, and Rappoport[96] indicated that there was a reduced amount of active polysomes in the developing rat brain (Fig. 9). Examination of the nuclear RNA in its capacity to enhance amino acid incorporation into ribosomal protein showed that there was a marked reduction in "mRNA-like" RNA in the brain during the growth of the rat. This was also noted in the nuclear RNA profiles from analytical ultracentrifugation which showed less 17 and 27S RNA in the adult relative to the profile in young rat brain.[96] Murthy demonstrated that there was a progressive decline in the concentration of polyribosomes in the young and adult rat brain.[46] His conclusions were derived using two approaches: (1) measuring the incorporation of phenylalanine in the presence and absence of poly U, and (2) comparing the sucrose density gradient distribution of

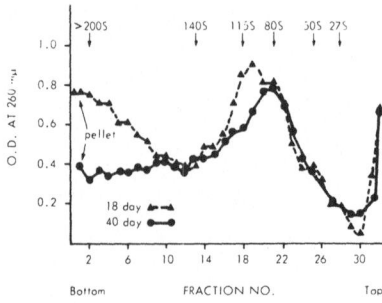

Fig. 9. Sucrose gradient sedimentation profile of ribosomes from 18- and 40-day rat brains, respectively. One milliliter of the above brain ribosomal preparation was layered over 26 ml of a linear sucrose gradient (5–25%), stratified over 3 ml of 50% sucrose on the bottom of the tube and the tubes were centrifuged at 63,500 g for 3 hr in a swinging bucket SW 25.1 rotor at 5°C.

ribosomal preparation which were separately treated with poly U and ribonuclease. This was consistent with the findings of Dellweg et al.[44] where they demonstrated that the number of ribosomes bound in polysome complexes decreased with increasing age of the rat.

Dellweg et al.[44] measured the quantitative and qualitative changes in tRNA's of young and adult rat brains. They showed that the amount of tRNA's increased during development, reaching a maximum at about the tenth week of life. They also demonstrated a second minor valyl-tRNA component in young rat brains in addition to the major valyl-tRNA. This minor component disappeared as the animals advanced to an age of 3 weeks. The other tRNA's tested (leucyl-tRNA, phenylalanyl-tRNA, lysyl-tRNA) did not show any differences in activity in the two age groups studied.[44]

Changes of various enzyme activities during development of the brain have been studied[97]; however, measurements of absolute amounts of enzyme protein have not been carried out. Whether an increase in enzyme activity during development actually signifies an increase in enzyme protein or is due to the action of an activator or a loss in some inhibitor, remains to be shown. In order to demonstrate that there is an actual increase in enzyme protein during development, NADase in the developing rat brain was chosen for this study. The activity of this enzyme is low at birth and increases progressively as the rat brain develops, reaching maximum activity at approximately 33 days of age and remaining constant at this level.[98,99] NADase was isolated from 35–40-day rat brain and purified 800-fold. The purified enzyme was used to generate rabbit antisera which was tested by the Ouchterlony[100] gel diffusion method and by immunoelectrophoresis.[101] Subsequently, the sera was used to determine the amount of NADase in treated homogenates of brains from rats at 10, 20, and 30 days using the radial immunodiffusion technique of Mancini et al.[102] It was found that there was a linear relationship between the increase in NADase activity with the increase in diameter of the radial immunodiffusion precipitant lines indicating an increase in protein in direct proportion to the increase in enzyme activity in the rat brain from 10 to 30 days of age.[99]

It has been generally recognized that cell nuclei are capable of synthesizing protein since isolated nuclei have been shown to incorporate amino acids into protein.[103] Recently, Burdman and Journey[103] have demonstrated protein synthesis in isolated rat brain nuclei. However, histones are known to be synthesized in the cytoplasm and then transferred to the newly synthesized DNA for the formation of chromatin.[17] The nonhistone proteins on the chromatin are presumed to be formed in the nucleus, although clear evidence for this has not been reported. However, reports that these proteins show a rapid turnover[104,105] and the implication that these proteins may play a role in the genetic control of protein synthesis[106–109] has recently aroused interest in regard to the synthesis of chromatin protein during development of the rat brain. In a preliminary study[54] two groups of rats, 10 and 60 days of age, were injected intraperitoneally with L-leucine-U-^{14}C and the appearance of the isotope in nonhistone chromatin protein was

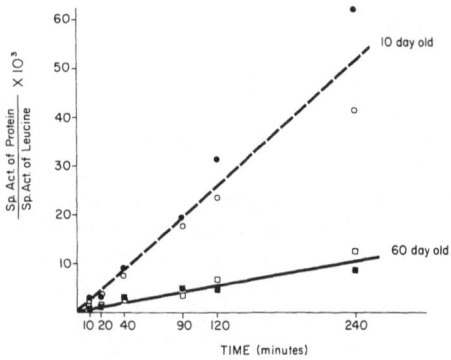

Fig. 10. Changes in relative specific activity of total brain proteins in 10- and 60-day rats during 4 hr after intraperitoneal injection of L-leucine-U-^{14}C. Data corrected for leucine pools in each age group.

followed at various intervals for 2 hr. Both histone and nonhistone protein formation were investigated, and the results are shown in Table II and Figs. 10 through 13. The composition of the chromatin isolated from rat brain from both ages is shown in Table II. The rate of change in the specific activity of total protein in the rat brain from both age groups, following injection of leucine, is shown in Fig. 10 in which the 10-day rats show a much greater

TABLE II

Composition of Rat Brain Chromatin
(weight ratio based on DNA)

Age (days)	Number of animals	DNA	RNA	Histones	Nonhistone protein
10	24	1.00	0.077 ± 0.015[a]	1.03 ± 0.09	0.63 ± 0.07
60	24	1.00	0.046 ± 0.007	1.00 ± 0.13	0.96 ± 0.13

[a] Average deviation from the mean.

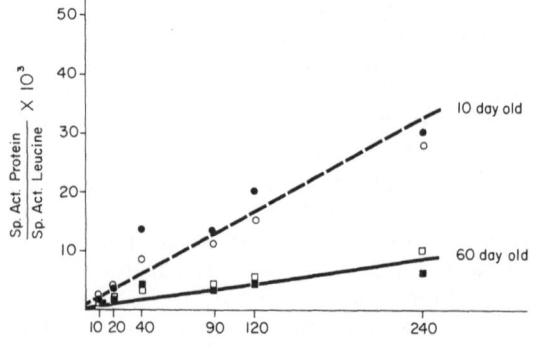

Fig. 11. Changes in relative specific activity of total chromatin proteins (histones plus nonhistone protein) in 10- and 60-day rat brains during 4 hr after intraperitoneal injection of L-leucine-U-^{14}C. Data corrected for leucine pools in each age group.

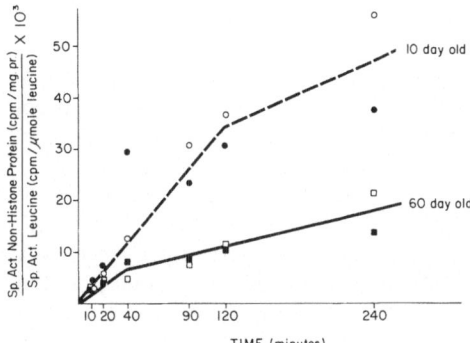

Fig. 12. Changes in relative specific activity of nonhistone proteins in 10- and 60-day rat brain chromatin during 4 hr after intraperitoneal injection of L-leucine-U-[14]C. Data corrected for leucine pools in both age groups.

rate of incorporation than the 60-day brain protein. The leucine pools in the 10 and 60-day rat brain were determined and found to be 0.61 and 0.17 μmoles of leucine per gram wet weight of brain, respectively. The rate of change in the specific activity of total chromatin protein (histones plus nonhistone protein) is shown in Fig. 11. Here too, the chromatin in the 10-day rat brain shows a much higher leucine uptake than that in the 60-day brain chromatin. Figure 12 illustrates that nonhistone protein incorporates leucine to a much greater extent than do the histones in both age groups, and that the nonhistone proteins in the 10-day brain chromatin incorporate leucine to a greater extent than in the 60 day. These results show that the formation of nonhistone protein occurs to a much greater extent (5 times) than the histones in both age groups (Fig. 13). The findings demonstrate the active synthesis of nonhistone proteins in the chromatin in the very young brain and young adult brain, suggesting that these proteins may have a role in the "transcription" reaction, possibly during the entire life span of the organism. The speculative role will be further discussed.

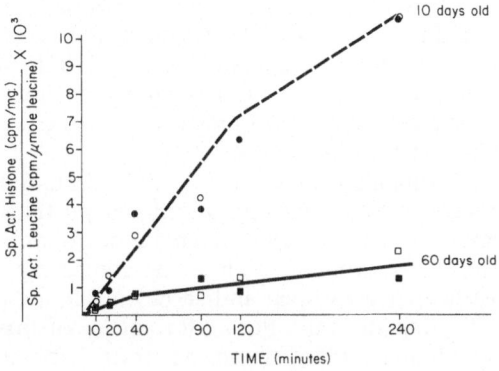

Fig. 13. Changes in relative specific activity of histones in 10- and 60-day rat brain chromatin during 4 hr after intraperitoneal injection of L-leucine-U-[14]C. Data corrected for leucine pools in both age groups.

VI. MYELINATION

It is well recognized that in all mammals there is a marked increase in the size of the brain of the adult in contrast to the brain of the newborn. Thus, in the rat, the brain increases from five- to ten-fold,[39,110] in the cat and rabbit four-fold,[7] and in man approximately three-fold.[111] This weight increase is due to (1) the outgrowth of the neuronal processes or neuropil, (2) the increase in the number of neuroglial cells, (3) progressive myelination, (4) increased vascularization, and (5) a considerable microneuron proliferation.[112] The extensive increase and complexity in the neuropil results in more extensive intraneuronal connections in the mature brain. This also reflects some phylogenetic differences in that the primates and man show more extensive neuropil in contrast to lower animals.[113] Haug[113] indicated that the increase in glia with maturation in all vertebrates is not phylogenetically significant, presumably due to the greater dependence of the neurons on the glia as sources of metabolites in the mature brain. To permit increase in myelination, the oligodendroglia increase in number and thus contribute to the gross weight of the brain during maturation. Concomitantly, there is a considerable increase in vascularization such as shown by Horstman.[114] In general, myelination has been shown in the human to begin first in the lower layers and subsequently in the upper layers of the brain.[115] Peters[116] has recently described the stages in the formation of the myelin sheaths. In the first stage, the axon becomes enclosed by a myelin-forming Schwann cell to form a mesaxon. Subsequently, the mesaxon elongates in a spiral manner around the enclosed axon until multiple turns are completed. The cytoplasm and the mesaxon then disappear to form a compact myelin sheath.[116] In spite of the dogma that neurons do not multiply after birth, it has recently been shown that microneurons multiply in certain regions of the brain for long intervals after birth. Using thymidine-^{3}H, Altman[117–121] demonstrated that neurogenesis continues in the ependymal and subependymal layers of the forebrain ventricles and in certain subpial areas of the rat brain. He showed that cell multiplication occurs at a very low rate in the white and gray matter in young adult rats including not only neuroglia, but also short-axon neurons known as microneurons. This was noted in the walls of the olfactory ventricles from which the microneurons migrated to the laminated olfactory bulb. He also noted proliferation of neuroblasts in the dentate gyrus and their differentiation into granular cells. It appears that hippocampal neurogenesis also continues for a long time after birth of mammals.

Animals such as the rat, rabbit, cat, and dog, are born with little if any myelin which then appears in the postnatal period as indicated by eye-opening, marked improvement in coordination of movement, and a characteristic EEG. Animals such as the guinea pig, cow, horse, and deer are born with their eyes open and show coordinated movement shortly after birth, reflecting the intrauterine formation of myelin in the brain before birth. The human infant is somewhere in between these two classes of animals, indicating a partial myelination *in utero* but not sufficiently complete to permit coordinated movement and optical focusing until some months after birth. Wells and Dittmer[122] have studied individual changes in lipids of the rat brain during development in an attempt to differentiate those lipids which are characteristically found in myelin from those which are considered to

TABLE III

The Lipid Composition of Developing Rat Brain[a,e]

Lipid[b]	3 days	6 days	12 days	18 days	24 days	42 days	180 days	330 days
Phosphoglycerides								
Phosphatidyl choline	14.72	14.82	20.38	24.38	24.79	24.95	24.65	24.89
Phosphatidyl ethanolamine	5.25	5.66	7.96	9.37	11.00	10.89	10.72	10.49
Phosphatidyl glycerol	0.12	0.20	0.16	0.20	0.29	0.27	0.28	0.29
Phosphatidyl inositol	1.21	1.38	1.59	1.86	2.04	2.17	2.20	2.30
Phosphatidyl serine	2.91	3.56	4.51	6.10	7.04	8.25	8.50	8.97
Phosphatidyl glycerol phosphate	0.10	0.13	0.20	0.16	0.16	0.17	0.06	0.04
Phosphatidic acid	0.14	0.21	0.26	0.39	0.70	1.03	1.31	1.36
Diphosphatidyl glycerol	0.19	0.21	0.34	0.52	0.57	0.68	0.60	0.55
Diphosphoinositide	0.01	0.05	0.05	0.16	0.15	0.19	0.21	0.20
Triphosphoinositide	0.03	0.04	0.05	0.14	0.17	0.24	0.41	0.39
Choline plasmalogen	0.04	0.06	0.07	0.09	0.16	0.22	0.35	0.34
Ethanolamine plasmalogen	2.19	2.75	4.73	7.02	11.3	13.5	13.0	13.2
Inositol plasmalogen	0[c]	2	2.05	0.08	0.13	0.13	0.13	0.11
Plasmalogenic acid	0.02	0.95	0.15	0.17	0.18	0.15	0.20	0.12
Ethanolamine phosphoglyceryl ether	0.18	0.23	0.38	0.69	1.02	1.06	1.04	1.13
Sphingolipids								
Sphingomyelin	0.23	0.26	1.04	2.15	3.19	3.62	3.70	4.10
Cerebroside	0[d]	0	2.3	5.8	10.3	18.6	21.8	22.5
Sulfatide	0.14	0.32	0.79	1.28	2.04	3.22	4.22	4.48
Ganglioside								
Hexosamine	0.31	0.73	0.84	0.96	1.01	1.06	1.15	1.13
Sialic acid	0.51	1.27	1.45	1.74	1.97	2.08	2.18	2.07

TABLE III (continued)

Lipid[b]	3 days	6 days	12 days	18 days	24 days	42 days	180 days	330 days
Others								
Cholesterol	10.7	12.6	22.6	32.2	38.3	39.5	40.2	40.6
Galactosyl-diglyceride	0.05	0.06	0.31	0.86	1.29	1.46	1.56	1.62
Brain weight (g)	0.40	0.61	1.11	1.34	1.55	1.63	1.85	1.86
Percentage recovery of phosphorus in combined fractions	97.0	96.5	98.0	98.5	98.0	97.0	97.0	98.0

[a] From Wells and Dittmer.[122] Reproduced by permission of the authors.
[b] Data for sterol ester, glycerides, and serine plasmalogen are given in the text.
[c] Lower level of detection would be on the order of 0.01 μmole/g.
[d] Lower level of detection would be on the order of 0.05 μmole/g.
[e] Concentration in μmoles/g wet weight.

reside within other membranes in the brain cells. Table III lists individual lipids and their changes in rat brains from 3 days of age up to 330 days of age. Table IV is a summary of changes in lipids and the concomitant morphological changes during the four phases of the developing rat brain. Wells and Dittmer[122] discuss their results on the basis of the percent of the concentration of lipid in the brain of the 180-day-old rat using the four growth periods proposed by McIlwain.[14] Since the prenatal period, designated by McIlwain[14] as Period I, was not studied by Wells and Dittmer,[122] it is omitted from Table IV. In Period II, myelin is absent in the rat brain and there is a marked outgrowth of axons and dendrites, represented in Table IV, in the 3 and 6-day-old-rats, primarily in the gray matter. During Period III, which is the 10–20 day interval, the brain doubles in weight. Myelination is initiated on approximately the thirteenth day after birth and increases rapidly through this period. This is reflected by the evoked potentials from the cerebral cortex and by coordinated movement of the young animals as represented in Table IV by rats 12 and 18 days of age. Period IV represents the period after weaning through adulthood during which myelination continues but at a reduced rate, in rats 24, 42, 180, and 330 days of age. Wells and Dittmer[122] categorized the lipids into five groups correlated with the above periods of development. The first group are lipids which change in nonmyelinated brain, including sterol esters and gangliosides. No other lipid had a pattern of change similar to these esters. They suggest that the sterol esters may be associated with cell proliferation in the fetus. The gangliosides, however, showed a 2.5-fold increase only during Period II, reaching 90% of the adult level in the twenty-fourth day. Since these changes are primarily

TABLE IV

Summary of Lipid and Morphological Changes[a]

Period	Age of rats studied (days)	Characteristic morphological changes	Description of lipid changes[b]				
			Class 1[c]	Class 2[d]	Class 3[e]	Class 4[f]	Class 5[g]
II	3 and 6	Axons and dendrites grow out. Myelin is virtually absent.	Sterol ester disappeared. Ganglioside increased from 24 to 60% of adult level.	On third day 10% the adult concentration found and little change occurred up to the sixth day.	Between third and sixth day these lipids increased 1.2- to 1.7-fold from a range of 17 to 34% the adult level.	These lipids increased from a range of 49–60 to 60–63% the adult level between third and sixth day.	Sulfatide and DPI[a] increased markedly between the third and sixth day from less than 5% the adult level.
III	12 and 18	Rapid phase of myelination.	Ganglioside reached 73% of adult level by twelfth day and increased to 83% by eighteenth day.	10–28% of the adult concentration reached on twelfth day and a 2- to 3-fold increase occurred by the eighteenth day.	In general these lipids increased to from 36 to 53% the adult level by the twelfth day and to 54–87% by the eighteenth day.	Concentrations continued steady increase with 84–99% of the adult level being reached by the eighteenth day.	Sulfatide and DPI increased 1.6- and 3.2-fold between twelfth and eighteenth day. PGP reached maximum concentration.

TABLE IV (continued)

Period	Age of rats studied (days)	Characteristic morphological changes	Description of lipid changes[b]				
			Class 1[c]	Class 2[d]	Class 3[e]	Class 4[f]	Class 5[g]
IV	24, 42, and 180	Decreased rate of myelination; maturation.	Ganglioside reached adult level by twenty-fourth day.	In general, increase continued from a range of 47–86% at the twenty-fourth day to 58–100% the adult level on the forty-second day. The exceptions (PA and TPI) showed greatest increase in last part of period.	Continued increase to concentrations of 82–98% the adult level by the twenty-fourth day; 97–100% of the adult level reached by forty-second day. Choline plasmalogen showed greatest increase in last part of period.	Adult levels reached by the twenty-fourth day.	Sulfatide and DPI increased in manner similar to class 2 lipids. PGP decreased in concentration to lowest levels found.

[a] From Wells and Dittmer.[122] Reproduced by permission of the authors.
[b] Abbreviations used: DPI, diphosphoinositide; PGP, phosphatidyl glycerol phosphate; TPI, triphosphoinositide; PA, phosphatidic acid.
[c] Lipids that appear to be primarily associated with changes in the nonmyelinated brain.
[d] Lipids that appear to be associated primarily with myelination.
[e] Lipids apparently associated with changes in both myelinated and nonmyelinated brain.
[f] Lipids that show no changes that can be directly related to morphological changes.
[g] Lipids that show marked changes during development not directly associable with morphological changes.

associated with cell growth during Periods II and III, their involvement in myelination is uncertain.

The second group of lipids, considered to be associated solely with myelination, shows an increase in Period III and Period IV. The maximal increase in these lipids, cerebroside, sphingomyelin, triphosphoinositide, phosphotidic acid, galactosyl diglyceride and inositol plasmalogen, was noted to be two-to-three-fold between the twelfth and eighteenth day (Period III).

The third group of lipids, ethanolamine plasmalogen, choline plasmalogen, diphosphatidyl glycerol, ethanolamine phosphoglyceryl ether, cholesterol, and plasmalogenic acid, showed a marked increase in concentration beginning in Period II, from 1.2- to 1.8-fold from the third to eighteenth day, and continued increasing through Periods III and IV. These lipids appear to be associated with the nonmyelin membranes in the brain.

The fourth group of lipids, which are not involved in myelination, consists of phosphotidyl choline, phosphotidyl ethanolamine, and phosphotidyl inositol. Phosphotidyl glycerol was considered by Wells and Dittmer[122] to be intermediate between lipid groups three and four. Phosphatidyl glycerol phosphate, diphosphoinositide and sulfatide did not fit into any of the lipid groups and the authors suggest that the first two may be intermediates in the formation of other lipids. Since 50% of the total lipid sulfur has been estimated to be in the myelin, the sulfatides appear to be associated, in part, with both myelinated and nonmyelinated structures. A study by McKhann et al.[123] has shown that $^{35}SO_4$ incorporation into sulfatides is very rapid in the rat brain at 10 days of age, reaching a maximum at 23 days (Periods III and IV), decreasing rapidly up to the age of 25 days, and then more gradually decline to adulthood.

Suzuki et al.,[124,125] studying the variations in the gangliosides in the myelin fraction of the developing rat brain, found the total ganglioside concentration was relatively constant at all ages. However, the relative proportions of the individual major gangliosides change as the animal matures. They also found that the quantity of monosialoganglioside in myelin was much higher (55.8 mole %) than in the whole brain (23.4 mole %) and this increased in the myelin fraction as the rat matures, reaching a concentration of 90 mole % in the 144-day rat and remaining at this level in the older rat (425 days).[124,125] Steric acid comprised 90% of the total fatty acids in this ganglioside. Dhopeshwarkar et al.,[126] studying the incorporation of acetate-1-^{14}C into fatty acids of the developing rat brain found that the uptake of intraperitoneally injected acetate was higher in the weanling rat brain than in the adult brain. The rate of incorporation of the acetate into the total lipids decreased linearly from 10 to 20 days, reaching minimal levels in the 27-day rat and decreasing slightly in the 50-day rat brain, indicating a considerable enrichment of lipids in the weanling stage. However, the palmitate to sterate ratio decreased and the oleate to sterate ratio increased in the older brains. These authors found that palmitic acid showed a high specific activity in every stage of development of the brain. The drop in specific activity of the fatty acid with increasing chain length from palmitic to docosahexaenoic was greater in the brains of the younger rats. The specific activity and percent distribution of the radioactivity in palmitic and palmitoleic acids were nearly equal, confirming the direct dehydrogenation of palmitate to form palmitoleic acid. A similar pattern was observed with steric and oleic acids. *De novo* synthesis of palmitic acid from acetate-^{14}C was confirmed.[126]

Salway et al.[127] have studied activity changes in the enzymes of phosphoinositide metabolism during development of the rat brain. The enzymes CDP-diglyceride inositol phosphatidate transferase, phosphatidyl inositol kinase, diphosphoinositide kinase, and triphosphoinositide phosphomonoesterase catalyze the following reactions in the order stated:

(I)   CDP-diglyceride + inositol → phosphatidyl inositol + CMP
(II)   Phosphatidyl inositol + ATP → diphosphoinositide + ADP
(III) Diphosphoinositide + ATP → triphosphoinositide + ADP
(IV)   Triphosphoinositide + H_2O → diphosphoinositide + Pi

The activity of all the enzymes was found to increase with age. Phosphatidyl inositol kinase (II) rose sharply before myelination started, reaching a maximum at about 6 days of age, while diphosphoinositide kinase (III) and triphosphoinositide phosphomonoesterase (IV) showed the most rapid increase in activity during myelination. CDP-diglyceride inositol phosphatidate transferase (I) showed a gradual rise in activity between 6 and 10 days, reaching a maximum in 25 days and then decreasing slightly in the 70-day rat. From these studies, the authors conclude that the metabolism of triphosphoinositide is associated with myelin formation.[127]

However, Hauser and his associates[128] reported that the activities of triphosphoinositide phosphomono- and phosphodiesterase, in brain acetone powder extracts from newborn and mature rats, showed no change. Eichberg and Hauser[129] using brain homogenates from developing rats found a progressive increase in the synthesis of triphosphoinositide and a decrease in the synthesis of diphosphoinositide during brain maturation. They suggest that the capability of homogenates from unmyelinated rat brains (from rats younger than 12 days) to metabolize diphosphoinositide and triphosphoinositide supports the concept that these substances undergo turnover at an extra myelin site.

VII. HORMONES

A. Thyroxine and Growth Hormone

The hormones from the thyroid gland, gonads, and the adrenal cortex have a definitive influence on both the growth and function of the brain.[130] Thyroidectomy of the neonatal rat profoundly decreases the rate of body and brain growth and causes alteration in myelination, which is later manifested in abnormal behavior. In addition, the perikarya of nerve cells is smaller and the number of dendrites and axons is reduced.[131] Hamburgh and Flexner[132] showed that thyroidectomy substantially decreases the activity of succinic dehydrogenase in the brain and also depresses the cholinesterase activity. When rats are given a therapeutic dose of thyroxine on the tenth day of thyroidectomy, succinic dehydrogenase activity is found to be at a normal level 15 days later. However, if the thyroxine therapy is instituted on the fifteenth day or later, the activity of this enzyme is not restored to normal. Geel and Timiras[133] reported that neonatal thryoidectomy causes a reduction of acetylcholinesterase and cholinesterase activities in the cerebral cortex and hypothalamus of the rat at 22 days of age. However, administration of physiological levels of thyroxine initiated on the sixth day of age restores the enzyme activities to normal and restores their ability to incorporate leucine into proteins.[134] Other investigators[135] have shown that neonatal thyroidectomy markedly decreases the amount of RNA and protein in the rat cerebellum and cerebral cortex. Garcia Argiz et al.[136] studied the effect of neonatal thyroidectomy on the synthesis of brain enzymes. They noted a significant decrease in the succinic dehydrogenase, glutamic decarboxylase, γ-amino butyric acid transaminase, Mg adenosine triphosphatase, and Na^+,K^+-adenosine triphosphatase in rat cerebral cortex. In the cerebellum, there is a substantial decrease in GABA transaminase and Na^+,K^+-adenosine triphosphates, while succinic dehydrogenase and glutamic decarboxylase decrease temporarily.[136] However, the administration of thyroxine to the neonatal thyroidectomized rats on the tenth day restores these enzyme activities to normal. Such therapeutic treatment, initiated on the fifteenth day in the neonatal thyroidectomized rats, has no restorative effects.[137]

Several investigators[138–141] observed that treatment of neonatal thyroidectomized rats with growth hormone (somatotrophin) partially restores the body weight of the cretinous animal. Additional treatment with thyroxine early in the postnatal period further enhances normal development in these animals. Eayrs[142] interpreted the action of these hormones by suggesting that incremental growth affected by growth hormone is due to its influence on the synthesis of proteins, while thyroxine is more specifically involved in the maturation phenomenon. Krawiec et al.[137] and Gomez et al.[143] also found that treatment of neonatally thyroidectomized rats with growth hormone restores brain and body growth, as assessed by body weight and by changes of the metabolic functions of the brain to normal, if the treatment is instituted at the age of 10 days. However, treatment with growth hormone on the fifteenth day does not ameliorate the cretinous condition in these animals. In addition, Gomez et al.[143] reported that growth hormone restores oxygen and glucose utilization and lactic acid production by brain slices from neonatal cretinous rats treated with growth hormone. Krawiec et al.[137] showed that the cerebrum of the hypothyroid rats has a significantly lower content of RNA than the normal controls. The RNA content in both the cerebral cortex and cerebellum of these cretinous rats is restored by treatment with thyroxine or growth hormone. Similarly, treatment with either hormone restores the levels of succinic dehydrogenase, γ-aminobutyric acid transaminase, and aspartate aminotransferase activities in both the cerebral cortex and the cerebellum. Treatment with the above hormones is effective only when instituted by the age of 10 days or earlier.[137]

Injection of thyroxine into normal neonatal animals elicites precocious eye opening and earlier maturation of the brain, as demonstrated by both maturelike behavior of the very young animals and by the evoked electro-encephalogram when the animals experience novel stimuli.[142,144] Rats treated with thyroxine at 3 days of age acquired a conditioned avoidance response faster than untreated controls, as evidenced by tests 16–18 days after injection.[10,144]

Recently Cocks et al.[145] found that brains from hyperthyroid rats incorporate a greater amount of radioglucose into amino acids than normal controls, and thyroidectomized rats showed a large decrease in this incorporation. These authors suggest that lack of thyroid hormone retards biochemical maturation, while excess hormone induces premature appearance of adult metabolic patterns in the young brain.

Injection of growth hormone into pregnant rats was noted by Zamenhof[146] to effectively increase the weight of the cerebellar hemispheres and the number of cells in the cortex of the offspring. A more recent report by this author and his collaborators[147] confirms the previous observed increase in the number of cortical neurons in the offspring of rats given prenatal treatment with growth hormone. Clendinnen and Eayrs[148] have also noted that, following prenatal treatment of pregnant dams with growth hormone, the offspring exhibited an enhanced cortically mediated behavior. This was probably the result of enlarged neurons which included enlargement of the perikarya and expansion of the dendritic fields. Injection of growth hormone into 7-day rats did not affect any changes at 28 days of age in either brain or body size relative to uninjected controls.[149] Gregory and Diamond[150] and Diamond[151] recently found that, in spite of the decreased size of brains in 33–35-day-old rats hypophysectomized at 4–5 days of age, no changes were evident in brain morphology and concluded that the presence of growth hormone in the postnatal period is not essential for normal brain growth. It is conceivable that hormone in the maternal milk or residual hormone in circulation may account for these observations.

It has been known for some time that that hypothyroidism in the very young rat, human, and other animals results in abnormal myelin deposition[152] which may be more or less corrected with the early treatment of thyroid hormone. A recent study by Walravens and Chase[153] showed that the cerebrosides and sulfatides (myelin lipids) decrease 50–60% and the cholesterol 23% in the brain of 18-day rats which have been thyroidectomized at birth. On the other hand, Cuaron et al.[154] reported that the lipid phosphorus content in rabbit brain was not modified by hypothyroidism. It appears that hypothyroidism inhibits the synthesis of lipids and myelin without affecting other membranes.[153]

Hamburgh and Bunge[155] and Hamburgh[156] showed that thyroxine enhanced myelin formation in cerebellar explants of neonatal rats and mice.

When these explants were cultured at 30°C without thyroxine, the differentiation of cells into neurons and glial elements was delayed but progressed gradually, while myelogenesis was almost completely suppressed. When thyroxine was added to these cultures at low temperature, some myelogenesis was observed, but to a much lesser degree than in explants kept at 36°C. Shifting the cultures from low (30°C) to optimal (36°C) temperature resulted in recovery of myelogenesis. When thyroxine was added to cultures kept at optimal temperature, it was consistently noted that there was an accelerated emergence (by 1–2 days) of myelin in these cultures in contrast to untreated controls.[156] These observations led the author to suggest that one of the stimuli for myelogenesis may be a metabolic component whose rate of formation is temperature dependent and enhanced by the presence of thyroxine.

B. Other Hormones

It has been reported that neonatal administration of estradiol[157,158] and of cortisol[159,160] hastened the functional development of the brain and spinal cord and the appearance of myelin in the rat brain.[161] When nominal doses of cortisol and estradiol were injected separately into rat pups, between the sixth and tenth days of age, and the brain and spinal cord excised on the twelfth day, there was an increase in the amount of cerebrosides in the spinal cord and cerebrum of estradiol-treated rats. In the cortisol-treated animals, an increase in the cerebrosides was found only in the spinal cord.[159] This suggested to the authors that these hormones stimulate premature myelination, thus affecting precocious functional maturation in these animals.[162]

DeVellis and Inglish[163] reported that hypophysectomy or adrenalectomy reduced the activity of glycerolphosphate dehydrogenase in the cerebral hemispheres and brain stem of adult rats. Injection of cortisol in either of these surgically treated rats restored the glycerolphosphate dehydrogenase activity. After hypophysectomy had been performed on male rats at 20 days of age, glycerolphosphate dehydrogenase activity was appreciably below that of the normal controls in both the cerebrum and brain stem at 30 days of age, although the protein content in these organs was not significantly different from normal controls. These findings indicate that the effect of hypophysectomy on the brain glycerolphosphate dehydrogenase is specific since other enzymes in the glycolytic pathway were not affected.[163]

Winick and Coscia[164] found that large amounts (1.0 mg) of cortisone acetate injected into rats produced a typical runting syndrome affecting both the rat body and brain. When the rats were sacrificed on the twenty-fifth day, a marked decrease in body weight, organ weights (including the brain), and organ content of protein, RNA, and DNA was noted. Howard[165] found that mice treated with cortisone at 2 or 3 days of age showed a reduction in brain size and in DNA and RNA formation.

VIII. NUTRITION

Infants from humans and animals, reared under restricted protein diet, not only exhibit a marked decrease in size and weight, but also manifest abnormal behavior.[166–173] A restricted protein diet is here defined as approximately one-third to one-fourth (8 % protein) of the protein content in a normal diet (25–27 % protein) for a pregnant animal which allows birth of a normal litter of viable offspring.[174] Offspring from dams with restricted protein intake during gestation exhibit a marked decrease in body weight (30 % less than controls) and a change in brain composition (10 % decrease in DNA and 20 % decrease in protein,[175] and later exhibit abnormalities in behavior.[176] Such deficits in weight and behavior are only partially reversible when the above animals are subsequently rehabilitated by an ample supply of protein.[166–172,175] Observations in humans[166] and rats[177] have shown the critical period in growth when protein restriction permanently alters the stature and behavior of the mature animal.[166–171] However, protein restriction during the period of lactation also induces abnormalities in behavior and stature. In general, protein-restricted diets maintained prior to and during initiation of myelination result in severe and irreversible abnormalities in both growth of the brain and behavior when these animals mature.[172,178] Since myelination is initiated in many animals during gestation, such as the human, guinea pig, sheep and pig, protein-restricted diets in this initial phase of growth would be expected to cause permanent changes in the development of the animal. Hence, gestation is considered to be the most vulnerable period for dietary restriction or deprivation.

In contrast to the abnormal behavior of offspring from nutritionally restricted pregnant dams, many studies have been reported with normal newborn animals maintained under altered conditions during lactation, but only two will be discussed here. Howard[165] reported that newborn mice removed from their mother for 2-3 hr daily during the lactation period showed a decrease in size at weaning but exhibited superior maze-solving ability in contrast to controls. This author indicated that increased activity experienced by the experimental mice when returned to their mothers was probably responsible for their enhanced behavior at postweaning in contrast to unexperienced controls. The observation that dams that have been food restricted during gestation yield offspring which have a significant decrease in both DNA and protein levels in the brain suggests that the systems responsible for the synthesis of these components have been altered. Examination of mammalian growth indicates two major phases in the development of an organ following differentiation. The first phase is the multiplication of cells, which is a reflection of an increase in DNA as well as protein. The second phase of growth is the enlargement of existing cells, indicating an increase in protein with very little change in DNA.

A recent study by van Marthens and Zamenhof[179] showed that ligation of one uterine horn of mature female rats increased the nutrient supply to the other horn. These females, mated 1 week after the operation, bore half of the normal number of offspring. The offspring, however, were heavier, had a higher cerebral weight, a higher total cerebral protein, and a higher total amount of cerebral DNA than the normal litter from sham-operated controls. The experimental offspring also had a higher number of neurons than the controls. These investigators suggested that ligation of a uterine horn supplied a greater abundance of nutrients to the limited number of embryos and thus influenced enhanced growth.

Benton et al.[178] investigated brain changes in rats nutritionally deprived during lactation by allowing 16–21 newborn rats to be nursed by one dam in contrast to control rats in which no more than 10 newborns were nursed by a single dam. A diminished body weight of approximately 50% and a 20% reduction in brain weight, total brain lipids, cholesterol, and phospholipids was found. Brain cerebrosides were more affected than other lipids, resulting in a 50% reduction in contrast to controls. When the deprived pups were allowed *ad libitum* food intake at 3 weeks of age, body weight, brain weight, and concentration of brain lipids became equal to those in the controls. A study by Culley and Lineberger[180] of young rats deprived of food by restricting their feeding from the dam to an interval each day long enough for each animal to gain only 0.5 g per day, showed significantly less DNA and RNA per brain than controls of the same age. They also found a deficit in brain weight, brain lipid, phospholipid, cholesterol, and cerebroside content. Adequate feeding of these animals did not restore these components to normal values. However, animals restricted in food for 11 days after birth and then fed *ad libitum*, showed a greater accrual of all the components in contrast to those rats which were restricted in food up to the seventeenth and sixtieth day after birth and then given adequate nutrition. This indicates that at 17 days of age, the rat has attained its adult complement of DNA (or cells) and that the ability of the rat brain to recover from food restriction is dependent upon the number and types of cells that it contains by 17 days of age. Undernourishment during the postnatal period causes a greater deficit of DNA in the cerebellum and least in the pons-medulla and intermediate change in the mid- and forebrain. This is consistent with the report that the rate of cell formation is high in the cerebellum and low in the medulla during the postnatal period, when only glia and microneurons are formed.[110,121] Long axon cells (macroneurons) are formed only during the prenatal period[121] and consequently are not affected by post-natal undernourishment.

Chase et al.[176] undernourished newborn rats by increasing the litter to 16 animals per dam in contrast to well-fed litters which consisted of 4 rats per day. When the rats were sacrificed at 18 days of age, the weights of both cerebrum and cerebellum were found to be significantly lower in the under-nourished rats than in the controls. The cerebellar DNA was found to be significantly lower while the cerebrum DNA was not significantly different from controls. The total protein was found to be greater in the cerebellum than the cerebrum which contained less protein and had smaller cells in the undernourished rat brain. On the other hand, the cerebellum showed a reduction in both DNA and protein suggesting that the cell number but not necessarily the cell size is reduced.

IX. INTERPRETATION OF GENE ACTIVATION

Existing information permits a reasonable, although limited, interpretation of cellular differentiation and growth on a molecular basis. Cell differentiation is the transition in cell character due to activation and repression of different genes within a group of cells each containing a homologous genetic repertoire.[1] Growth is the summation of cell proliferation and enlargement solely due to enhanced protein synthesis under genetic regulation.[181] The characteristics of cells, tissues, and organs are encoded in the genetic material or DNA in each organism in the form of a triplet nucleotide sequence within the primary structure of DNA.[182] In eukaryotic cells, the double-stranded DNA in the chromatin contains histones which stabilize the double helical structure.[183] In addition, chromatin also contains nonhistone (acidic) proteins and a small amount of RNA.[184] Chromatin from various sources has been shown to have similar composition, with variations mainly in RNA and nonhistone protein content as illustrated in Table V.[184]

Abundant evidence has accumulated to show that extraction of histones from chromatin enhances the template activity of the remaining DNA-nucleoprotein (DNA-nonhistone protein complex).[106,184–187] Additional histones combine ionically and nonspecifically with DNA, hence mask its template function in the formation of RNA by RNA polymerase.[188]

TABLE V

Chemical Compositions of Varied Chromatins[a]

Source of chromatin	Content, relative to DNA, of				Template activity (Percentage of DNA)
	DNA	Histone	Nonhistone protein	RNA	
Pea embryonic axis	1.00	1.03	0.29	0.26	12
Pea vegetative bud	1.00	1.30	0.10	0.11	6
Pea growing cotyledon	1.00	0.76	0.36	0.13	32
Rat liver	1.00	1.00	0.67	0.043	20
Rat ascites tumor	1.00	1.16	1.00	0.13	10
Human HeLa cells	1.00	1.02	0.71	0.09	10
Cow thymus	1.00	1.14	0.33	0.007	15
Sea urchin blastula	1.00	1.04	0.48	0.039	10
Sea urchin pluteus	1.00	0.86	1.04	0.078	20

[a] From Bonner et al.[184]

Chromatin isolated from differentiating tissues, used as templates for RNA synthesis, yields RNA's with varied composition at each stage of differentiation, reflecting the changes in genetic activation.[189,190] Recent reports have indicated that mammalian chromatin contains organ-specific "masked" DNA. Paul and Gilmour[106] found that if purified DNA is first combined with nonhistone protein before addition of histones, the reconstituted nucleoprotein is organ specific, exhibiting the same template activity in RNA synthesis as the original chromatin from which the DNA was extracted. These authors indicate that the nonhistone proteins are responsible for unmasking organ-specific DNA sequences. A recent study by the above investigators[107] compared the nature of the RNA formed by native chromatin and that formed from reconstituted chromatin from the same tissue. By competitive hybridization, they found both RNA's were similar, hence they concluded that, in contrast to histones which have a nonspecific repressive effect on transcription, nonhistone (acidic) proteins render specific regions of DNA available for transcription.[107]

Electrophoretic comparison of acidic proteins of chromatin from different animal tissues revealed that they are different from one tissue to another in the same species. The acidic proteins from the same tissue from various species also differ.[191]

Huang and Huang[192] have found that chick embryo chromatin contains acidic protein covalently linked to RNA. Using this RNA-acidic protein complex in reconstituting chromatin, the template activity of the chromatin yielded an RNA which resembled native nuclear RNA. When the RNA in this complex (RNA-acidic protein) was degraded (with zinc nitrate) within the chromatin, the resulting template yielded an RNA different from either nuclear RNA or original chromatin-primed RNA.[192] Reconstitution of chromatin, which resembled native chromatin, required the presence of a chromosomal RNA (not mRNA) for sequence specificity in the DNA-chromosomal protein interaction.[193]

It has been demonstrated that isolated chromatin can act as a template for purified E. coli RNA polymerase for the synthesis of RNA, but the available DNA in the chromatin as a nucleoprotein is only transcribed 5% or less of the DNA when the chromatin is stripped of histones, isolated and purified.[106,184–186] Recent studies by Stevenin et al.[62,194] have shown that less than 2% of the chromatin DNA is available for transcription in the rat brain. These findings indicate that the majority of the genes are repressed in the mammalian organism, whereas in bacteria, 80–100% of the genome is always available for transcription.[109] This repression is due to histones masking the genomes.

In contrast to bacteria, in which the Jacob–Monod concept of repressors interact with activator genes[195] has been well demonstrated, cells in eukaryotic organisms appear to have a different mechanism of controlling genetic activity.[109] Since histones are recognized as the genetic repressors in differentiated cells, it is difficult to explain how specific genomes are activated without implying the existence of genetic activators in order to explain temporal genetic activation during differentiation in the gastrula and during maturation of tissues and organs in the postembryonic metazoans. Using the concept of genetic activators, it is possible to formulate a speculative mechanism to account for the temporal and sequential activation of genes and to suggest which components in the chromatin may be the activators. As indicated in Table V, the composition of chromatin within differentiated cells is similar. The present information indicates that the histones are localized in the deep grooves on the double-stranded DNA. Because of the relatively low molecular weight of the histones (15,000–30,000),[196] the double-stranded DNA is triple-stranded with histone, thus presenting the genetic fibrillar strand which is intermittently covered with histone adjacent to an area of exposed double-stranded DNA (Fig. 14). As a working hypothesis, one can speculate that the exposed double helical DNA also represents repressed genomes since any tendency for the exposed double-coiled DNA to unwind (B in Fig. 14) will require energy to overcome the multitude of hydrogen bonds which are responsible for the double helical coil. In addition, the presence of histone on either side of the exposed double helical DNA will also stabilize the double helix (Fig. 14). Since RNA polymerase requires a double-stranded (native) DNA as a template, and since this enzyme synthesizes RNA by conservative transcription, using only a single strand of DNA, it is apparent that the double-stranded DNA (free of histone) has to be uncoiled into relatively parallel single strands to permit the RNA polymerase to associate with one of the strands during the synthesis of RNA. It is conceivable that the acidic proteins in association with chromatin RNA may act as gene "activators." Thus, the RNA may act as a complementary "read-out" by hydrogen binding to a specific DNA loci to be activated, and this RNA can bind with a specific portion on one of the DNA strands as specific genetic anticodon, and secondly, the protein attached to the RNA, representing a much greater bulk of this nucleoprotein particle, would interdigitate between the unwound strands of DNA, forming hydrogen bonds with bases from both strands and thus act as a wedge to maintain the strands separated and extended. This will permit the RNA-nucleoprotein (RNA-aP, Fig. 14) to bind in a complementary fashion with

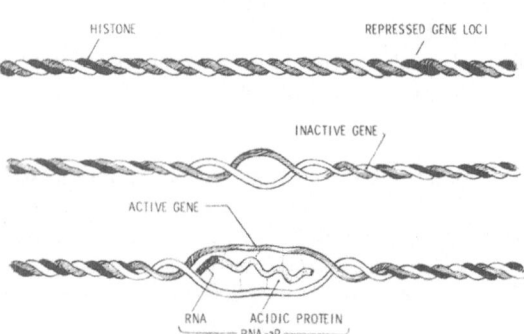

Fig. 14. A schematic illustration of the conceptual mechanism of gene activation by a chromatin containing RNA-acidic protein (RNA nonhistone protein) complex. (a) A double-stranded DNA (gray and white) with histones (black) at both ends; (b) Portion of DNA free of histones partially unwound; (c) Activated gene with unwound DNA strands stabilized by the RNA-acidic protein complex.

a requisite site on one of the two strands of DNA, while leaving the other strand available for association with the enzyme, RNA polymerase. This will permit assembling the substrates (nucleotide triphosphates) on the template for the formation of new RNA. It can be seen in Fig. 14 that the chromatin RNA-acidic protein is conceptualized as a gene activating complex. The postulated mechanism explains why the RNA polymerase requires double-stranded DNA for its template but uses only one strand in its transcription (conservative replication) since it is evident that attachment of RNA to a DNA strand makes this strand unavailable for association with the polymerase and thus the second strand is available for transcription. Bonner[184] originally conceived the possibility that chromatin RNA may serve as an anticodon for recognition of specific genome sites and thus select specific portions of DNA for transcription, although he did not speculate further as to the details of this interaction.

At present, other investigators such as Davidson[109] have come to the conclusion that, in contrast to the bacterial systems in which the Jacob-Monod[195] mechanism is responsible for genetic depression in differentiated cells, there is a need for an activator, in this review postulated to be the RNA-acidic protein which activates those genomes not repressed by histones.

ACKNOWLEDGMENT

This work was supported in part by a Robert A. Welch Foundation Grant, No. H-180 and a Public Health Service Grant, GM 12519.

X. REFERENCES

1. P. R. Gross, Biochemistry of differentiation, *Ann. Rev. Biochem.* **37**:631–660 (1968).
2. R. B. Scott and E. Bell, *in Molecular and Cellular Aspects of Development* (E. Bell. ed.), pp. 217–222, Harper and Row, New York (1967).
3. J. Paul, Molecular aspects of cytodifferentiation, *Adv. Comp. Physiol. Biochem.* **3**:115–172 (1968).
4. J. D. Ebert, *in The Neurosciences* (G. C. Quarton, T. Melnechuk, and F. O. Schmitt, eds.), pp. 241–247, The Rockefeller University Press, New York (1967).
5. J. D. Ebert and F. E. Samson, Gene expression, *Neurosci. Res. Prog. Bull.* **5**(3):227–303 (1967).
6. R. G. Spector, *in Neurochemistry* (C. W. M. Adams, ed.), pp. 239–252, Elsevier, Amsterdam (1965).
7. W. A. Himwich, Biochemical and neurophysiological development of the brain in the neonatal period, *Intl. Rev. Neurobiol.* **4**:117–158 (1962).
8. S. Skoglund, Growth and differentiation, *Ann Rev. Physiol.* **31**:19–42 (1969).
9. W. Wechsler, Developmental analysis of the fine structure of different nerve cell types, *Exp. Biol. Med.* **1**:153–169 (1967).
10. S. Schapiro and R. J. Norman, Thyroxine: effects of neonatal administration on maturation, development and behavior, *Science* **155**:1279–1281 (1967).
11. S. M. Crain, Development of electrical activity in the cerebral cortex of the albino rat, *Proc. Soc. Exp. Biol. Med.* **81**:49–51 (1952).

12. J. Bures, The ontogenetic development of steady potential differences in the cerebral cortex in animals, *Electroencephalog. Clin. Neurophysiol.* **9**:121–130 (1957).

13. L. B. Flexner, D. B. Tyler, and L. J. Gallant, Biochemical and physiological differentiation during morphogenesis, X. Onset of electrical activity in developing cerebral cortex of fetal guinea pig, *J. Neurophysiol.* **13**:427–430 (1950).

14. H. McIlwain, *Biochemistry and the Central Nervous System* 2nd ed. Little, Brown and Co., Boston (1959).

15. M. Atlas and V. P. Bond, The cell generation cycle of the eleven-day mouse embryo, *J. Cell. Biol.* **26**:19–24 (1965).

16. L. D. Hodge, T. W. Borun, E. Robbins, and M. D. Scharff, in *Biochemistry of Cell Division* (R. Baserga, ed.), pp. 15–37, C. C. Thomas, Springfield, Ill. (1969).

17. D. Gallwitz and G. C. Mueller, Histone synthesis *in vitro* by cytoplasmic microsomes from HeLa cells, *Science* **163**:1351–1353 (1969).

18. E. Robbins and T. W. Borun, The cytoplasmic synthesis of histones in HeLa cells and its temporal relationship to DNA replication, *Proc. Natl. Acad. Sci.* **57**:409–416 (1967).

19. E. K. Schandl and J. H. Taylor, Early events in the replication and integration of DNA into mammalian chromosomes, *Biochem. Biophys. Res. Commun.* **34**:291–300 (1969).

20. R. B. Painter and A. Schaefer, State of newly synthesized HeLa DNA, *Nature* **221**:1215–1217 (1969).

21. G. C. Mueller, Biochemical events in the animal cell cycle, *Fed. Proc.* **28**:1780–1789 (1969).

22. L. R. Gurley and J. M. Hardin, The metabolism of histone fractions, I. Synthesis of histone fractions during the life cycle of mammalian cells, *Arch. Biochem. Biophys.* **128**:285–292 (1968).

23. L. H. Kedes and P. R. Gross, Identification in cleaving embryos of three RNA species serving as templates for the synthesis of nuclear proteins, *Nature* **223**:1335–1339 (1969).

24. G. C. Mueller and K. Kajiwara, Actinomycin D and *p*-fluorophenylalanine inhibitors of nuclear replication in HeLa cells, *Biochim. Biophys. Acta* **119**:557–565 (1966).

25. J. J. Sisken, in *The Proliferation and Spread of Neoplastic Cells*, pp. 159–174, Williams and Wilkins, Baltimore (1968).

26. T. P. Brent, J. A. V. Butler, and A. R. Crathorn, Variations in phosphokinase activity during the cell cycle in synchronous populations of HeLa cells, *Nature* **207**:176–177 (1965).

27. E. Stubblefield and G. C. Mueller, Thymidine activity in synchronized HeLa cell cultures, *Biochem. Biophys. Res. Commun.* **20**:535–538 (1965).

28. Y. Hotta and H. Stern, in *Development* (E. Bell, ed.), rev. ed., pp. 187–194, Harper and Row, New York (1967).

29. H. Stern and Y. Hotta, Biochemical studies of male gametogenesis in liliaceous plants, *Current Topics Develop. Biol.* **3**:37–63 (1968).

30. J. Bukovsky and J. S. Roth, Some factors affecting the phosphorylation of thymidine by transplantable rat hepatomas, *Cancer Res.* **25**:358–364 (1965).

31. H. Tiedemann, Inducers and inhibitors of embryonic differentiation: Their chemical nature and mechanism of action, *Exptl. Biol. Med.* **1**:8–21 (1967).

32. T. Yamada, A chemical approach to the problem of the organism, *Adv. Morphogenes.* **1**:1–53 (1961).

33. T. Yamada, The inductive phenomenon as a tool for understanding the basic mechanism of differentiation, *J. Cell. Comp. Physiol. Suppl.* 1 ad **60**:49–64 (1962).

34. H. Tiedemann, K. Kesselring, U. Becker, and H. Tiedemann, Uber die induktions-fahigkeit von microsomen- und zellkernfraktionen aud embryonen und leber von huhnern, *Develop. Biol.* **4**:214–241 (1962).

35. A. Suzuki and I. Kawakami, Inductive effects of nuclei and their subfractions isolated from rat liver, *Embryologia* **8**:75–78 (1963).

36. J. Langman, in *The Structure and Function of Nervous Tissue* (G. H. Bourne, ed.), Vol. 1, pp. 33–65, Academic Press, New York (1968).
37. P. Glees and K. Meller, in *The Structure and Function of Nervous Tissue* (G. H. Bourne, ed.), Vol. 1, pp. 301–323, Academic Press, New York (1968).
38. J. Altman, in *Handbook of Neurochemistry* (A. Lajtha, ed.), Vol. 2, pp. 137–182, Plenum Press, New York (1969).
39. M. Winick and A. Noble, Quantitative changes in DNA, RNA, and protein during prenatal and postnatal growth in the rat, *Develop. Biol.* **12**:451–466 (1965).
40. F. L. Margolis, DNA and DNA-polymerase activity in chicken brain regions during ontogeny, *J. Neurochem.* **16**:447–456 (1969).
41. A. D. Bharucha and M. R. V. Murthy, personal communication (1969).
42. S.-C. Sung, DNA synthesis in the developing rat brain, *Canad. J. Biochem.* **47**:47–50 (1969).
43. D. A. Rappoport, R. R. Fritz, and J. L. Myers, in *Handbook of Neurochemistry* (A. Lajtha, ed.), Vol. 1, pp. 101–119, Plenum Press, New York (1969).
44. H. Dellweg, R. Gerner, and A. Wacker, Quantitative and qualitative changes in ribonucleic acids of rat brain dependent on age and training experiments, *J. Neurochem.* **15**:1109–1119 (1968).
45. L. M. Barbato, I. W. Barbato, and A. Hamanaka, Comparative study of incorporation of isotopically labeled precursors into nuclear and microsomal RNA of the developing brain, *Brain Res.* **9**:213–223 (1968).
46. M. R. V. Murthy, Kinetic aspects of ribosomal and messenger RNA formation in young rat brain, *Biochim. Biophys. Acta* **166**:115–123 (1968).
47. S. Yamagami, R. R. Fritz, and D. A. Rappoport, Biochemistry of the developing rat brain, VII. Changes in the ribosomal system and nuclear RNAs, *Biochim. Biophys. Acta* **129**:532–547 (1966).
48. Z. S. Tencheva and A. A. Hadjiolov, Characterization of rat brain ribonucleic acids by agar gel electrophoresis, *J. Neurochem.* **16**:769–776 (1969).
49. H. Adams, The incorporation of (^{14}C) orotic acid into the nuclear and microsomal RNA fractions of the developing rat cerebral cortex: Some effects of deoxycholate treatment, *J. Neurol. Sci.* **8**:171–181 (1968).
50. B. Daneholt and S.-O. Brattgärd, A comparison between RNA metabolism of nerve cells and glia in the hypoglossal nucleus of the rabbit, *J. Neurochem.* **13**:913–921 (1966).
51. E. Egyházi and H. Hydén, Biosynthesis of rapidly labeled RNA in brain cells, *Life Sci.* **5**:1215–1223 (1966).
52. H. R. Mahler, W. J. Moore, and R. J. Thompson, Isolation and characterization of ribonucleic acid from cerebral cortex of rat, *J. Biol. Chem.* **241**:1283–1289 (1966).
53. S. C. Bondy and S. Roberts, Developmental and regional variations in ribonucleic acid synthesis on cerebral chromatin, *Biochem. J.* **115**:341–349 (1969).
54. R. R. Fritz and D. A. Rappoport, unpublished data.
55. R. G. Roeder and W. J. Rutter, Multiple Forms of DNA-dependent RNA polymerase in eukaryotic organisms, *Nature* **224**:234–237 (1969).
56. S. H. Barondes, Studies with an RNA polymerase from brain, *J. Neurochem.* **11**:663–669 (1964).
57. S. Furusawa and D. A. Rappoport, unpublished data.
58. P. Mandel and S. Edel-Harth, Free nucleotides in the rat brain during postnatal development, *J. Neurochem.* **13**:591–595 (1966).
59. T. Itoh and J. H. Quastel, Ribonucleic acid biosynthesis in adult and infant rat brain *in vitro*, *Science* **164**:79–80 (1969).
60. S. Yamagami and D. A. Rappoport, unpublished data.
61. G. R. Dutton, A. T. Campagnoni, H. R. Mahler, and W. J. Moore, Studies on the labeling patterns of RNA from cerebral cortex nuclei, *J. Neurochem.* **16**:989–997 (1969).

62. J. Stevenin, P. Mandel, and M. Jacob, Relationship between nuclear giant-size dRNA and microsomal dRNA of rat brain, *Proc. Natl. Acad. Sci.* **62**:490–497 (1969).

63. L. Lim and D. H. Adams, Microsomal components in relation to amino acid incorporation by preparations from the developing rat brain, *Biochem. J.* **104**:229–238 (1967).

64. D. H. Adams and M. E. Fox, Some studies on rat brain microsomes in relation to growth and development, *Brain Res.* **12**:157–164 (1969).

65. M. R. V. Murthy and D. A. Rappoport, Biochemistry of the developing rat brain, VI. Preparation and properties of ribosomes, *Biochim. Biophys. Acta* **95**:132–145 (1965).

66. S. Navon and A. Lajtha, The uptake of amino acids by particulate fractions from brain, *Biochim. Biophys. Acta* **173**:518–531 (1969).

67. T. C. Johnson, Cell-free protein synthesis by mouse brain during early development, *J. Neurochem.* **15**:1189–1194 (1968).

68. M. L. Vahvelainen and S. S. Oja, The uptake and incorporation into protein of (^{3}H) tyrosine by slices prepared from developing rat brain cortex, *Brain Res.* **13**:227–233 (1969).

69. S. S. Oja, Studies on protein metabolism in developing rat brain, *Annales Academiae Scientiarum Fennicae, Series A, V. Medica* **131**:1–81 (1967).

70. T. C. Johnson and M. W. Luttges, The effects of maturation on *in vitro* protein synthesis by mouse brain cells, *J. Neurochem.* **13**:545–552 (1966).

71. R. J. Schain, M. J. Carver, J. H. Copenhaver, and N. R. Underdahl, Protein metabolism in the developing brain: Influence of birth and gestational age, *Science* **156**:984–985 (1967).

72. B. W. Moore and V. J. Perez, in *Physiological and Biochemical Aspects of Nervous Integration* (F. D. Carlson, ed.), pp. 343–359, Prentice-Hall, Englewood Cliffs, N.J. (1968).

73. H. Hydén and B. S. McEwen, A glial protein specific for the nervous system, *Proc. Natl. Acad. Sci.* **55**:354–358 (1966).

74. B. S. McEwen and H. Hydén, A study of specific brain proteins on the semi-micro scale, *J. Neurochem.* **13**:823–833 (1966).

75. B. S. McEwen, in *Physiological and Biochemical Aspects of Nervous Integration* (F. D. Carlson, ed.), pp. 361–381, Prentice-Hall, Englewood Cliffs, N.J. (1968).

76. L. Levine, in *The Neurosciences* (G. C. Quarton, T. Melnechuk, and F. O. Schmitt, eds.), pp. 220–230, Rockefeller University Press, New York (1967).

77. L. C. Mokrasch, Incorporation *in vitro* of ^{14}C amino acids and ^{14}C-palmitate into the rat brain proteolipids, *Fed. Proc.* **22**:300 (1963).

78. C. B. Klee and L. Sokoloff, Mitochondrial differences in mature and immature brain, *J. Neurochem.* **11**:709–716 (1964).

79. C. B. Klee and L. Sokoloff, Amino acid incorporation into proteolipid of myelin *in vitro*, *Proc. Natl. Acad. Sci.* **53**:1014–1021 (1965).

80. M. Kies, E. B. Thompson, and E. C. Alvord, Jr., The relationship of myelin protein to experimental allergic encephalomyelitis, *Ann. N.Y. Acad. Sci.* **122**:148–160 (1965).

81. R. F. Kibler, R. Shapira, S. McKneally, J. Jenkins, P. Selden, and F. Chow, Encephalitogenetic protein: Structure, *Science* **164**:577–580 (1969).

82. H. C. Ranch and S. Raffel, Immunofluorescent localization of encepholitogenic protein in myelin, *J. Immunol.* **92**:452–455 (1964).

83. E. C. Alvord, Jr., in *The Central Nervous System* (O. T. Bailey and D. E. Smith, eds.), pp. 52–70, Williams and Williams, Baltimore (1968).

84. E. Mehl and F. Wolfgram, Myelin types with different protein component in the same species, *J. Neurochem.* **16**:1091–1097 (1969).

85. G. R. Dutton and S. Barondes, Microtubular protein: Synthesis and metabolism in developing brain, *Science* **166**:1637–1638 (1969).

86. B. S. Wenger, Brain and nerve proteins: Functional correlates, *Neurosci. Res. Prog. Bull.* **3**(6), 1–53 (1965).

87. H. P. Freidman and B. S. Wenger, Adult brain antigens demonstrated in chick embryos by fractionated antisera, *J. Embryol. Exptl. Morphol.* **13**:35–43 (1965).

88. D. J. McCallion and J. Langman, An immunological study on the effect of brain extract on the developing nervous tissue in the chick embryo, *J. Embryol. Exptl. Morphol.* **12**: 107–118 (1964).

89. S. C. Bondy and S. V. Perry, Incorporation of labeled amino acids in the soluble protein fraction of rabbit brain, *J. Neurochem.* **10**:603–609 (1963).

90. M. R. V. Murthy and D. A. Rappoport, Biochemistry of the developing rat brain, V. Cell-free incorporation of L-(1-^{14}C) leucine into microsomal protein, *Biochim. Biophys. Acta* **95**:121–131 (1965).

91. S. Gelber, P. L. Campbell, G. E. Deibler, and L. Sokoloff, Effects of L-thyroxine on amino acid incorporation into protein in mature and immature rat brain, *J. Neurochem.* **11**:221–229 (1964).

92. K. Suzuki, S. R. Korey, and R. D. Terry, Studies on protein synthesis in brain microsomal system, *J. Neurochem.* **11**:403–412 (1964).

93. D. H. Adams and L. Lim, Amino acid incorporation by preparations from the developing rat brain, *Biochem. J.* **99**:261–265 (1966).

94. T. C. Johnson and G. Belytschko, Alteration in microsomal protein synthesis during early development of the mouse brain, *Proc. Natl. Acad. Sci.* **62**:844–851 (1969).

95. D. A. Rappoport and R. R. Fritz, *in Protides of the Biological Fluids* (H. Peeters, ed.), pp. 53–62, Elsevier, Amsterdam (1966).

96. S. Yamagami, R. R. Fritz, and D. A. Rappoport, unpublished data.

97. D. Richter, *in Regional Development of the Brain in Early Life* (A. Minkowski, ed.), pp. 137–156, F. A. Davis, Philadelphia (1967).

98. R. M. Burton, The pyridine nucleotide and diphosphopyridine nucleotidase levels of the brain of young rats, *J. Neurochem.* **2**:15–20 (1957).

99. J. Myers and D. A. Rappoport, unpublished data.

100. O. Ouchterlony, Antigen—antibody reactions in gels, *Acta Pathol. Microbiol. Scand.* **26**: 507–515 (1949).

101. J. J. Scheidegger, Une micro-methode de l'immuno-electrophorese, *Intern. Arch. Allergy Appl. Immunol.* **7**:103–110 (1955).

102. G. Mancini, A. O. Carbonara, and J. F. Heremans, Immunochemical quantitation of antigens by single radial immunodiffusion, *Immunochem.* **2**:235–254 (1965).

103. J. A. Burdman and L. J. Journey, Protein synthesis in isolated nuclei from adult rat brain, *J. Neurochem.* **16**:493–500 (1969).

104. L. S. Hnilica, *in Developmental and Metabolic Control Mechanisms and Neoplasia*, pp. 273–295, Williams and Wilkins Co., Baltimore (1965).

105. J. Bonner, M. E. Dahmus, D. Fambrough, R. C. Huang, K. Marushige, and D. H. Y. Tuan, The biology of isolated chromatin, *Science* **159**:47–56 (1968).

106. J. Paul and R. S. Gilmour, Organ-specific restriction of transcription in mammalian chromatin, *J. Mol. Biol.* **34**: 305–316 (1968).

107. R. S. Gilmour and J. Paul, RNA transcribed from reconstituted nucleoprotein is similar to natural RNA, *J. Mol. Biol.* **40**:137–139 (1969).

108. K. Marushige and J. Bonner, Template properties of liver chromatin, *J. Mol. Biol.* **15**: 160–174 (1966).

109. E. H. Davidson, *Gene Activity in Early Development*, Academic Press, New York (1968).

110. K. R. Brizzee, J. Vogt, X. Kharetchko, Postnatal changes in glia/neuron index with a comparison of methods of cell enumeration in white rat, *Progr. Brain Res.* **4**:136–149 (1964).

111. J. L. Conel, "Postnatal Development of the Human Cerebral Cortex," Vols. 1–6, Harvard University Press, Cambridge, Mass. (1939–1963).

112. J. Altman, in *The Neurosciences* (G. C. Quarton, T. Melnechuk, and F. O. Schmitt, eds.), pp. 723–743, Rockefeller University Press, New York (1967).

113. H. Haug, in *Structure and Function of the Cerebral Cortex* (D. B. Tower and J. P. Schade, eds.), pp. 28–34, Elsevier, Amsterdam (1960).

114. E. Horstmann, in *Structure and Function of the Cerebral Cortex* (D. B. Tower and J. P. Schade, eds.), pp. 59–63, Elsevier, Amsterdam (1960).

115. P. I. Yakovlev, Morphological criteria of growth and maturation of the nervous system in man, *Res. Publ. Assoc. Res. Nervous Mental Disease* **39**:3–46 (1962).

116. A. Peters, in *The Structure and Function of Nervous Tissue* (G. H. Bourne, ed.), Vol. 1, pp. 141–186, Academic Press, New York (1968).

117. J. Altman, Autoradiographic study of degenerative and regenerative proliferation of neuroglia cells with tritiated thymidine, *Exptl. Neurol.* **5**: 302–318 (1962).

118. J. Altman, Autoradiographic investigation of cell proliferation in the brains of rats and cats, *Anat. Record* **145**: 573–591 (1963).

119. J. Altman and G. D. Das, Autoradiographic and histological evidence of postnatal hippocampal neurogenesis in rats, *J. Comp. Neurol.* **124**: 319–335 (1965).

120. J. Altman, Proliferation and migration of undifferentiated precursor cells in the rat during postnatal gliogenesis, *Exptl. Neurol.* **16**:263–278 (1966).

121. J. Altman and G. D. Das, Autoradiographic and histological studies of postnatal neurogenesis, I. A. longitudinal investigation of the kinetics, migration and transformation of cells incorporating tritiated thymidine in neonate rats, with special reference to postnatal neurogenesis in some brain regions, *J. Comp. Neurol.* **126**:337–389 (1966).

122. M. A. Wells and J. C. Dittmer, A comprehensive study of the postnatal changes in the concentration of the lipids of developing rat brain, *Biochemistry* **6**:3169–3175 (1967).

123. G. M. McKhann, R. Levy, and W. Ho, in *Regional Development of the Brain in Early Life* (A. Minkowski, ed.), pp. 189–199, F. A. Davis, Philadelphia (1967).

124. K. Suzuki, S. E. Poduslo, and W. T. Norton, Gangliosides in the myelin fraction of developing rats, *Biochim. Biophys. Acta* **144**:375–381 (1967).

125. K. Suzuki, J. F. Poduslo, and S. E. Poduslo, Further evidence for a specific ganglioside fraction closely associated with myelin, *Biochim. Biophys. Acta* **152**:576–586 (1968).

126. G. A. Dhopeshwarkar, R. Maier, and J. F. Mead, Incorporation of (1-^{14}C) acetate into the fatty acids of the developing rat brain, *Biochim. Biophys. Acta* **187**:6–12 (1969).

127. J. G. Salway, J. L. Harwood, M. Dai, G. L. White, and J. N. Hawthorne, Enzymes of phosphoinositide metabolism during rat brain development, *J. Neurochem.* **15**:221–226 (1968).

128. G. Hauser, J. Eichberg, S. M. Gompertz, and M. Ross, in *First International Meeting of the International Society for Neurochemistry*, p. 93, Strasbourg, France (1967).

129. J. Eichberg and G. Hauser, Polyphosphoinositide biosynthesis in developing rat brain homogenates, *Ann. N.Y. Acad. Sci.* **165**:784–789 (1969).

130. H. J. Campbell and J. T. Eayrs, Influence of hormones on the central nervous system, *Brit. Med. Bull.* **21**:81–86 (1965).

131. J. T. Eayrs and G. Horn, The development of cerebral cortex in hypothyroid and starved rats, *Anat. Record* **121**:53–61 (1955).

132. M. Hamburgh and L. B. Flexner, Biochemical and physiological differentiation during morphogenesis, XXI. Effect of hypothyroidism and hormone therapy on enzyme activities of the developing cerebral cortex of the rat, *J. Neurochem.* **1**:279–288 (1957).

133. S. E. Geel and P. S. Timiras, Influence of neonatal hypothyroidism and of thyroxine on the acetylcholinesterase and cholinesterase activities in the developing central nervous system of the rat, *Endocrinol.* **80**:1069–1074 (1967).

134. S. E. Geel, T. Valcana, and P. S. Timiras, Effect of neonatal hypothyroidism and of thyroxine on L-(^{14}C) leucine incorporation in protein *in vivo* and the relationship of ionic levels in the developing brain of the rat, *Brain Res.* **4**:143–150 (1967).

135. J. M. Pasquini, B. Kaplun, C. A. Garcia Argiz, and C. J. Gomez, Hormonal regulation of brain development, I. The effect of neonatal thyroidectomy upon nucleic acids, protein and two enzymes in developing cerebral cortex and cerebellum of the rat, *Brain Res.* **6**:621–634 (1967).

136. C. A. Garcia Argiz, J. M. Pasquini, B. Kaplun, and C. J. Gomez, Hormonal regulation of brain development, II. Effect of neonatal thyroidectomy on succinate dehydrogenase and other enzymes in developing cerebral cortex and cerebellum of the rat, *Brain Res.* **6**:635–646(1967).

137. L. Krawiec, C. A. Garcia Argiz, C. J. Gomez, and J. M. Pasquini, Hormonal regulation of brain development, III. Effects of triiodothyronine and growth hormone on the biochemical changes in the cerebral cortex and cerebellum of neonatally thyroidectomized rats, *Brain Res.* **15**:209–218 (1969).

138. H. M. Evans, M. E. Simpson, and R. I. Pencharz, Relation between the growth promoting effects of the pituitary and the thyroid hormone, *Endocrinol.* **25**:175–182 (1939).

139. T. N. Salmon, Effect of pituitary growth substance on the development of rats thyroidectomized at birth, *Endocrinol.* **29**:291–296 (1941).

140. R. O. Scow and W. Marx, Response to pituitary growth hormone of rats thyroidectomized on the day of birth, *Anat. Record* **91**:227–236 (1945).

141. R. O. Scow, M. E. Simpson, C. W. Asling, C. H. Li, and H. M. Evans, Response by the rat thyroparathyroidectomized at birth to growth hormone and to thyroxine given separately or in combination, I. General growth and organ changes, *Anat. Record* **104**:445–463 (1949).

142. J. T. Eayrs, Protein anabolism as a factor ameliorating the effects of early thyroid deficiency, *Growth* **25**:175–189 (1961).

143. C. J. Gomez, N. E. Ghittoni, and J. M. Dellacha, Effect of L-thyroxine or somatotrophin on body growth and cerebral development in neonatally thyroidectomized rats, *Life Sci.* **5**:243–246 (1966).

144. S. Schapiro, Metabolic and maturational effects of thyroxine on the infant rat, *Endocrinol.* **78**:527–532 (1966).

145. J. A. Cocks, R. Balazs, and J. T. Eayrs, The effect of thyroid hormones on the biochemical maturation of the rat brain, *J. Biochem.* **111**: 18p (1969).

146. S. Zamenhof, Stimulation of cortical-cell proliferation by the growth hormone, III. Experiments on albino rats, *Physiol. Zool.* **15**:281–292 (1942).

147. S. Zamenhof, J. Mosley, and E. Schuller, Stimulation of the proliferation of cortical neurons by prenatal treatment with growth hormone, *Science* **152**:1396–1397 (1966).

148. B. G. Clendinnen and J. T. Eayrs, The anatomical and physiological effects of prenatally administered somatotrophin on cerebral development in rats, *J. Endocrinol.* **22**:183–193 (1961).

149. M. C. Diamond, R. E. Johnson, C. Ingham, and B. Stone, Lack of direct effect of hypophysectomy and growth hormone on postnatal rat brain morphology, *Exptl. Neurol.* **23**:51–57 (1969).

150. K. M. Gregory and M. C. Diamond, Effects of early hypophysectomy on brain morphogenesis in the rat, *Exptl. Neurol.* **20**:394–402 (1968).

151. M. C. Diamond, The effects of early hypophysectomy and hormone therapy on brain development, *Brain Res.* **7**:399–406 (1968).

152. M. Hamburgh, E. Lynn, and E. P. Weiss, Analysis of the influence of thyroid hormone on prenatal and postnatal maturation of the rat, *Anat. Record.* **150**:147–162 (1964).

153. P. Walravens and H. P. Chase, Influence of thyroid on formation of myelin lipids, *J. Neurochem.* **16**:1477–1484 (1969).
154. A. Cuaron, J. Gamble, N. B. Myant, and C. Osorio, The effect of thyroid deficiency on the growth of the brain and on the deposition of brain phospholipids in foetal and newborn rats, *J. Physiol.* (*London*) **168**:613–630 (1963).
155. M. Hamburgh and R. P. Bunge, Evidence for a direct effect of thyroid hormone on maturation of nervous tissue grown *in vitro*, *Life Sci.* **3**:1423–1430 (1964).
156. M. Hamburgh, Evidence for a direct effect of temperature and thyroid hormone on myelinogenesis, *Develop. Biol.* **13**:15–30 (1966).
157. L. M. Heim and P. S. Timiras, Gonad-brain relationship: Precocious brain maturation after estrodiol-rats, *Endocrinol.* **72**:598–606 (1963).
158. A. Vernadakis and P. S. Timiras, Effect of estrodiol on spinal cord convulsions in developing rats, *Nature* **197**:906 (1963).
159. A. Vernadakis and D. M. Woodbury, Effect of cortisol on the electroshock seizure thresholds in developing rats, *J. Pharmacol. Exp. Ther.* **139**:110–113 (1963).
160. A. Vernadakis and D. M. Woodbury, Effects of cortisol and diphenylhydantoin (Dilantin) on spinal cord convulsions in developing rats, *J. Pharmacol. Exp. Ther.* **144**:316–320 (1964).
161. J. J. Curry and L. M. Heim, Brain myelination after neonatal administration of oestradiol, *Nature* **209**:915–916 (1966).
162. R. Caspar, A. Vernadakis, and P. S. Timiras, Influence of estrodiol and cortisol on lipids and cerebrosides in the developing rat brain and spinal cord of the rat, *Brain Res.* **5**:524–526 (1967).
163. J. DeVellis and D. Inglish, Hormonal control and glycerolphosphate dehydrogenase in the rat brain, *J. Neurochem.* **15**:1061–1070 (1968).
164. M. Winick and A. Coscia, Cortisone-induced growth failure in neonatal rats, *Pediat. Res.* **2**:451–455 (1968).
165. E. Howard, Effects of corticosterone and food restriction on growth and on DNA, RNA, and cholesterol contents of the brain and liver in infant mice, *J. Neurochem.* **12**:181–191 (1965).
166. B. F. Chow and C.-J. Lee, Effect of dietary restriction of pregnant rats on body weight gain of the offspring, *J. Nutr.* **82**:10–18 (1964).
167. C.-J. Lee and B. F. Chow, Protein metabolism in the offspring of underfed mother rats, *J. Nutr.* **87**:439–443 (1965).
168. A. M. Hsueh, C. E. Agustin, and B. F. Chow, Growth of young rats after differential manipulation of maternal diet, *J. Nutr.* **91**:195–200 (1967).
169. C.-J. Lee and B. F. Chow, Metabolism of proteins by progeny of underfed mother rats, *J. Nutr.* **94**:20–26 (1968).
170. B.-N. Blackwell, R. Q. Blackwell, T. T. S. Yu, Y.-S. Weng, and B. F. Chow, Further studies on growth and feed utilization in progeny of underfed mother rats, *J. Nutr.* **97**:79–84 (1968).
171. M. Simonson, R. W. Sherwin, J. K. Anilane, W. Y. Yu, and B. F. Chow, Neuromotor development in progeny of underfed mother rats, *J. Nutr.* **98**:18–24 (1969).
172. A. N. Davidson and J. Dobbing, Myelination as a vulnerable period in brain development, *Brit. Med. Bull.* **22**:40–44 (1966).
173. M. Winick and P. Rosso, The effect of severe early malnutrition on cellular growth of human brain, *Pediat. Res.* **3**:181–184 (1969).
174. J. W. Millen, *The Nutritional Basis for Reproduction*, C. C. Thomas, Springfield, Ill. (1962).
175. S. Zamenhof, E. van Marthens, and F. L. Margolis, DNA (cell number) and protein in neonatal brain: Alteration by maternal dietary protein restriction, *Science* **162**: 322–323 (1968).

176. H. P. Chase, W. F. B. Lindsley, Jr., and D. O'Brien, Undernutrition and cerebellar development, *Nature* **221**:554–555 (1969).

177. M. Winick, Malnutrition and brain development, *J. Pediat.* **74**:667–679 (1969).

178. J. W. Benton, H. W. Moser, P. R. Dodge, and S. Carr, Modification of the schedule of myelination in the rat by early nutritional deprivation, *Pediatrics* **38**:801–807 (1966).

179. E. van Marthens and S. Zamenhof, Deoxyribonucleic acid of neonatal rat cerebrum increased by operative restriction of litter size, *Exptl. Neurol.* **23**:214–219 (1969).

180. W. J. Culley and R. O. Lineberger, Effect of undernutrition on the size and composition of the rat brain, *J. Nutr.* **96**:375–381 (1967).

181. N. Kretchmer and R. E. Greenberg, Some physiological and biochemical determinants of development, *Adv. Pediatrics* **14**:201–251 (1966).

182. J. D. Watson, *Molecular Biology of the Gene*, W. A. Benjamin, New York (1965).

183. G. Zubay, in *The Nucleohistones* (J. Bonner and P. Ts'o, eds.), pp. 95–107, Holden-Day, San Francisco (1964).

184. J. Bonner, M. E. Dahmus, D. Fambrough, R.-C. Huang, K. Marushige, and D. Y. H. Tuan, The biology of isolated chromatin, *Science* **159**:47–56 (1968).

185. G. P. Georgiev, Histones and the control of gene action, *Ann. Rev. Genetics* **3**:155–180 (1969).

186. K. B. Smith, R. B. Church, and B. J. McCarthy, Template specificity of isolated chromatin, *Biochemistry* **8**:4271–4277 (1969).

187. G. P. Georgiev, The nature and biosynthesis of nuclear ribonucleic acids, *Progr. Nucleic Acid Res. Mol. Biol.* **6**:259–351 (1967).

188. V. G. Allfrey, V. C. Littau, and A. E. Mirsky, On the role of histones in regulating RNA synthesis in the cell nucleus, *Proc. Natl. Acad. Sci.* **49**:414–421 (1963).

189. R. B. Church and B. J. McCarthy, RNA synthesis in regenerating and embryonic liver, I. The synthesis of new species of RNA during regeneration of mouse liver after partial hepatectomy, *J. Mol. Biol.* **23**:459–475 (1967).

190. A. O. Pogo, V. G. Allfrey, and A. E. Mirsky, Evidence for increased DNA template activity in regenerating liver nuclei, *Proc. Natl. Acad. Sci.* **56**:550–557 (1966).

191. J. E. Loeb and C. Creuzet, Electrophoretic comparison of acidic proteins of chromatin from different animal tissues, *FEBS Letters* **5**:37–40 (1969).

192. R. C. C. Huang and P. C. Huang, Effect of protein-bound RNA associated with chick embryo chromatin on template specificity of the chromatin, *J. Mol. Biol.* **39**:365–378 (1969).

193. I. Bekhor, G. M. Kung, and J. Bonner, Sequence-specific interaction of DNA and chromosomal protein, *J. Mol. Biol.* **39**:351–364 (1969).

194. J. Stevenin, J. Samec, M. Jacob, and P. Mandel, Determination de la fraction du genome codant pour les RNA ribosomiques et messagers dans le Cerveau du rat adulte, *J. Mol. Biol.* **33**:777–793 (1968).

195. F. Jacob and J. Monod, Genetic regulatory mechanisms in the synthesis of protein, *J. Mol. Biol.* **3**:318–358 (1961).

196. L. S. Hnilica, Proteins of the cell nucleus, *Progr. Nucleic Acid Res. and Mol. Biol.* **7**:25–106 (1967).

197. E. Stubblefield, in *The Proliferation and Spread of Neoplastic Cells*, pp. 175–193, Williams and Wilkins Co., Baltimore (1968).

Chapter 14

AGING

Robert L. Herrmann

Department of Biochemistry
Boston University School of Medicine
Boston, Massachusetts

I. INTRODUCTION

The study of the phenomenon of cellular aging is severely limited by the meager amount of basic information available, and nowhere is this more evident than in the area of the neural sciences. The present review is thus a very limited one, and some of the data considered have of necessity been more of developmental than aging significance. For broader information on aging, the reader is referred to the excellent review of Strehler[1] and to a recent symposium volume.[2]

II. MORPHOLOGY OF AGING BRAIN

A. Histological Changes

The decrease of cell number in the central nervous system with old age has been considered one of the few well-established events of the aging process. Ganglion cells of the central nervous system show no mitotic activity, so dead cells are not replaced.[3] Losses have been observed in Purkinje cells in the cerebellar cortex[4] and in the neuron population of the cerebral cortex in aging rats.[5] Similar results have been reported in measurements of aging human brain.[6,7] An increase in the percentage of neurons with satellite cells has also been reported.[8] These results, which are corroborated by the recent studies reported by Bondereff,[9] seem to be in disagreement with studies of variation of DNA concentration with age, to be considered in a later section.

B. Alterations in Subcellular Constituents

Studies of subcellular constituents are complicated by the difficulty of following the life history of single organelles. Dempsey[10] described degenerating mitochondria in old tissues, and Andrew[11] has observed an age-related shift from filamentous to granular appearance of mitochondria of Purkinje cells of the mouse cerebellum. The question to be answered is whether these differences are directly caused by aging or are rather due to the altered environment of the aging cell, since mitochondria have been shown to be very susceptible to environmental influences. Palay and Palade[12] have reported changes in endoplasmic reticulum associated with a decrease in the number of granular, RNA-containing structures of the reticulum called Nissl bodies. These structures, which may be related to the Golgi apparatus, may change because of age-related changes in the cellular milieu, or their presence may reflect an actual decline in the protein-synthetic capability of the cell.

A further morphological change is seen in the accumulation of pigmented inclusion bodies which have been called lipofuscin granules in nerve and muscle tissue. That these are distinct from Nissl substance has been shown by Bethe and Fluck.[13] The histochemical characterization of lipofuscin and its experimental formation by various stress conditions has been reported by Sulkin and Srivanij.[14] These results could be interpreted to indicate that lipofuscin accumulation is not a fundamental event in the aging of nervous tissue but rather is the result of various types of insult to the organism in the course of age. The nature and possible origin of lipofuscin will be considered in a later section.

III. METABOLISM OF AGING BRAIN

A. Nucleoprotein Changes

1. Structure

Age differences in the thermal melting of DNA have been reported by Hahn and Fritz[15] for rat liver. These studies have been extended to mouse brain deoxynucleoprotein by Kurtz and Sinex.[16] The T_m of these preparations from brain nuclei were found to decrease regularly for animals from 3 days to 13 months of age and thereafter, to increase for animals up to 30 months of age. The T_m of deoxynucleoprotein from 30-month-old mice was found to be 8°C higher than the preparation from 13-month-old animals. The protein-to-DNA ratio varies in a roughly similar manner, decreasing from 3 days to 5 months, remaining constant until 13 months, and increasing thereafter. The change in T_m as a function of age is interpreted as more than an effect of the variation in protein content of deoxynucleoprotein. That a qualitative aspect is present is indicated by an examination of the variation of T_m as a function of the ratio of the log of protein to DNA. Separate straight lines of differing slopes are obtained for deoxynucleoprotein from young and old mice; in addition, through the middle age range, the ratio of total protein to DNA remains constant while T_m varies. The conclusion of the authors is that a qualitative component exists in the protein-mediated stabilization of the DNA helix with age.

2. *Histone and Nonhistone Protein*

Kurtz and Sinex[16] have observed that mouse brain chromatin has a constant histone composition as a function of age over the range from 3 days to 30 months. Conversely, nonhistone content varies by severalfold from high levels in young and old to low levels in mature brain. These workers suggest that the factor or factors determining the melting behavior of brain chromatin may reside in the nonhistone fraction. This result is of considerable interest since the histone fraction has generally been considered the most likely mammalian counterpart of the repressor proteins of bacterial systems.[17,18]

3. *DNA Concentration and Turnover*

Burger[19] has reported an apparent increase in the DNA of very old human brain due to loss of cytoplasm. Conversely, Cammermayer[20] has observed a loss of chromatin in brain nuclei with age. More recently, Yajima observed no changes in DNA concentrations as a function of age in diencephalon, brain stem, cerebrum, and cerebellum.[21] The most definitive study to date is that of Hahn,[22] who has measured DNA, RNA, and protein concentrations in five fractions of brain as a function of age for a large number of Wistar strain rats. Large changes occur during the growth period, DNA and RNA decreasing after the first month and protein increasing up to about 6 months of age. DNA and RNA decrease only slightly during adult and senescent periods and this decrease is largely restricted to the hemispheres. It is suggested that the loss of neurons which has been found with histological methods may be compensated for by an increase in other cells such as glia. Alternatively, losses may be restricted to a very few cell types, such as the Purkinje cells of the cerebellum, or the pyramidal cells of the third cortical lamina, as borne out by the cytophotometric measurements of Wulff and Freshman[23] and of Wayner *et al.*[24] Hahn concludes that age-related loss in brain function due to cellular aging in surviving neurons may be at least as important as the quantitative loss of neurons. Such cellular aging could be caused in part by alterations in the structure of the nucleic acids.

B. Alterations in Other Systems

1. *RNA Synthesis and Concentration*

The effect of age upon synthesis of RNA has been studied by Wulff, Quastler, and Sherman[25] by measurement of the incorporation of tritiated cytidine into various mouse tissues. The labeled precursor was found to be incorporated with more efficiency in old rather than young mouse liver. This result was interpreted to indicate more rapid turnover of RNA in older animals, perhaps as a result of feedback effects at the genetic level. More recently, Wulff *et al.*[26] have observed that the incorporation of precursor is profoundly affected by the amount injected, indicating that the age differences may be due to differences in pool sizes of intermediates in the various tissues. Also, considerable variation was found to exist from tissue to tissue. In contrast to the results with liver, tissues of the central nervous system from young mice incorporated more cytidine-^{3}H than those from old. However, upon addition of unlabeled cytidine, the age-related differences in incorporation of cytidine-^{3}H into spinal cord, lower motoneurons, and cerebellar Purkinje cells were diminished by virtue of a relatively greater increase in incorporation of precursor in old rather than in young mice. Studies of RNA concentration in rat brain reported by Hahn[22] were mentioned previously in conjunction with DNA studies.

2. Protein Synthesis

The incorporation of unlabeled amino acids into nerve cells has been measured by several groups of workers. Palladin, Belik, and Kracko[27] observed lower rates of incorporation in old animals and the reverse was found by Oeriu.[28] A similar study was carried out by Jakoubek et al.[29] for isolated spinal motoneurons, but with variation in time of labeling. The latter workers found no significant changes with age in the turnover rates of proteins between adolescent and old animals. As pointed out by these authors, incorporation experiments depend heavily upon the relationship between the specific activities of the proteins and their precursors, upon the heterogeneity of nervous tissues, and upon the different turnover rates of different proteins in the tissues. Judging by present results, it would seem desirable to have methods for the examination of the turnover rates of individual proteins of nerve tissue, especially since Lajtha and Toth[30] have shown higher uptake of amino acids into young rat brain, making the precursor–product relationship very uncertain.

3. Energy Production

Differences between brain mitochondria from young and old rats has been reported by Weinbach and Garbus.[31] The rate of phosphorylation was unchanged with α-ketoglutarate, malate, or succinate as substrate, but with β-hydroxybutyrate, the rate decreased by 60%, though the P/O ratio was unaffected. The P/O ratio of mitochondria from aged animals was shown to decrease upon storage for several days, whereas those from young animals were unaffected.

C. Alteration in Specific Enzyme Levels

There has been little work done on age-related changes in specific enzyme levels in brain. Developmental changes have been reported for rat brain neuraminidase by Carubelli[32] and for acid and alkaline DNase by Sung.[33] Of possible importance to aging in brain is the observation by the latter author of a very high ratio of alkaline DNase to acid DNase in cerebellum in adult rats but not in young animals.

Reiner[34] has studied carbohydrate metabolism in rat brain homogenates and observed a decline in brain oxygen consumption with animals over 2 years of age. Anaerobic glycolysis was also found to decrease.

D. Accumulation of Certain Tissue Components

The accumulation of lipofuscin granules with age was mentioned as one morphological aspect of aging. The chemical nature of lipofuscin pigment has been studied by Strehler and his co-workers.[35] Chromatographic behavior on silicic acid columns indicates that the pigment may in part represent autoxidation products of cephalin and perhaps other unsaturated lipids. It is suggested that lipofuscin results from accumulation and autoxidation of lipid components of lysosomes. Its significance for aging is uncertain.

IV. POSSIBLE MECHANISMS OF AGING

A variety of theories of cellular aging have been proposed. For a detailed discussion, see the review by Strehler.[1] Three general theories may be briefly outlined. The first group concerns instability of critical cellular structures or functions, cytoplasmic or nuclear. Within this category are found those theories which invoke changes in regulatory mechanisms at the genetic level. Hahn[36] has proposed such a mechanism to explain the increased thermal stability of DNA as a function of age. Thus, repressor proteins have been postulated to become irreversibly bound to their sites on DNA, leading to cellular dysfunction. An alternate view, that of Wulff, Quastler, and Sherman[37] suggests that with senescence, alterations occur in the primary structure of DNA. These are transcribed to messenger RNA and ultimately translated into altered proteins. Since some of these altered proteins would represent inactive enzymes, the pool sizes of endproducts of pathways in which they were involved would decrease. This would in turn lead to a derepression of enzyme synthesis, leading to an increase in the synthesis of RNA and nonfunctional enzymes and a corresponding waste to the cell.

A second group of aging theories is based upon the idea of aging as a normal or planned extension of differentiation and development. Thus, aging may be a genetic propensity of the organism, programmed into the genetic complement of the cell, perhaps as a result of evolutionary selection. A theory of this type has been proposed by Medvedev.[38] The third group of theories is based upon the interdependence of or antagonism between various cellular components. These interactions have been conceived at the level of the immune systems and also at the level of hormone function.

The impression of this reviewer is that much more data are needed in order to construct a meaningful hypothesis for the aging mechanism. Probably no single factor is primary. As Pearl once suggested many years ago,[39] aging is a phenomenon which involves the total structural–functional potential of the cell.

V. REFERENCES

1. B. L. Strehler, *Time, Cells, and Aging*, Academic Press, New York (1962).
2. *Aspects of the Biology of Aging*, Symposium of the Society for Experimental Biology, Number XXI, Academic Press, New York (1967).
3. C. P. Leblond and B. E. Walker, Renewal of cell populations, *Physiol. Rev.* **36**:255–276 (1956).
4. T. Inukai, On the loss of Purkinje cells, with advancing age, from the cerebellar cortex of the albino rat, *J. Comp. Neurol.* **30**:229 (1919).
5. H. Kuhlenbeck, Senile changes in the brain of Wistar Institute rats, *Anat. Rec.* **88**:441 (1944).
6. H. Brody, Organization of the cerebral cortex. III. A study in aging in human cerebral cortex, *J. Comp. Neurol.* **102**:511–556 (1955).
7. W. Andrew and M. A. Bari, Some aspects of age changes in the spinal cord compared with those in other parts of the nervous system, *in Proc. of the Fifth International Cong. of Neuropathology*, Excerpta Medica Foundation, New York, 1965. International Congress Ser. No. 100, pp. 518–525.
8. H. M. Wahal and H. H. Riggs, Changes in the brain associated with senility, *Arch. Neurol. Psychiat. Chicago* **2**:151 (1960).
9. W. Bondereff, Histophysiology of the aging nervous system, *in Advances in Gerontological Research* (B. L. Strehler, ed.), Vol. 1, Academic Press (1964), p. 1.
10. E. W. Dempsey, Mitochondrial changes in different physiological states, *in Ciba Foundation Colloquia on Aging* (G. E. W. Wolstenholme and E. C. P. Millar, eds.), Vol. 2: *Aging in Transient Tissues*, Little, Brown, Boston (1956), pp. 100–102.
11. W. Andrew, comments *in The Biology of Aging* (B. L. Strehler *et al.*, eds.), Publ. No. 6, Am. Inst. Biol. Sci., Washington (1960), p. 37.
12. S. L. Palay and G. E. Palade, The fine structure of neurons, *J. Biophys. Biochem. Cytol.* **1**:69 (1954).
13. A. Bethe and M. Fluck, Uber das gelbe Pigment der Ganglienzellen, seine kolloidchemischen and topographischen Beziehungen zu andern Zells-trukturen und eine elekive Methode zu seiner Darstellung, *Z. Zellforsch* **27**:211 (1937).
14. N. M. Sulkin and P. Srivanij, The experimental production of senile pigments in the nerve cells of young rats, *J. Gerontol.* **15**:2 (1960).
15. H. P. von Hahn and E. Fritz, Age-related alterations in the structure of DNA. III Thermal stability of rat liver DNA related to age, histone content and ionic strength, *Gerontologia, Basel* **12**:237 (1966).
16. D. I. Kurtz and F. M. Sinex, Age-related differences in the Association of brain DNA and nuclear protein, *Biochim. Biophys. Acta* **145**:140 (1967).
17. R. C. Huang and J. Bonner, Histone, a suppressor of chromosomal RNA synthesis, *Proc. Nat. Acad. Sci. U.S.* **48**:1216 (1962).
18. R. L. Herrmann, Gene interactions in lower organisms as models for development and aging, *J. Gerontol.* **22**:Part II, 9 (1967).
19. M. Burger, Der desoxyribonucleursaure- und ribonuclensaure-gehalt des menschlichen Gehirns im Laufe des Levens, *J. Altenforsch.* **12**:133 (1958).
20. J. Cammermayer, Cytological manifestations of aging in rabbit and chinchilla brains, *J. Geront.* **18**:41 (1963).
21. A. Yajima, The nucleic acid content of the brain tissue of rats as influenced by age, *Tohuku J. Exp. Med.* **85**: 259 (1965).
22. H. P. von Hahn, Distribution of DNA and RNA in the brain during the life span of the albino rat, *Gerontologia, Basel* **12**: 18 (1966).
23. V. J. Wulff and M. Freshman, Age-related reduction of RNA content of rat cardiac muscle and cerebellum, *Arch. Biochem. Biophys.* **95**:181 (1961).

24. M. J. Wayner, V. J. Wulff, and M. Piekielniak, Ribonucleic acid content of tissues of rats of various ages, *J. Gerontol.* **17**:455 (1962).

25. V. J. Wulff, H. Quastler, and F. G. Sherman, The incorporation of ^{3}H-cytidine in mice of different ages, *Arch. Biochem. Biophys.* **95**:548 (1961).

26. V. J. Wulff, H. Quastler, F. G. Sherman, and H. V. Samis, The effect of specific activity of ^{3}H-cytidine on its incorporation into tissues of young and old mice, *J. Gerontol.* **20**:34 (1965).

27. A. V. Palladin, Y. V. Belik, and L. I. Kracko, Rate of protein renewal in the brain in states of stimulation and inhibition at different ages of the test animal, *Biokhymia* **22**:359 (1957).

28. S. Oeriu, Proteins in development and senescence, *in Advances in Gerontological Research* (B. L. Strehler, ed.), Vol. 1, Academic Press, New York (1964), pp. 23–78.

29. B. Jakoubek, E. Gutmann, J. Fischer, and A. Babicky, Rate of protein renewal in spinal motoneurons of adolescent and old rats, *J. Neurochem.* **15**:633 (1968).

30. A. Lajtha and J. Toth, The brain barrier system. II. Uptake and transport of amino acids by the brain, *J. Neurochem.* **8**:216 (1961).

31. E. C. Weinbach and J. Garbus, Age and oxidative phosphorylation in rat liver and brain, *Nature* **178**:1225 (1956).

32. R. Carubelli, Changes in rat brain neuraminidase during development, *Nature* **219**:955 (1968).

33. S.-C. Sung, Deoxyribonucleases from rat brain, *J. Neurochem.* **15**:477 (1968).

34. J. M. Reiner, The effect of age on carbohydrate metabolism of tissue homogenates, *J. Gerontol.* **2**:315 (1947).

35. D. D. Hendley, B. L. Strehler, M. C. Reporter, and M. V. Gee, Further studies on human cardiac age pigment, *Fed. Proc.* **20**:298 (1961).

36. H. P. Hahn, A model of "regulatory" aging of the cell at the gene level, *J. Gerontol.* **21**:291 (1966).

37. V. J. Wulff, H. Quastler, and F. G. Sherman, An hypothesis concerning RNA metabolism and aging, *Proc. Nat. Acad. Sci. U.S.* **48**:1373 (1962).

38. Zh. A. Medvedev, The nucleic acids in development and aging, *in Advances in Gerontological Research* (B. L. Strehler, ed.), Vol. 1, pp. 181–206, Academic Press, New York (1964).

39. R. Pearl, S. L. Parker, and B. M. Gonzalez, Experimental studies on the duration of life, VII. The Mendelian inheritance of duration of life in crosses of wild type and quintuple stocks of *Drosophila melanogaster*, *Am. Naturalist* **57**:153 (1923).

Chapter 15

HIBERNATION

A. V. Palladin and N. M. Poljakova

Institute of Biochemistry of the Ukrainian Academy
of Sciences,
Kiev, U.S.S.R.

I. INTRODUCTION

Biologically, hibernation is an adaptation of the animal organism to the lowered temperature of its environment which has developed in the process of evolution.

Hibernating animals, such as the ground squirrel or chipmunk (*Citellus citellus*), the marmot (*Marmota marmota*), the hamster (*Cricetus cricetus*), the common dormouse (*Glis glis*), and the hedgehog (*Erinaceus*) possess imperfect thermal regulation. This group of mammals, which, with respect to thermal regulation, occupies a position between the warm-blooded (homiothermal) and cold-blood (poikilothermal) animals, forms a group of heterothermal animals.

In warm-blooded or homiothermal animals the body temperature is fairly constant, and does not depend on the ambient temperature.

Cold-blooded or poikilothermal animals are incapable of regulating their body temperature, which depends on the temperature of their environment.

Heterothermal animals fall into a state of hibernation in autumn under conditions of a temperate climate. In the wakeful state they maintain a constant body temperature and differ little from warm-blooded animals; in the state of hibernation, however, their body temperature is lowered and depends to a great extent on the environmental conditions.

Hibernation is a state of prolonged profound dormancy of the animal, which may be interrupted from time to time. Every 6–10 days the animal wakes up for a short time (several hours), and then it again falls asleep.

II. GENERAL CHARACTERISTICS OF HIBERNATION

A. Physiological State of Animals During Hibernation and Awakening

The state of hibernation is characterized by extremely low physiological activity (respiration, circulation, excretion), an abrupt drop in the rate of metabolism, and a marked slowing down in the activity of the central nervous system.

1. Temperature of Hibernating Animals

The dependence of the body temperature of hibernating animals on the environmental conditions holds only within a certain temperature range. Below a definite boundary, low temperature may serve as a stimulus that awakens the animal. The low temperature evoking the awakening was called the "minimum temperature" by Pfluger. It differs for various species, fluctuating from 2.3 to 4°C.[1] According to other investigators, the body temperature of hibernating animals may fall to 0°C.

The awakening of hibernating animals is a complicated and fairly long process. The awakening and warming of the body occurs at different rates in various species. The body temperature does not rise uniformly on awakening: the anterior part of the body is warmed more rapidly than the posterior. Awakening passes through a number of states—profound sleep, light sleep, sleepiness, and wakefulness.

Hibernating animals may readily be awakened artificially by raising or lowering the ambient temperature. Artificial awakening is attended by a sharp rise of all physiological functions, including the activity of the central nervous system.

2. Respiration

Respiration is greatly slowed down during hibernation. Whereas a ground squirrel performs 60–140 respirations per minute in the wakeful state, one respiration in the course of several minutes may be observed in the hibernating ground squirrel.[2]

3. Circulation

Circulation is also extremely slow during hibernation. According to Valentin's data,[3] the heart of a hibernating animal contracts once in 3–4 min, sometimes even more rarely. Owing to the slowing down of cardiac activity, there is a considerable fall in blood pressure during hibernation, the pressure in the carotid artery equalling 70–72 mm Hg.

4. Blood

In hibernating hamsters the quantity of total protein in the blood increases by 22%, and blood albumins increase by 27%; there is also an increase in the β-globulin content of the serum. The absence of changes in the α-albumin content parallels the considerable fall in γ-globulin content during hibernation.[4]

Svihla[5,6] in experiments on ground squirrels, Suomalainen[7] in experiments on hedgehogs, and Denyes[8] in experiments on hamsters found an increase in the clotting time during hibernation. Zain-ul-Abedin[9] in experiments on lizards found a sharp increase in clotting time in the state of hibernation. The increase in clotting time is due, in the main, to the decrease in blood prothrombin (Svihla).

The quantity of formed elements of the blood diminishes during hibernation, the number of leukocytes being sharply reduced.

5. Weight

Loss of weight during hibernation was studied in detail by Valentin,[10] who found that during hibernation animals lose on the average 0.27% of their weight per day; the main loss in weight coincides with the periods of awakening. During the state of profound sleep the loss of weight is extremely small.

The capacity for regeneration is lowered during hibernation. Thus, after removal of part of the liver, regeneration of the part was completed by the eleventh day in wakeful ground squirrels, while in hibernating animals regeneration ceased after 10 days.[11]

B. Metabolism in Hibernating Animals During Hibernation and Awakening

1. Respiratory Metabolism

A study of respiratory metabolism in hibernating animals showed qualitative and quantitative alterations during hibernation, a sharp fall in oxygen consumption and CO_2 excretion and a lowered respiratory quotient being noted. The respiratory quotient was found to vary in different hibernating animals. The more profound the hibernation, the less intense the respiratory metabolism and the lower the respiratory quotient.

2. Carbohydrate-Phosphorus Metabolism

Investigations of carbohydrate metabolism during hibernation have long since established (as early as by Claude Bernard) that the muscles and liver of the hibernating animals are fairly rich in glycogen, which rapidly vanishes on awakening; the glycogen content is lower than during the wakeful period in the skeletal muscles and liver, while it is higher in the myocardium.[12] The lactic acid content is also greatly decreased in the myocardium and skeletal muscles of the ground squirrel. The quantity of inorganic phosphate and ATP diminishes, while the creatine phosphate content is considerably increased.[13]

Immediately after artificial awakening (during the first 7.5 min) the glycogen content falls, in comparison with the hibernation period, in all tissues. During the following 15–30 minutes the glycogen content progressively increases. The ATP content of the myocardium and skeletal muscles remains unchanged as compared with the wakeful period. During awakening, the inorganic phosphate concentration, which decreased during hibernation, is perceptibly raised. The phosphocreatine content is correlated with the changes in glycogen: with intensifying glycolysis its content is raised (initial period of awakening), with intensifying glycogenolysis it falls (remote period of awakening).[14]

Haberey[15] studied the labeled carbon content of the exhaled carbon dioxide after intraperitoneal injection of labeled glucose into a ground squirrel and a garden dormouse and found that oxidation of labeled glucose was less intense in December–March than in June–October; the inference was drawn that in hibernating animals in December–March the path of direct oxidation of glucose is less important.

During hibernation the hexokinase activity in the ground squirrel liver is sharply attenuated. In the period of awakening the hexokinase activity rises in ground squirrels with the body temperature.[12]

Changes in carbohydrate metabolism during hibernation are also indicated by the fall in the sugar content of the blood. The blood sugar is greatly reduced (by 25 % on the average) in hibernating marmots as compared to the content in wakeful animals. In ground squirrels, the blood sugar content also diminishes, but to a lesser extent.[13] During awakening the sugar content of the blood increases slightly.[14]

3. Nitrogen Metabolism

There have been few investigations on nitrogen metabolism in hibernating animals, although they go back to the middle of the last century.[15] Research into nitrogen metabolism was conducted in greater detail by Nagai[16] at the beginning of this century. He found that during hibernation there was a considerable increase in the amino acid nitrogen excreted with the urine and a great decrease in the excretion of urea and ammonia nitrogen, which indicates disturbance in the amino acid deamination processes during hibernation. Since amino acid deamination occurs in the animal organism chiefly through oxidation, the increased excretion of amino acids, is, apparently, due to an attenuation of oxidation processes during hibernation.

The urea content in the whole blood, serum, and muscles of the European hedgehog is higher in the hibernating than in the wakeful animal, which is explained by decrease in the capacity of the kidneys to excrete nitrogen metabolism products during hibernation.[17]

III. NERVOUS SYSTEMS DURING HIBERNATION

A. General State of Nervous System During Hibernation

The state of the nervous system of hibernating animals differs considerably from that of the same animals in the wakeful state. The nervous system is in a state of profound inhibition. In the period of deep sleep, spontaneous electric activity of the cerebral cortex ceases in mammals.[22,23]

The excitability of the peripheral nervous system is greatly attenuated during hibernation. This was first pointed out by Valentin,[24] who found that the rate of spreading of excitation along the nerves is only one meter per second during hibernation. These data were confirmed by Horvath[25] in investigations on marmots.

The changes in the state of the peripheral nervous system are also pointed out by Merzbacher,[26] who established that after cutting out part of a nerve in hibernating animals, no anatomical or physiological manifestations of degeneration were observed in the remaining part of the nerve. The process of degeneration sets in immediately after excitation of the animal, and when it falls asleep ceases once more.

The nerves of hibernating animals can tolerate a lower temperature than those of nonhibernating animals. According to Chatfield[27] the nerves of hibernating hamsters continue functioning at a temperature of 3.4°C, while rat nerves cease functioning at about 9°C.

The vegetative nervous system is of great importance in hibernation of mammals. In the opinion of Hess[28] and Bruman,[29] hibernation is regulated by the vegetative nervous system, the tonus of the sympathetic nervous system falls during hibernation, and the parasympathetic influences predominate.

B. Chemical Composition and Metabolism of the Brain in Hibernating Animals

1. Phosphorus Compounds

A study of the changes in brain composition during hibernation showed that the content of nitrogenous matter and nonprotein nitrogen in the brain is unaltered during hibernation; during prolonged and deep sleep there is a rise in the total phosphorus in the brain, while the quantity of acid-soluble phosphorus is somewhat reduced. At the same time the lipoid phosphorus increases somewhat.

Studying the content of some substances separately in the gray and the white matter of the cerebral hemispheres and in the cerebellum during hibernation, Feinschmedt and Ferdman[30] showed that the total quantity of nitrogenous matter remains unchanged in all sections; the residual nitrogen content and the ratio of residual nitrogen to total nitrogen decreases to a greater extent in the gray matter than in the white matter and cerebellum in hibernating ground squirrels.

During hibernation the total phosphate content in the gray and white matter of the brain diminishes in ground squirrels; it is the same in the cerebellum, but to a lesser degree. The acid-insoluble phosphorus is reduced more in the gray matter than in the white.

A study of the content of creatine phosphoric acid, creatine, and ATP in the brain of hibernating ground squirrels showed that a great part of the creatine in the frozen brain is in the form of creatine phosphoric acid in the bound state. The ATP content is higher in the brain of hibernating animals than in that of wakeful animals.[31]

2. Carbohydrate Metabolism

Studying the histochemical topography of glycogen in the brain of hibernating animals (European hedgehogs) Stark[32] found that, as the animal falls asleep and becomes unconscious, glycogen is accumulated selectively and is, evidently, synthesized chiefly in the neurons of nonspecific formations of the brain stem (reticular nucleus of the mesencephalon roof, medial thalamus); in the cytoplasm of neurons of other brain divisions the polysaccharide concentration is almost unchanged.

The selective accumulation of glycogen in the neurons of nonspecific formations of the brain stem permits the inference that these divisions play the greatest part in maintaining homeostatic cerebral equilibrium.[33]

During awakening the quantity of glycogen sharply decreases; at this time it is the chief source of energy.

With the aim of determining the rate of glycogen metabolism in the brain during hibernation, investigations were conducted on the intensity of radio-active carbon incorporation in the brain glycogen in wakeful and hibernating ground squirrels. These researches showed that glycogen metabolism was attenuated during hibernation.[34]

The lactic acid content in the brain (frozen) of hibernating ground squirrels is lower than in that of the wakeful animals.

On studying the intensity of brain tissue respiration in ground squirrels during wakefulness and hibernation, Nechiporenko[35] found that the intensity of brain respiration was lower in the hibernating animals than in the wakeful ones. It was found at the same time that the capacity for utilizing glucose, which is known to be the basic substrate of brain respiration, is also lowered during hibernation. This is, evidently, due to the fall in activity of the enzyme systems of respiration.

In vitro determination of oxygen absorption and phosphorylation in the brain of nonhibernating (rats and guinea pigs) and hibernating mammals (hamsters and ground squirrels) showed that while the level of respiration and phosphorylation in the brain of nonhibernating animals undergoing artificial cooling remained constant, distinct changes in the oxidating phosphorylation were noted in the hibernating animals in experiments with succinate as a substrate. Separation of phosphorylation from respiration was noted during hibernation: respiration was intensified, while the ATP content remained unaltered. These changes are typical only for brain tissue, and are not observed in the liver, skeletal muscles, and myocardium.[36]

3. Nucleic Acids and Their Metabolism

Investigation of the nucleic acid content in the brain of hibernating ground squirrels[37] showed absence of distinct changes in the RNA, phosphoprotein, and phospholipid of the brain; in the spinal cord the difference in content of these substances was pronounced, especially in phospholipids and RNA, the content of which was elevated during hibernation.

Weill, Mandel, and Kayser,[38] studying the effect of hibernation and summer wakefulness on the relative weight of the organs in ground squirrels and on the contents of protein, RNA, and DNA in the organs, also found that the nucleic acid content in the brain was almost the same during hibernation and wakefulness, as was the case with the protein content; the RNA/DNA ratio is also unaltered.

Along with determinations of the nucleic acid content, investigations were also conducted on the intensity of their renewal by studying the rate of incorporation of radioactive phosphate. Skvirskaya and Silich[37] found considerable differences in the intensity of incorporation of radioactive phosphorus in RNA, phosphoproteins, and phospholipids of the brain and spinal cord in wakeful and hibernating, as well as in artificially awakened animals. The specific activity (impulses per milligram of phosphorus of the given fraction) for the various phosphorus-containing compounds in the brain and spinal cord of hibernating ground squirrels was many times lower than during wakefulness, and was zero for some of them. Thus, during hibernation RNA, phosphoprotein, and phospholipid metabolism is lower than in the wakeful state.

Interesting data were obtained on artificially awakened animals. On awakening, the intensity of radioactive phosphorus incorporation in nucleic acids, phosphoproteins, and phospholipids increased, but did not come up to the level of the wakeful animals.

The use of labeled phosphorus also permitted determining the fact that penetration of phosphorus from the blood into the tissues, in particular nerve tissue, is extremely retarded during hibernation.

The activity of certain nucleic metabolism enzymes and phosphoprotein phosphatase was also studied. In the ground squirrel, the ribonuclease activity is reduced during hibernation in the brain to a lesser degree in the spinal cord. The phosphoprotein phosphatase activity in the brain is also considerably attenuated during hibernation, and reaches its normal value during artificial awakening. Similar changes in the phosphoprotein phosphatase activity are also observed in the spinal cord, but are considerably less pronounced.

4. Protein Metabolism of Brain

Investigating protein metabolism in the brain during hibernation, which is characterized by profound inhibition of central nervous system activity, and comparing it with metabolism in the wakeful state and the artificial awakening, which is accompanied by a relatively rapid enhancement of all physiological functions including those of the brain, Belik[39] determined the rate of radioactive methionine incorporation in brain proteins of the ground squirrel.

Investigations showed that during hibernation the intensity of radio-active sulfur incorporation in brain proteins, in other words the renewal of brain proteins, was at an extremely low level; in some experiments, when the animals were in a state of torpor, protein renewal was practically nil.

During artificial awakening of ground squirrels, which is attended by a sharp rise in nervous activity, the intensity of brain protein renewal was 25 times as high on the average as in hibernating animals.

With respect to brain protein renewal, wakeful ground squirrels occupied an intermediate position: the specific radioactivity of the brain proteins was several times higher than in hibernating animals but considerably lower than in the artificially awakened.

With the aim of checking whether these differences in the rate of incorporation of radiomethionine in the brain proteins are due to differences in the permeability of the blood–brain barrier in various functional states, the specific radioactivity of the acid-soluble fraction of brain tissue was determined during hibernation, wakefulness, and artificial awakening. This fraction may be regarded as a kind of "metabolic pool," from which the brain cells receive material for the synthesis and renewal of their component parts. These determinations revealed the highest radioactivity of the acid-soluble fraction in hibernating animals; in the wakeful and artificially awakened animals it was considerably lower.

It may thus be considered that the differences in intensity of brain protein metabolism in hibernating ground squirrel on the one hand, and in wakeful and artificially awakened squirrels on the other hand, are due to different intensities in the processes of brain protein renewal in these states.

5. Content of Amino Acids in Brain

Mandel[40] found that in the brain of garden dormouse the total protein content calculated per gram of DNA phosphorus was 16% lower during hibernation than during wakefulness. He determined the free amino acid composition and found that the Thr and Tyr of the brain were higher during hibernation than in the awakened dormouse. Glu, and especially Asp, were lower, while $Glu-NH_2$ and, to a lesser degree, GABA were higher, which may be explained in the author's opinion both by retardation of Krebs' cycle and as a result of removing excess ammonia. On the whole, the alterations observed in the brain during hibernation reflect the effect on metabolism in the brain of both sleep and hunger.

The changes in the content of Glu and GABA in the brain during hibernation and awakening of ground squirrels were also studied by Mikhailo-vich,[41] who drew other conclusions: that during profound hibernation Glu and GABA and, particularly, glutamine in the brain are lower than in the wakeful state. On awakening from hibernation, the Glu and glutamine content are unchanged, while the GABA level is somewhat lowered.

According to the data of Emirbekov,[42] the glutamic acid content of the cerebral hemispheres of ground squirrels increases by 7.5 %, as compared to wakeful animals, at the onset of hibernation, when the animal is in a sleepy, torpid state; in the cerebellum the quantity of Glu is somewhat lowered. The Asp concentration is almost unaltered in the cerebral hemispheres, but increases 3.5 times in the cerebellum. The GABA content in the cerebral hemispheres and the cerebellum increases sharply during hibernation. A week after the onset of profound sleep the content of the above-mentioned amino acids decreases in both brain divisions as compared with the moment of falling asleep. The quantity of Asp in the cerebellum and of GABA in the cerebral hemispheres is almost the same as the quantity in the wakeful animal, while the quantity of GABA in the cerebellum is twice as high as in the wakeful animal.

Two months of hibernation lead to further changes in the Glu and GABA contents and to an increase in the Asp content in the investigated brain divisions. On awakening from prolonged hibernation, the content of amino acids in ground squirrel does not immediately return to normal. The Glu decarboxylase activity in the brain of hibernating ground squirrels is raised by 30 %.

6. Cholinesterase and Acetylphosphatase

According to the data of Robinson and Bradley,[43] the cholinesterase activity of the brain is depressed by 14 % during hibernation in ground squirrels, and the sensitivity to cholinesterase inhibitors is raised.[44]

On studying the acetylphosphatase and ATPase activities in the brain stem, in the cerebral cortex, in the vicinity of the Ammon's horn, and in the thalamus zone of golden hamsters during hibernation and wakefulness, it was found that acetylphosphatase in various brain divisions was more active during hibernation; the activity of the enzyme varied in various brain divisions, but was always higher than during wakefulness.[45]

C. Renewal of Proteins of Subcellular Structures of Brain Tissue During Hibernation and Awakening

More and more attention has been devoted recently to the study of the biochemistry of intracellular structures, in particular to metabolism within these structures. With the aim of determining the effect of hibernation on metabolism in various intracellular structures of brain tissue, Belik[46] studied protein metabolism in various subcellular structures of brain tissue by the intensity of radioactive methionine incorporation in the proteins of the nuclei, heavy and light mitochondria, microsomes, and the soluble cytoplasmic fraction of the brain tissue of hibernating, wakeful, and artificially awakened ground squirrels.

The investigations showed that the specific radioactivity of the proteins of all investigated intracellular fractions of the brain tissue was lowest in hibernating ground squirrels; in the artificially awakened it was about two orders higher than in the hibernating animals, and in the wakeful animals several tens of times higher than in the hibernating, but 15–20 % lower than in the artificially awakened animals.

Within each group of animals investigated the incorporation of radioactive methionine went on most actively in the microsome proteins, and in the proteins of the soluble cytoplasmic fraction. The intensity of incorporation of radioactive methionine in the proteins of nuclei was somewhat lower, and it was still lower in the case of proteins of the mitochondria, especially the heavy ones.

Determination of the radioactivity of the acid-soluble fraction in these experiments also revealed the highest level of radioactivity in hibernating ground squirrels; in the wakeful and artificially awakened animals it was lower.

Thus, metabolism (renewal) of protein substances in the brain of ground squirrels is lowered during hibernation; furthermore, during hibernation, wakefulness, and artificial awakening, alterations are observed in protein metabolism, not only on investigating the rate of renewal of summary brain proteins, but also on studying the protein metabolism of the various intracellular fractions. The lowest metabolism is typical of the state of hibernation in the case of both the summary brain proteins and the proteins of the various intracellular structures. Metabolism is considerably higher in the wakeful state and still higher during artificial awakening. On the other hand, the proteins of the various intracellular structures incorporate labeled amino acids, or in other words are renewed, with different intensities. In all investigated functional states of ground squirrels (hibernation, wakefulness, and artificial awakening) metabolism was found to be most intense in the microsome proteins, less intense in the nuclear proteins, and slowest in the mitochondrial proteins.

Of particular interest is the discovery of higher intensity of metabolism of the summary brain proteins and the proteins of various intracellular fractions of the brain tissue in artificially awakened ground squirrels (when the nervous system passes into the state of excitation) as compared with that of wakeful animals. The question of the intensity of protein renewal during excitation of the central nervous system is more difficult to decide with other animal species, and various scientists studying protein metabolism during excitation of the central nervous system induced by drugs or other factors, such as electric stimulation, have obtained contradictory results.

Investigations on hibernating animals during hibernation and artificial awakening elucidate the influence of the principal physiological states of the central nervous system—excitation and inhibition—on the process of metabolism of various substances in the brain (proteins, phosphoproteins, nucleic acids, phospholipids, and glycogen). On interpreting the results, however, consideration should also be given to the effect of hypothermia on the central nervous system and its metabolism.

IV. THE INFLUENCE OF HYPOTHERMIA ON METABOLISM IN THE BRAIN

Since one of the factors which can influence brain metabolism in hibernating mammals is low body temperature, the data on the effect of hypothermia on the mammalian organism, in particular on the brain, are of great interest.

Investigations by Vladimirov[47] have shown that a fall of body temperature to 25–20°C in rabbits does not give rise to disorders in the energy metabolism of the brain; the oxidation processes and phosphorylation are retarded (by three times); glycolysis is inhibited to a lesser extent; the rate of hexosophosphate formation is slightly decreased, and so is oxygen consumption.

Use of labeled phosphorus showed that hypothermia depresses renewal of the phosphorus of phosphoproteids and, especially, of phospholipids.

The study of protein metabolism by determining the rate of incorporation in brain proteins of labeled Met, Gly, and Tyr in dogs subjected to hypothermia showed a strikingly large diminution of the intensity of amino acid renewal: at a body temperature of 20°C the Tyr renewal rate in the brain proteins is 20 times less than at normal body temperature. The conclusion can be drawn that the functional activity of the nerve cells ceases during hypothermia owing to string inhibition of the biochemical processes directly connected with metabolism of protein and lipoid substances.[48] Of the various divisions of the central nervous system (cerebral cortex, region of the rhomboid fossa, cerebellum, medulla oblongata, and spinal cord) protein metabolism is most strongly disturbed in the cerebellum.[49]

According to Maur,[50] the general glucose metabolism is disturbed during hypothermia, and it takes the form, mainly, of lactic acid formation (through pyruvic acid); in his opinion, the alteration of oxidation metabolism of carbohydrates in hypothermia is similar to that of hibernating animals during hibernation.

Hypothermia in dogs induces accumulation in the brain tissue of glycogen and ammonia, and a slight fall in the ATP and Glu.[51] The activity of the enzymes cholinacetylase and cholinesterase is lowered in the rat brain during hypothermia; this affects the acetylcholine content.[52]

In vitro experiments on isolated guinea pig brain tissue in hypothermia showed a decrease in creatine phosphate synthesis and an increase in inorganic phosphate. All this[53] indicates a fall of the rate of penetration of glucose into the cells and its oxidation during hypothermia, and a decrease of energy-rich phosphorus compounds in the cells.

It can thus be seen that in some cases the change in the content of some substances and their metabolism during hibernation of mammals and during hypothermia of nonhibernating mammals is the same; in other cases the processes of metabolism in the brain take different pathways during hibernation and hypothermia.

V. REFERENCES

1. D. Ferdman and O. Feinschmidt, Der Winterschlaf, in *Ergebnisse der Biologie*, Vol. 8. pp. 2–6, Springer, Berlin (1932).
2. A. Horvath, *Zbl. Med. Wiss.* **55**:865–870 (1872).
3. G. Valentin, *Moleschotts Untersuchungen* **7**:39 (1860).
4. F. E. South and H. Jeffay, *Proc. Soc. Exp. Biol. Med.* **98**:885–887 (1958).
5. A. Svihla, H. Bowman, and R. Pearson, *Science* **115**:272 (1952).
6. A. Svihla, H. Bowman, and R. Ritenour, *Science* **115**:306–307 (1952).
7. P. Suomalainen and E. Lehto, *Arch. Soc. Fen. "Vanamo"* **6**:94 (1952).
8. A. Denyes and J. D. Carter, *Nature* **190**:450–451 (1961).
9. M. Zain-ul-Abedin and B. Katorski, *Canad. J. Physiol. Pharmacol.* **44**:505–507 (1966).
10. G. Valentin, *Moleschotts Untersuchungen* **5**:11–19 (1858).
11. J. F. Thomson, R. L. Straube, and D. E. Smith, *Comp. Biochem. Physiol.* **5**:297–305 (1962).
12. P. D. Dvornikova, *Biokhim. Zhurn.* **15**:85–91 (1940).
13. M. L. Zimny, *J. Cell. Compar. Physiol.* **48**:371–394 (1956).
14. M. L. Zimny and R. Gregory, *Amer. J. Physiol.* **195**:233–236 (1958).
15. P. Haberey, *Compt. Rend. Soc. Biol.* **159**:2051–2054 (1965).
16. G. M. Davydova and N. G. Stepanova, *Zh. Evolyuts. biokhimii. i Fiziol.* **1**:32–37 (1965).
17. O. Fainshmidt and D. Ferdman, *Nauk. Zap. Ukr. biokhim. In-ty* **6**:65–74 (1933).
18. P. Raths, *Z. Biol.* **112**:282–299 (1961).
19. G. Valentin, *Moleschotts Untersuchungen* **2**:1–12 (1857).
20. H. Nagai, *Z. Allg. Physiol.* **9**:243–250 (1909).
21. R. Kristoffersson, *Suomalais, Tiedeakat. Toimituks* **45** (1961).
22. C. P. Lyman and P. O. Chatfield, *Science* **117**:533–534 (1953).
23. P. O. Chatfield, C. P. Lyman, and D. P. Purpura, *Clin. Neurophysiol.* **3**:225 (1951).
24. G. Valentin, *Moleschotts Untersuchungen* **10**:526 (1870).
25. A. Horvath, *Verh. med.-phys. Ges. in Würzburg* **13**:60–67 (1879).
26. L. Merzbacher, *Erg. Physiol.* **3**:214 (1904).
27. P. O. Chatfield, A. F. Battista, C. P. Lyman, and J. P. Garcia, *Am. J. Physiol.* **155**:179–185 (1948).
28. W. K. Hess, *Ueber die Wechselbeziehungen Zwischen physischen and vegetativen Functionen*, Zürich (1925).
29. F. Bruman, *Z. vergl. Physiol.* **10**:419–430 (1929).
30. O. Fainshmidt and D. Ferdman, *Nauk. Zap. Ukr. biokhim. In-ty* **6**:75–84 (1933).
31. D. L. Ferdman and P. D. Dvornikova, *Biokhim. zhurn.* **15**:69–83 (1940).
32. M. B. Shtark, *Dan, USSR* **153**:1216–1219 (1963).
33. A. Oksche, *Za. Zellforsch. Mikroscop. Anat.* **54**:307 (1961).
34. A. V. Palladin, in *Comparative Neurochemistry* (D. Richter, ed.), pp. 131–137, Pergamon Press, Oxford (1964); in *Voprosy biokhimii nervnoi sistemy*, pp. 149–155, Naukova Dumka, Kiev (1965). (Problems of the Biochemistry of the Nervous System.)
35. Z. Yu. Nechiporenko, *Ukr. biokhim. zhurn.* **18**:77–86 (1946).
36. G. Vincendon, R. Bidet, R. Jund, P. Mandel, and Ch. Kayser, *Bull. Soc. Chim. Biol.* **47**:929–944 (1965).
37. E. B. Skvirs'ka and T. P. Silich, *Ukr. biokhim. zhurn.* **27**:385–393 (1955).
38. J. D. Weill, P. Mandel, and Ch. Kayser, *Bull. Soc. Chim. Biol.* **39**:1395–1407 (1957).
39. Ya. V. Belik and L. S. Krachko, *Ukr. biokhim. zhurn.* **33**:684–692 (1961).
40. P. Mandel, Y. Godin, J. Mark, and Ch. Kayser, *J. Neurochem.* **13**:535–536 (1966).
41. Lj. T. Mihailović, Lj. Kržalić, D. Čupić, and B. Beleslin, *Experientia* **21**:100–101 (1965).
42. É. Z. Émirbekov, *DAN USSR* **179**:1485–1486 (1968); *Dok. Biol. Sci.* (English translation, Consultants Bureau) Vol. 179, p. 230).

43. J. D. Robinson and R. M. Bradley, *Nature* **197**:389–390 (1963).
44. J. F. Scaife and D. H. Campbell, *Nature* **182**:1739–1739 (1958).
45. L. C. Mokrasch, *Am. J. Physiol.* **199**:950–954 (1960).
46. Ya. V. Belik, *in Tret'ya Vsesoyuznaya konferentsiya po Biokhimii nervnoi systemy*, pp. 39–45, Yrevan (1963).
47. G. E. Vladimirov, *in Voprosy Biokhimii Nervnoi Systemy*, pp. 247–257, Kiev (1957).
48. G. E. Vladimirov, *in Konferentsiya po Primeneniyu Radioizotopov v Nauke*, pp. 5–9, Moskva (1958).
49. V. I. Nikulin, *Eksper. zirurgiya* **1**:55–60 (1957).
50. J. M. Maur, D. M. McComiskcy, J. W. Haynes, and J. R. Beaton, *Canad. J. Biochem. Physiol.* **40**:1427–1438 (1962).
51. M. S. Gayevskaya, E. A. Nosova, *in Tret'ya Vsesoyuznaya Konferentsiya po Biokhimii Nervnoi Systemy*, pp. 421–430, Yrevan (1963).
52. L. N. Ponomarenko, *Byull. Eksper. Biol. Med.* **54**:47–50 (1962); *Bull. Exp. Biol. Med.* (USSR) English Trans. Vol. 54, No. 12 (1964), p. 1351.
53. P. Joanny, J. Corrio, and J. J. Papy, *Compt. Rend. Soc. Biol.* **160**:1295–1299 (1966).

Chapter 16

THE CHEMISTRY AND BIOLOGY OF NERVE GROWTH FACTOR

Isaac Schenkein

Irvington House Institute
Department of Medicine
New York University School of Medicine
New York, New York

I. INTRODUCTION

The "field" of nerve growth factor (NGF) has developed out of a series of investigations performed some 20 years ago by Bueker.[1-3] His experiments, done within the framework of the "center–periphery" problem, dealt with magnitude of peripheral innervation and its causal relationship to the relative size of the "center," whence the innervating neurons had their origin.[4-10] Prior to Bueker's unorthodox procedures, most investigators had used implantation or ablation techniques in order to produce the desired increases or decreases of peripheral fields. Bueker departed from these, implanting tumor masses in the body wall of the study object (the chick embryo) instead of the usual homologous or heterologous structures. The importance of the limb periphery as a factor essential for the differentiation of the lateral motor column in the limb segments of the spinal chord was known, as was the fact that the "stimulus," i.e., altered size of periphery, was not species-specific.[10] Successful implantation of tumor masses (sarcoma 180) in lieu of the hind limb periphery quickly revealed important deviations from the previously observed relationships between size of periphery and development of center.[1-3] Bueker observed that:

1. The tumor became "innervated" by the peripheral nerves which normally form the lumbosacral plexus.
2. There was marked hypertrophy and hyperplasia of the spinal ganglia on the experimental side, although the periphery (i.e., the tumor mass) at the time of observation was decidedly smaller than the contralateral control side.
3. The lateral motor neurons of the experimental side were significantly smaller than those of the other side.
4. The tumor maintained its cellular autonomy.

5. Successful implants of rhabdo-myosarcomas did not produce these effects, an observation which was later extended by Bueker's survey of mouse neoplasias, which showed that only certain sarcomas were effective, all carcinomas investigated being negative.[11–13]

These results were unexpected and Bueker's tentative explanation was that: "Histochemical properties of sarcoma 180 (which) favor sensory innervation and hence enlargement of spinal ganglia." The increased size of the center (spinal ganglia) was thus due to "histochemical properties" of the innervated mass rather than its size *per se*.[12] In 1951, Levi-Montalcini and Hamburger confirmed Bueker's original findings and stated that they were in agreement with his hypothesis.[14] These investigators, however, added a number of observations which proved pivotal to the further development of the problem.[15] These were:

1. The ingrowth of nerve fibers into the tumor mass produced enlargement of the sympathetic ganglia, to a far greater degree than that of the spinal ganglia.
2. The fibers entering the tumor mass, which had their origin in the sympathetic ganglia, came only from one type of ganglionic cell, i.e., the late-differentiating mediodorsal cells.
3. There appeared a "field effect," i.e., cells not directly connected with innervation of the tumor were also involved in the general picture of the enlarged center.

It was therefore felt that the implanted tumors (sarcomas 180 and 37 were used) produced: "A specific growth promoting agent which selectively stimulated the growth of some type of nerve fibers but not of others."

Elegant proof that a growth-promoting factor was involved came in 1952 from the work of Levi-Montalcini and Hamburger. They conclusively showed that extraembryonic transplants of the tumor (allantoic grafts) produced bilaterally the same hypertrophy and hyperplasia of both spinal and sympathetic ganglia, though the transplanted tumor mass was not accessible to direct innervation. The concept of a diffusable growth factor first enunciated by these investigators now had a firm foundation.[15,16]

II. CELL-FREE PREPARATIONS

In 1953 Bueker and Hilderman were able to confirm these observations but reported that a preparation of cell-free material from sarcoma 180 obtained by Berkefeld filtration failed to show any effect on either spinal or sympathetic ganglia when injected into the yolk sac of chick embryos. This failure became subsequently more understandable after studies by Cohen *et al.*[17] revealed that little or no active material could be obtained from sarcomas that had not been transplanted and allowed to grow in the chick embryo prior to the attempted isolation of the growth factor. This remarkable requirement in the isolation of active material from sarcoma 180 has never been elucidated, but may well be related to the observations made later by Burdman *et al.*[18] that considerable NGF activity was detectable in the sera of children with neuroblastomas, though virtually no active material could be obtained from these tumors. The year 1954 saw several further developments. An *in vitro* bioassay became available after Levi-Montalcini *et al.*[19] showed that small fragments of sarcoma 180, when incubated adjacent to explanted spinal or sympathetic ganglia of the 7–8-day-old chick embryo in a firm plasma clot, elicited a remarkable halolike dense outgrowth of neurites from the ganglion.[19] This characteristic assay and its semiquantitative evaluation (1 − 4 + outgrowth) became the cornerstone of virtually all subsequent work with cell-free fractions, a major reason being that the specificity of the *in vivo* observations made with tumor fragments was maintained with it, i.e., only spinal and sympathetic ganglia responded. That same year Cohen *et al.*[17] isolated from the microsomal fraction of sarcoma 180 a nucleoprotein fraction (260 mμ/280 mμ of 1.78) that with the newly introduced *in vitro* bioassay produced the same typical response as that obtained with fragments of the tumor. The biological activity proved to be nondialyzable and heat labile. It showed poor solubility below *p*H 5 and above *p*H 9, an observation that has since gained in significance in view of the recently reported behavior of a large molecular weight particle as NGF and its dissociation into subunits at both acid and alkaline *p*H. Cohen rapidly took the purification scheme further, showed that the biological activity was resistant to both DNase and RNase and considered it to be protein or "related" to protein, since considerable loss of activity occurred as a result of proteolysis by enzymes. The purified material proved labile to heat and acid, but showed the remarkable property of being stable to alkali (0.1 N for 1 hr at room temperature), as well as to 6 M urea.[20] It was during these investigations (to be more precise, in the use of whole snake venom as a source of phosphodiesterase) that the striking discovery was made that crude snake venom itself on a w/w basis contained far more biological activity than the most purified fraction from either sarcoma 180 or 37.[21] Cohen developed a purification scheme for the activity from snake venom which resulted in the isolation of a fraction which again showed many characteristics of protein and appeared homogeneous in the ultracentrifuge (single boundary with an S_{20} of 2.2).[22] Heat and acid lability were again manifest, as was loss of biological activity upon proteolysis. Like the protein from sarcoma 180, the material proved stable to alkali as well as 6 M urea. Other experiments showed that purification of the growth factor from crude venom was coincident with the loss of a variety of enzymatic activities present in the crude venom.[22,23] Proteolytic activity was an outstanding exception in this respect, the purified material possessing 75% of the protease activity of the starting material. This finding is of interest, since in the mouse submaxillary gland, which subsequently became the major source for NGF, proteolytic activity is closely associated with NGF or, as is claimed by Varon *et al.*, is an integral activity of one of the subunits which these workers believe form the NGF particle.[24] The *in vivo* biological activity of the purified fraction from snake venom was studied by Levi-

Montalcini et al.[25] It was demonstrated that injection of this material into the yolk sac of the chick embryo provoked both an enlargement of sensory and sympathetic ganglia and an aberrant distribution of sympathetic nerves in viscera and blood vessels. Enlarged sympathetic ganglia were in a more advanced state of differentiation as attested to by the presence of many neurofibrils in the cytoplasm of the experimental sympathetic neurons and their paucity in the control ganglia.[25] These phenomena were identical to the ones observed as a result of implantation of tumor masses in the chick embryo and the conclusion was drawn that the agent released by the tumor was similar or even identical to that isolated from snake venom. Injections of purified material obtained from tumors never proved fully successful with the in vivo experiments, presumably because only very small amounts were available for experimentation.

The finding that the male mouse submaxillary gland is an even better source for NGF than is snake venom, was made by Cohen in 1960 on the basis of a reasoned relationship of organ homology between the poison glands of reptiles and the salivary glands of mammals.[26] The method of purification devised by him yielded a protein that was homogeneous by ultracentrifugation and a mol. wt. of 44,000 was estimated on the basis of an S_{20} value of 4.32. The specific activity (the definition of the biological unit, B.U., as the minimum quantity of protein that resulted in $4+$ biological activity with the in vitro assay was made in Ref. 19) was 0.015 μg on the basis of a protein determination by the method of Lowry et al. Other characteristics were again quite similar to those of the material isolated from snake venom or tumor, i.e., acid and heat lability, alkali stability, and a 50% and 95% loss of biological activity after treatment with trypsin and chymotrypsin respectively. A further important observation made by Cohen et al., was that the purified material was antigenic.[26] The in vitro biological activity of 0.2 μg NGF protein from the mouse was completely abolished by 1×10^{-3} ml of antiserum obtained from rabbits that had been injected in the footpad with the antigen suspended in Freund's adjuvant. It was also noted that 1 ml of this antiserum could abolish the activity (when tested in vitro) of 800 B.U.'s of the material obtained from snake venom. In contrast to this was the finding that 16,000 units of a commercially obtained snake venom antiserum were needed to abolish the activity of one B.U. (i.e., 0.015 μg) of the NGF from the mouse gland.

An extensive series of biological experiments were performed by Levi-Montalcini et al. with the purified fraction from the mouse gland.[27] It was shown that both sensory and sympathetic ganglia of human fetuses responded in the in vitro assay in the same fashion as did ganglia from the chick embryo, i.e., outgrowth of a dense halo of neurites. Daily injections of newborn mice with this material (300 B.U./gram body weight/up to 3 weeks) resulted in a marked hypertrophy of the sympathetic ganglia, with average volume increases of 300% above the controls for the superior cervical ganglion. It was noted that cellular hypertrophy was of greater importance than the increase in cell numbers. In general, the effects were considered similar to those provoked by the factor from sarcoma 180 or snake venom. An outstanding exception was the absence of response of the sensory ganglia of the mouse. Other observations indicated that NGF activity (as measured by the *in vitro* assay) could be elicited from sera of mice, those of male animals being consistently higher in NGF titers than those of the female. This sexual dimorphism (see later part of this review) was observed to extend to the sympathetic nervous system itself and the statement was made that the sympathetic neurons in male mice are considerably and consistently larger and stain more intensely with basic dyes than females of the same size.[27]* Yet removal of the submaxillary or sublingual glands of both adult and weanling mice did not result in detectable size differences of the sympathetic ganglia. In extension of Cohen's observation on the antigenicity of the NGF from the mouse gland, Levi-Montalcini and Booker[28] showed that the antiserum had remarkable and apparently specific effects on ganglionic cells of the neonatal mouse. Injections of the antiserum produced a near total destruction of the sympathetic neurons. It was observed that the rabbits that were "producing" the antiserum also showed an extensive destruction of cells of their sympathetic nervous system. These basic observations on the *in vivo* action of the mouse NGF and an antiserum to it have resulted in a series of investigations on the general biological and more special pharmacological aspects of the changes induced in both the "hypersympathetic" and in the "immunosympathectomized" animals. A later part of this review will discuss some of the salient data.

The publication by Cohen of methods for the purification of NGF from tumors, snake venoms, and the mouse submaxillary gland which resulted in proteins to which several powerful criteria of homogeneity could be applied, was in no way the end of the problem of the identification of a defined chemical(s) with the nerve growth promoting activity. As can be seen from Table I, there have been isolated since 1960, often from apparently unrelated starting materials, and in some cases with considerable criteria of purity, a variety of proteins which each elicit an *in vitro* response not distinguishable from that provoked by either the tumor fragment or the purified proteins obtained by Cohen from snake venom or the mouse gland. In some cases no experimental work has been done beyond the observation that cell-free fractions showed the specific biological activity either *in vitro* or *in vivo*. (See Table IB.) In others, (Table IA) more extensive investigations were made. Molecular weights assigned to the various factors vary widely. S_{20} values range from 1.5S to 7.1S from which molecular weights were estimated, ranging from 14,000 to 140,000. With the exception of data published by Banks *et al.* for highly purified NGF from snake venom,[29] no direct values have been reported for either the partial specific volume or the diffusion constant of the isolates. Factors like pH, concentration dependence, and hydrodynamic shape of the molecule may be at the root of some of the differences of the estimated molecular weights. The existence of monomer–

---

* In a later publication Levi-Montalcini reverses this observation by stating that the sympathetic nerve cells from male and female mice compare in size.[52]

TABLE IA

Source	Authors	B.U.^{a} (in protein)	Characteristics
Tumor	Cohen[20]	250 μg	$E_{260}/E_{280} = 1.78$
Venom	Cohen[22]	3×10^{-1} μg	$S_w = 2.2$ mol. wt. 20,000
Venom	Banks[29]	1×10^{-8} μg	$S_w = 2.85$ mol. wt. 40,000 Other forms of 24,000 and 38,000
Venom	Zanini[87]	1×10^{-4} μg	Mol wt. ranging from 24,000 to 30,000
M.S.G.^{b}	Cohen[26]	1.5×10^{-2} μg	$S_w = 4.22$ mol. wt. 44,000
M.S.G.	Schenkein[35]	3×10^{-2} μg	$S_w = 2.41$ and 4.24 A + C = AC
M.S.G.	Levi-Montalcini[85,86]		Mol. wt. 36,500 as minimum, on basis of quantitative amino acid analysis.
M.S.G.	Angeletti[36]	1×10^{-2} μg	Subunits mol. wt. 44,000 = 130,000
M.S.G.	Varon[32–32]	1.3×10^{-2} μg	Subunits alpha, beta, gamma Each 30,000. Complex $S_w = 7.1 = 140,000$
M.S.G.	Schenkein[37]	1×10^{-2} μg	NGF "A" ⎱ Heterogeneous
		1×10^{-8} μg	NGF "B" ⎰ fraction
M.S.G.	Zanini[87]	1×10^{-3} μg	Subunits mol. wt. 14,000 = 28,000
		1×10^{-4} μg	Either 14,000 or 28,000 plus B.S.A.

^{a} Biological Unit.
^{b} Mouse submaxillary gland.

dimer and multimer relationships among identical as well as among non-identical subunits have been inferred from data of analytical gel electrophoresis and "molecular-sieving" procedures. The recently published data by Varon et al.[30–32] and by Smith et al.[33] describing a "system" of particles that are required for the full expression of the biological activity will be discussed in some detail here, since they represent a drastic departure from the simple "one protein–one factor" idea initially brought forward by the work of Cohen et al., though Schenkein et al., in 1962[34,35] and Angeletti in 1967[36] had already indicated that the NGF activity was related to multiple particles or that it could exist in several molecular weight forms that were each stable enough to be isolated and studied as such.

In a sequence of papers[30–32] Varon et al. and Smith et al.[33] have claimed that:

1. The NGF activity is related to a protein particle with an estimated mol. wt. of some 140,000 on the basis of an S_{20} value of 7.1.

TABLE IB

Source	Authors	No physico chemical characteristics but activity demonstrated
Chick embryo	Bueker et al.[55,56]	(Low level in several organs)
Spinal ganglia	Bueker et al.[55,56]	From chick embryo
Sympathetic ganglia	Levi-Montalcini[57]	From human, mouse, rat, cat, cow
Granuloma tissue	Levi-Montalcini[60]	After injection of carrageenin
Serum	Levi-Montalcini[27]	Mice
Saliva	Levi-Montalcini[27]	Mice (low level)
Urine	Levi-Montalcini[27]	Mice (low level)
Serum	Burdman[18]	Children with neuroblastoma. Tumor itself no activity.
Amphibian	Winnick and Greenberg[58]	
Teleost fish	Shulman-Weis[59]	
Frog	Winnick and Greenberg[58]	

2. The particle is dissociable into three different types of subunits each of mol. wt. 30,000 ($S_{20} = 2.5$); the dissociation occurring as a function of pH, i.e., below pH 5 and above pH 8. (Although an earlier paper from this group gave the pH limits as being 3.8 and 10.3.)

3. Of the three types of subunits (referred to as alpha, beta, and gamma) two (alpha and gamma) display electrophoretic heterogeneity.

4. All three subunits are necessary for the full expression of the biological activity. The beta subunit by itself displays (on a protein basis) 25% of the biological activity of the fully assembled particle.

Each of the three subunits can be isolated by ion-exchange chromatography, and it was shown that spontaneous reassembly occurs at neutral pH. There are four different types of alpha subunits (alpha 1, 2, 3, 4) and three different types of gamma subunits (gamma 1, 2, 3). Recombination of any one alpha and any one gamma subunit with the beta subunit produces a 7S species with the same physicochemical properties and specific biological activity as the material isolated directly as the 7S particle by gel filtration at neutral pH. (1 B.U. = 0.015 μg protein.) These investigators furthermore find preparations of NGF to be mixtures of the 7S species (different alpha and gamma subunits), all with the same biological activity, and liken it to the enzyme aspartic transcarbamylase, in that only one of the NGF subunits is biologically active, though its level of activity is modified by interaction with the inactive subunits. It should also be mentioned that Varon et al. have stated that by their method of purification (a procedure that involves gel filtration on Sephadex G-100, G-150 and chromatography on DEAE-cellulose) the 7S particle represents some 2% of the total soluble protein of the mouse submaxillary gland and has 80% of the expressed activity of the gland homogenate. The yield per gland in terms of B.U.'s was 14 times that reported by Cohen et al. No activity could be detected in proteins from the gland with mol. wt. of either 20,000 or 44,000 which were those assigned to the purified materials from snake venom and mouse gland respectively when purified by the method of Cohen et al.

Finally, Green et al.,[24] have found that the gamma subunit possesses a potent esterase activity on α-N-benzoyl-L-arginine ethyl ester. The enzymatic activity largely disappears upon reassociation of this gamma subunit with the alpha and beta subunits to form the original 7S particle with NGF activity. Whether or not this enzyme is identical with, or related to, the estero-proteases from the mouse submaxillary gland described by other workers [92,93,95] remains to be determined.

The differences in specific activities of NGF's isolated from snake venom and/or mouse salivary glands reported by different investigators are even more dramatic, at least when taken on the basis of the minimum amount of protein needed for the specific in vitro response.

Recently, Banks et al. have demonstrated that with a relatively simple procedure, NGF can be obtained from the venom of Vipera russelli with a specific activity of 1×10^{-8} μg/B.U.[29] Two grams of venom yielded 4 mg of final material or 2×10^{11} B.U./g of crude. Earlier, however, Cohen et al. (see above) had reported that 25% of the NGF activity of 1 g of crude venom from Agkistrodon piscivorus could be isolated in 5.2 mg of protein with a specific activity of 3×10^{-1} μg/B.U.[21,22] This gives 6.5×10^5 B.U.'s as total yield per gram of this venom. The venom of V. russelli on the basis of a comparison done by Cohen is some four times as active as Ag. piscivorus on a w/w basis. The V. russelli venom should then yield 2.6×10^6 B.U./g by the procedure of Cohen, in sharp contrast to the data reported by Banks et al. It is clear that specific activities and yields of these purified NGF's from the two snake venoms are of an entirely different order of magnitude. Similar differences can be seen to exist in the material from the mouse submaxillary gland.

Schenkein et al.[37] have reported on the apparent existence of two "response levels" of NGF ("A" and "B"). When isolated by a modified procedure and tested with the standard in vitro assay of Levi-Montalcini et al., NGF "A" showed full biological activity at 3×10^{-2} μg/B.U., a value similar to that reported by Cohen et al. (mol. wt. 44,000) or that reported by Varon et al. (mol. wt. 140,000). NGF "B" became apparent upon serial tenfold dilutions of the solution that gave NGF "A." In terms of protein concentration, NGF "B" was active at doses from 2×10^{-8} to 2×10^{-10} μg/ml, a value similar to that reported by Banks et al. for the material from snake venom. Two tentative explanations were forwarded for this diphasic dose-response curve. One was based on the hypothesis that both NGF "A" and "B" coexisted as proteins in the fraction from which serial tenfold dilutions were made. An amplification mechanism at the level of the responding nerve cells was postulated, since it could be calculated that at a protein concentration of 1×10^{-9} $\mu g/ml$ there were present only extremely small numbers of protein molecules though a large number of neurites were growing out from the explanted chick embryo ganglion. The alternate possibility considered that NGF "B" is not a protein, but is the result of the possibly catalytic transformation of one or more of the components of medium 199 which was used as the solvent for the serial tenfold dilution.[38] This process would then allow for significant numbers of the newly formed substance to be present even after the serial dilutions. It, of course, requires that one (or more) of the materials present in the original solution be the transformer or catalyst.

The question arises as to how the large differences in specific activity and molecular weight can be resolved for the same, and in terms of target tissue, apparently highly specific biological activity, especially in view of the fact that for several of the isolated proteins powerful criteria of purity have been presented. There are as yet no clear answers. It is possible that the different methodologies used (mainly gel filtration versus ethanol precipitation, followed by salting out with ammonium sulfate and use of ion-exchangers) play an important role in determining the final characteristics of the growth factor. It is also possible that none of the isolated proteins are the "real" NGF but merely serve as carrier for an as-yet unidentified substance, an idea suggested by several authors.[29,38,39] NGF activity in a protein from the mouse gland with 0.02 μg/B.U. could be due to contamination of that protein by the actual factor. It would, for example, take a chemically undetectable impurity of NGF "B" to account for the specific activity of NGF preparations where 1 B.U. = 0.01–0.1 μg. The same would hold true for the NGF from snake venom prepared by the method of Banks et al. (with 1 B.U. = 1×10^{-8} μg) when compared with the factor from snake venom isolated by the method of Cohen (1 B.U. = 1×10^{-1} μg) and is in fact suggested by these authors.[29] The concept of "carriage" of an NGF of small molecular weight by various proteins would have to be reconciled with both the observed whole or partial loss of biological activity after proteolysis and the abundant immunological data. These will be considered in a later section of this review.

III. METABOLIC EFFECTS OF NGF

Cohen, in his publication of the method of purification of NGF from snake venom also reported some important observations of the metabolic effects of this NGF on the ganglia which had been incubated with it.[21–23]

The morphological phenomenon (i.e., the outgrowth of neurites) was completely inhibited by iodoacetate (final concentration was 1×10^{-4}M) but not by fluoride, cyanide, or dinitrophenol, at concentrations from 10^{-4}M to 10^{-2}M. It was also found that the presence of either glucose or mannose in the growth medium was an absolute requirement. These sugars could not be replaced by fructose, arabinose, ribose, galactose, gluconic or glucoronic acid, malic, fumaric, succinic, or α-keto glutaric acid. Some growth was observed with either lactate or pyruvate as replacement. A general requirement for amino acids in the growth medium was inferred from an experiment which demonstrated that the analog p-fluorophenyl-alanine was inhibitory, but that growth was initiated upon addition of phenylalanine. Careful measurements furthermore indicated a shift of ganglionic carbohydrate metabolism to a nonoxidative pathway, since differential labeling experiments with glucose showed a 50% increase in the appearance of the C_1 of glucose as CO_2. Other experiments with radioactive precursors showed that incorporation of ^{14}C-labeled lysine into protein was 50% higher than in control ganglia. RNA synthesis was increased by 35%. No measurable change was noted in de novo synthesis of DNA. Angeletti et al. working with the purified factor from the mouse submaxillary gland, found essentially the same picture.[40,41] Their metabolic studies were extended by the use of the antibiotics puromycin and actinomycin D.[42] The data obtained suggested that the stimulation of protein synthesis provoked by the NGF was dependent on a primary effect on nuclear RNA synthesis. Toschi et al.,[43] however, found no apparent differences in the RNA obtained from spinal ganglia incubated with NGF with the RNA from controls when examined chromatographically or by sedimentation analysis.[43] Burdman found some increases in the total RNA content of sensory ganglia treated with NGF, though the differences were small.[44] The same author found that base composition expressed as $(G + C)/(A + U)$ ratio was not influenced by NGF.

The effect of NGF on lipid biosynthesis has been studied by Liuzzi et al.,[45] Angeletti et al.,[46] and Liuzzi and Foppen.[47] It was shown by these workers that NGF markedly stimulates the incorporation of acetate-1,2-^{14}C into lipids of sensory ganglia (in vitro), and that the increased rate of lipogenesis was dependent upon the stimulation of DNA-primed RNA synthesis. An hypothesis was forwarded[45] that NGF induces enzymes concerned with the synthesis of fatty acids in the responding nerve cells via stimulation of the synthesis of mRNA. An interesting finding was that when higher than optimal quantities of NGF were used, the incorporation of acetate was inhibited. Liuzzi and Foppen[47] have reported the isolation of a steroidlike compound from the nonsaponifiable lipids of the sensory ganglia of the chick embryo. The synthesis of this as-yet unidentified material was diminished as a result of the addition of mouse salivary gland NGF to the culture medium. It was pointed out by these authors that this was the first observation of an inhibitory rather than stimulatory action of the NGF. Gandini-Attardi et al.[48] have reported that profiles of soluble proteins, obtained by plotting specific radioactivities of fractions from sucrose gradients from either embryonic sensory or from mature sympathetic ganglia incubated with NGF for 7 hr showed significant differences when compared to the soluble proteins from control ganglia. DEAE-cellulose fractionation of these materials indicated that the proteins emerging in the "acidic" region of the chromatogram had higher specific radioactivity when obtained from ganglia incubated with NGF, though the staining pattern after analytical acrylamide gel electrophoresis showed no differences. The possible influence of NGF on amino acid transport across the neuronal membrane was studied by Levi-Montalcini et al.[49] Use was made of the nonmetabolizable analog α-aminoisobutyric acid-1-^{14}C and it was shown that NGF at concentrations from 0.05 μg/ml to 0.5 μg/ml had no effect on the intracellular accumulation of label. A parallel experiment with radioactive leucine done under the same conditions confirmed the previously reported enhancement of incorporation into protein of the precursor in cultures to which NGF was added. It was concluded from these data that the enhanced protein synthesis was not due to increased levels of intracellular amino acids.

Among the most remarkable effects of NGF on ganglionic metabolism ranks the following observation by Crain et al.[50] Incubation of dorsal root ganglia from the 8-day-old chick embryo with NGF resulted in the frequent occurrence of a unique structure of highly ordered mosaics of cytoplasmic granules, which when examined with the electron microscope appeared to approach a crystalline organization. It was speculated that the paired rows of granules represent an early stage in the formation of additional cisternae of the endoplasmic reticulum or that the "crystals" are related to the development of neurotubules. Electron micrographs have shown that the structures are already apparent after only 30 min of incubation of the ganglia with NGF. Crain et al.[50] feel that one is dealing here with an extremely rapid development of cell organelles from cytoplasmic granules triggered or augmented by NGF.

The available data do not yet permit an understanding of either mechanism or site of action of the NGF, especially, since (as previously discussed) there is considerable uncertainty as to the chemical nature of NGF. What is certain is that, with either the fraction from snake venom or with that from the mouse, the process of nerve growth stimulation is dependent on an energy source (glucose or mannose) and is accompanied by a shift of glucose metabolism to a pathway which is presumably the pentose shunt. Cyanide is not inhibitory, and therefore a noncytochrome oxidase-linked oxidative process must be involved. Distinction between net synthesis and turnover was, however, not unambiguous. In general, interpretation of the metabolic data is complicated by the fact that ganglia are not pure cultures of nerve cells and that a variable amount of necrosis occurs in the central part of the ganglion during incubation. Experiments by Levi-Montalcini et al. have, however, indicated that for long-term in vitro cultures of these nerve cells, the presence of NGF in the medium is obligatory, if cells are to be maintained.[51,52] The study performed by A. I. Cohen et al.[53] on the requirement for NGF in cultures of embryonic sensory and sympathetic ganglia is in full agreement with Levi-Montalcini's observations. Rieske[54] has come to the same conclusions in her studies with cell cultures of the trigeminal ganglion of the 10-day-old chick embryo.

IV. SOURCES OF NGF

The sources from which the various NGF's have been isolated or from which tissue fragments or cell-free preparations were derived can be divided into the following categories:

1. Certain mouse sarcomas.[1-3,11-14,16,17] (Isolation only possible after passage of the tumor on the chick embryo.)
2. Secretions of the poison glands of reptiles.[21,22,29]
3. Submaxillary glands of mice.[26,27]
4. Spinal and sympathetic ganglia of chick embryos and mammals.[55-57]
5. Spinal-axial region of tadpoles.[58]
6. Spinal-axial region of bony fishes.[58,59]
7. Sera from children with neuroblastomas, though not from the tumor tissue itself.[18]
8. Granuloma tissue obtained after injection of carrageenin into subcutaneous tissue of mammals.[28,60]
9. Certain preparations of thrombin.[61]

The peculiarity concerning the obligatory passage as grafts of sarcoma 180 on the chick embryo, prior to isolation of an active NGF fraction has already been mentioned, as was the possible relationship of this to the observation that neuroblastomas are as such not a source, but that sera of children with these neoplasias do show considerable NGF activity. The submaxillary gland of the mouse presents a different but equally interesting facet in that its histological structure exhibits a sexual dimorphism, which is closely echoed by a variety of proteins that can be isolated from it, including the NGF. It was Cohen who first observed that the male mouse had on a w/w basis 7–8 times as much NGF as the female, and that in the male, NGF made its appearance only around onset of sexual maturity of the animal.[23,26] Subsequently, these data were correlated with intrinsic differences in both structure and function of the gland as a function of sex or sexual maturation. Caramia et al.[62] established in 1962 that castration of male mice resulted in a marked reduction of the NGF content in the submaxillary gland. They also demonstrated that testosterone replacement therapy returned NGF to its original level. Testosterone also "masculinized" the mature female's gland, which as a result showed levels of NGF comparable to those of the male. Bueker et al.[63] showed that a spontaneous increase in NGF levels of the female gland occurs during pregnancy, lasting throughout lactation. They hypothesized that this might be a "route" of administration of NGF from mother to offspring. Two further points of interest can be made here: One is that the ganglia themselves have small but measurable quantities of NGF as established for the chick embryo by Bueker et al.[55,56] for sympathetic ganglia of several mammals by Levi-Montalcini et al.,[47] and for a variety of fishes and amphibians by Winnick et al. and by Shulman-Weis.[58,59] The other point is the observation made by Levi-Montalcini et al.[60] that cell-free fractions from granuloma tissue obtained by the subcutaneous injection of carrageenin into guinea pig, monkey, or rabbit have NGF activity; Levi-Montalcini in fact has suggested that NGF is produced everywhere by cells of the mesenchymal type, but that it is concentrated by the salivary gland and used for normal growth and maintenance by the nerve cells. The problem of the site of synthesis of the mouse NGF has, however, not been solved; reports in the literature are conflicting. In 1960[27] it was reported that sialoadenectomy had little or no effect on adult or weanling mice both in terms of the size of neuronal cells of the sympathetic system as well as by measurement of the titer of NGF in the serum of the animals. However, in 1962[62] and in 1964[64] in studies using a different approach it was stated that removal of the salivary gland in mice does result in decrease and even total disappearance of the NGF from the blood of these animals.[62]

A recent observation by Hoffman et al.[61] (who have devised a novel bioassay for NGF by cultivating sympathetic cells directly on the surface of sliced acrylamide gels) indicates that three zones of NGF activity can be resolved on the gels. An additional discovery was that commercial thrombin preparations show significant NGF activity, observations which deserve careful evaluation. The morphological criterion of growth accepted by these workers is not the presence of a halo of neuritic outgrowth from explanted ganglia, but rather those seen by Levi-Montalcini in her extensive studies with trypsin-dissociated ganglionic cells, i.e., presence of neuritic processes.

It is clear then that there are already known a significant variety of sources for NGF, or rather for NGF activity, and it is by no means certain that the same molecule is involved in all cases. How all these factors can be fitted into some unified pattern is not apparent, and constitutes an interesting problem in comparative biochemistry.

V. IMMUNOLOGY AND PHARMACOLOGY

Cohen's discovery of the antigenicity of the NGF[26] from mouse submaxillary glands has already been mentioned. Injection of 1 mg of the purified preparation with Freund's adjuvant into the footpad of the rabbit resulted after 4 weeks in an antiserum that showed a specifically precipitating antibody which completely inhibited the in vitro biological activity. Incubation of the antigen-antibody complex with alkali for 1 hr followed by neutralization, gave a solution that with the in vitro assay showed 50% of the activity prior to complex formation.

Up to this point there was no evidence that the NGF played a role in the normal economy of the animal. Cohen[26] reasoned, however, that if such a factor were present in the general circulation of the animal, it would be partially or even entirely removed by injection of the antiserum, which might result in an observable physiological and/or morphological phenomenon. In fact daily subcutaneous injections of the antiserum into newborn mice resulted in a dramatic destruction of the cells of the sympathetic ganglia. The conclusion was reached that either in the mouse's circulation there is a protein that is similar or identical in structure to the NGF isolated from its salivary gland, and which is vital for these cells, or there occurs an interaction of the antibody with the antigen or antigenlike material on the surface of the neuronal cells resulting in the cytotoxic effect.

Both the *in vitro* and the *in vivo* observations were rapidly confirmed and expanded in an extensive study by Levi-Montalcini *et al.*[28] It was shown that injection of the antiserum into rats, rabbits, and kittens resulted in a near total (97–99%) destruction of the sympathetic ganglia, the cellular damage proving irreversible. Inspection of the sympathetic ganglia at different head and trunk regions of these animals revealed that all ganglia were affected in the same way, allowing the data presented in the study to be based on cell counts of the superior cervical ganglia.

Sabatini *et al.*[65] have done an electron microscope study on the early effects of the antiserum on the fine structure of sympathetic neurons. It was observed that the cytolytic process was faster *in vitro* than *in vivo*. *In vitro*, after a few hours of incubation with the antiserum, the ganglia showed changes that were interpreted as evidence for a cytotoxic antigen-antibody reaction.

The lack of toxic effects in animals that received antiserum even over prolonged periods of time has made it possible to raise colonies of mice and rats deprived of their sympathetic nervous system.[66,67] Such animals have been excellent materials for a variety of studies.

The first pharmacological observations seem to be due to Crain *et al.*[68] in a study of catecholamine levels of mouse sympathetic ganglia that were enlarged as a result of *in vivo* treatment with NGF. These ganglia showed a 300%–400% increase in norepinephrine levels over controls. Epinephrine levels remained barely detectable. No changes were noted for either amine in brain, adrenal gland, spleen, and liver.

In 1962, Levi-Montalcini *et al.*[69] reported that the norepinephrine content of heart, spleen, and lung from both rats and mice was drastically lowered in animals treated with the antiserum to complete immunosympathectomy. No change was observable in brain tissue. A similar picture was present when the enzyme monoamine oxidase was the measured parameter.

These two basic studies have been followed by a series of observations made in different laboratories which are in the main confirmatory.[70–75]

Refractoriness to the effect of the antiserum was found when treatment was started several days postnatally rather than immediately after birth.[74] Inherent absence of response to antiserum was observable in cells of the peripherally located adrenergic ganglia innervating either vas deferens or uterus.[75] An increased level of urinary excretion of norepinephrine in immunosypathectomized animals was noted after amphetamine injection.[74,76] Vogt[77] believes that this represents norepinephrine from the unaffected cells of the prevertebral ganglia.

Brody[78] has studied cardiovascular responses following immunological sympathectomy. No changes were noted in heart rate or body temperature, but a mild ptosis of the eyelids and miosis were seen. Studies with the ganglionic stimulant DMPP* indicated that no vasoconstriction or dilation occurred as a result of sympathetic stimulation and the conclusion was reached that the immunosympathectomy effectively abolished vasomotor function of the fibers of the sympathetic nervous system.

Klingman[79] injected antiserum in gravid mice and presented evidence that the antibody crosses the placenta, resulting in a partial immuno-sympathectomy of the fetus. Previously, Levi-Montalcini had reported that suckling mice of mothers receiving antiserum became immunosympathec-tomized.[60]

Hamberger et al.[80,81] have studied the effect of immunosympathec-tomy on the tissue distribution of monoamines in rats using the fluorescence technique of Falck. They showed that peripheral effector organs, normally innervated by fibers from the superior cervical and stellate ganglia, showed complete loss of the typical fluorescence attributable to norepinephrine. Fluorescent adrenergic nerve termini resistant to the antiserum were found in the vas deferens. No changes were observed in hypothalamus and adrenal medulla.

Olson[82] has observed that mice treated with NGF showed increased density of the adrenergic ground plexus in iris, submaxillary gland, parotid, blood vessels and intramural ganglia of the G.I. tract. Normally noninner-vated tissues were also found to contain a considerable number of adrenergic termini. Of special interest was the striking increase of adrenergic termini in various types of autonomic ganglia.

Of more than passing interest is the observation made by Shulman-Weiss[83] of the significant reduction in cell number of the spinal ganglia of the zebra fish after treatment with antiserum. The appearance of the remaining cells was normal, an observation also made by Zaimis et al.[75] who showed that the coeliac ganglia of rats after antiserum treatment was reduced in size by loss of cell numbers, the residual cells appearing normal. Shulman-Weis also notes that the effect she has obtained in the zebra fish was the first instance of an *in vivo* response to the antiserum (to mouse gland NGF) by spinal rather than sympathetic ganglia. The same author also demonstrated that injection of NGF (isolated from mouse submaxillary glands) into embryos and young larvae of *Ambystoma maculatum* resulted in a significant enlargement of the spinal ganglia.[84]

VI. DISCUSSION

Central to the problem of understanding the mechanism of action of NGF is the question of its unambiguous chemical identification. Corollary questions are those of the possible relatedness of the NGF's from the different sources and the attendant immunological observations. If indeed only one specific protein molecule is the actual NGF from, for example, the mouse submaxillary gland, then one is faced with explaining the array

* DMPP: 1,1-dimethyl-4 phenyl-piperazinium iodide.

of proteins with NGF activity, which at least in their molecular weights, vary in a manner that cannot be easily explained in terms of a single small fundamental unit. A recent paper by Zanini, Angeletti, and Levi-Montalcini[87] claims the identity of the NGF from the mouse with a unique protein on immunochemical grounds. These investigators state that the method of purification of NGF from mouse gland first published by Cohen[26] (which leads to the 44,000 mol. wt. species) has been repeatedly confirmed, but that a modification (first step is gel filtration!) has given a protein that appeared homogeneous by chromatographic, electrophoretic, and ultra-centrifugational criteria. Its molecular weight is, unfortunately, not reported, but it appears that this preparation can under "appropriate conditions" be separated into forms with different molecular weights that are multiples of a basic unit whose molecular weight was estimated by gel filtration to be about 14,000.*

A perusal of the specific biological activities reported for the various NGF's from the mouse gland (see Table IA) indicates that almost all require similar quantities of protein for the B.U. i.e., 1×10^{-2} μg. Exceptions are the previously discussed NGF "B"[37] and the observation by Zanini et al. that the above-mentioned fundamental unit of 14,000 or its dimer of 28,000 has a specific activity of 1×10^{-3} μg/B.U. which can be further reduced by the addition of albumin to 1×10^{-4} μg/B.U.† (It appears that other inert proteins also sharply increase the specific activity of the NGF.)[87] It is quite remarkable that so many "different" proteins can be isolated, which virtually all have the same specific activity, and on this basis it is not easy to accept the thesis that there is one unique protein which is the NGF.

An alternate approach, in which the various isolated proteins are regarded as being "carriers" for NGF, has already been mentioned. This hypothesis requires the biosynthesis by either rabbit or goat of a specific antibody to several proteins which each have the presumably small molecular weight NGF noncovalently bound as hapten. There is some precedent for this in the system of Plescia et al.[88] who have demonstrated that electro-static complexes of methylated bovine serum albumin with denatured DNA produce DNA-specific antibodies and in the studies by Green et al.[89] of hapten-carrier relationships.

On the other hand, if, as is claimed by Zanini et al.[87] the NGF activity is uniquely related to a specific protein, then the question arises of how the rabbit antibody against the mouse protein can so profoundly influence the in vivo growth and differentiation of neuronal cells of fish, chick embryo, rabbit, mouse, rat and kitten, a list to which the human and the salamander[84] must be added if the growth stimulating effects of the antigen (NGF) are considered, rather than the destructive effects of the antibody.

---

* An earlier paper[36] from the same laboratory gave the fundamental unit as being 20,000 and 40,000 for the factor from snake venom and mouse gland respectively.

† This same paper reports the specific activity of NGF preparation from two venoms (*Bothrops jararaca*; *Crotalus adamanteus*) as being 10^{-4} μg/ml, a 1000-fold improvement over previous reports, except that of Banks et al.[29] (See above.)

Winnick and Greenberg[58] have claimed that the NGF from goldfish, tadpoles, human, chicken, and mouse show immunological identity by the double immunodiffusion technique when challenged by antiserum to mouse NGF. Weak if any cross reactions are reported by others.[26,87] Zanini et al.[87] show that their purified NGF (mol. wt. 14,000 = 28,000) has a single precipitin line on immunoelectrophoresis. It appears, however, that when samples of NGF from different stages of the purification scheme were challenged with the antiserum obtained against the pure fraction, there appeared differences in the immunoelectrophoretic pattern which were considered evidence for different states of molecular aggregation of the NGF, or of unspecific interaction with other macromolecules.

Finally, there is the purely speculative possibility that the endogenously present proteolytic activity of mouse salivary gland homogenates[37,93,95] plays an important role in the various isolation procedures leading to NGF or possibly even the several other factors that have been isolated or reported as present (epidermal growth factor,[90] thymolytic factor,[61,91] spinal tubule factor,[61] granulocytosis factor,[94] a factor stimulating growth of rat hepatoma cells in culture[96] and a purified protein whose effect on embryonic muscle tissue *in vitro* was interpreted as "dedifferentiative" changes[92]). This last factor has both esterase and peptidase activity which can be inhibited by phenylmethanesulfonyl fluoride, an inhibitor of tryptic activity. The inhibited protein loses its biological activity, and Attardi et al.,[93] the investigators of this material, hypothesize that its action on muscle tissue is a result of the enzymatic activity, via release of biologically active peptides. The reports of persistent esteroproteolytic activity in NGF preparations from both venom and mouse gland[22-24,37] as well as that of NGF activity in preparations of human and bovine thrombin[61] could be due to contaminants that are unrelated to the NGF (be it molecule or activity), or the thrombin, but could conceivably also be inherently related to NGF and/or other reported biologically active fractions from the mouse gland. NGF itself may in fact be an enzyme whose substrate we are all searching for.

VII. REFERENCES

1. E. D. Bueker, Implantation of tumors in the hind limb field of the embryonic chick and the developmental response of the lumbo-sacral nervous system, *Anat. Rec.* **102**:369–390 (1948).
2. E. D. Bueker, Hypertrophy of spinal ganglion cells in chick embryos after the substitution of mouse sarcoma 180 for the hind limb periphery, *Anat. Rec.* **100**:735 (1948).
3. E. D. Bueker, Innervation of tumors transplanted in the hind limb of the embryonic chick, *in Year Book of the American Philosophical Society*, pp. 185–186 (1950).
4. E. D. Bueker, Intracentral and peripheral factors in the differentiation of motor neurons in transplanted lumbo-sacral spinal chords of chick embryos, *J. Exp. Zoology* **93**:99–120 (1943).
5. V. Hamburger and E. L. Keefe, The effects of peripheral factors on the proliferation and differentiation in the spinal chord of chick embryo, *J. Exp. Zool.* **96**:223 (1944).

6. E. D. Bueker, The influence of a growing limb on the differentiation of somatic motor neurons in transplanted lumbo-sacral spinal chords of chick embryos, *J. Exp. Zoology* **93**:99–129 (1945).

7. V. Hamburger, Development of the nervous system, *Ann. N.Y. Acad. Sci.* **55**:117–132 (1952).

8. E. D. Bueker, Limb ablation experiments of the embryonic chick and its effect as observed on the mature nervous system, *Anat. Rec.* **97**:2 (1947).

9. V. Hamburger, The mitotic patterns in the spinal chord of the chick embryo and their relation to histo-genetic processes, *J. Comp. Neurol.* **88**:221 (1948).

10. V. Hamburger and R. Levi-Montalcini, Proliferation, degeneration and differentiation in the spinal ganglia of the chick embryo under normal and experimental conditions, *J. Exp. Zool.* **111**:457 (1949).

11. E. D. Bueker, Hypertrophy and hyperplasia of sympathetic and spinal ganglia in the chick embryo induced by sarcoma I, *Anat. Rec.* **112**:2 (1952).

12. E. D. Bueker and H. L. Hilderman, Gross stimulatory effects of mouse sarcomas 1, 37 and 180 on spinal and sympathetic ganglia of chick embryos as contrasted with effect of other tumors, *Cancer* **6**:397–415 (1953).

13. E. E. Bueker, Screening tumors (*in vivo*) for their effect on the growth of spinal and sympathetic ganglia of the embryonic chick, *Cancer Res.* **17**:190–199 (1957).

14. R. Levi-Montalcini and V. Hamburger, Selective growth stimulatory effects of mouse sarcomas on the sensory and sympathetic nervous system of the chick embryo, *J. Exp. Zool.* **116**:321–362 (1951).

15. R. Levi-Montalcini, Effects of mouse tumor transplantation on the nervous system, *Ann. N.Y. Acad. Sci.* **55**:330–343 (1952).

16. R. Levi-Montalcini and V. Hamburger, A diffusable agent of mouse sarcoma producing hyperplasia of sympathetic ganglia and hyperneurotization of the chick embryo, *J. Exp. Zool.* **123**:233–288 (1953).

17. S. Cohen, R. Levi-Montalcini, and V. Hamburger, A nerve growth stimulating factor isolated from sarcoma 37 and 180, *Proc. Nat. Acad. Sci.* **40**:1014–1018 (1954).

18. J. A. Burdman and M. N. Goldstein, Long term tissue culture of neuroblastomas. *In vitro* studies of an NGF in sera of children with neuroblastomas, *J. Nat. Cancer. Inst.* **33**:123–133 (1964).

19. R. Levi-Montalcini, H. Meyer, and V. Hamburger, *In vitro* experiments on the effects of mouse sarcoma 180 and 137 on the spinal and sympathetic ganglia of chick embryo, *Cancer Res.* **14**:49–57 (1954).

20. S. Cohen and R. Levi-Montalcini, Purification and properties of a nerve growth factor isolated from mouse sarcoma 180, *Cancer Res.* **17**:15–20 (1957).

21. S. Cohen and R. Levi-Montalcini, A nerve growth stimulating factor from snake venom, *Proc. Nat. Acad. Sci.* **42**:571–574 (1956).

22. S. Cohen, Purification and metabolic effects of a nerve growth promoting protein from snake venon, *J. Biol. Chem.* **234**:1129–1137 (1959).

23. S. Cohen, A nerve promoting protein, in *The Chemical Basis of Development* (W. McElroy and B. Glass, eds.) pp. 665–667, Johns Hopkins Press, Baltimore (1958).

24. L. A. Greene, E. M. Shooter, and S. Varon, Subunit interaction and enzymatic activity of mouse 7S NGF, *Biochem.* **8**:3735–3741 (1969).

25. R. Levi-Montalcini and S. Cohen, *In vitro* and *in vivo* effects of a nerve growth stimulating agent isolated from snake venom, *Proc. Nat. Acad. Sci.* **42**:695–699 (1956).

26. S. Cohen, Purification of a nerve growth promoting protein from mouse salivary gland and its neurocytotoxic antiserum, *Proc. Nat. Acad. Eci.* **46**:302–311 (1960).

27. R. Levi-Montalcini and B. Booker, Excessive growth of the sympathetic ganglia by a protein isolated from the salivary gland, *Proc. Nat. Acad. Sci.* **46**:373–384 (1960).

28. R. Levi-Montalcini and B. Booker, Destruction of the sympathetic ganglia in mammals by an antiserum to the nerve growth promoting protein, *Proc. Nat. Acad. Sci.* **46**:384–394 (1960).

29. B. E. C. Banks, D. V. Banthorpe, A. R. Berry, H. S. Davies, S. Doonan, D. M. Lamont, R. Shipolini, and C. A. Vernon, Purification of NGF from three snake venoms, *Biochem. J.* **108**:157 (1968).

30. S. Varon, J. Nomura, and E. M. Shooter, Subunit structure of a high mol. wt. form of the NGF from mouse submaxillary gland, *Proc. Nat. Acad. Sci.* **57**:1782 (1967).

31. S. Varon, J. Nomura, and E. M. Shooter, The isolation of the mouse NGF protein in a high molecular weight form, *Biochemistry* **6**:2202 (1967).

32. S. Varon, J. Nomura, and E. M. Shooter, Reversible dissociation of the mouse NGF protein into different subunits, *Biochemistry* **7**:1296 (1968).

33. A. P. Smith, A. Varon, and E. M. Shooter, Multiple forms of the nerve growth factor protein and its subunits, *Biochemistry* **7**:3259 (1968).

34. Isaac Schenkein and E. D. Bueker, Dialyzable co-factor in nerve growth promoting protein from mouse salivary gland, *Science* **137**:433–434 (1962).

35. Isaac Schenkein and E. D. Bueker, The nerve growth factor as two essential components, *Ann. N.Y. Acad. Sci.* **118**:171–182 (1964).

36. P. Angeletti, P. Calissano, J. S. Chen, and R. Levi-Montalcini, Multiple molecular forms of the NGF, *Biochim. Biophys. Acta* **147**:180–182 (1967).

37. Isaac Schenkein, M. Levy, E. D. Bueker, and E. Tokarsky, NGF of very high yield and specific activity, *Science* **159**:640–643 (1968).

38. Isaac Schenkein, E. D. Bueker, and M. Levy, Some aspects of the chemistry of NGF, Conference on Hormones in Development; University of Nottingham, England (in press). Sept. 1968.

39. S. Varon, Conference on The Biology of Neuroblastoma, *J. Ped. Surg.* **3**:123 (1968).

40. P. Angeletti, A. Liuzzi, R. Levi-Montalcini, and D. Gandini-Attardi, Effect of a NGF on glucose metabolism by sympathetic and sensory nerve cells, *Biochim Biophys. Acta* **90**:445–450 (1964).

41. P. Angeletti, A. Liuzzi, and   . Pochiari, Carbohydrate metabolism in embryonic sensory ganglia, *in 1st Int. Meeting Neurochemistry* (Abstract of Communications), Strasbourg, France (1967).

42. P. Angeletti, D. Gandini-Attardi, G. Toschi, M. L. Salvi, and R. Levi-Montalcini, Metabolic aspects of the effect of NGF on sympathetic and sensory ganglia: Protein and RNA synthesis, *Biochim. Biophys. Acta* **95**:111–120 (1965).

43. G. Toschi, E. Dore, P. Angeletti, R. Levi-Montalcini, and Ch. deHaen, Characteristics of labeled RNA from spinal ganglia of chick embryo and the action of a specific growth factor (NGF), *J. Neurochem.* **13**:539–544 (1966).

44. J. Burdman, Early effect of an NGF on RNA content and base ratios of isolated chick embryo sensory ganglia neuroblasts in tissue culture, *J. Neurochem.* **14**:367–371 (1967).

45. A. Liuzzi, P. Angeletti, and R. Levi-Montalcini, Metabolic effects of a specific NGF on sensory and sympathetic ganglia: Enhancement of lipid biosynthesis, *J. Neurochem.* **12**:705–708 (1965).

46. P. Angeletti, A. Liuzzi, and R. Levi-Montalcini, Stimulation of lipid biosynthesis in sympathetic and sensory ganglia by a specific NGF, *Biochim. Biophys. Acta* **84**:778–781 (1964).

47. A. Liuzzi and F. Foppen, Sterol-like compound from sensory ganglia, *Biochem. J.* **107**:191 (1968).

48. D. Gandini-Attardi, P. Calissano, and P. Angeletti, Protein synthesis in embryonic sensory ganglia, *Brain Res.* **6**:367–370 (1967).

49. R. Levi-Montalcini and P. Angeletti, The action of nerve growth factor on sensory and sympathetic cells, *in Organogenesis* (R. L. deHaen and H. Ursprung, eds.), pp. 187–198, Holt, New York (1965).

50. S. M. Crain, H. Benitez, and A. E. Vatter, Some cytologic effects of salivary NGF on tissue cultures of peripheral ganglia, *Ann. N.Y. Acad. Sci.* **118**:206–231 (1964).

51. R. Levi-Montalcini and P. Angeletti, Essential role of the NGF in the survival and maintenance of dissociated sensory and sympathetic embryonic nerve cells *in vitro*, *Develop. Biol.* **7**:653–659 (1963).

52. R. Levi-Montalcini, The NGF. Its mode of action on sensory and sympathetic nerve cells, Harvey Lecture Delivered April 15, 1965.

53. A. I. Cohen, E. C. Nicol, and W. Richter, NGF requirement for development of dissociated embryonic sensory and sympathetic ganglia in culture, *Proc. Soc. Exp. Biol. Med.* **116**: 784–789 (1964).

54. E. Rieske, Einflusz eines Specifischen Nerven Wachstum Faktors (NGF) auf Zell Culturen des Ganglion Trigeminale, *Z. Zellforschung* **95**:546–567 (1969).

55. E. D. Bueker, Isaac Schenkein, and Joan Bane, The problem of distribution of NGF, *Anat. Rec.* **136**:172 (1960).

56. E. D. Bueker, Isaac Schenkein, and Joan Bane, The problem of distribution of a NGF specific for spinal and sympathetic ganglia, *Cancer Res.* **20**:1220–1228 (1960).

57. R. Levi-Montalcini and P. Angeletti, Growth control of the sympathetic system by a specific protein factor, *Quart. Rev. Biol.* **3**:99 (1961).

58. M. Winnick and R. Greenberg, Appearance and localization of a nerve growth promoting protein during development, *Pediatrics* **35**:221 (1965).

59. J. Shulman-Weis, The occurence of NGF in teleost fishes, *Experientia* **24**:736 (1968).

60. R. Levi-Montalcini and P. Angeletti, Biological properties of a nerve growth promoting protein and its antiserum, *in 4th International Neurochemical Symposium*, Pergamon Press, New York (1960).

61. H. Hoffman and J. M. McDougall, Some biological properties of proteins of the mouse submaxillary gland as revealed by growth of tissue on electrophoretic acrylamide gels, *Exp. Cell Res.* **51**:485–503 (1968).

62. F. Caramia, P. Angeletti, and R. Levi-Montalcini, Experimental analysis of the mouse submaxillary gland in relationship to its NGF content, *Endocrinology* **70**: 915–922 (1962).

63. E. D. Bueker, P. Weis, and I. Schenkein, Sexual dimorphism of the mouse submaxillary gland and its relationship to nerve growth stimulating protein, *Proc. Soc. Exp. Biol. Med.* **118**:204–207 (1965).

64. R. Levi-Montalcini and P. Angeletti, Hormonal control of the NGF content in the submaxillary gland of mice, *in Salivary Glands and their Secretions* (L. M. Sreebny and J. Meyers, eds.), pp. 129–141 (1964).

65. M. T. Sabatini, A. Pellegrino, and E. ReRobertis, Early effects of A.S. against NGF on fine structure of sympathetic neurons, *Exp. Neurol.* **12**:4 (1965).

66. R. Levi-Montalcini and S. Cohen, Biological properties of a nerve growth promoting protein and its antiserum, Fourth International Neurochemical Symposium, pp. 362–377, Pergamon Press, New York (1960).

67. E. Zaimis, The immunosympathectomized animal: A valuable tool in physiological and pharmacological research, *J. Physiol. (London)* **177**:35–36 (1964).

68. S. M. Crain and R. Wiegand, Catecholamine levels of mouse sympathetic ganglia following hypertrophy produced by salivary gland NGF, *Proc. Soc. Exp. Biol. Med.* **107**:663 (1961).

69. R. Levi-Montalcini and P. Angeletti, Noradrenaline and monoamine oxidase contents in immunosympathectomized animals, *Int. J. Neuropharmacol.* **1**:161–164 (1962).

70. G. Klingman, Effect of immunosympathectomy on catecholamine levels and dopa decarboxylase activity in the peripheral tissue of rat, *Fed. Proc.* **23**:455 (1964).

71. G. Klingman, Catecholamine levels and dopa decarboxylase activity in peripheral organs and adrenergic tissues in the rat after immunosympathectomy, *J. Pharmacol. Exp. Ther.* **148**:14-21 (1965).

72. G. Klingman, The effect of immunosympathectomy on the superior cervical ganglion of the rat, *Fed. Proc.* **24**:132 (1965).

73. G. Klingman and J. D. Klingman, Effects of immunosympathectomy on the superior cervical ganglion and other adrenergic tissues of the rat, *Life Sci.* **4**:2171-2179 (1965).

74. R. Levi-Montalcini and P. Angeletti, Immunosympathectomy, *Pharmacol. Rev.* 619-628 (1966).

75. E. Zaimis, E. L. Berk, and B. A. Callingham, Biochemical and functional changes in the sympathetic nervous system of rats treated with NGF antiserum, *Nature* **206**:1220-1222 (1965).

76. A. Carpi and O. Oliverio, Urinary excretion of catecholamines in immunosympathecto-mized rats. Balance phenomenon between the adrenergic and the noradrenargic system, *Int. J. Neuropharmacology* **3**:427-431 (1964).

77. M. Vogt, Source of noradrenaline in the immunosympathectomized rat. *Nature* **204**:1315-1316 (1964).

78. M. J. Brody, Cardiovascular responses following immunological sympathectomy, *Circulation Res.* **15**:161-167 (1964).

79. G. Klingman, *In utero* immunosympathectomy of mice, *Int. J. Neuropharmacology* **5**:163-170 (1966).

80. B. Hamberger, R. Levi-Montalcini, K. A. Norberg, and F. Sjöquist, Changes in the cellular distribution of monoamines induced by immunosympathectomy, *Pharmacologist* **6**:2 (1965).

81. B. Hamberger, R. Levi-Montalcini, K. A. Norberg, and F. Sjöquist, Monoamines in immunosympathectomized rats, *Int. J. Neuropharmacology* **4**:91-95 (1965).

82. L. Olson, Outgrowth of sympathetic adrenergic neurons in mice treated with a NGF, *Z. Zellforschung* **81**:155-173 (1967).

83. J. Shulman-Weis, Analysis of the development of the nervous system of the zebrafish, Brachydanio Rerio. II. The effect of NGF and its antiserum on the nervous system of the zebrafish, *J. Embryol. Exp. Morph.* **19**:121-135 (1968).

84. J. Shulman-Weis, The effects of nerve growth factor on the spinal ganglia of *Ambystoma maculatum*, *J. Exp. Zool.* **170**:481-488 (1969).

85. R. Levi-Montalcini, L. Tentori, G. Vivaldi, P. Angeletti, and G. B. Marini-Bettolo, Composizione in ammino acidi del fattore di crescita del sistema nervoso (NGF), *Gazz. Chim. Ital.* **95**:333-337 (1965).

86. M. L. Salvi, P. Angeletti, and L. Frati, Frazionamento delle proteini solubili della ghiandola sotto-mascelare del topo. Localizazione di alcune componenti biologicamente attive, *Il Farmaco* **20**:12-21 (1965).

87. A. Zanini, P. Angeletti, and R. Levi-Montalcini, Immunochemical properties of the nerve growth factor, *Proc. Nat. Acad. Sci.* **61**:835-842 (1968).

88. O. J. Plescia, W. Braun, and N. C. Palczuk, Production of antibodies to denatured DNA, *Proc. Nat. Acad. Sci.* **52**:279-285 (1964).

89. I. B. B. Green, W. E. Levine, B. Paul, and B. Benacerraf, The relationship of the hapten to the carrier in the induction and specificity of the immune response, in *Nucleic Acids in Immunology* (O. J. Plescia and W. Braun, eds.), Springer-Verlag, New York (1968).

90. S. Cohen, Isolation of a mouse submaxillary gland protein accelerating incisor eruption and eyelid opening in the newborn animal, *J. Biol. Chem.* **237**:1555-1562 (1962).

91. E. D. Bueker and I. Schenkein, Effects of daily subcutaneous injections of nerve growth stimulating protein fractions on mice during postnatal to adult stage, *Ann. N.Y. Acad. Sci.* **118**:183-205 (1964).

92. D. G. Attardi, M. J. Schlesinger, and S. Schlesinger, Submaxillary gland of mouse: Properties of a purified protein affecting muscle tissue *in vitro, Science* **156**:1253–1255 (1967).

93. D. G. Attardi, R. Levi-Montalcini, and P. Angeletti, Submaxillary gland of mouse: Effects of a fraction of tissues of mesodermal origin *in vitro, Science* **150**:1307 (1965).

94. P. Angeletti, M. L. Salvi, F. Capani, and L. Frati, Granulocytosis inducing factor from the mouse submaxillary gland, *Biochim. Biophys. Acta* **111**:344–346 (1965).

95. I. Schenkein, M. Boesman, E. Tokarsky, L. Fishman, and M. Levy, Proteases from mouse submaxillary gland, *Biochem. Biophys. Res. Comm.* **36**:156–165 (1969).

96. A. Grossman, K. P. Lele, J. Sheldon, I. Schenkein, and M. Levy, The effect of esteroproteases from mouse submaxillary gland on the growth of rat hepatoma cells in tissue culture, *Exptl. Cell Res.* **54**:260–263 (1969).

Chapter 17

THE ACTION OF THYROID HORMONES

Louis Sokoloff

Section on Developmental Neurochemistry
Laboratory of Cerebral Metabolism
National Institute of Mental Health
U.S. Department of Health, Education and Welfare
Public Health Service
Bethesda, Maryland

I. INTRODUCTION

The thyroid hormones produce profound and diverse effects on the physiological, metabolic, and biochemical processes of most cells and tissues of the mammalian organism. Although the tissues of the central nervous system are among those affected, their responses to the hormones are quite atypical. First of all, the sensitivity of the brain to the thyroid hormones is limited only to its period of growth, development, and maturation; once fully developed to maturity, the brain no longer exhibits any of the usual effects which can be attributed to a direct action of the hormones. Second, there is a strong morphogenetic component in the effects of the thyroid hormones on the developing central nervous system. Most other tissues, no matter how greatly altered functionally and biochemically, undergo relatively little anatomical change under the influence of these hormones. The developing mammalian brain, however, undergoes a marked morphological reorganization, indeed differentiation, which appears to be dependent on the action of the thyroid hormones. In many ways, the effects of the thyroid hormones in the mammalian central nervous system are more like those seen in thyroxine-induced amphibian metamorphosis than in other mammalian tissues.

Despite the dissimilarities of their effects in the various tissues, it is likely, or at least most fruitful to assume that the diverse effects of such structurally specialized chemical substances reflect a single, common, fundamental biochemical mode of action. In order, therefore, to appreciate and to comprehend more fully the mechanisms of the influences of the thyroid hormones on the central nervous system, it is necessary to review their chemistry and their actions on metabolic and biochemical processes in general.

II. CHEMICAL NATURE OF THYROID HORMONES

The designation thyroid hormones is restricted to thyroxine and triiodothyronine because they are the only two compounds with classical thyroid hormone activity known to be synthesized in the gland and secreted into the circulation. Both are present in the blood, the triiodothyronine normally contributing less than 5% to the total blood hormonal iodine content.[1] Other thyroactive compounds with related chemical structure, for example, the tetroiodo- and triiodoacetic and propionic analogues, have been isolated from peripheral tissues, but these are probably metabolites of the hormones which have retained biological activity.

The chemical structures of the thyroid hormones are depicted in Fig. 1. The hormones are iodinated diphenyl ether derivatives of L-alanine with a phenolic group in the 4'-position. The exact structural requirements for thyromimetic activity have not been fully and precisely defined.[2,3] The alanine structure is not essential for biological activity; analogues with other side chains, such as acetic, propionic, or pyruvic acid, retain activity although the relative potency in the various biological actions of the hormones may be altered.[2,3] The D-alanine analogue, D-thyroxine, has only a fraction of the biological activity of the L-thyroxine, but it is metabolized and excreted at least three times more rapidly and distributed quite differently to the tissues.[4] In almost all *in vitro* actions of thyroxine in cell-free and purified enzyme systems, D-thyroxine is nearly as effective as L-thyroxine. The side chain probably plays no role in the specific chemical mechanism of hormonal action. It may influence biological activity only by its effects on protein binding, particularly in plasma, storage, distribution, metabolic degradation, and excretion of the hormones or their analogues.

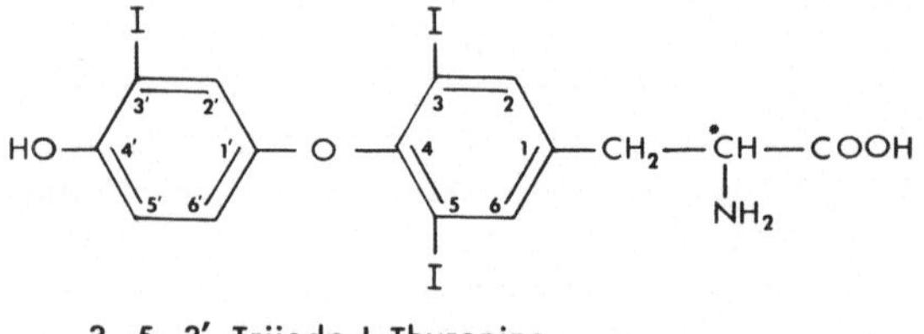

3, 5, 3′-Triiodo-L-Thyronine

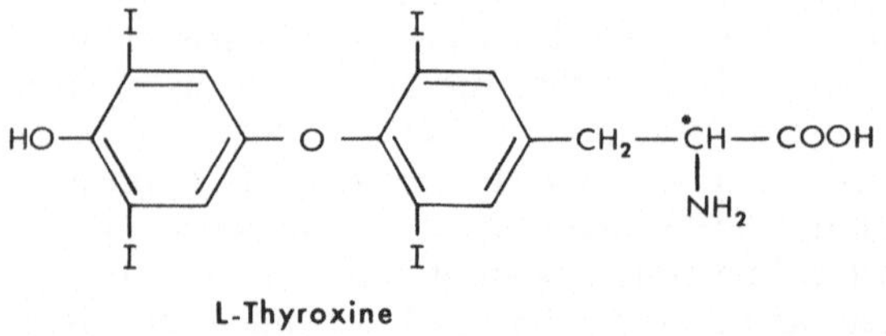

L-Thyroxine

Fig. 1. Chemical structure of the thyroid hormones, L-thyroxine and L-triiodothyronine. Asterisk indicates an asymmetric carbon.

The iodine atoms *per se* are not essential; bromo-, fluoro-, chloro-, and nitro- analogs exhibit activity, though with decreasing effectiveness in that order.[2,3] Alkyl substitution in the 3 and 5 positions results in markedly reduced activity intermediate between the chloro- and nitro-analogues; alkyl substitution in the 3' and/or 5' positions, however, leads to enhanced activity, even greater than that of the comparably iodinated molecules.[2,5] Substitution at the 3 and 5 positions is essential, and iodine substituents there are associated with the greatest amount of activity. 3,5-Diiodothyronine is slightly active, but thyronine and 3',5'-diiodothyronine are completely devoid of activity.[3] Substitution at either or both of the 3' and 5' positions is required for maximal activity. Triiodothyronine is reputed to be more active than thyroxine. The difference may reflect extracellular mechanisms rather than structure-function relationships in the specific mode of action intracellularly. Triiodothyronine is less firmly bound to plasma proteins than thyroxine,[6-8] and a given dose then leads to a higher free hormone concentration in the blood and more rapid distribution to the tissues.

The phenolic structure of the thyroid hormones has been suspected to be of fundamental importance to their functional activity. Niemann and Mead[9] studied a variety of isomers of thyroxine with a phenolic hydroxyl group displaced to various positions of the outer ring. Only those isomers capable of oxidation to a quinoid structure were active, and Niemann and Mead[9] postulated that thyromimetic activity is correlated with the oxidation-reduction potential of the system as a whole. More recently, however, O-alkyl derivatives, in which the phenolic structure is converted to an ether structure, have been found to be active,[2,3,10] and Jorgensen et al.[11] have observed considerable thyromimetic activity with 3,5-diiodo-4-(2',3'-dimethylphenoxy)-DL-phenylalanine, an analogue with no substituent at the 4' position and incapable of being metabolically converted to a phenolic structure. The functional role of the phenolic structure, if any, remains, therefore, obscure.

The ether structure is essential although a sulfur can be substituted for the oxygen bridge with retention of some activity.[2,12] Whether it need be a diphenyl ether is uncertain but it appears unlikely since benzyloxy derivatives of diiodotyrosine exhibit slight degrees of thyromimetic activity.[2]

The relatively long latency in the biological activity of the thyroid hormones has led to the common belief that they are not the active agents but must first be chemically converted to the active compounds. Almost all active analogues or derivatives of the hormones have at one time or another been investigated as the putative agents, and all have failed to achieve the distinction. Recent developments indicate, however, that the latent period is more apparent than real and that the effects exhibiting a latent period are delayed secondary consequences of earlier actions of the hormones. Effects on RNA and protein metabolism have been observed much earlier than the effects on metabolic rate.[13,14] In fact, recent studies have demonstrated that triiodothyronine stimulates protein synthesis with essentially no latent period at all.[14]

III. GENERAL BIOLOGICAL, PHYSIOLOGICAL, AND METABOLIC EFFECTS OF THYROID HORMONES

The thyroid hormones have broad and diffuse effects which ramify into almost every aspect of bodily function and metabolism. Extensive studies of these effects and their underlying biochemical mechanisms have led to a voluminous literature which can only briefly and selectively be considered in this volume. There are a number of excellent reviews available for more detailed and comprehensive treatment of the subject.[6,15–21]

A. Energy Metabolism

Perhaps the earliest discovery concerning the functions of the thyroid was in regard to its effects on oxygen consumption and heat production.[22] In mammals thyroid deficiency is associated with a marked decrease in oxygen consumption; complete thyroidectomy can lead eventually to as much as a 50% decrease in basal metabolic rate (BMR). The hypometabolism is reversed by the administration of thyroid hormones which, if administered in excess, can drive the BMR above normal. The stimulation of oxygen consumption occurs after a latent period of several hours following the administration of a dose of hormone; the latent period with triiodothyronine is shorter than with thyroxine, probably because of its lesser binding to plasma proteins and greater availability for diffusion into the cells. All tissues except mature brain, testis, and spleen[23] participate in the augmentation of oxidative metabolism, and all classes of foodstuffs, carbohydrate, protein, and lipid, are the substrates for the increased oxygen utilization.[24] Thyroid hormones cause rapid depletion of liver glycogen stores,[25–27] partly because of increased carbohydrate utilization but also because of the potentiation of the glycogenolytic effects of epinephrine.[28] Synergism with epinephrine may also contribute to the mobilization of fatty acids from adipose tissue and the elevated plasma free fatty levels in hyperthyroidism.[29,30] Serum cholesterol levels are inversely related to thyroid activity even though its synthesis is stimulated by thyroid hormones.[31]

The increased oxidative metabolism does not appear to lead to comparable increases in available energy for physical work. This is most evident in muscle which becomes noticeably inefficient in the hyperthyroid state. The increment in total body oxygen consumption associated with a given load of muscular effort is markedly increased in hyperthyroidism,[14] the additional energy apparently dissipated as heat. It has been suggested that the role of the thyroid is primarily in heat production and thermoregulation by inducing nonenergy conserving, heat-producing, calorigenic shunts in electron transport, for example, NADPH-cytochrome reductase.[32,33] The effects of thyroid hormones on oxidative metabolism are, in fact, limited to warm-blooded animals and are not seen in poikilothermic species. Increased thyroid activity is unquestionably an important component of the mechanisms of acclimation to cold,[34] but it is doubtful whether so limited a function is consistent with the numerous other peripheral actions of thyroid hormones.

B. Nitrogen Metabolism

The thyroid hormones produce prominent and, in some ways, paradoxical effects on nitrogen metabolism. Protein synthesis is stimulated in all tissues, which also respond with increased oxygen consumption.[35] In the growth-retarded, hypothyroid, immature mammal this effect may be anabolic and lead to positive nitrogen balance and stimulation of growth,[6] but excessive doses become catabolic and retard growth despite the continued stimulation of protein synthesis.[6,36] Thyroid hormones also stimulate protein synthesis in fully grown animals,[35] but urinary nitrogen excretion is usually markedly increased.[6,15,22,24,37] With large doses, negative nitrogen balance and weight loss ensue. The apparent paradox appears to be the result of a relative nutritional insufficiency associated with the increased energy metabolism. Adequate caloric intake restores normal nitrogen balance,[24] and magnesium and vitamin supplementation has been reported to reverse the catabolic effects of excessive thyroid hormone administration.[38] Hyperthyroidism is also characterized by a creatinuria which probably reflects a decreased creatine uptake by muscle tissue.[39] Creatine phosphate deficiency may cause contracting muscle to be more dependent on ATP generation from oxidative phosphorylation and might account for the increased oxygen consumption per given amount of work in hyperthyroid muscle.[24]

C. Growth and Metamorphosis

In adult mammals the action of thyroid hormones is most prominently reflected in their effects on metabolic rates. In young mammals, thyroid hormones also profoundly influence growth and development, and thyroid deficiency in the immature leads to stunted growth and incomplete maturation. Growth and development can be reinstituted by adequate thyroid replacement therapy, provided treatment is initiated early enough. Mild juvenile hyperthyroidism may lead to accelerated growth and the achievement of a greater than projected body stature; severe hyperthyroidism generally retards growth, probably because of the relative nutritional insufficiency.[38]

The most dramatic example of the effects of thyroid hormones on developmental processes is their action in amphibian metamorphosis.[40] The metamorphosis of the tadpole into an adult frog is absolutely dependent on thyroid hormones. It is a period of phenomenal structural reorganization of the entire organism into essentially a new phenotype. The process is obviously one of revision of gene expression, but the mechanism by which thyroid hormones initiate it is unknown. Protein and RNA synthesis are both increased, and tissue levels of various enzymes are drastically altered.[41,42] The characteristic stimulation of metabolic rate by thyroid hormones in warm-blooded animals is, however, absent in the cold-blooded amphibian during metamorphosis or at any other time.

IV. BIOCHEMICAL ACTIONS OF THYROID HORMONES

The relative simplicity of the chemical structure of the thyroid hormones and the conspicuousness of their biological effects have encouraged intense efforts to elucidate their biochemical mechanism of action. These studies have uncovered myriad biochemical effects, all capable of explaining some but none consistent with all of the biological actions of the hormones. In no case has a biochemical effect been precisely and unequivocally defined at the molecular level. The multiplicity of effects has confounded rather than facilitated attempts to establish the primary mode of action. Numerous ostensibly unrelated biochemical processes and pathways are accelerated or slowed; a variety of enzyme activities are activated, induced, or repressed; and even *in vitro* a number of purified enzymes, which catalyze distinctly different classes of reactions and operate by obviously different mechanisms, are either stimulated or inhibited by thyroid hormones. Many of these effects, particularly those observed following *in vivo* administration, reflect secondary or even more remote consequences of earlier more specific biochemical actions of the hormones and are clearly cellular adaptations to changed intracellular conditions arising from earlier effects. The biochemical effects of the thyroid hormones and the theories of their mechanism of action have been comprehensively examined in a number of reviews.[6,16,19,21,43]

A. Effects on Electron Transport

Because of their prominent effects on oxygen consumption, the effects of thyroid hormones on the respiratory electron transport chain have received major attention. Thyroxine itself has been considered an electron carrier which could undergo oxidation–reduction in its phenolic ring to and from a quinoid structure[9]; it has even been proposed as a component of an additional site in the electron transport chain where phosphorylation is coupled to oxidation.[44] Experimental evidence in support of this hypothesis is lacking and, in fact, it appears unlikely in view of the demonstrated biological activity of analogues which lack the 4'-phenolic hydroxyl group and cannot, therefore, form quinoid structures.[2,3,10] Components of the mitochondrial respiratory chain, such as cytochrome c[45,46] and ubiquinone,[47] and the microsomal NADPH-cytochrome c reductase system[32] are increased in hyperthyroidism and reduced in hypothyroidism, but these changes appear to be secondary adaptations to the changes in metabolic rate rather than the cause.[46,47] Lindberg et al.[48] have recently concluded, on the basis of the effects of thyroxine on a variety of mitochondrial reactions, that its actions on oxidative phosphorylation and mitochondrial structure are secondary to a direct interaction of the hormone with the flavin-linked electron transport site in the respiratory chain.

B. Effects on Oxidative Phosphorylation

Dinitrophenol, a substituted phenol like the thyroid hormones, is the classical uncoupler of oxidative phosphorylation. It dissociates electron transport from the phosphorylation process so that electron transport no longer drives ATP synthesis.[49] The uptake of inorganic phosphate into ATP is inhibited, and, in fact, the opposite reaction, ATP hydrolysis, is stimulated. Electron transport is released by this uncoupling, and oxygen consumption is increased. The net result is that the P/O ratio is markedly depressed. The thyroid hormones have similar though not identical effects on oxidative phosphorylation.[16] Mitochondria isolated from thyroxine-treated animals or preincubated with thyroid hormones *in vitro* exhibit depressed P/O ratios.[16,50-53] The oxygen consumption, however, is not increased by the thyroid hormones *in vitro*; it may, in fact, be reduced. ATPase is stimulated but only weakly compared to the effect of dinitrophenol. Furthermore, relatively high doses or concentrations are required to produce these effects, and, in contrast to dinitrophenol, thyroid hormones depress oxidative phosphorylation only in intact mitochondria and not in actively phosphory-lating submitochondrial particles.[54]

Recently, Hoch[55,56] observed almost immediate effects of minute doses of thyroid hormones on rat liver mitochondrial respiratory control. In normal tightly coupled mitochondria, respiration is regulated by the level of phosphate acceptors and is restricted by low levels of ADP. Thyroid hormones loosen the coupling and reduce the effectiveness of this control; in effect oxygen consumption is stimulated, but only in the presence of less than optimal concentrations of ADP. Human skeletal muscle mitochondria from hyperthyroid patients show the same deficiency.[57] Such an effect suggests some interaction with the coupling reactions between oxidation and phosphorylation, but it does not necessarily mean uncoupling. Uncoupling implies energy wastage which may be compatible with the calorigenesis and reduced muscle efficiency associated with hyperthyroidism but cannot be reconciled with the stimulation of energy-requiring reactions, such as protein synthesis, growth, and maturation, by thyroid hormones. A change in respiratory control is not inconsistent with such effects; it does not by itself explain them, but it may be an indication of some mitochondrial-hormone interaction, other than uncoupling, which leads to profound changes in cellular functions. Uncoupling may, however, be of significance in extreme thyrotoxic states when the concentrations of the thyroid hormones reach relatively enormous levels.[58]

C. Effects on Mitochondrial Structure

Closely related to and, perhaps, responsible for the effects on oxidative phosphorylation are the actions of thyroid hormones on mitochondrial structure. Low concentrations of hormone, much lower than required to depress the P/O ratio, cause mitochondria to swell.[59-62] Mitochondria normally exhibit swelling-contraction cycles,[62] but hypothyroidism reduces and hyperthyroidism enhances their fragility and tendency to swell.[60,61] Mitochondria from tissues which are physiologically thyroxine-insensitive do not swell in response to thyroxine.[63] The morphological change is due to water accumulation and reflects functional and structural alterations of the mitochondrial membrane.[62,64] Uncoupling of oxidative phosphorylation is not the cause; dinitrophenol, a more potent uncoupler, not only does not cause swelling but reverses the swelling induced by thyroid hormones.[60,62] Lehninger[62] believes that uncoupling by thyroid hormones is secondary to the swelling mechanism. Thyroactive compounds and other swelling agents, such as Ca^{2+}, inorganic phosphate, glutathione, and a variety of other substances, cause the release from the mitochondrial membrane of a protein-bound fatty acid derivative, U factor, which causes mitochondrial swelling, uncouples oxidative phosphorylation, and stimulates mitochondrial ATPase activity.[62] ATP, which reverses the swelling effect, causes the incorporation of the U factor into the cardiolipin and phosphatidic acid fractions of the mitochondrial membrane.[62]

It is clear that thyroid hormones produce profound alterations in the cytostructural organization and function of the mitochondria. These have led to extensive speculation that changes in mitochondrial permeability may underlie the physiological actions. Against this view is the relative non-specificity of the swelling effect; numerous other naturally occurring agents also induce mitochondrial swelling and yet have no semblance of thyromimetic activity.[62] The effect of thyroid hormones on mitochondrial structure is real and relevant, but it probably represents not the cause but a consequence of the fundamental biochemical mechanism of action of the hormones. It provides strong evidence, however, that the mitochondria are the primary locus of that action.

D. Effects on Catecholamine Metabolism

The thyroid hormones potentiate the action of the adrenergic amines. Cardiovascular and metabolic responses to a given dose of epinephrine are more intense and prolonged in hyperthyroidism and reduced in hypothyroidism.[28,30,65–68] Epinephrine has a calorigenic action qualitatively like that of the thyroid hormones, and many of the physiological and behavioral manifestations of hyperthyroidism resemble those of increased sympathetic nervous activity and can be ameliorated by adrenolytic drugs. There have been reports that thyroid hormones lower catechol-O-methyl transferase[69] and monamine oxidase[70] activities, but there has been controversy concerning the significance of these effects to the overall metabolic degradation and excretion of the catecholamines.[71,72] Catecholamines are, however, largely inactivated by uptake and binding in sympathetic nerve endings within the tissues; thyroid hormones have been reported to diminish this process.[73] Brewster et al.[74] have argued that the actions of the thyroid hormones are mediated entirely by the sympathetic amines. It is likely that these amines initiate or exaggerate some of the effects attributed to the thyroid hormones, but sympathetic blockade by means of dibenzyline or guanethidine does not eliminate or even alter their most fundamental metabolic actions.[75,76] The sympathetic nervous system and the adrenergic amines may modulate but do not mediate the effects of the thyroid hormones.

E. Effects on Specific Enzyme Activities

Thyroid hormones exert effects on a number of specific enzymes and the chemical reactions which they catalyze. Wolff and Wolff[43] have listed almost 70 enzyme activities which are influenced by thyroid hormones *in vivo* and more than 30 which are affected by the hormones *in vitro*. Most of those altered only *in vivo* probably reflect secondary changes and may bear no special relationship to the specific molecular mechanism of hormonal action. Some of the actions *in vitro*, however, are on isolated, purified enzymes. Unfortunately, the classes of enzymes represented are so diverse that no common denominator on which to base a single mechanism of action can be recognized. The nearest approach to this goal has been achieved by Wolff and Wolff[77] who found that thyroxine inhibits a large number of zinc-containing, NAD^+-linked dehydrogenases. Among these are the dehydrogenases which catalyze the oxidation of glutamate, malate, triosephosphate, lactate, and alcohol. The effect on glutamate dehydrogenase has been found to be accompanied by a dissociation of the enzyme protein into smaller subunits.[78,79] Thyroxine forms complexes with a number of metal ions, but its action on the dehydrogenases does not appear to be related to complex formation with the Zn^{2+} contained in them.[43,77] The effect on alcohol dehydrogenase has been attributed to the noncompetitive inhibition by thyroid hormones of the binding of the nicotinamide portion of the NAD^+ cofactor to the enzyme.[80] The action of thyroid hormones on this class of dehydrogenases may eventually prove to be a model for specific chemical action of the thyroid hormones, but it is at present inadequate to explain the multiplicity of their physiological and biochemical effects.

The search for a single mechanism of action is confounded by the effects on other classes of enzymes. Creatine phosphokinase is strongly inhibited by thyroid hormones.[81,82] The mechanism is unknown, but this effect may have important implications for the accumulation and utilization of energy stores in muscle. Thyroxine inhibits a number of pyridoxal phosphate-dependent enzymes, but the effect appears to be due to pyridoxal phosphate deficiency.[83] Since pyridoxal phosphate is the cofactor in the various transamination and amino acid decarboxylation reactions, this action could be of importance in the nervous system by virtue of its effects on γ-aminobutyric acid metabolism and the synthesis of the biogenic amines. Thyroxine forms a highly insoluble complex with Mg^{2+}[84] and inhibits a number of Mg^{2+}-dependent enzyme activities through this action.[43] Indeed, thyroxine administration causes a rapid fall in serum Mg^{2+} levels, and magnesium deficiency causes catabolic effects similar to those observed in hyperthyroidism.[85] Mg^{2+} deficiency also causes uncoupling of oxidative phosphorylation, and magnesium reverses the uncoupling action of thyroxine both *in vivo* and *in vitro*.[86,87] Some of the debilitating effects of thyrotoxicosis may, in fact, be secondary to absolute or relative magnesium deficiency and can be reversed by magnesium administration.[38,85,86]

One of the most dramatic effects of thyroid hormones is the induction of mitochondrial L-α-glycerophosphate dehydrogenase activity.[88] Greater than tenfold increases in enzyme activity can be achieved in liver and kidney following a single dose of hormone *in vivo*. There are no direct effects on the enzyme *in vitro*. Only the flavin-linked mitochondrial enzyme is induced; the soluble NAD^+-linked α-glycerophosphate dehydrogenase is unaffected. The two enzymes and α-glycerophosphate and its oxidation product, dihydroxyacetone phosphate, constitute a shuttle mechanism by which the hydrogen of extramitochondrial NADH is transported into the mitochondria and made available to the respiratory electron transport chain. It is a major pathway for the reoxidation of NADH produced extramitochondrially during aerobic glycolysis. The mechanism and consequences of the enzyme induction are unknown, but it is quantitatively the most prominent physiological or biochemical action of thyroid hormones yet described in mammalian tissues.

F. Effects on Protein and Nucleic Acid Synthesis

Recent developments have led to an intriguing new and unifying concept of the mechanism of action of the thyroid hormones. Numerous studies have established that thyroid hormones stimulate protein synthesis.[14,35,36,43,58,89-93] Since structural and enzymatic proteins are so fundamentally involved in all aspects of cellular morphology, metabolic activity, and function, an effect on protein synthesis could readily ramify into all the diverse and multiple actions ascribed to the thyroid hormones. The administration of thyroid hormones to normal animals stimulates the incorporation of amino acids into protein in tissues, such as liver, kidney, and heart, which also respond with increased metabolic rates; no effects on protein synthesis are observed in mature brain, testis, or spleen, all tissues which retain normal rates of oxygen utilization in hyperthyroidism[23,35] (Fig. 2). The increased metabolic rate appears to be secondary to the effect on protein synthesis; it follows the stimulation of protein synthesis by a number of hours[91] and is prevented or abruptly reversed by inhibitors of protein synthesis.[94,95] Contrary to most effects of thyroid hormones *in vivo*, the stimulation of protein synthesis occurs with essentially no latent period.[14] Hypothyroidism leads to reduced rates of protein synthesis[58,91] which are restored to normal by thyroid hormone replacement therapy.[90,91]

The effects on protein synthesis are not only observed *in vivo* and in cell-free tissue preparations from treated animals but can also be reproduced in cell-free preparations from normal animals by thyroid hormones added *in vitro*.[58,89] Stimulations occur with thyroxine concentrations as low as 10^{-7}M and increase with increasing hormone concentration until an optimum is reached at about 4×10^{-4}M; above this concentration the effect abruptly reverses from stimulation to inhibition.[58] The reversal may reflect the binding of Mg^{2+} by high concentrations of thyroxine and may be of significance in severe thyrotoxic states.[58] The mechanism of the stimulation of protein synthesis has been extensively studied in the *in vitro* system and has been found to be a translational effect at the level of the transfer of *t*RNA-bound amino acid to ribosomal protein.[96,97] It is exerted primarily on the

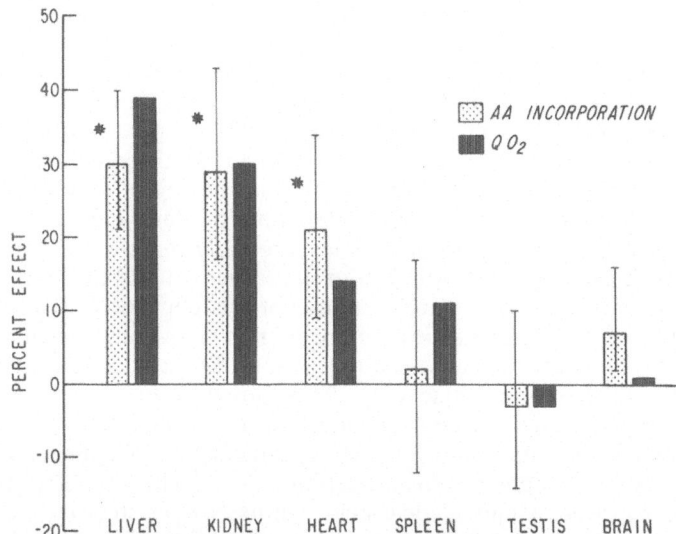

Fig. 2. Organ distribution of the effects of hyperthyroidism on *in vivo* amino acid incorporation in protein (AA) and tissue oxygen consumption (Q_{O_2}). Asterisk indicates statistical significant effect ($P < 0.05$). The means $\pm$ S.E. are indicated. The data on protein synthesis are taken from Michels *et al.*[35]; the data on oxygen consumption are from Gordon and Heming.[23]

elongation or completion of the nascent polypeptide chain; initiation of new protein chains appears to be secondarily stimulated, perhaps, because accelerated completion and release of existing polypeptide chains free the ribosomes and template RNA more rapidly for recycling.[98] The effect is not limited to a single specific species of protein but is observed in a large variety of tissues synthesizing different types of protein[98]; in fact, it can be elicited in the synthesis of artificial polypeptides directed by synthetic messenger polyribonucleotides.[99]

The thyroid hormones do not, however, stimulate protein synthesis directly. They first interact with mitochondria in an energy-dependent reaction leading to some product or consequence which is actually responsible for the enhanced ribosomal protein-synthesizing activity.[93] The identity of the products of the thyroxine-mitochondrial reaction has not yet been established, nor is there any convincing evidence that it is a single unique compound rather than a readjustment of concentrations of existing substances, but it is known that the protein synthesis stimulating activity is heat stable, dialyzable, and acid labile.[100] Mitochondria from thyroxine-insensitive tissue, such as mature brain, are incapable of reacting with thyroid hormones to produce this stimulating activity,[101,102] a deficiency which may be the basis of the lack of responsiveness of these tissues to the hormones.

When administered *in vivo*, thyroid hormones also stimulate nuclear RNA polymerase activity and the synthesis of ribosomal and, perhaps, also messenger RNA.[13] These effects result ultimately in increased cellular contents of functional ribosomes and, therefore, also result in a second basis for increased rates of protein synthesis.[36,91] The effects on nucleic acid synthesis cannot, however, be reproduced by thyroid hormones *in vitro*, are completely prevented *in vivo* by mild fasting of the animal, and appear several hours after the initial mitochondria-dependent cytoplasmic stimulation of protein synthesis.[14] They probably represent a secondary, positive feedback, cellular adaptation to the earlier cytoplasmic stimulation of protein synthesis. An initial mitochondria-dependent cytoplasmic stimulation of the existing cellular protein synthesizing machinery is followed by a nuclear-mediated cellular adaptative increase in the amount of protein synthesizing machinery.[14,100] Although secondary, the nuclear-mediated response may represent the mechanism for the selective transcription of genetic information which is so obviously displayed in thyroid hormone-induced amphibian metamorphosis.[41,42] The actions of thyroid hormones on protein and nucleic acid synthesis offer thus far the most comprehensive explanation of their multiple physiological and biochemical effects.

V. ACTIONS OF THYROID HORMONES IN THE NERVOUS SYSTEM

It will be noted from the previous discussion that the consequences of the biological actions of the thyroid hormones in mammalian tissues are, in general, confined to the metabolic or biochemical domain. In most tissues their role is manifested only in alterations of the rates and balances of metabolic and chemical processes with little or no apparent influence on the structure or specific functional activity of the cells. The situation is quite different in the central nervous system. The thyroid hormones decisively influence the postnatal growth, development, and maturation of the mammalian brain. Their effects in this tissue are in many respects more like those which occur in amphibian metamorphosis than in other mammalian tissues. They are reflected mainly in the structural and functional reorganization and differentiation which characterize maturation rather than in mere changes in metabolic rates. The uniqueness of the nervous system in this regard may be related to the timing of the action of thyroid hormones. In most mammalian tissues the thyroid hormones are active all through life. In the brain, responsiveness to the hormones is confined only to the period of development and maturation, a time when the plasticity of the tissue is great, and profound structural and functional changes normally occur. Once maturation of the brain is achieved, however, the thyroid hormones exert little, if any, further effect.

A. Anatomical Effects

The elegant studies of Eayrs and his associates, which have done so much to clarify the action of thyroid hormones on the morphogenesis of the nervous system, have recently been reviewed by Eayrs.[103] These investigations were carried out in rats, which are particularly suitable for developmental studies because cerebral cortical differentiation and maturation occur almost entirely postnatally. Thyroidectomy at birth leads to impairment of normal histiogenesis which, if allowed to progress beyond a critical developmental age, becomes permanent and irreversible.[103–105] Brain water content and brain weight are low, myelinization is slow and deficient, perikarya are small in size and more closely aggregated, and axonal and dendritic densities are reduced. The neuronal nuclei are not particularly affected, and partly because of the reduction in perikaryonal cytoplasm, brain DNA concentration is increased and RNA content decreased.[106–108] The perikarya are more densely packed because of a reduction in the amount of axonal and dendritic material between them. There is, in fact a paucity of axodendritic connections in the neuropil which may be of relevance to the functional deficiencies in cretinism.[105] Eayrs[103] has pointed out that impairment in axonal development is more prominent in layer 4 of the cerebral cortex which contains the specific afferent plexuses associated with the thalamic sensory projection system.

All the morphological defects in the nervous system of the hypothyroid animal represent failures in development and do not occur in the mature nervous system. They can be prevented or reversed in the immature hypothyroid animal by thyroid hormone replacement therapy, but only if that therapy is begun early enough.[103] Excessive thyroid hormone administration may even accelerate the maturation process.[103] Hamburgh and Bunge[109,110] have recently reported an effect of thyroxine on myelinogenesis in cerebellar cortex grown in tissue culture; thyroxine added directly *in vitro* to the culture accelerated the onset and rate of myelinogenesis.

B. Neurophysiological and Behavioral Effects

Associated with the morphological defects in the cretinous nervous system are concomitant disturbances in the development of electrophysiological and behavioral activities.[103] The electroencephalograms of neonatally thyroidectomized rats remain low in amplitude,[111] and the appearance of evoked responses is delayed.[112] When it finally appears, the evoked response is characterized by an abnormally long latency and duration and a low amplitude.[112] Innate behavioral responses, such as startle, righting, and placing reactions, are also delayed in their appearance.[103] There are severe decrements in maze performance and learning behavior, and the impairment is not merely a sluggishness in response but an increase in errors and a reduced rate of error correction with repeated trials.[103,113,114] The cretinous rat does not learn well from experience and, like its human counterpart, can be considered mentally retarded. The electrophysiological and behavioral abnormalities can as readily be corrected by thyroid hormone therapy as the morphological changes, but the same time limitations exist. All the defects caused by neonatal thyroidectomy are completely or partially reversible, provided therapy is begun before the critical postnatal age, 24 days in the rat[103]; the later the onset of therapy, the less complete is the recovery.

If thyroidectomy is delayed until maturity has been achieved, or even after the critical period has passed, then morphological effects are negligible and abnormalities in electrical and behavioral functions are completely reversible.[103] Excitability of the central nervous system is reduced, as evidenced by a rise in electroshock threshold.[115] The amplitude of the EEG is lowered, and there may be a fall in the alpha frequency. Behaviorally, the adult hypothyroid is slow and sluggish, and although there is an appearance of mental retardation, the defect is one of slowness of response rather than inability to learn. In severe myxedema mental disturbances may occur ranging from coma to psychosis. Excess thyroid produces opposite effects. Hyperthyroidism increases CNS excitability[115,116] which is manifested not only in electroshock thresholds and reflex activity, but also in emotional and affective states.

Eayrs[103] has hypothesized that the physiological and behavioral changes caused by thyroid dysfunction in the immature animal reflect the effects on axodendritic relations in the neuropil and functional interactions of the neuronal elements. The functional disturbances in the mature animal are less clearly understood and have been attributed to unspecified metabolic changes, either in the CNS or elsewhere in the body.

C. Metabolic Effects

The action of thyroid hormones on metabolic rate also shows an age dependence in the central nervous system. Fazekas et al.[117] have studied the effects of thyroxine administration on the metabolic rate of the cerebral cortex of the rat from birth to early adulthood. Cerebral cortical O_2 consumption in the rat is normally quite low at birth and remains so for approximately the first 10 days of life; it then rises in a typical S-shaped growth type of curve until it reaches the adult level at approximately 45 days of age (Fig. 3). Thyroxine administration to newborn animals shifts this curve to the left, resulting in an earlier and steeper rise of the curve and a more rapid achievement of the normal adult level. Once that level is reached, however, further administration of thyroxine is without effect. The shape of the curve for oxygen consumption parallels and unquestionably reflects the structural and functional maturation of the nervous system. A number of cerebral cortical enzymes including the oxidative enzymes, succinic dehydrogenase and cytochrome oxidase, show a similar time course of development.[118] Thyroid hormones cause, therefore, an earlier and more rapid metabolic as well as morphological and functional maturation of the brain and stimulate cerebral oxygen consumption only during the period of growth and development of the brain. Thyroid hormones appear to have no direct

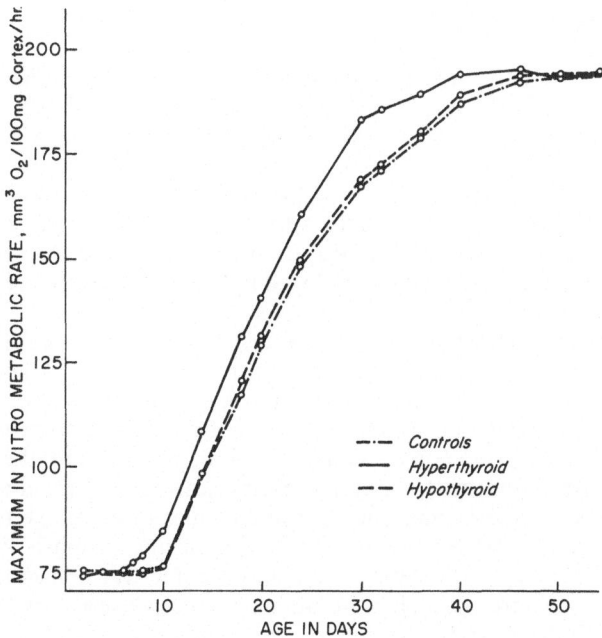

Fig. 3. Effects of hyperthyroidism and hypothyroidism on rat cerebral cortical oxygen consumption at various age from birth to maturity. The illustration is redrawn from the report of Fazekas et al.[117]

effects on cerebral metabolic rate in the fully mature brain. Cerebral oxygen consumption is clearly normal in adult hyperthyroidism,[23,117,119–121] but in human adult hypothyroidism one study reported it to be depressed[122] and another found it to be normal.[120] The apparent discrepancy is probably due to differences in severity and duration of the hypothyroidism. It is likely that simple deficiency of thyroid hormone does not in itself lower the cerebral metabolic rate, but when it is allowed to progress to myxedema, then secondary effects, perhaps, originating in other tissues, may intervene to cause lowering of the energy metabolism of the brain. Hypothyroidism present at birth or before a critical time in the course of brain development, if untreated, almost certainly leads to low metabolic rates in adulthood. This seems likely not only because of the impaired morphological and functional maturation of the brain[103] but also because of the failure of oxidative enzymes such as succinic dehydrogenase to develop normally in such circumstances.[118]

The difference in sensitivities of mature and immature brain to thyroid hormones is not a reflection of the operation of the blood–brain barrier. Thyroid hormones do traverse the blood–brain barrier of adult animals, perhaps, not freely, but rapidly enough for the amounts in brain to reflect sustained differences in the levels circulating in the blood.[123,124]

D. Biochemical Effects

Normal maturation is associated with changes in DNA, RNA, protein, and specific enzyme concentrations and quantities in the brain. There may be some increase in the total DNA content of the brain, probably due to glial multiplication, but the tremendous growth of perikaryonal cytoplasm, proliferation of axonal and dendritic processes, and myelin deposition lead to dilution of the DNA in the nuclei and a decline in DNA concentration.

RNA and protein contents and concentrations of the brain rise, however. All these changes are delayed and reduced in magnitude by thyroid deficiency at or shortly following birth.[106–108]

Postnatal hypothyroidism also lowers or impedes the development of a number of enzyme systems. Although phospholipid deposition is apparently unaffected,[125] the formation of cholesterol, cerebrosides, and sulfatides is markedly reduced in the brains of neonatally thyroidectomized rats.[126] These lipids constitute major components of myelin and their deficient synthesis is consistent with the impaired myelinization present in postnatal hypothyroidism. The normal increase in enzymes of oxidative metabolism, such as succinic dehydrogenase and cytochrome oxidase, is similarly affected.[118] Glutamic dehydrogenase and alanine transferase activities do not appear to be influenced, but glutamic decarboxylase activity is markedly inhibited.[108] The latter enzyme is responsible for γ-aminobutyric acid (GABA) formation, and its reduced development in hypothyroidism is consistent with the low brain GABA content[127] and deficient nerve-ending proliferation[108] which have also been observed.

The fact that the most prominent effects of thyroid hormones in brain are confined to the period of its most rapid growth and development suggests an action primarily on biosynthetic processes. Recent studies have, in fact, suggested the action of thyroid hormones on protein synthesis as the underlying basis of their effects in brain.[128] The rate of protein synthesis is normally three to four times greater in immature than in mature brain,[101,102] a difference undoubtedly reflecting the perikaryonal growth, axonal and dendritic proliferation, and myelinization which are occurring during the process of maturation. Like its effects on cerebral oxygen consumption, thyroxine stimulates protein synthesis in the immature but not the mature brain.[101,102] These effects on protein synthesis are observed in vivo[35,129,130] and also in vitro in cell-free brain preparations[101,102] as shown in Table I. The incorporation of leucine-^{14}C into protein was assayed in vitro with cell-free preparations of rat liver or brain as indicated in the table. The thyroxine was added in vitro directly to the reaction mixtures to a final concentration of 6.5×10^{-5}M. The details of the reaction mixture and assay conditions have been previously described.[101,128]

The subcellular components of the in vitro protein synthesis assay system used in the studies of Gelber et al.[101] and Klee and Sokoloff[102] were mitochondria, microsomes, and cell sap. Crossover experiments with all possible combinations of these three subcellular elements derived from

TABLE I

Effects of L-Thyroxine Added *In Vitro* on the Incorporation of Leucine-^{14}C into Protein in Cell-Free Preparations of Brain and Liver from Infant and Adult Rats[a]

Tissue	Animal age (days)	Protein specific activity		L-Thyroxine effect	
		Control	Thyroxine[b]		
		(cpm/mg protein)		Δ cpm/mg	Percentage[c]
Liver	40–50 (3)	39	61	+22	+57
	15–16 (2)	32	45	+13	+42
Brain	40–50 (7)	12	9	−13[b]	−21
	15–16 (6)	44	50	+6[b]	+16

[a] From data of Gelber *et al.*[101]
[b] Denotes statistically significant effect ($P < 0.05$) as determined by method of paired comparison.
[c] Mean of individual percentage effects, not percentage difference between the means. The number of experiments from which the means are derived are indicated by the numbers in parentheses.

mature and immature brain have clearly established that functional differences in their mitochondria are responsible for the difference in thyroxine sensitivities of mature and immature brain[102,128] (Fig. 4).

In Figs. 4 and 5 protein synthesis and the effects of thyroxine on the rate were assayed *in vitro* in cell-free preparations as in the experiments summarized in Table I. The final thyroxine concentration was 6.5×10^{-5}M. The cell-free

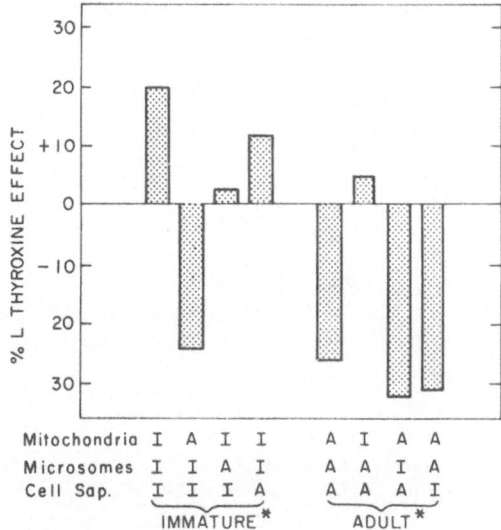

Fig. 4. The role of mitochondria in the different effects of thyroxine on protein synthesis in mature brain as compared to immature brain. I indicates that the homogenate fraction is obtained from immature brain. A indicates that the homogenate fraction is obtained from adult brain. Asterisk indicates at least two components derived from sources as indicated. Data from Klee and Sokoloff.[102]

system contained mitochondria, microsomes, and cell sap derived from immature brain (I), mature brain (A), or mature liver (L). Immature rats were 15–16, and mature rats were 40–50 days of age. The effects of thyroxine on protein synthesis were examined in all possible combinations of these three cell fractions derived from mature brain and from immature brain or liver. Note that stimulation by thyroxine is observed when the mitochondria are either from immature brain (Fig. 4) or from liver (Fig. 5) and never with mature brain mitochondria, regardless of the source of the cell sap and microsomes.

It will be recalled from the previous discussion that mitochondria are involved in the mechanism of the stimulation of protein synthesis by thyroid hormones. A preliminary interaction between mitochondria and the thyroid hormone yields a product or consequence which is responsible for the enhancement of ribosomal protein biosynthetic activity.[14,93] Immature brain mitochondria share with the mitochondria from other thyroxine-sensitive tissues the ability to participate in this interaction. The mitochondria of mature brain have apparently lost this ability, and this may be the basis of the loss of thyroxine sensitivity.

The action of thyroid hormones on protein synthesis goes far to explain many of their major effects in the central nervous system. As has been found in other tissues[91] and in the body as a whole,[94,95] the effects of thyroid hormones on metabolic rate are secondary to those on protein synthesis. Changes in RNA content of the tissues[13,106] can reflect cellular adaptations to the altered rate of protein synthesis.[14] The influence of thyroid hormones on the morphological and functional development of the brain[103] and the time course of the appearance of various enzymatic activities[118] are fully

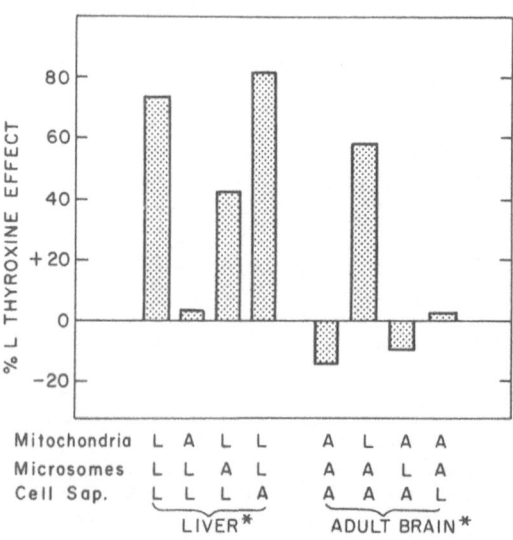

Fig. 5. The role of mitochondria in the different effects of thyroxine on protein synthesis in mature brain as compared to liver. L indicates that the homogenate fraction is obtained from liver. A indicates that the homogenate fraction is obtained from adult brain. Asterisk indicates at least two components derived from sources as indicated. Data from Klee and Sokoloff.[102]

consistent with the effects on protein synthesis. Finally, the change in thyroxine sensitivity with maturation of the brain and its implications in regard to the critical periods of development can at present be explained only in relation to the role of mitochondria in the mechanism of the thyroxine stimulation of protein synthesis.[100,128]

ACKNOWLEDGMENTS

The author wishes to express his appreciation to Mrs. Lucille Wilfand and Mrs. Barbara Franke for their bibliographic assistance and their aid in the preparation of the manuscript.

VI. REFERENCES

1. J. Wynn, Organic iodine constituents in human serum, *Arch. Biochem. Biophys.* **87**:120–124 (1960).
2. H. A. Selenkow and S. P. Asper Jr., Biological activity of compounds structurally related to thyroxine, *Physiol. Rev.* **35**:426–474 (1955).
3. T. C. Bruice, N. Kharasch, and R. J. Winzler, A correlation of thyroxine-like activity and chemical structure, *Arch. Biochem. Biophys.* **62**:306–317 (1956).
4. D. F. Tapley, F. F. Davidoff, W. B. Hatfield, and J. F. Ross, Physiological disposition of D- and L-thyroxine in the rat, *Am. J. Physiol.* **197**:1021–1027 (1959).
5. C. M. Greenberg, B. Blank, F. R. Pfeiffer, and J. F. Pauls, Relative activities of several 3'- and 3':5'-alkyl and aryl thyromimetic agents, *Am. J. Physiol.* **205**:821–826 (1963).
6. R. Pitt-Rivers and J. R. Tata, *The Thyroid Hormones*, Pergamon Press, London (1960).
7. J. R. Tata and C. J. Shellabarger, An explanation for the difference between the responses of mammals and birds to thyroxine and triiodothyronine, *Biochem. J.* **72**:608–613 (1959).
8. K. Sterling and M. Tabachnick, Determination of the binding constants for the interaction of thyroxine and its analogues with human serum albumin, *J. Biol. Chem.* **236**:2241–2243 (1961).
9. C. Niemann and J. F. Mead, The synthesis of DL-3,5-diiodo-4-(3',5'-diiodo-2' hydroxyphenoxy)-phenylalanine, a physiologically active isomer of thyroxine, *J. Am. Chem. Soc.* **63**:2685–2687 (1941).
10. K. Tomita, H. A. Lardy, D. Johnson, and A. Kent, Synthesis and biological activity of O-methyl derivatives of thyroid hormones, *J. Biol. Chem.* **236**:2981–2986 (1961).
11. E. C. Jorgensen, N. Zenker, and C. Greenberg, Thyroxine analogues. III. Antigoitrogenic and calorigenic activity of some alkyl substituted analogues of thyroxine, *J. Biol. Chem.* **235**:1732–1737 (1960).
12. C. R. Harington, Synthesis of a sulfur-containing analogue of thyroxine, *Biochem. J.* **43**:434–437 (1948).
13. J. R. Tata and C. C. Widnell, Ribonucleic acid synthesis during the early action of thyroid hormones, *Biochem. J.* **98**:604–620 (1966).
14. L. Sokoloff, P. A. Roberts, M. M. Januska, and J. E. Kline, Mechanisms of stimulation of protein synthesis by thyroid hormones *in vivo*, *Proc. Natl. Acad. Sci.* **60**:652–659 (1968).
15. D. Marine, *in Glandular Physiology and Therapy*, pp. 315–333, Am. Med. Assoc., Chicago (1935).
16. F. L. Hoch, Biochemical actions of thyroid hormones, *Physiol. Rev.* **42**:605–673 (1962).

17. R. Pitt-Rivers and W. R. Trotter (eds.), *The Thyroid Gland*, Vol. I, Butterworths, London (1964).

18. J. H. Means, L. J. DeGroot, and J. B. Stanbury, *The Thyroid and Its Diseases*, 3rd ed., McGraw-Hill, New York (1963).

19. S. B. Barker, Mechanism of action of thyroid hormone, *Physiol. Rev.* **31**:205–243 (1951).

20. S. B. Barker, Peripheral actions of thyroid hormones, *Fed. Proc.* **21**:635–641 (1962).

21. J. Wolff and R. C. Goldberg, in *Biochemical Disorders in Human Disease* (R. H. S. Thompson and E. J. King, eds.), pp. 289–351, Academic Press, New York (1957).

22. A. Magnus-Levy, Über den respiratorischen Gewechsel unter dem Einfluss der Thyroidea sowie unter verschiedenen patholischen Zustanden, *Berl. Klin. Wchnschr.* **32**:650–652 (1895).

23. E. S. Gordon and A. E. Heming, The effect of thyroid treatment on the respiration of various rat tissues, *Endocrinology* **34**:353–360 (1944).

24. W. M. Boothby and I. Sandiford, The total and the nitrogenous metabolism in exophthalmic goiter, *J. Am. Med. Assn.* **81**:795–800 (1923).

25. I. A. Mirsky and R. H. Broh-Kahn, The effect of experimental hyperthyroidism on carbohydrate metabolism, *Am. J. Physiol.* **117**:6–12 (1936).

26. R. Sternheimer, The effect of a single injection of thyroxine on carbohydrates, protein, and growth in the rat liver, *Endocrinology* **25**:899–908 (1939).

27. S. D. Burton, E. Robbins, and S. O. Byers, Effect of hyperthyroidism on glycogen content of the isolated rat liver, *Am. J. Physiol.* **188**:509–513 (1957).

28. K. R. Hornbrook, P. V. Quinn, J. H. Siegel, and T. M. Brody, Thyroid hormone regulation of cardiac glycogen metabolism, *Biochem. Pharmacol.* **14**:925–926 (1965).

29. C. Rich, E. L. Bierman, and I. L. Schwartz, Plasma nonesterified fatty acids in hyperthyroid states, *J. Clin. Invest.* **38**:275–278 (1959).

30. M. Vaughan, An *in vitro* effect of triiodothyronine on rat adipose tissue, *J. Clin. Invest.* **46**:1482–1491 (1967).

31. K. Fletcher and N. B. Myant, Influence of the thyroid on the synthesis of cholesterol by liver and skin *in vitro*, *J. Physiol.* **144**:361–372 (1958).

32. A. H. Philips and R. H. Langdon, The influence of thyroxine and other hormones on hepatic TPN-cytochrome reductase activity, *Biochim. Biophys. Acta* **19**:380–382 (1956).

33. V. R. Potter, Possible biochemical mechanisms underlying adaptation to cold, *Fed. Proc.* **17**:1060–1063 (1958).

34. J. S. Hart, Metabolic alterations during chronic exposure to cold, *Fed. Proc.* **17**:1045–1054 (1958).

35. R. Michels, J. Cason, and L. Sokoloff, Thyroxine: effects on amino acid incorporation into protein *in vivo*, *Science* **140**:1417–1418 (1963).

36. J. R. Tata, in *Actions of Hormones on Molecular Processes* (G. Litwack and D. Kritchevsky, eds.), pp. 58–131, John Wiley, New York (1964).

37. H. F. Müller, Beitrage zur kenntniss der Basedowischen Krankheit, *Deutsches Arch. f. Klin. Med.* **51**:335–412 (1893).

38. S. N. Gershoff, J. J. Vitale, I. Antonowicz, M. Nakamura, and E. E. Hellerstein, Studies of interrelationships of thyroxine, magnesium, and vitamin B_{12}, *J. Biol. Chem.* **231**:849–854 (1958).

39. C. D. Fitch, R. Coker, and J. S. Dinning, Metabolism of creatine-1-C^{14} by vitamin E-deficient and hyperthyroid rats, *Am. J. Physiol.* **198**:1232–1234 (1960).

40. J. F. Gudernatsch, Feeding experiments on tadpoles. II. A further contribution to the knowledge of organs with internal secretion, *Am. J. Anat.* **15**:431–480 (1914).

41. P. P. Cohen, Biochemical aspects of metamorphosis: transition from ammontelism to ureotelism, *The Harvey Lectures* **60**:119–154 (1964–1965).

42. E. Frieden, Thyroid hormones and the biochemistry of amphibian metamorphosis, *Recent Progr. Hormone Res.* **23**:139–194 (1967).

43. E. C. Wolff and J. Wolff, *in The Thyroid Gland* (R. Pitt-Rivers and W. R. Trotter, eds.), Vol. I, pp. 237–281, Butterworths, London (1964).

44. R. D. Dallam and R. B. Howard, Thyroxine-enhanced oxidative phosphorylation of rat liver mitochondria, *Biochim. Biophys. Acta* **37**:188–189 (1960).

45. D. L. Drabkin, Cytochrome C metabolism and liver regeneration. Influence of thyroid gland and thyroxine, *J. Biol. Chem.* **182**:335–349 (1950).

46. H. M. Klitgaard, Effect of thyroidectomy on cytochrome C concentration of selected rat tissues, *Endocrinology* **78**:642–644 (1966).

47. S. Pedersen, J. R. Tata, and L. Ernster, Ubiquinone (coenzyme Q) and the regulation of basal metabolic rate by thyroid hormones, *Biochim. Biophys. Acta* **69**:407–409 (1963).

48. O. Lindberg, H. Löw, T. E. Conover, and L. Ernster, *in Biological Structure and Function* (T. W. Goodwin and O. Lindberg, eds.), Vol. II, pp. 3–24, Academic Press, London (1961).

49. W. F. Loomis and F. Lipmann, Reversible inhibition of the coupling between phosphorylation and oxidation, *J. Biol. Chem.* **173**:807–808 (1948).

50. G. F. Maley and H. A. Lardy, Metabolic effects of thyroid hormones *in vitro*. II. Influence of thyroxine and triiodothyronine on oxidative phosphorylation, *J. Biol. Chem.* **204**:435–444 (1953).

51. F. L. Hoch and F. Lipmann, The uncoupling of respiration and phosphorylation by thyroid hormones, *Proc. Natl. Acad. Sci.* **40**:909–921 (1954).

52. G. F. Maley and H. A. Lardy, Efficiency of phosphorylation in selected oxidations by mitochondria from normal and thyrotoxic rat livers, *J. Biol. Chem.* **215**:377–388 (1955).

53. H. G. Klemperer, The uncoupling of oxidative phosphorylation in rat liver mitochondria by thyroxine, triiodothyronine, and related substances, *Biochem. J.* **60**:122–135 (1955).

54. C. Cooper and A. L. Lehninger, Oxidative phosphorylation by an enzyme complex from extracts of mitochrondria. I. Span β-hydroxybutyrate to oxygen, *J. Biol. Chem.* **219**:489–505 (1956).

55. F. L. Hoch, Rapid effects of a subcalorigenic dose of L-thyroxine on mitochondria, *J. Biol. Chem.* **241**:524–525 (1966).

56. F. L. Hoch, Early action of injected L-thyroxine on mitochondrial oxidative phosphorylation, *Proc. Natl. Acad. Sci.* **58**:506–512 (1967).

57. L. Ernster, D. Ikkos, and R. Luft, Enzymic activities of human skeletal muscle mitochondria: a tool in clinical metabolic research, *Nature* **184**:1851–1854 (1959).

58. L. Sokoloff and S. Kaufman, Thyroxine stimulation of amino acid incorporation into protein, *J. Biol. Chem.* **236**:795–803 (1961).

59. D. F. Tapley, C. Cooper, and A. L. Lehninger, The action of thyroxine on mitochondria and oxidative phosphorylation, *Biochim. Biophys. Acta* **18**:597–598 (1955).

60. D. F. Tapley, The effect of thyroxine and other substances on the swelling of isolated rat liver mitochondria, *J. Biol. Chem.* **222**:325–339 (1956).

61. D. F. Tapley, Mode and site of action of thyroxine, *Proc. Mayo Clinic* **39**:626–636 (1964).

62. A. L. Lehninger, Water uptake and extrusion by mitochondria in relation to oxidative phosphorylation, *Physiol. Rev.* **42**:467–517 (1962).

63. D. F. Tapley and C. Cooper, Effect of thyroxine on the swelling of mitchondria isolated from various tissues of the rat, *Nature* **178**:1119 (1956).

64. G. E. Paget and J. M. Thorp, An effect of thyroxine on the fine structure of the rat liver cell, *Nature* **199**:1307–1308 (1963).

65. E. Goetsch, Newer methods in the diagnosis of thyroid disorders: pathological and clinical, *New York J. Med.* **18**:259–267 (1918).

66. O. Thibault, Action renforçatrice de la thyroxine sur l'effet inhibiteur de l'adrénaline sur l'intestin de Lapin isolé, *Compt. Rend. Soc. Biol.* (*Paris*) **142**:499–504 (1948).

67. T. S. Danowski, A. C. Heineman Jr., J. V. Bonessi, and C. Moses, Hydrocortisone and/or desiccated thyroid in physiologic dosage. XIV. Effects of thyroid hormone excesses on pressor activity and epinephrine responses, *Metabolism* **13**:747–752 (1964).

68. H. E. Swanson, Interrelationships between thyroxine and adrenalin in the regulation of oxygen consumption in the albino rat, *Endocrinology* **59**:217–225 (1956).

69. A. D'Iorio and J. Leduc, The influence of thyroxine on the *O*-methylation of catechols, *Arch. Biochem. Biophys.* **87**:224–227 (1960).

70. M. H. Zile, Effect of thyroxine and related compounds on monamine oxidase activity, *Endocrinology* **66**:311–312 (1960).

71. T. S. Harrison, Adrenal medullary and thyroid relationships, *Physiol. Rev.* **44**:161–185 (1964).

72. R. P. Zimon, E. V. Flock, G. M. Tyce, S. G. Sheps, and C. A. Owen Jr., Effect of thyroid hormones on metabolism of DL-norepinephrine by isolated rat liver, *Endocrinology* **80**:808–814 (1967).

73. R. J. Wurtman, I. J. Kopin, and J. Axelrod, Thyroid function and the cardiac disposition of catecholamines, *Endocrinology* **73**:63–74 (1963).

74. W. R. Brewster Jr., J. P. Isaacs, P. F. Osgood, and T. L. King, The hemodynamic and metabolic interrelationships in the activity of epinephrine, norepinephrine, and the thyroid hormones, *Circulation* **13**:1–20 (1956).

75. A. Surtshin, J. K. Cordonnier, and S. Lang, Lack of influence of the sympathetic nervous system on the calorigenic response to thyroxine, *Am. J. Physiol.* **188**:503–506 (1957).

76. W. Y. Lee, D. Bronsky, and S. S. Waldstein, Studies of thyroid and sympathetic nervous system interrelationships. II. Effects of guanethidine on manifestations of hyperthyroidism, *J. Clin. Endocrinol. Metab.* **22**:879–885 (1962).

77. J. Wolff and E. C. Wolff, The effect of thyroxine on isolated dehydrogenases, *Biochim. Biophys. Acta* **26**:387–396 (1957).

78. J. Wolff, The effect of thyroxine on isolated dehydrogenases. II. Sedimentation changes in glutamic dehydrogenase, *J. Biol. Chem.* **237**:230–235 (1962).

79. J. Wolff, The effect of thyroxine on isolated dehydrogenases. III. The site of action of thyroxine on glutamic dehydrogenase, the function of adenine and guanine nucleotides, and the relation of kinetic to sedimentation changes, *J. Biol. Chem.* **237**:236–242 (1962).

80. K. McCarthy, W. Lovenberg, and A. Sjoerdsma, The mechanism of the inhibition of horse liver alcohol dehydrogenase by thyroxine and related compounds, *J. Biol. Chem.* **243**:2754–2760 (1968).

81. B. A. Askonas, Effect of thyroxine on creatine phosphokinase activity, *Nature* **167**:933–934 (1951).

82. S. A. Kuby, L. Noda, and H. A. Lardy, Adenosinetriphosphatecreatine transphosphorylase. III. Kinetic Studies, *J. Biol. Chem.* **210**:65–82 (1954).

83. A. Horvath, Inhibition by thyroxine of enzymes requiring pyridoxal-5-phosphate, *Nature* **179**:968 (1957).

84. H. Lardy, *in The Thyroid, Brookhaven Symposium in Biology*, No. 7, 1954, Brookhaven National Laboratory, pp. 90–101, Upton, New York (1955).

85. J. J. Vitale, D. M. Hegsted, M. Nakamura, and P. Connors, The effect of thyroxine on magnesium requirement, *J. Biol. Chem.* **226**:597–601 (1957).

86. J. J. Vitale, M. Nakamura, and D. M. Hegsted, The effect of magnesium deficiency on oxidative phosphorylation, *J. Biol. Chem.* **228**:573–576 (1957).

87. S. H. Mudd, J. H. Park, and F. Lipmann, Magnesium antagonism of the uncoupling of oxidative phosphorylation by iodo-thyronines, *Proc. Natl. Acad. Sci.* **41**:571–576 (1955).

88. Y. P. Lee, A. E. Takemori, and H. Lardy, Enhanced oxidation of α-glycerophosphate by mitochondria of thyroid-fed rats, *J. Biol. Chem.* **234**:3051–3054 (1959).

89. L. Sokoloff and S. Kaufman, Effects of thyroxine on amino acid incorporation into protein, *Science* **129**:569–570 (1959).

90. O. Stein and J. Gross, Effect of thyroid hormone on protein biosynthesis by cell-free systems of liver, *Proc. Soc. Exptl. Biol. Med.* **109**:817–820 (1962).

91. J. R. Tata, L. Ernster, O. Lindberg, E. Arrhenius, S. Pedersen, and R. Hedman, The action of thyroid hormones at the cell level, *Biochem. J.* **86**:408–428 (1963).

92. J. R. Tata, *in Proceedings of the Second International Congress of Endocrinology*, London, 1964, International Congress Series No. 83 (S. Taylor, ed.), pp. 46–56, Excerpta Medica Foundation, Amsterdam (1965).

93. L. Sokoloff, *in Proceedings of the Second International Congress of Endocrinology*, London, 1964, International Congress Series No. 83 (S. Taylor, ed.), pp. 87–94, Excerpta Medica Foundation, Amsterdam (1965).

94. J. R. Tata, Inhibition of the biological action of thyroid hormones by actinomycin D and puromycin, *Nature* **197**:1167–1168 (1963).

95. W. P. Weiss and L. Sokoloff, Reversal of thyroxine-induced hypermetabolism by puromycin, *Science* **140**:1324–1326 (1963).

96. L. Sokoloff, S. Kaufman, P. L. Campbell, C. M. Francis, and H. V. Gelboin, Thyroxine stimulation of amino acid incorporation into protein. Localization of stimulated step, *J. Biol. Chem.* **238**:1432–1437 (1963).

97. L. Sokoloff, P. L. Campbell, C. M. Francis, and C. B. Klee, Thyroxine stimulation of amino acid incorporation into ribosomal protein, *Biochem. Biophys. Acta* **76**:329–332 (1963).

98. R. L. Krause and L. Sokoloff, Effects of thyroxine on initiation and completion of protein chains of hemoglobin *in vitro*, *J. Biol. Chem.* **242**:1431–1438 (1967).

99. L. Sokoloff, C. M. Francis, and P. L. Campbell, Thyroxine stimulation of amino acid incorporation into protein independent of any action on messenger RNA synthesis, *Proc. Natl. Acad. Sci.* **52**:728–736 (1964).

100. L. Sokoloff, *in Proceedings of the Third Kettering Symposium* (A. Pietro, M. R. Lamborg, and F. T. Kenney, eds.), pp. 345–367, Academic Press, New York (1968).

101. S. Gelber, P. L. Campbell, G. E. Deibler, and L. Sokoloff, Effects of L-thyroxine on amino acid incorporation into protein in mature and immature rat brain, *J. Neurochem.* **11**:221–229 (1964).

102. C. B. Klee and L. Sokoloff, Mitochondrial differences in mature and immature brain. Influence on rate of amino acid incorporation into protein and responses to thyroxine, *J. Neurochem.* **11**:709–716 (1964).

103. J. T. Eayrs, Endocrine influence on cerebral development, *Arch. de Biologie (Liège)* **75**:529–565 (1964).

104. J. T. Eayrs and S. H. Taylor, The effect of thyroid deficiency induced by methyl thiouracil on the maturation of the central nervous system, *J. Anat. (London)* **85**:350–358 (1951).

105. J. T. Eayrs, The cerebral cortex of normal and hypothyroid rats, *Acta Anat.* **25**:160–183 (1955).

106. S. E. Geel and P. S. Timiras, The influence of neonatal hypothyroidism and thyroxine on the ribonucleic acid and deoxyribonucleic acid concentrations of rat cerebral cortex, *Brain Res.* **4**:135–142 (1967).

107. J. M. Pasquini, B. Kaplum, C. A. Garcia Argiz, and C. J. Gomez, Hormonal regulation of brain development. I. The effect of neonatal thyroidectomy upon nucleic acids, protein, and two enzymes in developing cerebral cortex and cerebellum of the rat, *Brain Res.* **6**:621–634 (1967).

108. R. Balàzs, S. Kovacs, P. Teichgräber, W. A. Cocks, and J. T. Eayrs, Biochemical effects of thyroid deficiency on the developing brain, *J. Neurochem.* **15**:1335–1349 (1968).

109. M. Hamburgh and R. P. Bunge, Evidence for a direct effect of thyroid hormone on maturation of nervous tissue grown *in vitro, Life Sci.* **3**:1423–1430 (1964).

110. M. Hamburgh, Evidence for a direct effect of temperature and thyroid hormone on myelinogenesis *in vitro, Developmental Biol.* **13**:15–30 (1966).

111. P. B. Bradley, J. T. Eayrs, and K. Schmalbach, The electroencephalogram of normal and hypothyroid rats, *Electroenceph. Clin. Neurophysiol.* **12**:467–477 (1960).

112. P. B. Bradley, J. T. Eayrs, A. Glass, and W. Heath, The maturational and metabolic consequences of neonatal thyroidectomy upon the recruiting response in the rat, *Electroenceph. Clin. Neurophysiol.* **13**:577–586 (1961).

113. J. T. Eayrs and W. A. Lishman, The maturation of behavior in hypothyroidism and starvation, *Brit. J. Animal Behavior* **3**:17–24 (1955).

114. J. T. Eayrs and S. Levine, Influence of thyroidectomy and subsequent replacement therapy upon conditioned avoidance learning in the rat, *J. Endocrinol.* **25**:505–513 (1963).

115. D. M. Woodbury, Effect of hormones on brain excitability and electrolytes, *Recent Progr. Hormone Res.* **10**:65–107 (1954).

116. P. S. Timiras, D. M. Woodbury, and S. L. Agarwal, Effect of thyroxine and triiodothyronine on brain function and electrolyte distribution in intact and adrenalectomized rats, *J. Pharmacol. Exptl. Therap.* **115**:154–171 (1955).

117. J. F. Fazekas, F. B. Graves, and R. W. Alman, The influence of the thyroid on cerebral metabolism, *Endocrinology* **48**:169–174 (1951).

118. M. Hamburgh and L. B. Flexner, Biochemical and physiological differentiation during morphogenesis. XXI. Effect of hypothyroidism and hormone therapy on enzyme activities of the developing cerebral cortex of the rat, *J. Neurochem.* **1**:279–288 (1957).

119. L. Sokoloff, R. L. Wechsler, R. Mangold, K. Balls, and S. S. Kety, Cerebral blood flow and oxygen consumption in hyperthyroidism before and after treatment, *J. Clin. Invest.* **32**:202–208 (1953).

120. P. Scheinberg, Cerebral circulation and metabolism in hyperthyroidism, *J. Clin. Invest.* **29**:1010–1013 (1950).

121. W. Sensenbach, L. Madison, S. Eisenberg, and L. Ochs, The cerebral circulation and metabolism in hyperthyroidism and myxedema, *J. Clin. Invest.* **33**:1434–1440 (1954).

122. P. Scheinberg, E. A. Stead Jr., E. S. Brannon, and J. V. Warren, Correlative observations on cerebral metabolism and cardiac output in myxedema, *J. Clin. Invest.* **29**:1139–1146 (1950).

123. D. H. Ford and J. Gross, The metabolism of I^{131}-labeled thyroid hormones in the hypophysis and brain of the rabbit, *Endocrinology* **62**:416–436 (1958).

124. D. H. Ford and J. Gross, Central nervous system–thyroid interrelationships, *Brain Res.* **7**:329–349 (1968).

125. A. Cuaron, J. Gamble, N. B. Myant, and C. Osorio, The effect of thyroid deficiency on the growth of the brain and on the deposition of brain phospholipids in foetal and newborn rabbits, *J. Physiol. (London)* **168**:613–630 (1963).

126. P. Walravens and H. P. Chase, Influence of thyroid on formation of myelin lipids, *J. Neurochem.* **16**:1477–1484 (1969).

127. A. E. Ramirez de Guglielmone and C. J. Gomez, Influence of neonatal hypothyroidism on amino acids in developing brain, *J. Neurochem.* **13**:1017–1025 (1966).

128. L. Sokoloff, Action of thyroid hormones and cerebral development, *Am. J. Dis. Children* **114**:498–506 (1967).

129. L. Schneck, D. H. Ford, and R. Rhines, The uptake of S^{35}-L-methionine into the brain of euthyroid and hyperthyroid neonatal rats, *Acta Neurol. Scand.* **40**:285–290 (1965).

130. S. Geel, T. Valcana, and P. S. Timiras, Effect of neonatal hypothyroidism and of thyroxine on L-[^{14}C] leucine incorporation in protein *in vivo* and the relationship to ionic levels in the developing brain of the rat, *Brain Res.* **4**:143–150 (1967).

Chapter 18

PROTEIN TURNOVER

A. Lajtha and N. Marks

New York State Research Institute
for Neurochemistry and Drug Addiction
Ward's Island, New York, N.Y.

I. INTRODUCTION

Regulation of protein composition, distribution, and metabolism in tissues is a fundamental process essential to life. Alterations in protein metabolism play a crucial role not only during the physiological functioning of the nervous system but also during cellular differentiation and during pathological changes. Although our knowledge of synthetic mechanisms is rather advanced, there is less known about mechanisms of breakdown, and nothing about how synthesis and breakdown are linked together. Knowledge of both processes is required before exact schemes for "turnover" and its regulation can be formulated. The mechanisms for degradation were discussed in our review (in this volume) and the manner in which these might be integrated with synthetic mechanisms is summarized in schematic fashion in Fig. 1.

The following were the main questions that interested early investigators of protein turnover: (a) Are some, or many, of the cerebral proteins metabolically active? (b) How does the pattern and mechanism of brain protein turnover compare to other tissues (especially as brain is an example of a nonregenerative tissue)? (c) What is the relationship of protein metabolism to function or to the alterations resulting from administration of drugs, hormonal and nutritional factors, or changes in patterns of behavior?

Protein metabolism has many ramifications, as reviewed elsewhere in the Handbook and in recent symposia (see Section VI). The present chapter focuses attention on the very concept of turnover and questions related to its measurement. Studies not specifically cited here can be found in the bibliographies of Lajtha[1-3] and Oja[5]; other sources include the reviews of Hyden,[22] Richter[2g,10], Waelsch,[23] Waelsch and Lajtha,[24] Palladin,[25] Neuberger and Richards,[4b] Miller,[4d] Droz,[1k,2c] Munro,[4a,e] Piha,[216] S. Roberts,[1q,2a] and Altman.[1q,75] It must be emphasized at the outset that the complexity of obtaining pure brain fractions and proteins, together with the assumptions invoked in calculating turnover, at best yield results that are approximations.

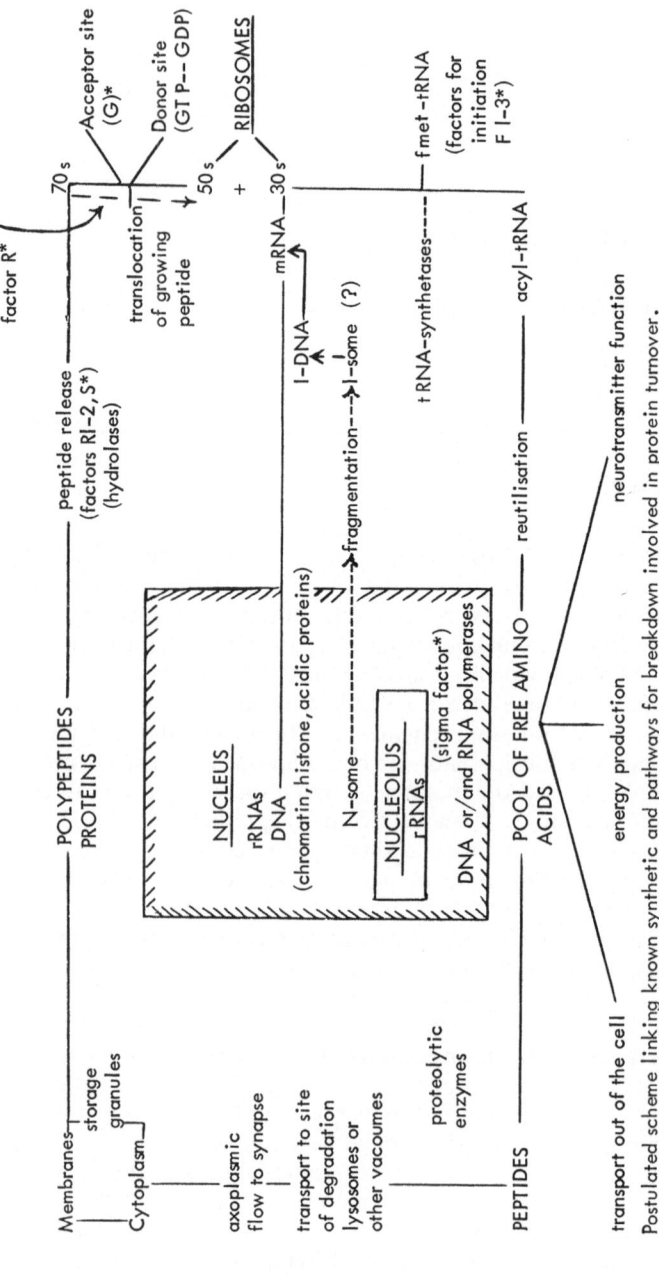

Fig. 1. Postulated scheme linking known synthetic and degradative pathways involved in protein turnover. Factors which influence synthetic mechanisms in microorganisms are indicated by asterisks. In mammalian tissues the major rRNA's are 40S, 60S, forming a combined 80S particle. The sigma factor is that associated with DNA-dependent RNA polymerase.

II. TURNOVER IN WHOLE BRAIN

A. Concepts and Terminology

The term turnover denotes the overall metabolism of proteins, encompassing both synthesis and degradation. In relatively simple systems (microorganisms), the rate of synthesis can be evaluated from the incorporation of labeled precursors, and degradation from the decay of radioactivity. Under steady-state conditions, the two rates are equal and both represent the rate of turnover. It is important to realize that steady-state conditions cannot always be assumed, especially as far as the specific protein or a specific structure is concerned. Turnover is best evaluated by noting the appearance and durability of a particular protein species, by convention expressed as the half-life $(t_{1/2})$. Most studies on turnover and its regulation have been performed with systems less complex than brain and it may be premature to extrapolate from such findings (e.g., microorganisms) to nerve tissue. The lack of lysosomal or mitochondrial particulates in bacteria further handicaps comparison of this nature. Mandelstam[30] observed some reduction in protein breakdown in rapidly growing bacterial colonies; in contrast, others have reported rapid breakdown in nongrowing colonies.[31-34] If similar conditions pertain to mammalian tissues, the quantity of protein at a given time is determined by differential alteration of rates of breakdown or synthesis. In practice, studies of turnover should be accompanied by an attempt to evaluate the relative contribution of these two processes. It is possible that in mammalian cells the two processes (breakdown and synthesis) are linked in an unknown manner such that alteration of one unavoidably affects the rate of the other (by a possible complex feedback mechanism, Fig. 1; also see discussion in Section VI). Basically, all methods attempting to evaluate turnover, or its synthetic or breakdown components, rely on rates of change utilizing three well-defined methods:

I. Administration of labeled precursors in single doses (pulse labeling). A variant of this procedure is to administer a second but different label at a later period, known as double-label procedure.

II. Continuous administration of isotope. This can be achieved by feeding, by infusion, or direct perfusion of the isolated head (Seta, Mitsunobu, and Lajtha, unpublished).

III. Study of kinetics of enzyme change. There are four variants for this method which are capable of giving information on rates of synthesis or breakdown, namely: (a) rate of increase of enzyme (hormones, stress, etc.), (b) restoration of enzyme activity after irreversible inhibition, (c) rates of decay following enzyme induction, and, (d) rate of decay of enzyme after inhibition of protein synthesis.

Although different sets of assumptions apply to the analysis of data in all three methods, mathematical expressions can be derived for an approximation of the half-life or relative rates of degradation in comparison to synthesis. Early studies clearly established the metabolic lability of brain proteins, but early investigators frequently failed to recognize the assumptions implied in the experimental method. These studies were previously discussed by Lajtha,[3] especially the pioneering efforts performed in the period 1940–1950, when labeled materials first become available. The three methods above are described briefly, together with their implied assumptions according to the tenets advanced by Reiner,[27,77] Buchanan,[35] Swick,[36,52] Koch,[37] more recently by Lajtha,[3] Schimke,[4g] and Schimke and Doyle.[9] In general, studies in brain encompass fractions with more than one population of proteins, with different rates of synthesis or breakdown. The departures from linearity for decay or incorporation of label observed imply the summation of one or two exponential functions.[27] At a first approximation, leaving aside all other assumptions, the rates of decay for components A_1 and A_2 can be determined according to the first-order law

$$-\frac{dC}{dt} = A_1 e^{-k_1 t} + A_2 e^{-k_2 t} \tag{1:1}$$

or in the case of a one-component system this can be simplified to

$$\ln \frac{C_0}{C_t} = kt \tag{1:2}$$

where k is the first-order rate constant, C_0 the composition at time zero (i.e., 100%), and C_t at time t. The half-life ($t_{1/2}$) is when $C_t = 50\%$, reducing the first term to $\ln 2$ or 0.693. If an average half-life of 14 days (336 hr) is assumed for brain proteins, $k = 2.06 \times 10^{-3}$ hr^{-1}. This would represent a conversion rate of approximately 0.2% hr^{-1}. For such small values it can be assumed that $\ln (1 - \delta) \simeq \delta$, where δ represents the fractional value converted at time t, or $C_t = C_0(1 - \delta)$. Substitution of this value in equation (1:2) gives the expression $\ln (1 - \delta) = -kt \simeq -\delta$. Since it is known that $k = \ln 2/t_{1/2}$, the equation can be simplified to

$$\frac{t \ln 2}{24C} = t_{1/2} \text{ (days)} \tag{1:3}$$

or

$$\frac{2.89 \times t}{C} = t_{1/2} \tag{1:4}$$

where C is the per cent conversion per hour (i.e., $t = 1$). An example for the calculation of C is provided in the legend to Table II. In addition to exponential labeling patterns cited for brain in this review, similar observations are reported on exponential decay of intracellular proteins, including arginase,[9,38,48] tyrosine transaminase,[39] tryptophan pyrrolase,[40] catalase,[41] serine dehydratase,[42] etc.

Method I

A single administration of label results in a rapid and precipitous decline in free pools, as illustrated in Fig. 2. The rates of change of precursor pools will depend on the route of injection, and in brain, on the age of the animal. In our own studies on myelin turnover,[43] intracisternal injection led to a rapid decline of Leu-^{14}C in the acid-soluble material of rat brain. Depending

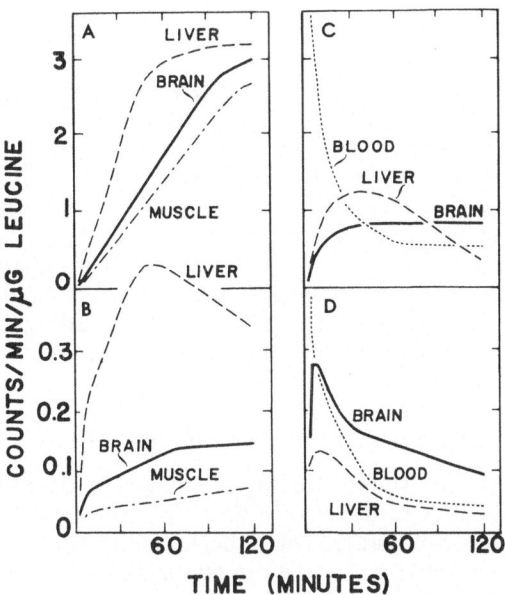

Fig. 2. Change in specific activity of Leu-^{14}C in newborns (A, C) and adult mice (B, D) with time in the trichloracetic acid-soluble pool (C, D) or the insoluble protein (A, B). Leu was determined by a microbiological method. Results show that intravenous injection in adults or intraperitoneal in newborns is accompanied by a rapid decay of the pool of radioactive leucine in blood. The change in pool size in liver is included for comparison. Incorporation is linear with time up to 1.5 hr in newborn brain but increases slowly after 10 min in the adult, reaching a maximum at 1 hr.

on the concentration of the amino acid introduced, and its rate of conversion to other intermediates, the rate constants for transport to the precursor pool can be estimated. Knowledge of the specific activity of the precursor is essential if the rate of incorporation into the protein-bound form is to be determined. In pulse-labeling methods, consideration should be given to aspects of the precursor which include penetration of label (route of injection, and rate-limiting factors concerned with transport, and knowledge of the specific activity of the precursor pool(s). It can be inferred from many studies that there exists a functional heterogeneity of the precursor pool.[335] There are no studies in brain describing the kinetics of formation of a single protein relative to such precursor amino acid pools. Mathematical models invoked in the case of other tissues involve a "multicompartmental" analysis of considerable complexity[5,44,335] (Section IIA, C).

Metabolic conversion of the labeled amino acids (e.g., rapid in the case of Leu,[45,118,285] Glu[1j]; slow with Tyr,[5,46] Val[1,46] should also be taken into account. The use of carboxyl-labeled amino acids has been suggested as an alternative since radioactive CO_2 is eliminated more rapidly[268]; the end-products after decarboxylation do not add to the radioactivity of the precursor pool (except for the fixation of a limited amount of CO_2 into Glu, Asp, and Glu·NH$_2$).[52,286]

The following aspects of the product should be considered. (a) Heterogeneity of turnover for different proteins. It cannot always be assumed that the radioactivity measured is representative of all the proteins in the sample. Heterogeneity is related also to different anatomical regions. (b) Reutilization of labeled amino acids produced by breakdown. (c) Interconversion of proteins. (d) Redistribution by processes of axoplasmic flow. (e) The merits of short- versus long-term experiments for measurement of the short-lived versus the longer lived components.

All the above have important implications for the evaluation of altered metabolic patterns consequent to conditioning or training experiments.

A method to determine whether proteins are turning over at dissimilar rates is the double isotope technique.[49] Giving two isotopes, ^{14}C and ^3H, separated by time periods enables an assessment of relative rates; ^{14}C will represent the decay time point, and ^3H the initial incorporation rate, if the animal is sacrificed shortly after the second administration. In double isotope experiments, assumptions are made concerning the identical rates of synthesis at the time of the two isotope administrations, and the fact that the different proteins being compared are synthesized from the same amino acid pool.[338]

Method II (Continuous Administration)

Infusion or feeding of animals with a radioactive diet enables measurement of turnover of longer lived proteins, since they are exposed to pools fully equilibrated with label for comparatively long periods of time (assuming the initial period before staturation of the pool can be ignored). Swick et al.[36,52] devised a method using carbonate-^{14}C fed to rats. The isotope is rapidly incorporated into arginine via the urea cycle, with 90% of the radioactivity in the guanidino carbon. The specific activity of the precursor pool can be easily determined from that of the excreted urea, since the guanidino carbon of Arg forms its carbon source. The rate of degradation of the protein can be calculated from the same formula as Eq. (1:1), where C_0 and C_t are the specific activities of arginine and urea respectively, and k is the first-order constant. More strictly, the relationship using other labeled materials, when reutilization over long time periods can occur, is

$$\frac{C_p}{C} = \frac{k_2 t}{(1 - e^{-k_1 t})} - \frac{k_2}{k_1} \qquad (1:5)$$

where C_p and C are respectively the specific activities of the protein-bound and free amino acid at time t, and k_1 and k_2 are respectively the fractional rates of incorporation into protein or breakdown. This equation is difficult to solve unless $k_1 \gg k_2$ and t is sufficiently long.[50] Infusion studies permit some simplification since the pool size can be considered as relatively constant (particularly in shorter-term experiments when the amount of amino acid introduced is small compared to total flux). If pool sizes are constant and if c and C represent total amounts per gram of free and protein-bound amino acid, then $k_1 c = k_2 C$, or

$$\frac{k_1}{k_2} = \frac{C}{c} = R \qquad (1:6)$$

(the equation can be solved if Rk_2 is substituted for k_1). The half-life of rat muscle protein with i.v. infusion of Lys-^{14}C derived from this equation was 6 days for periods of 6–10 hr ($R = 50$).[51]

Method IIIa (Kinetics of Enzyme Change)

This approach does not of necessity require isolation of a specific component in a homogeneous state (enzyme or protein) or the use of isotopes with all their attendant assumptions. The method has not found much application in brain but has been exploited by Schimke and others[9] to study enzyme changes in liver. Such studies are relevant to unexplained findings in brain related to irreversible inhibition of acetylcholinesterase (and its subsequent restoration) and the effect of inhibitors of protein synthesis. In the case of method IIIc it is assumed that following an induction to a higher level of enzyme activity (by increase in enzyme proteins), synthesis (S) will subsequently follow zero-order kinetics on withdrawal of the agent, but that decay of enzyme activity (breakdown of the protein) follows nonlinear kinetics expressed by first-order constant k [Eq. (1:1)]. If enzyme (E) is expressed as content per organ weight (no appreciable change in mass of the organ is observed during the experimental period), then the model can be expressed according to the equation:

$$\frac{dE}{dt} = S - kE \qquad (1:7)$$

At any time that steady-state conditions prevail ($dE/dt = 0$) then

$$E = \frac{S}{k} \qquad (1:8)$$

i.e., the level of enzyme protein is determined by the relative rates of synthesis (S) or degradation k. An example of this method would be induction of tryptophan pyrrolase, or tyrosine transaminase to new levels of higher activity by suitable agents (steroids, etc.). Such studies assume that the inductive agent ceases activity promptly on withdrawal, and that the change in activity (increase or decay) is due to an actual change in enzyme protein and not simply activation (release from bound forms, cofactors, etc.). Schimke et al.[9] overcame this objection by precipitating the enzyme under study with its specific antiserum. In rats, change from a low protein diet (8%) to conditions of starvation led to a twofold increase in activity attributable to decreased rate of breakdown. Conversely, change from a high protein (70%) to a low (8%) led to a decrease in total arginase resulting from both a decreased rate of synthesis and an increased rate of degradation. Schimke[9] concluded that differences in arginase levels with rats maintained on an 8% or 70% protein diet (two different steady-state conditions) were due to differences in rates of synthesis. However, during acute changes (starvation) the level of enzyme is determined by the rate of breakdown.

The major assumption in method IIIa is that the new steady-state condition resulted from an initial de novo synthesis of protein (substantiated in part if measured after isolation with antiserum). Methods utilizing known inhibitors of protein synthesis (IIId) generally assume that rates of degradation are unaffected. This is not always the case since puromycin in addition to inhibition of synthesis will also abolish breakdown of tyrosine trans-aminase,[39] or glutamate dehydrogenase[53]; it can also directly inhibit a degradative enzyme such as purified brain arylamidase *in vitro*.[1m] This raises the important question of whether inhibitors in tissues could inhibit the synthesis of degradative enzymes and thus indirectly affect turnover; this provides an alternative mechanism linking synthesis to breakdown. Inhibition of a purified enzyme *in vitro* (e.g., brain peptidase),[54] by puro-mycin suggests that its effect on memory retention *in vivo* may be more complex than hitherto supposed. In method *IIIb* (irreversible inhibition), it is assumed that restoration of activity is a result of new enzyme synthesis. The assumption in this case is that there is no reactivation of enzyme-inhibitor complex.

In summary, no single method is completely free of assumptions and/or limitations. A variety of methods may be required for the study of turnover of different proteins or components of the nervous system. Whenever possible, the use of more than one approach is suggested before proposing models for mechanisms of turnover (particularly when rates of synthesis or degradation are differentially affected).

B. Incorporation Studies in Brain with Pulse Labeling

Method I (Pulse Labeling)

Early investigations could not take into consideration all the problems cited above, and were preoccupied with questions of labile (or dietary) proteins versus inert (or structural) materials.[55,56] The major indices of protein change in early studies were cytological techniques or measurement of nitrogen balance in living animals. Borsook and Keighley in 1935[57] first proposed the concept that tissue proteins were subject to a "continuing metabolism" with dietary proteins serving as the major precursor. This concept was confirmed and extended by the use of labeled precursors, leading to the general concept introduced by Schoenheimer[8] for the dynamic state of body constituents.

Although cytological observations earlier this century implied alteration in brain protein as a result of electrical stimulation of application of drugs (see review of Hyden[22]), attempts to measure turnover in brain were ignored until the mid-1950's (see review of Lajtha[3]). Up to that point some skepticism was expressed concerning the lability of brain proteins, as re-inforced by negative results of early incorporation experiments with iso-topes.[58–60] Greenberg *et al.*[62] concluded that transport was an important factor since brain homogenates could incorporate labeled amino acids *in vitro*. Subsequent repetition of earlier experiments with i.v. or i.p. or isotopes of higher specific activity[3] demonstrated incorporation into brain protein. Gaitonde and Richter[63] showed that intracerebral injection of Met-^{35}S led to the appearance of ^{35}S in the available brain pools and other body compartments. These early studies established two facts: (a) a large fraction of cerebral proteins is in the dynamic state and undergoes rapid synthesis and breakdown; (b) the turnover rate is not homogeneous, some proteins turning over very rapidly, others very slowly.

Lajtha et al.[3,64] and Gaitonde and Richter[63] reported half-lives of 10–22 days, with a mean of 14; others[47,65,66] reported half-lives of 6–9 days for the more active fractions. These findings were extended in a number of subsequent studies[5,67–69] confirming the high metabolic activities of cerebral proteins and the apparent heterogeneity of metabolic rates (Table I). A number of important considerations following from the early findings were not always clearly realized in later studies. The following are among the most important: (a) for a calculation of half-lives of protein from rates of incorporation of labeled amino acids, the specific activity of the amino acid incorporated is needed. This is never measured directly and is frequently omitted; consequently the assumptions used for calculation are often

TABLE IA

Half-Life Values for Brain Proteins[a]

Area	$t_{1/2}$ (days)	Comment (methods)	References
Rat brain	14	^{35}-Met (I)	63
Rat brain	9	^{14}-C (I)	66
Rabbit brain	6	^{14}-C (I)	65
Rat brain	10–22	Feeding ^{14}-C diet (II)	96,314
Goldfish brain	10	15-min pulses (I)	268
	25	30 min	
	33	4 hr	
Cerebral cortex (rat)	2 (3)	1-hr pulse (I)	67
	11 (26)	6-hr pulse	
	31 (37)	12-hr pulse	
Basal ganglia (rat)	3 (5)	1-hr pulse	
	20 (26)	6-hr pulse	
Cerebellum (rat)	2.5 (3)	1-hr pulse	
	14 (14)	12-hr pulse	
Pons medulla (rat)	20 (28)	6-hr pulse	
	35 (28)	12-hr pulse	
Cerebral cortex (rat)	4–6	Tyr-^{3}H (adult)	5
Cerebral cortex (rat)	4–14	^{14}C perfusion (II)	b
Cerebral cortex slices	144–176	Adults,	b
(in vitro)	6	1 day old	

[a] Half-lives are quoted to the nearest significant figure since all values are approximations dependent on the method and assumptions used (Section II). Individual papers should be consulted for full experimental details.

[b] The perfusion data in Table II, and the in vitro study in Table IV are included for comparison. The nature of the labeled amino acid is indicated in col. 3; in time studies of Piha et al.[67] a mixture of amino acids from Chlorella hydrolyzed proteins was employed; the values in parentheses refer to soluble proteins in supernatant fractions and the other values to insoluble protein.

TABLE IB

Half-Life Values for Proteins in Different Brain Fractions in Selected Fractions and Components[a]

Fraction or component	$t_{1/2}$ (days)	Comment and method	References
I.			
Nuclei	24	Tritiated water	71
Mitochondria	16	in diet (II)	71
Microsomes	13		71
Ribosomes	12	[14]H (I)	68
Mitochondria, synaptosomes Synaptic membranes, soluble vesicles, and mitochondrial fractions	20–24		68, 95
II.			
Components			
S-100	16	^{3}H (I)	263
Histones	54–117	^{14}Leu (I)	216
Histones	5		*
Neurotubule (mouse)	4	5-day-old (I)	259, 260
Myelin—basic protein	21	^{3}H (I)	276
—basic protein	45	i/c (I)	*
—basic protein	14–21	i/c (I) ^{14}Lys	43
—proteolipid	25	i/c (I)	*
—Wolfgram fraction	30	i/c (I) ^{14}Lys	43
—Folch–Lees	10–25	i/c (I) ^{14}Lys	43
Glycoprotein	1–4	^{14}C-glucosamine (I) Newborns	266
	12.5	Adults	

[a] Details are similar to those of Table IA. The asterisks refer to unpublished data of Bondy (histones), M. Smith (basic protein and proteolipid) (personal communications). i/c denotes intracisternal injection.

erroneous. As an example, the free amino acid pool at the nerve endings has little relationship to the proteins at nerve endings transported from the cell body by axoplasmic flow. (b) A half-life of 14 days means a fractional turnover of about 0.2% per hour; this alteration or the smaller change seen at very short pulses may not be representative of the total (Table I). (c) Exposure to short pulses (2 hr) leads to relatively long half-lives (0.2% conversion/1 hr). This can be interpreted as showing that proteins with a significantly higher than average rate of turnover do not form a very large percentage of total cerebral proteins. There is no evidence whether turnover of one protein is uniform in all areas or structures of brain, or is uniform during development; also, interconversion of proteins is an additional consideration.

Most later studies of cerebral protein metabolism investigated special problems such as developmental changes, regional distribution, and drug effects, and only a few were concerned with the measurements of half-lives in whole brain. Maurer[65] and Schultz et al.[66] estimated by autoradiography a half-life of 17 days for glial proteins, with a half-life of the most active fraction, ganglion cells, of less than 1 day. Oja[5] found in adults a half-life of tyrosine incorporation of 4–6 days. Some of the half-life measurements in specific brain regions discussed below gave results similar to those enumerated above. It must be concluded that previous measurements *in vivo* leave a lot to be desired; measurement of half-lives using better-defined methods are clearly indicated.

C. Long-Term Incorporation Studies

Method II

Exposure of cerebral proteins to short pulses will fail to detect long-lived proteins, which require longer periods to acquire sufficient label for detection and measurement. The majority of *in vitro* experiments, or the testing of the effects of drugs of training *in vivo*, are essentially short-term incorporation experiments. There are very few whole brain studies involving exposure to a constant level of label for extended periods (feeding, infusion, perfusion).

Thompson and Ballou[70] exposed young rats to a constant level of tritium oxide from conception to 6 months of age and exposed adult animals for 4 months; then they measured the subsequent decay of radioactivity. About 50% of the brain tritium showed a half-life of 16 days, the other 50%, a half-life of 150 days. No correction was made for the percentage of cerebral ^3H present in nucleic acids and lipids, although it can be expected that DNA and cholesterol, for example, form a significant proportion of the more stable components. Buchanan,[35] after continuous feeding of labeled carbon (feeding yeast grown on labeled sucrose), found a half-life of 10 days for one-third of the brain and a total average of 59 days for defatted brain. In these experiments, again, not only proteins but many other organic constituents were labeled. Khan and Wilson[71] fed tritiated water from conception to weaning and found a half-life of 16 days for mitochondria and 24 days for nuclei. Approximately 21% of the nuclear tritium was not replaced; this corresponded to the DNA level of the nuclei.

In our experiments, ^{14}C-labeled lysine was fed from conception to 60 days of age, after which the decay of Lys-^{14}C in the proteins was measured.[314] In these experiments the bulk of cerebral proteins (at least 90%) decayed, with a half-life of between 10 and 20 days; only about 2% of the label remained in the brain after 150 days. At the present time somewhat similar experiments are in progress in our laboratory (Dunlop and Lajtha, in preparation), which involve feeding labeled amino acids throughout growth—labeling all cerebral proteins and measuring the half-life of individual fractions from the decay of label. These experiments seem to indicate that the major portion, if not all, of the cerebral proteins are in a dynamic state, and furthermore, that the portion with relatively slower turnover is only a small fraction of the total.

TABLE II

Half-Lives of Rat Brain Proteins Determined by an Infusion Method[a]

Amino acid ^{14}C	Time of infusion (hr)	Conversion (percent/hr)[b]	$t_{1/2}$ (days)
Arg	0–2	0.5	5.8
	2–3	0.25	11.5
Lys	0–3	0.51	5.7
	0–6	0.65	4.4
Val	0–9	0.22	13.0
Tyr	0–6	0.57	5.0
Leu	0–3	0.8	3.6
	3–6	0.58	5.0

[a] Data based on the studies of Seta, Mitsunobu, and Lajtha (unpublished). Infusion via the femoral vein and artery of adult rat. Rate of infusion 1.164 ml/hr in medium containing 0.9% NaCl, heparin, and 0.7–1.6 mμmoles of amino acid (0.2–0.45 μCi) depending on the amino acid. The trichloracetic pool size of brain(s) expressed as cpm/μmole of amino acid, and averaged between consecutive time points. The PCA and TCA-insoluble fractions (Sp) expressed as cpm/μmole of protein-bound amino acid (increment of change between time periods). In the case of Val the average pool size S was 2910, and Sp was 64.8, yielding the conversion rate 0.22. When substituted in equation (1 : 4) (expressed as parts per 100) the value for $t_{1/2}$ was 13 days. Pool size was corrected for metabolic conversion by a chromatographic method.

[b] $[(Sp_2 - Sp_1)100]/[0.5(S_1 + S_2)] = \%$ conversion/hr.

Infusion studies now in progress in our laboratory (Seta, Mitsunobu) offer the potential of studying turnover during sustained periods of constant specific activity of plasma amino acids. Results summarized in Table II show a range of 0.25–0.8% protein renewal per hour based on knowledge of the specific activity of the precursor pool; that the half-lives were lower indicated a more rapid turnover compared to most other studies quoted in Table I. The method for calculating half-lives based on the specific activity of the precursor pool and that of the protein-bound amino is appended to the legend. Correction for the metabolic conversion of the different amino acids was made by a paper chromatographic method, which could be subject to some error; we estimate that this could modify results by a factor of two.

D. Regional Differences in Turnover

The structural complexity of the nervous system needs no elaboration here. The various morphological units forming functional compartments and differing in their metabolic and enzymatic composition have already been discussed in previous volumes of this handbook. Regional heterogeneity of protein metabolism was, among others, confirmed by autoradiographic and histochemical methods and supplemented by quantitative data with subcellular system.[6,75,79,97] The difficulties associated with autoradiography or cytological techniques for both quantitation and localization are well known, and are discussed in Vol. 2 of the Handbook. The chief problem in both techniques is retaining recognizable ultrastructure after fixation and diffusion of soluble components from sites of reaction. Despite such limitations, early studies of incorporation *in vivo* with thio-amino acids or tritiated leucine substantially confirm the concept of lability and heterogeneity of brain proteins and different structures. Early literature is reviewed by Lajtha,[3] and Friede.[6] Autoradiographic studies (reviewed recently by Droz[1k,2c]) in different species are in broad agreement: the highest grain densities occur over perikarya of the neurons (some 60-fold higher than glial cells of white matter). In general the grain densities appear to be similar to the distribution seen after Nissl staining, which supports earlier cytological studies (see Hydén[22]). Specific areas demonstrate exceptionally high grain patterns; these include Ammon's horn, fascia dentata, habenula, choroid plexus, and some but not all Purkinje cells of cerebellum (see Friede[6]). In general, the nucleolus and nuclear membranes show higher incorporation compared to the cytoplasm and proximal dendrites, which may be linked to the role these structures play in formation of the ribosome, or production of *m*RNA, especially in relation to the regulation of rates of protein synthesis (see Munro[4e]). Quantitative autoradiographic findings based on grain density suffer from inadequate knowledge concerning the specific activity of the precursor pool(s), and actual sites of synthesis. Transport of perikaryon proteins by axoplasmic flow may mask gray/white matter differences and account for data not entirely consistent with studies on isolated tissues.[1i,2j] Of interest in regard to possible sites of transport, Klika and Lodin[72] reported high uptake of Met-^{35}S in blood vessels and arachnoidal cells, with maximal grain densities in clusters of meningiothelial cells. Heterogeneity of uptake in different areas was confirmed in studies by Ford et al.,[73,151] who showed relatively high uptakes into ventral horn motor cells compared to the tissue as a whole. A more detailed anatomical separation was done by Merei and Gallyas,[74] who autographically analyzed the incorporation of methionine in 38 areas of the brain.

Earlier studies from this laboratory[3,24] measured quantitatively the incorporation of Lys-^{14}C in different areas of the monkey brain (Table III). Incorporation varied with time of exposure, with initially higher rates in cortex and lowest rates in the older phylogenetic regions; however, the half-lives calculated on the basis of the specific activity of the respective pools as actual Lys residues incorporated into available protein lysine were relatively uniform. Calculated turnover time depended on the time of exposure, and half-lives were in each case lower in the shorter time experiments.

TABLE III

Half-Life Times of Proteins in Monkey Brain Areas[a]

Areas of brain	Half-life time in days calculated from different time points		
	5′	10′	45′
Cord	10.6	9.5	18.5
White	3.9	4.0	8.7
Medulla–pons	6.6	6.7	13.7
Thalamus–hypothalamus	6.1	7.8	16.8
Cerebellum	5.4	7.6	14.1
Cerebral cortex	5.5	6.7	12.7

[a] Data based on that of Lajtha, Furst, and Waelsch.[3,78] Single systemic injection (Method Ia) of Lys-^{14}C using adult macaque monkeys.

The half-life for white matter was from 4 to 9 days, and for cerebral cortex 5 to 13 days, for animals killed between 5 and 45 min. Such determinations were confirmed in the studies of Piha and collaborators,[67] where the calculated half-life of each area was lower in 6-hr experiments ($t_{\frac{1}{2}} = 3$–8 days) than in experiments over 6 days ($t_{\frac{1}{2}} = 14$–35 days, Table IA). In shorter experiments the cortex was more active compared to cerebellum than in longer term experiments. In miniature pigs near birth Schain et al.[76] found a fairly uniform rate of incorporation in the three areas investigated (pons-medulla, cerebellum, and cortex). In rabbits injected with Tyr-^{3}H, Blomstrand et al.[146] observed higher incorporation of Tyr-^{3}H in cerebellum with lower incorporation into caudal brain stem, hippocampus, and cranial brain stem areas. A field that remains to be explored is regional differences resulting from functional change. Water deprivation results in depletion of many neurohypophyseal hormones, and incorporation studies with intracerebral injection of Cys-^{35}S show increased rates of incorporation into octapeptides and their precursors in the median eminence and supraoptic nuclei.[1p] This aspect, and other regional changes resulting from altered sensory input are discussed in Section VI, E.

In summary, very little quantitative data are available concerning turnover in different brain areas, especially in fine anatomical structures. Qualitative studies are limited by the diversity of cellular structure, and quantitative studies in relation to actual sites of turnover cannot easily be undertaken with methods currently available (cytological or autoradiographic). More recent studies have emphasized data based on analyses in isolated systems, intact structures such as slices of spinal cord or different brain areas, cell suspensions, etc., or in subcellular fractions. Such data are examined and reviewed below according to the relevant structure based on studies both in vivo and in vitro.

III. TURNOVER IN STRUCTURAL ELEMENTS

A. Spinal Cord

The distinct morphological and functional differences of spinal cord from other regions of the CNS offer a valuable tool for investigation of protein metabolism. In the few studies available, incorporation rates following systemic injection of labeled amino acids are lower than brain[3,84,85] although higher rates are reported with intracisternal injection.[89] The spinal cord represents a very good model for studying some aspects of axoplasmic flow since intracord injection of labeled amino acids is followed by an outflow of labeled materials to the ventral spinal roots, as detected by autoradiographic[1k] and direct counting procedures.[3] Studies by Ochs[2j] in cats show two components associated with the flow of labeled material following intracord injection of Leu-^{3}H: analysis of 3-mm segments of the ventral root gave a fast flow estimated at 930 mm/day; the slow phase of axoplasmic flow had the rate of only a few millimeters per day. Direct injection into the L7 dorsal root ganglia with analysis of the sciatic nerve gave a more satisfactory indication of fast transport in the form of a crest of activity found to be 410 mm/day in adult cats. Analysis of ventral roots after cord injection showed highest activity in the soluble, compared to nuclear, mitochondrial fractions: separation on Sephadex columns and by isoelectric focusing revealed that the fast components were associated with small peptides or free leucine, while the slow components accounting for the bulk of radioactivity were associated with high molecular weight proteins (68,000–450,000). Weiss[28] in early studies concluded that material transported down the axon was to replace axoplasmic or synaptic components destroyed by degradation. Our studies[81] show the presence of potent hydrolytic enzymes in brain fractions and peripheral nerve trunks, and in ventral cord preparations of different species,[82] which could contribute to the pool of free amino acids or peptides, observed by Ochs as associated with fast component of axoplasmic flow. Important questions that remain unanswered in connection with axoplasmic flow are: (1) the precise mechanisms for conveying or transporting material down the axon; (2) the nature of the transported material; and (3) the site(s) of action and ultimate fate of the transported material. The exceptionally high rates for axoplasmic flow further underline the questions concerning the destination and fate of transported material. Some components may be transferred to muscle receptors[83] but it is doubtful whether the majority of proteins or other large macromolecules can pass the end terminal motor plates or dendritic membranes. Rapid transport would create intolerable intraaxonal pressures if considerable breakdown or reverse flow did not occur. In studies on breakdown in lobster ventral cord and in peripheral nerve preparations, we concluded that breakdown processes play a significant role in cell economy.[81]

Incorporation in spinal cord can be demonstrated by systemic injection[84,85] although this is less likely to reveal somatofugal flow since uptake readily occurs in Schwann or glial cells.[1e,2j] There is some indication that protein metabolism in spinal roots and ganglia differs from spinal cord based on variable sensitivity to DFP poisoning.[1w]

Recent studies by Globus et al.[86] show rapid incorporation of tritiated glycine applied to cat spinal cord through a triple-barreled intracellular micropipet. Rapid fixation in formalin revealed rapid labeling of all axonal collaterals with a marked increase following antidromic stimulation. It was concluded that rapid labeling of the adjacent neurons, but not glial cells, was due to an outflow from axon collaterals with release into the synaptic cleft followed by postsynaptic uptake and incorporation.

Studies in the giant Mauthner neurons obtained from spinal cord of fish are consistent with the existence of axoplasmic flow in these axons. Incorporation *in vitro* was linear with time up to 4 hr with a pronounced inhibition by puromycin and acetoxycyloheximide but not with chloramphenicol or actinomycin.[7c,11b] Absence of inhibition by the latter two antibiotics would exclude a major contribution by mitochondria to turnover in the axon. The highest incorporation occurred in the microsomal supernatant fractions (70%), with lower incorporation into the nuclear (19%), crude mitochondria (14%), and myelin fractions (4%).[7c,87,88] Determinations of the base ratios obtained from axonal RNA suggest that it is in part derived from ribosomal RNA although free ribosomes have yet to be isolated from peripheral nerve trunks. In relation to turnover, the interdependence of the Mauthner neurons and other spinal cord compartments is unknown.

1. *Cerebrospinal Fluid*

Many pathological conditions or experimentally induced diseases are characterized by a pronounced rise in levels of protein in the spinal fluid, but attempts to relate specific conditions with one particular protein species or enzyme have proved disappointing and contradictory.[90,91] Most proteins in the CSF are accounted for by albumin (58%), β-globulins (20%), and immunoglobulins (10%), with the total concentration only 0.5% compared to serum; these levels are subject to large and unspecific changes during disease.[90,92] Studies with labeled proteins indicate that CSF proteins are derived largely from serum and that turnover can be accounted for by transport into and out of this compartment.[93] In diseases such as multiple sclerosis, tuberculosis, meningitis, allergic and inflammatory conditions, there is a large increase in albumin derived from serum;[94] in multiple sclerosis a 30–40% increase in immunoglobulin is derived from an extravascular source, i.e., net synthesis in the CNS-neuroaxis.[90] The ability of brain to initiate formation of a new species of protein such as immunoglobulins could represent a survival mechanism in response to noxious stimulus. The maintenance of elevated levels in disease may at times be the result of a new equilibrium in turnover. Difficulties of evaluating the vascular contribution to the immunoglobulin pools preclude exact computation of its synthetic rate in brain or the CSF compartments. Appeltauer and Saa[85] observed higher incorporation into CSF proteins when Lys-^{14}C was introduced into the subarachnoid space than on intraperitoneal injection with a slow decay in radioactivity. Protein synthesis by the CSF itself is unlikely owing to the absence of the necessary components to complete synthesis and turnover. It has been postulated that production of "undesirable" proteins by net synthesis or degradation could result in the release of toxic factors, leading to pathological conditions. As an example, we have proposed that the degradation of basic (encephalitogenic) proteins following lipolysis may be involved in the etiology of EAE.[96]

B. Neurons and Glia

The heterogeneity of brain protein turnover indicated the necessity to separate distinct morphological cell types and study their properties in isolated preparations. Several important considerations apply to such studies: (a) the cellular damage resulting from isolation procedures, particularly to cellular processes; (b) the homogeneity and morphological identity of such preparations; (c) optimal conditions for studying environmental influences. Useful biochemical data can be acquired by this approach, although it must be anticipated that overall turnover must involve an integration, with considerable interdependence between all the cells.

1. Preparative Procedures

Earlier methods reviewed by Roots and Johnson,[20a] which utilized micromanipulation, were limited in scope by the quantity of material available and by the time needed for preparation. Bulk preparation of cells is now favored; this procedure involves disruption of tissue by chemical, enzymatic, or mechanical means, followed by filtration and gradient centrifugation on sucrose or ficoll gradients.[20b] Many such preparations are subject to considerable contamination, particularly in the case of the 'glial-enriched' material.[146] The methods available were devised in an empirical manner, aimed at separating cells with minimum damage to membranes and cellular processes; they range from disruption in the presence of hypotonic buffer,[146] acetone-glycerol,[147] proteolytic enzymes,[148,163] polyvinylpyrrolidone (PVP), serum albumin, and ficoll.[149] Treatment with proteolytic enzymes could be injurious to studies involving membrane or transport properties. It must be stressed that glial cells have diverse morphology with possible consequences to metabolic patterns; neuron/glial ratios also vary in different brain regions, or as a result of development; in some pathological conditions such as tumors the tissue is mainly of glial origin.[6] Cells from newborns lack the dendritic complexity that accompanies development and may be prone to less damage during isolation; preparations from older animals contain also a higher percentage of contamination resulting from myelin, capillaries, blood vessels, etc.

2. In Vivo

Studies on the sites of protein turnover must take into account redistribution of radioactivity by axoplasmic flow (transport of proteins from the soma to the synaptic terminals), particularly since the only portion of the neuron isolated is the nerve-cell body shorn of its cellular processes. Recent scanning micrographs of hypoglossal perikarya show only the base stubs of dendrites with surfaces covered by spherical or ovoid particles.[150] An additional consideration is the relative rates of transport of precursor to glial and neuronal cells and relative size of the pool of free amino acids. Ford and Rhines[151] observed a larger pool of free lysine in neurons in comparison to the adjacent neuropil, with higher levels in brain cells compared to spinal ganglia and the ventral/dorsal roots. Globus et al.[86] reported rapid labeling of neurons with iontophoretic application of Gly-^{14}C to spinal cells (Section III, A). Preliminary studies by Hamberger suggest[146] a higher concentrative uptake of amino acids by glial-enriched fractions which would support earlier concepts that glial cells regulate the "milieu" around nerve cells.[152]

Following systemic injection of Met-^{35}S in rats, Jakoubek et al.[153] observed lower incorporation into adult compared to senescent neurons isolated from spinal segments of rats. The half-lives estimated by quantitative autoradiography or by direct measurement of radioactivity were 2 days for the shorter lived and 27–38 days for the longer lived components (Fig. 5). Blomstrand and Hamberger[154] observed maximal incorporation at 4 hr in cells prepared by a bulk procedure following intravenous injection of Leu-^{14}C: the decay of radioactivity was rapid in neuron-enriched compared to glial fractions (Fig. 3A). The rates of incorporation were significantly lower than cells prepared by the bulk procedure of Johnson and Sellinger[155] illustrated in the same figure. The latter observed higher specific activities in glial fractions in comparison to neurons in an 18-day-old rat at 5 or 20 min but not at 10 min following a very short intrathecal pulse of Phe-^{14}C. This biphasic incorporation was attributed to a sequential involvement of different glial cells (oligodendrocytes and astrocytes). Incorporation was related to age, being higher in neurons at 5 and 10 days, but lower than glial fractions at 43 days of age. Blomstrand and Hamberger[154] estimated the half-lives of neurons and glial from brain as 15 and 30 hr and from spinal cord neurons as 20 hr although these rates are considerably lower than the generally accepted values (Section II). Autoradiographic studies usually show a higher incorporation in brain regions richest in neurons, but as noted, this method has limitations in respect to localization studies.[1k] Studies on incorporation of precursors for nucleic acid in association with zonal-gradient procedures for separation of cells are consistent with a more intensive protein metabolism in neurons.[156] Studies on subcellular fractions show high rates of incorporation in the microsomal, supernatant, and mitochondrial fractions of cells,[155,157] with rates for these fractions markedly higher in neurons compared to glial fractions.[157] In terms of specific protein components, Sellinger et al.[149] and Dutton and Barondes[158] reported a rapidly turning-over component in soluble fractions of neuronal perikarya from immature animals that was similar in characteristics to neurotubular protein. At 10 days, this acidic component represented 70% of the total labeled protein in

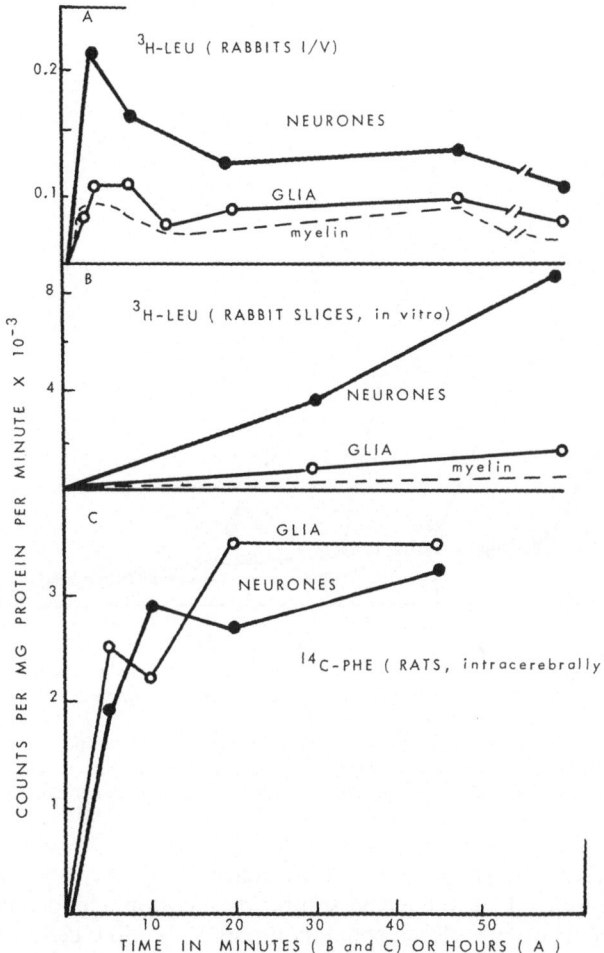

Fig. 3. Neuron–glial differences based on studies *in vivo* from cells obtained by sieving procedures and gradient centrifugation after i.v. injection of label to adult rabbits (A) or after incubation of slices (B), or sieving 18-day-old rat brain in presence of PVP and albumin (C). Radioactivity was measured in the TCA-insoluble material and results are expressed as change in specific radioactivity following a single pulse dose of about 100 μC Leu-^{3}H (A), or 6.25 μC Phe-^{14}C (C), or incubated in presence of Leu-^{3}H (B). Data in the figure are based on that of Blomstrand and Hamberger (A, B),[154] and Johnson and Sellinger (C).[155] The figure demonstrates higher incorporation in rabbit preparations in neurons compared to glia at short time intervals, and a biphasic response in rat preparations *in vivo* with higher incorporation in glia at 5 and 20 min. The dotted lines indicate other cell fractions used in comparison studies.

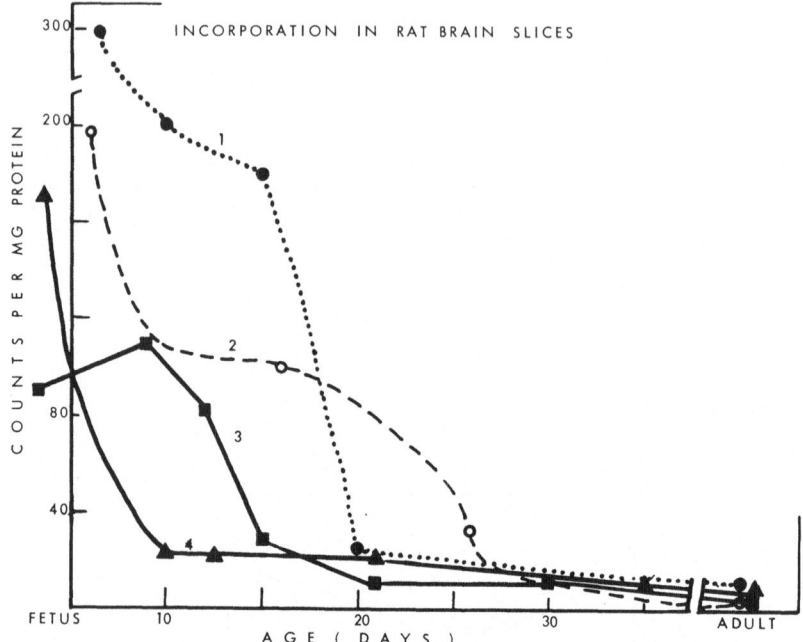

Fig. 4. Comparison of incorporation rates into mouse (curve 1) and rat brain slices (2–3) and the change with age. Data for (1) are from Ito and Arimatsu[233] who used an amino acid-^{14}C mixture of hydrolyzed *Chlorella* protein; for (2), Dunlop *et al.*[107] who used ^{14}C and ^{3}H labeled Lys and Arg; for (3), Orrego and Lipmann[100] who used Val-^{14}C; and for (4), Mokrasch and Manner[101] who used ^{14}C algal protein hydrolysate. In all cases except that of (3), incorporation fell immediately at birth and continued to drop with age.

soluble perikaryal fractions.[149] Burdman[159] observed a higher incorporation of Leu-^{14}C into the acid soluble deoxyribonucleoproteins, residual and soluble proteins of neuronal compared to glial cells. Exposure of neuronal cells to tritiated leucine for short periods led to rapid labeling of the chromatin (acidic) protein, but at longer periods of exposure other components acquired equal or higher radioactivity.

3. *In Vitro*

Incubation of rabbit brain homogenates or separated cell suspensions led to a threefold higher incorporation into the neuronal in comparison to the glial-enriched fractions.[157,164] The optimal conditions for incorporation into unfractionated cell suspensions appear to be close to those described for slices (Section IIIC). Glial cells appear to be more sensitive to K^+ than neurons since incorporation is optimal at 5 mM; high K^+ concentrations (105 mM) lead to inhibition of incorporation for Leu-[14]C.[146,160] The differential effect on glial cells may be related to its proposed role for controlling the milieu of neurons.[146,152] Subsequent fractionation of the cells showed exceptionally higher incorporation into the microsomal and supernatant fractions of neurons, but not into nuclear or mitochondrial fractions, when compared to glial particulates. Slices afford a good opportunity for studying neuronal–glial relationships *in vitro* since the unique spatial relationships are retained. Under optimal conditions for incubation of slices (10 mM K^+, 120 mM Na, O_2, glucose, etc., see Section III, C), neurons were two- to six-fold more active in terms of their protein metabolism. Neuronal–glial differences of five- to six-fold were observed with Leu as the label, but were lower in the case of Phe, Gly, and Glu. Neurons were slightly more sensitive to the inhibitory influences of cycloheximide and NAD, but no differences were detected with actinomycin D. It will be noted that labeling patterns of subcellular fractions *in vitro*, particularly in the case of mitochondria, are different from those *in vivo*. Such differences may find explanation in relative rates of transport of precursor in the whole animal compared to isolated systems. Preliminary studies by Blomstrand[146] indicate some differences in enzyme content [monoamine (MAO) and cytochrome oxidase] in the rabbit but no clear difference in other species; cytochrome was higher and MAO lower in rabbit neuronal preparations. Heterogeneity of turnover in glial cells could mask differences observed in all cited studies. Lovtrup-Rein[161] in studies of protein synthesis in isolated rat brain nuclei observed an incorporation ratio with Leu-^{14}C for astrocytes–neurons of 1.75 (30–60 min incubation in media containing K^+, Na and Mg^+, and glucose) in contrast to values for oligodendro- and microglia–neurons of 0.1. Burdman and Journey[162] observed neuron–astrocyte ratios of 1.2 but a oligodendro–neuron ratio of 0.1 for incorporation of Leu-^{14}C. These and all neuron–glia ratios cited above must be regarded with caution in the absence of precise morphological criteria for defining the cells in question. As noted, different ratios of cells could arise as a result of species, age, and regional differences in different areas of the brain. Relative rates of protein metabolism may be related to levels of nucleic acids and concentration of different groups of specific proteins. Glial cells are characterized by high levels of S-100 acidic protein, and these are subject to turnover with a half-life of about 14 days (Section VI, A). Exposure of rabbits to X-ray irradiation led to some neuron–glia differences in rates of incorporation, highly dependent on the time of sampling tissue. Hypoxia for 3 hr resulted in decreased incorporation in glial cells; at 12 hr incorporation in neuronal fraction from slices was increased (see Section VI).

C. Slices

The merits of studying isolated tissue with intact cellular organization in a controlled environment are emphasized in other chapters of the Handbook.[1f] Tissue slices are known to incorporate labeled precursors into their protein, proteolipid, and RNA moieties.[98–105] Studies on incorporation in brain slices were initiated by Lindan et al.[99] Huttonen[106] and Dunlop[107]

TABLE IV

Half-Lives of Proteins in Rat Brain Slices

Amino acid-^{14}C	Age	Incorporation (μM/mg/ hr $\times 10^{-3}$)	Rate^b (%/hr)	$t_{1/2}$ (days)
Lys	Adult	0.074	0.020	144
Val	Adult	0.085	0.017	170
Leu	Adult	0.151	0.018	160
Ile	Adult	0.093	0.023	176
Lys	1 day	1.79	0.50	5.8
Ile	1 day	1.98	0.53	5.4
Leu	1 day	3.88	0.50	5.8

^{a} Data from Dunlop and Lajtha (unpublished). Rat brain slices incubated in medium at pH 7.4, containing HEPES (25 mM), Na 132 mM, K 8 mM, MgSO$_4$ 1.3 mM, Ca 2.6 mM, Cl 130 mM, glucose 12 mM, for 1 hr at 37°C. The medium concentration of amino acid was 0.8–1.0 mM (53–1250 cpm/ $\mu M \times 10^{-3}$). At these levels 90% of the free amino acids in the slice was derived from the medium.

^{b} Conversion was calculated from lower ratio of counts in protein-bound amino acid and that of the free pool and substituted in Eq. (1:4) as in Table II.

showed that such incorporation is highly sensitive to alteration in environmental conditions. In general, conditions known to increase cellular respiration result in a decreased incorporation of labeled amino acids into slice proteins; for example, an increase in the level of potassium in the media, decreased calcium, addition of an acidic amino acid such as glutamate or electrical pulses.[1f,107–110] Mase et al.[111] observed a 40% decrease in the incorporation of Gly-^{14}C into brain slices with an 18-fold increase in potassium concentration: in the absence of magnesium (Mg) there was a 14% increase. The K^+–Na^+ ratio appeared to be a critical factor in such studies, since reduction of Na^+ content also had some effect. Recent studies show an optimal K–Na ratio of 8:132 mM.[106,107] Sodium concentration could be varied considerably but maximal incorporation occurred at a concentration of 1.3 mM or 2.6 mM Ca (in the presence of Mg). Variation with buffer anion or pH [HEPES-buffer is reported to be more effective than tris or glycyl-glycine (Table IV)] suggests the possibility of transport-linked phenomena, as substantiated by the absence of effect on slices preloaded with amino acids (incubation in media with high amino acid); the inhibition (20–30%) with addition of ouabain further suggests the involvement of ATPase.[106]

1. Electrical Pulses

Application of electrical pulses to brain slices is known to affect the movement of ions, glucose consumption, respiration, and the incorporation of labeled amino acids into protein of slices.[1f,100] The effects of the stimulatory amino acids were inhibitory; Orrego and Lipmann[100] observed a 50% reduction in incorporation with addition of glutamate, D-aspartate, homocysteate or, DL-aminoadipate. These authors did not attribute the decreased incorporation either to tissue damage or to a change in rates of breakdown as judged by the protein levels measured within the slice. Inhibition of incorporation with the application of electrical pulses was recently confirmed by C. T. Jones and Banks;[109] they observed a biphasic response for incorporation of Val-^{14}C, with enhancement for the first 5-min period, followed by inhibition. Pulses are known to cause multiple effects on tissue components: in addition to an increase in oxygen consumption, there is an alteration in the influx and efflux of ions and amino acids, leading to a depletion of energy-rich phosphates[1f] which in turn may indirectly affect rates of transport or the specific activity of the precursor pools. During the biphasic response, a short period of increased incorporation could result from a short-lived influx of amino acids and could be followed by inhibition that is due to the subsequent prolonged efflux. Recent studies by D. A. Jones and McIlwain[337] confirm that stimulation of neocortical slices (cerebral cortex, pyriform-olfactory lobe of guinea pig) with a variety of electrical pulses was either without effect or decreased incorporation into protein. It might be expected that stimulation of slices (pulses, K$^+$, low Ca^{2+}, acidic amino acids such as Glu) would lead to a depletion of high energy phosphates and that this is correlated with inhibition on incorporation. However, mild stimulation of the lateral olfactory lobe[337] was without effect on tissue phosphocreatine, K$^+$, or amino acid incorporation. Jones and McIlwain have calculated from the level of Leu in tissue (10% of brain protein) and its rate of turnover (1.2 nmoles/mg protein/hr) that it would only require 1% or less of the estimated production of high energy phosphate in tissues (300–700 μmoles/g tissue/hr). It would appear that under conditions of excitation $\sim$P is not the limiting factor. Based on our experience on change in proteolytic enzymes in frog sciatic nerve,[81] there is no marked change in breakdown on stimulation. The relationship of excitation to turnover (synthesis and breakdown) is an important but unresolved question.

Slices are frequently used to study change in specific subcellular fractions or protein components. These studies are commented upon in the relevant sections, but it should be stressed that factors that affect slice metabolism (ions, oxygen, substrate, medium concentration of amino acids, etc.) may affect rates of incorporation into different components. Optimal conditions must first be established for the preparation in question before direct causal effects can be demonstrated. Among the important considerations are the age of the animal (Fig. 4, Section V), regional variations in composition of the slice, and influence of amino acids in the medium on pool size (see below). It can be seen from the data in Table IV that rates of conversion in adult rats are 0.02%/hr or ten-fold lower than comparable rates *in vivo* (Table I). The rates in younger animals are comparable to those observed *in vivo*, particularly in the infusion experiments of Table II. In similar studies, Jones

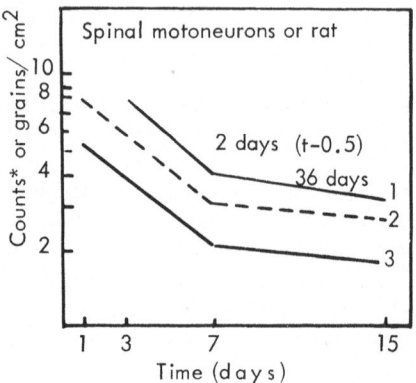

Fig. 5. Protein turnover in 6-week (line 3), or adolescent rats (lines 1 and 2) following the intravenous injection of Met-^{35}S and separation of spinal motoneurons by hand microdissection. The half-lives were estimated by direct counting procedures or quantitative autoradiography. (Data taken from Jakoubek and Gutmann[114d]).

and McIlwain[337] observed a rate for Val-^{14}C of 0.4 nmoles/mg protein/hr in adult guinea pig slices corresponding to about a $t_{\frac{1}{2}}$ of 20 days.

Hypoxia *in vivo* is reported to affect subsequent incorporation of labeled amino acid into slices of rabbit brain[112] and other tissues.[113] Oxygen deprivation can result in both morphological and histochemical changes (neuronal necrosis, microglial proliferation, and perivascular astrocytic swelling).[114] The effects of hypoxia vary with the degree of exposure to atmosphere containing 8 % oxygen : at 3 hr there is a decrease in incorporation of Leu-^{14}C into slices, but at 12 hr there is a slight increase. The effects of hypoxia were also observed in subcellular fractions prepared from slices; incorporation into neuroglial enriched preparations is preferentially increased at 3 hr compared to a decrease in neurons at 12 hr (see Section VI). In all other fractions at 3 hr there was a decreased incorporation. Jakoubek *et al.*[115] observed a 20% reduction in incorporation of labeled leucine in slices prepared from rats treated with ACTH but no significant change in animals subjected to stress (except at 17% increase in the free amino acid pool).

2. Amino Acid Pools

Studies on the kinetics of incorporation into slices and other preparations have indicated the presence of "functional pools" available within the tissue for purposes of protein synthesis and turnover.[5,116–121,335] An exception was the extensor digitorum longus muscle of rats, which incorporated labeled materials directly from the extracellular pool.[119] Studies in lymph node cells, ascites cells, diaphragm, liver, and brain point to a complex situation, where transport of material to the intracellular pool is a rate-limiting step in turnover studies.[120–123] At very high external concentrations, Folbergrova[102] reported some inhibition of incorporation which became less on the omission of acidic amino acids. In contrast, Huttonen[106] observed stimulation with amino acid mixtures when they were added at lower concentrations. At medium concentrations equivalent to that of the pool size in adult rat brain, Dunlop et al.[107] observed a pronounced inhibition even with omission of Glu and Asp; however, addition of only essential amino acids at 0.06% that in brain pools resulted in a stimulation of incorporation. Dunlop et al.[107] concluded that the variation in rates of incorporation reflected changes in transport to the precursor pool, which indirectly affect synthesis. A discussion of the manner in which precursor pools affect turnover in brain slices is premature in the absence of definitive knowledge concerning the slice spaces and compartments.[5,20d] Jones and McIlwain observed little effect on incorporation on addition of a mixture of 16 amino acids, or amino acids added individually. Except for a short lag period during the first 5–15 min of incubation (as also observed by C. T. Jones and Banks) incorporation was linear with time for 45 min. This lag phase could represent the time required to establish equilibrium with the intracellular precursor pool. Jones and McIlwain[337] were unable to accurately measure rates of incorporation based on the radioactivity changes in the medium unless the release of amino acids is taken into account. Proteolysis of slices is quite pronounced under these conditions and is related to the medium concentration of amino acids.[1q] Studies in collagen formation in vitro, in which proline is the major precursor, clearly demonstrate an important relationship between proline uptake and incorporation into collagen. In fetal rat calvaria the data on proline utilization based on the medium concentration, and tissue spaces (inulin) fitted a mathematical model involving functional compartmentation; an intracellular pool divided into a protein precursor pool and an equilibrated pool.[44] Intracellular transport is represented as a unidirectional process that leads into a rapidly turning over protein precursor pool from which synthesis originates. This pool equilibrates with the rest of the intracellular space, i.e., the equilibrated pool (efflux takes place from the latter pool).

The influence of hormones or drugs in vivo has some counterpart in the studies with slices. Clemens and Korner[117] recently observed a peak incorporation of Leu-^{14}C in liver slices in the presence of amino acids, six-fold higher than in plasma, with further stimulation on addition of growth hormone to the medium. The stimulatory effect of growth hormone in vitro was not observed in slices prepared from hypophysectomized rats; such results indicate the relative stability of control mechanisms altered as a result of hormonal imbalance (see Section VI, C). Jefferson and Korner[124] reported that livers perfused in situ with medium containing mixtures of amino acids in multiples of their concentration in normal rat plasma increased the incorporation of Phe-^{14}C or Lys. The presence of 11 amino acids (Arg, Asn, Ile, Leu, Lys, Met, Phe, Pro, Thr, Trp, Val) at 10 times their plasma level gave maximum incorporation, and subsequently increased the ability of ribosomes isolated from the slice to incorporate labeled amino acids. They concluded that amino acids play a role in regulation of synthesis, possibly by stimulating the ability of mRNA to become attached to ribosomes. Experiments in vivo by S. Roberts and Morelos[118] provide evidence that the availability of free amino acids (as affected by dietary changes) affects cerebral protein synthesis. In brain homogenates, Peterson and McKean[121] observed inhibition of 90% incorporation of Tyr-^{14}C by 1 mM Phe, and considerable inhibition of Ile-^{14}C by Leu. Interpretation of those effects, or the effects of drugs, is hindered by lack of concomitant measurements of uptake (transport) and associated mechanisms (levels of ATP, etc.). The influence of pool size on turnover, and the effect of drugs on incorporation of labeled materials, appear to be determined by factors related to transport, the critical periods of development (e.g., myelination), and in the case of some drug effects, the level of ATP. Agrawal et al.[125] reported greater inhibition of Met-^{35}S incorporated by Phe in 18-day-old (during period of myelinization) than in 8-day, or adult rats. They proposed that decreased myelin synthesis seen in hyperphenylalaninaemia or phenylketonuria may be due to an alteration in the free amino acid pool during the vulnerable periods of development. In the case of drugs, some stimulation of incorporation into ribosomal preparations has been reported for GABA,[2] but inhibition by phenothiazines[126]; in slices, inhibition was reported for ouabain, chloramphenicol, impramine, and prothiadene,[103,127] but not chlor-promazine.[103] Narcotics interfere to a limited extent with incorporation in ribosomal preparations and with uptake of free amino acids in slices,[127] and with incorporation into slices or cell suspensions as described by Clouet elsewhere in the Handbook.[1x]

D. Peripheral Nerve

Despite the importance attached to mechanisms of axoplasmic flow, very little information is available concerning synthetic or degradative processes within the axon itself. The spatial separation of perikarya and synaptic regions raises the question whether the axon has the capacity to synthesize proteins or other materials to meet the immediate demands involved in the transmission of nerve impulses, or recovery after excitation. Some years ago, several groups independently reported restoration of acetylcholinesterase activity along peripheral nerves following irreversible inactivation by organophosphorus inhibitors.[130,131] The lack of a proximo-distal gradient in the rate of restoration of enzyme along the nerve favored the concept of its local synthesis.[2h] Studies with transsected hypoglossal nerves of cat treated with puromycin supported the conclusion that this membrane-bound enzyme was synthesized locally in the axon.[1i,2h] Despite this evidence, difficulties in measurement of all the necessary factors for protein synthesis have led to some debate concerning the mechanisms of synthesis in peripheral nerve trunks. With sensitive techniques, it was possible to demonstrate the presence of RNA in axons, and incorporation of radioactivity into axonal proteins.[7c] To exclude the contribution of the Schwann or satellite cells or of myelin to turnover, many studies utilized axons denuded of these elements.[2h] The absence of a detectable ribosomal species within peripheral nerve axons does not support the concept of local protein synthesis by known mechanisms, although the presence of a membrane-bound form resisting normal isolation procedures cannot be excluded.[7c,2h] Protein turnover within the axon may be essential for the maintenance of "functional structures" associated with membrane excitability, and of the systems required for its recovery after depolarization, particularly the related transport systems. Although autoradiographic studies amply confirm the importance of axoplasmic flow as a major source of protein for distal areas, these mechanisms are presumably superimposed on protein supplied by local synthesis.

Local protein synthesis presupposes a supply of amino acids, which could be derived from three separate sources: (a) axoplasmic flow from the nerve cell body; (b) active transport across the axolemma;[144,145] and (c) local proteolysis.[1q] Our own studies indicate proximodistal gradients for a number of amino acids and proteolytic enzymes.[81,82] Proteolytic enzymes could be derived from organelles transported down the axon or could be intrinsic to the axoplasm or associated with its membrane. The variable nature of amino acid pools in different species, particularly in arthropods, suggests multiple functions for these components. In addition to serving as precursors for synthesis, amino acids could possibly act as neurotransmitters, as sources of energy, or as factors in osmoregulation.[81] The fate of proteins transported down the axon or synthesized locally is unknown. It has been speculated that such components replenish the "functional pools" of physiologically active components in areas contiguous with the membrane and the excitatory process.[2h] Certain proteolytic enzymes in nerve axons may serve an important role in the elimination of unwanted proteins by breakdown to amino acids, which are more easily transported across the synaptic endings, or indirectly in the release of physiologically active proteins (see Chapter 2, this volume). The role of proteolytic enzymes in the integrity of specific axonal elements, notably neurostenin and neurofibrillar or neurotubular proteins, is unknown, but is worthy of future exploration.

1. In Vivo

In the few studies available with mammalian tissue, incorporation of labeled amino acids was demonstrated in peripheral nerve protein, but at rates considerably lower than in the CNS.[85,132,133] Lajtha et al.[132] observed higher radioactivity in proximal than in distal regions of rat sciatic nerve 2 days after a pulse of labeled amino acids. Such evidence is consistent with current concepts for axoplasmic flow, as further confirmed by autoradiographic studies of Droz, and others.[2c,7e,134] Incorporation of labeled amino acids into peripheral nerve after transsection would imply local synthesis, in addition to the supply by axoplasmic flow.[135] Appeltauer and Saa[85] observed different labeling patterns in rat tibial nerve (from the L5 spinal route) dependent on the route of injection. Intraperitoneal injection of Lys-^{14}C, 30 min–30 hr before sacrifice led to a low incorporation in comparison to intrathecal injection, with the highest radioactivity in the distal areas. The higher proximal labeling on intrathecal injection was attributed to local protein synthesis with more rapid penetration of precursor to the relevant pools. In amphibian nerve, Caston and Singer[136] reported rates of incorporation equal to those of the brain and liver, with most of the radioactivity present in the soluble supernatant fractions. The variable rates of incorporation observed by these authors after intraperitoneal injection appeared to be related to the nutritional status of the animal; for example, during starvation, amino acids appeared to be diverted for energy production rather than for synthesis of macromolecules or formation of amino acyl-tRNA's (see Section VII, A).

2. In Vitro

Peripheral nerve is fairly stable, as shown by the incorporation of labeled precursor, continuing for several hours on incubation in the presence of glucose and oxygen.[137,138] In the absence of oxygen, or on the addition of inhibitors of metabolism (KCN, DNP) or the addition of diphtheria toxin, there is a marked reduction in rates of incorporation and protein turnover.[138] Age-related differences were reported in rat sciatic and optic nerves, and chicken peripheral nerve, with a fall in protein turnover with age.[139,140] Following a period of Wallerian degeneration, for 91 days, the subsequent incorporation of labeled precursors into soluble protein fractions was markedly increased.[1i,2o] E. Koenig[2h] reported a puromycin-sensitive incorporation of tritiated leucine in the IX cranial nerve roots of rabbits, but no inhibition by chloramphenicol or actinomycin D, with the implication that a contribution of mitochondrial protein synthesis was not involved. A study of turnover in axoplasm showed higher incorporation rates in comparison to the membranous elements.[2h,140]

Despite the significance attached to metabolic changes linked to functional events, few studies exist on the relation of protein turnover to membrane exitability.[141–143,81] Fisher et al.[141,142] observed an increased incorporation, with electrical stimulation for 10 min, of labeled amino acids or uridine when introduced into axons by microinjection techniques. Incorporation into TCA-insoluble fractions was further increased if amino acids were injected following an initial period of stimulation for 10 min; incorporation was inhibited in the presence of actinomycin D. The authors suggested that the membrane is a site for synthesis because incorporation was higher in sheath than in axoplasmic material. The reported interference with the action potential by intra-axonal injection of some proteolytic enzymes (Chapter 2, this volume) indicates the possible influence of breakdown processes on the integrity of axonal intrastructure and on membrane excitability. Recently, we explored the possible relationship of neutral proteinase and some aminopeptidases to excitability.[81] Stimulation of frog sciatic nerve, sectioned at the level of the spinal cord, but still retaining circulation and still connected to muscle, led to a significant change in pools of free amino acids (20%) and ammonia, but no change in activity of the proteolytic enzymes. Data on turnover for peripheral nerve are insufficient to permit any conclusions except to note that if electrical pulses increase incorporation, but not breakdown, this form of physiological activity may be accompanied by a net synthesis of new protein. Prolonged stimulation could lead to an exhaustion of precursors followed by activation or release of proteolytic enzymes; Ungar[14] reported increase in nonprotein nitrogen under such conditions in the isolated nerve of frog and rat.

IV. TURNOVER IN SUBCELLULAR ORGANELLES

A. Mitochondria

The demonstration that purified mammalian mitochondria are autonomous with respect to protein turnover *in vitro* (incorporation in the absence of cytoplasmic factors and ATP) has focused interest on their biogenesis and function.[2e,21,166] Early preparation from brain were subject to considerable contamination (myelin, synaptosomes, lysosomes, etc.) and skepticism was voiced concerning an independent protein turnover; however, many doubts were dispelled by the isolation of intramitochondrial DNA, and ribosomes, and lack of inhibition in the presence of ribonuclease.[68,166,167] Mitochondria are major sites for energy production, and their metabolism in most tissues is affected by the administration of selected drugs or hormones, or by exercise.[168,169] The presence of a distinctive DNA species in mitochondria has implications for the potential coding capacity of these organelles to form their own structural proteins and content of enzymes. Data based on electron microscopy or separation on cesium-ethidium bromide gradients indicate 1–2 molecules per mitochondrion; assuming about 250 mitochondria per cell, this represents 0.15% of that present in the nucleus.[165,170,171] The informational content of mitochondrial DNA is claimed to be insufficient to code for all mitochondrial components, and interaction with the nuclear genetic system may be essential for biogenesis. Turnover of mitochondrial DNA itself may reflect the rates of assembly and degradation of these organelles.

1. *In Vivo*

In vivo liver studies show a higher incorporation of labeled precursors into mitochondrial DNA compared to the nucleus, with the ratio of activities dependent on the rate of tissue growth.[171] In slow-growing tissues, mitochondrial DNA turned over more rapidly with a shorter half-life than fast-growing tissues (young rat liver or in hepatomas).[171,166] In *Neurospora*, pulse labeling experiments with choline imply replication by actual division of preexisting mitochondria, but insufficient information is available to decide whether similar mechanisms exist in mammalian cells.[172]

Fletcher and Sanadi[173] estimated the renewal rate of liver mitochondria as 10 days based on the kinetics of decay following systemic injection of

Met-^{35}S for cytochrome c and other protein components. This value has been confirmed by Bailey et al.[174] although lower values of 4–6 days were reported by Swick et al.[175] if correction is made for reutilization of amino acids formed by degradation. They used Arg labeled in the guanidine position since the probability of reincorporation of protein-derived, isotopic guanidine is small, especially if followed by administration of isotopic carbonate.[4g,166] The half-lives of two inducible enzymes, alanine and ornithine aminotransferase (E.C. 2.6.1.2 and 1.3; treatment with corticosterone) ranged from 0.7 to 1.0 day. The evidence from these experiments suggests that all mitochondria are in a dynamic state and that alteration of the enzyme complement is independent of mitochondrial genesis.

The presence of distinct mitochondrial populations in brain indicates differences in their formation and function.[177,178] D'Monte et al[2e] observed in electron micrographs different-sized mitocondria in selected areas of a continuous sucrose gradient used for fractionation of crude mitochondria from rat brain. Small mitochondria were located in the synaptosomal region and were distinct from those of the heavier mitochondria present in the pellet. Such observations are consistent with the three functional populations of rat brain mitochondria described by Neidle et al.[177]: (a) those containing acetyl-CoA synthetase and glutamic dehydrogenase; (b) those containing monoamine oxidase, succinic dehydrogenase (SDH), and glutaminase-1; and (c) particles in the synaptic ending region of the gradient characterized by relatively high levels of monoamine oxidase and SDH. Salganicoff and Koeppe[179] reported that some 75% of brain SDH was associated with synaptosomal mitochondria, but these observations are at variance with histochemical data of Hajos and Kerpel-Fronius,[180] who observed a significantly higher somal activity. Evidence for mitochondrial heterogeneity together with their associated compartments for tricarboxylic acid cycle intermediates or amino acids[1r,181] could influence protein turnover (transport of amino acids to the relevant precursor pools). Recent reports show the presence in brain and liver mitochondria of proteolytic enzymes and peptide hydrolases at levels sufficient to account for known rates of protein turnover.[2e] Mitochondria are also subject to destruction by sequestration within vacuoles, resulting in formation of mitochondrial debris and lipofuscin pigments.[1h]

Estimates on the decay of radioactive mitochondrial proteins of rat brain obtained from feeding experiments are consistent with the half-life of approximately 16 days.[71] Similar values were observed in the single-pulse experiments of Von Hungen et al,[68] with half-lives of 20–26 days, but longer than comparable experiments with liver preparations.[95,166] In our own studies, intracisternal injection of lysine-^{14}C in rats led to rapid labeling of all mitochondria present in the different cellular fractions separated on continuous sucrose gradients[2e] (Fig. 6). Incorporation into synaptosomal mitochondria could indicate transport from the nerve cell body as proposed by Barondes,[7e,182] who observed a time-related appearance of monoamine oxidase as the marker enzyme (see Section IV, C).

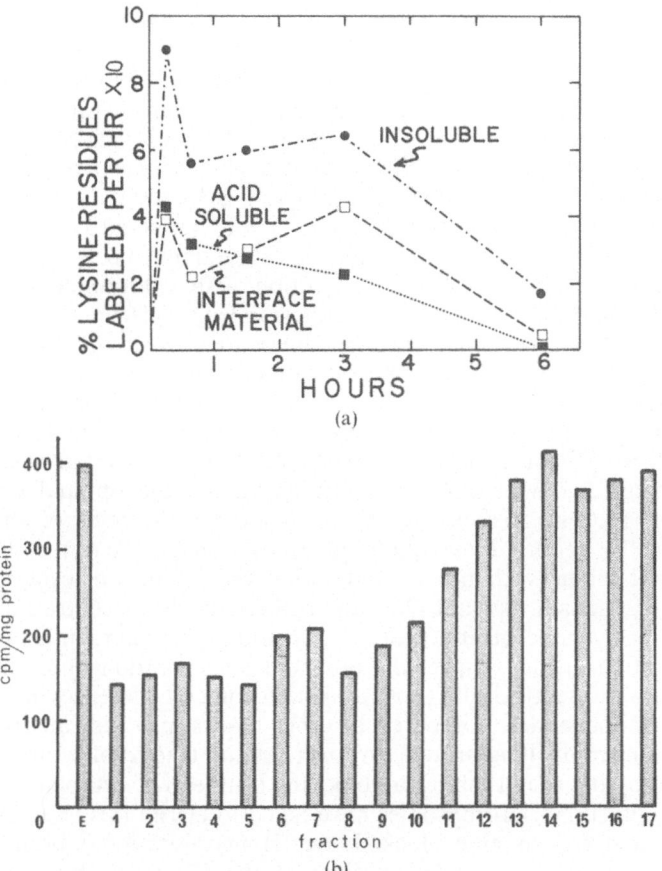

Fig. 6. (a) Turnover of protein and proteolipid fractions of mitochondrial fraction E (below) separated by the method of Eng.[20] Results are expressed as the percentage of Lys labeled (× 10) in protein by the method described by D'Monte, Mela, and Marks.[43] (b) Amino acid incorporation into mitochondrial subfractions. A total of 1.0 μCi of Lys-^{14}C was injected intracisternally into each rat, which was sacrificed after 30 min. Crude mitochondria were fractionated on a continuous sucrose gradient as described elsewhere.[2e,43] Fractions 12–16 represent crude myelin and fractions 5–12 synaptosomal fractions.

2. In Vitro

Several groups have reported the capacity of isolated crude or purified mitochondrial preparations to incorporate radioactive precursors.[166,176,184–187] Studies in liver and in brain by Campbell et al.[167] and by Bachelard[183] failed to demonstrate any requirement for added cytoplasmic factors, or for a source of energy, although Campbell et al.[167] reported some stimulation with addition of α-ketoglutarate, GABA, and some neurohormones. The effect of chloramphenicol and other inhibitors on turnover of mitochondrial proteins has led to considerable speculation and conflicting data.[184–189] Gordon and Deanin[184] failed to observe any inhibition in brain mitochondria with chloramphenicol but noted a marked effect with acetoxycycloheximide. These results are in contrast with those of Cunningham and Bridgers[185] who observed a 30% reduction of incorporation with the addition of chloramphenicol; Bosman and Hemswoth[186] reported a 60% inhibition with chloramphenicol but a virtual absence of any effect with cycloheximide. Aside from the question of inhibitor/protein concentrations, the differential sensitivity to antibiotics may indicate the presence of different mitochondrial populations. Alternatively, relative inhibition could be related to the purity of the fraction, the method of preparation, and the conditions used for incubation. Cunningham and Bridgers,[185] for example, observed inhibition with chloramphenicol in a medium enriched with potassium. In studies on the optimal conditions for incorporation of labeled amino acids and on the effect of antibiotics, Goldberg[169] reported maximum incorporation at 100 mM sodium and 10 mM potassium, with marked inhibition on addition of cycloheximide, insulin, or ouabain, but not chloramphenicol. High concentrations of potassium or low concentrations of calcium and magnesium tended to inhibit incorporation. In this respect the observed incorporation rates with mitochondria were identical with those of synaptosomes. Differences in sensitivity of brain mitochondria compared to other tissues may indicate alternative mechanisms controlling protein turnover. Attempts to resolve this confusing situation by the use of inhibitors blocking synthesis of cytoplasmic in contrast to mitochondrial ribosomes have failed to clarify this issue. Although one such material, emetine, blocked Leu-^{3}H incorporation in brain slices by 95%, it cannot be concluded easily that the 5% insensitive component (inhibited by chloramphenicol) is exclusively present in mitochondria.[190] Such explanations must overcome a number of assumptions concerning proteins with relative short half-lives, heterogeneity of mitochondria, penetrability of inhibitor to all areas of the slice, etc. As noted by Mahler et al.,[190] uncertainties are introduced in a tightly coupled system implied in protein turnover, when one component is blocked at the expense of another.

Subfractionation of mitochondria into its membrane and soluble components has provided some information on mechanisms of biogenesis.[166] For mitochondria of brain, the half-life was 18 days for the water-soluble components, 31 days for insoluble material, in contrast to 6 and 9 days respectively for soluble and insoluble fractions of liver, and 8 days for both soluble and insoluble fractions of kidney.[176] We investigated, in rat brain mitochondria, incorporation of Lys-^{14}C given intracisternally, hydrolase content of purified mitochondria, and protein composition of the isolated inner and outer membranes. Fragmentation of mitochondria was based on methods used for liver (phospholipase, large-amplitude swelling with the

TABLE V

Incorporation of L-Lysine-^{14}C into Rat Mitochondrial Membranes[a]

	Specific radioactivity	
	In vivo[b]	In vitro[c]
Subfraction E	474	7712
Inner membrane (IM)	549	8444
Outer membrane (OM)	407	3289
Soluble matrix	140	116

[a] IM and OM prepared by phospholipase treatment of purified mitochondria (fraction E) after separation on a continuous sucrose gradient.
[b] 1.0 μC lysine-^{14}C injected intracisternally 40 min prior to sacrifice.
[c] 2.5 μC lysine-^{14}C added to mitochondria suspended in the media of Campbell et al.[167]

addition of albumin and digitonin[178]. Rat brain membranes from mitochondria incorporated over 70% of the label compared to the soluble matrix fractions, with significantly higher incorporation into the inner membranes (Table V). Inner and outer membrane fractions displayed differences based on selected enzymes and in protein composition.[2e] Mitochondria free of morphological contamination showed a small but significant proteolytic activity (neutral and acid proteinases, arylamidase, and aminopetidase.[2e,178] Based on a number of assumptions, this trace activity would indicate a half-life for inner and outer membrane fractions of 4–8 days. Previously Brunner and Neupert[191] estimated a half-life for liver mitochondrial membranes of 3–11 days, with the outer membrane turning over at a higher rate. The presence of proteolytic enzymes could ensure complete autonomy with respect to synthesis and breakdown. Studies on protein turnover in mitochondria imply integrity of structure for finite periods rather than constant replacement by de novo function. As a further conclusion, heterogeneity of half-lives for different proteins suggests that in their origins, the outer membrane is different from the inner membrane. There is evidence that the outer membrane is of cytoplasmic origin and the inner is formed within the mitochondrion by endogenous processes. Evidence for the heterogeneity of turnover also exists for the incorporation of phospholipid precursors, with a higher incorporation of choline and serine into the outer compared to the inner membrane.[2e] The relationship between phospholipid and protein turnover in membranes represents an important aspect, since these components are intimately linked in structure and function. The complex structure of the mitochondrion produces several different compartments, with possible consequences to protein turnover. Currently there is considerable interest in the morphology of mitochondria, since the inner membrane is capable of forming various structural configurations with two extremes: (a) a contracted form characterized by a highly folded appearance and decreased matrical volume; (b) an expanded structure with the inner membrane compartment adjacent to the outer membrane. The presence of "contractible" or "structural" protein elements underlines the importance of protein replacement in this organelle. Work *et al.*[165] have suggested that liver mitochondria contain sufficient DNA for formation of its structural protein, but that a number of catalytic proteins (enzymes) are derived from extramitochondrial sources. A corollary to the heterogeneity of turnover rates is the possibility that the mitochondrion is neither formed nor degraded as a single entity.

The role of mitochondria in protein turnover in brain is unknown. Mitochondria could be an important source of energy in the supply of ATP for acyl-*t*RNA production; furthermore, their localization in synaptosomes is particularly intriguing, since they may form a significant source of protein for this organelle.

B. Synaptosomes

The synaptosome is a structure that includes pre- and postsynaptic membranes involved in the storage and release of putative transmitter substances. Synaptosomes or nerve-ending particles must be regarded as an operational entity resulting from the severing of pre- from postsynaptic attachments during isolation, followed by a sealing up of the membranes to form a small bag of about 0.5–1 μ diameter. The presence of cytoplasm and mitochondria within synaptosomes has consequences for the interpretation of mechanisms of protein turnover. Importance is attached to the question of whether the synaptosome can meet its local metabolic requirements or is dependent on a supply from the nerve cell body (axoplasmic transport) or associated cyctoplasmic components.[7e,192,182]

1. *In Vivo*

Lajtha *et al.*[193] were among the first to observe an incorporation of amino acids into fractions enriched with synaptosomes. This fraction prepared by a variety of procedures[20] has been the subject of intensive investigation.[195,202–205] The importance of axoplasmic transport is emphasized by the data of Barondes[2e] who observed a time-related appearance of radioactivity in synaptosomal soluble fractions in contrast to extracts of whole brain (Fig. 7). Von Hungen *et al.*[68] studied the kinetics of turnover following intraventricular injection of Leu-^{3}H prior to sacrifice at 4 hr–12 weeks and observed a half-life for synaptosomes or their fractions (soluble protein, synaptic vesicles and membranes) of about 20 days (Fig. 8). Synaptosomal mitochondria exhibited two decay rates for radioactivity consistent with half-life of 10 and 20 days, almost like nonsynaptosomal mitochondria (Table IB). More recently, several groups have attempted to fractionate synaptosomes

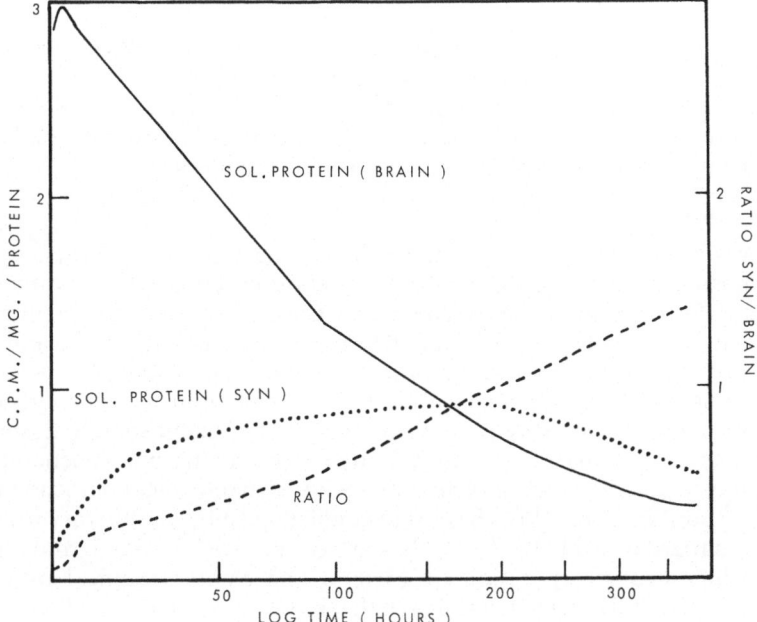

Fig. 7. Change in specific activity in soluble protein of whole brain compared to synaptosomes of mice brain injected intracerebrally with Leu-^{14}C. Data based on that of Barondes.[24] The ratio of activities showing a time-related appearance of radioactivity in nerve ending particles is indicated by the dotted line: the delayed appearance is attributed to redistribution by processes of axoplasmic flow.

Fig. 8. Time course of Leu-^{3}H label in synaptosomes and in synaptosomal mitochondria following an intraventricular injection. The data for cell body mitochondria are included for comparison. Actual $t\frac{1}{2}$ values are included in Table IB. Data taken from Von Hungen et al.[68] Both mitochondrial populations show two populations of proteins with different half-lives.

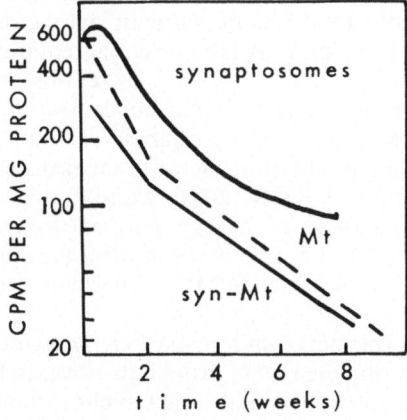

into their components by alternative schemes in an attempt to exclude cross-contamination by nonsynaptosomal membranes or other material (detergent or sonication procedures), but results in terms of turnover are highly variable. Some reports suggest that vesicle protein is subject to very low rates of turnover.[194,195] This question of heterogeneity can be decided only when better preparative procedures are available, assuming that valid methods for separating synaptic complexes from the axon are established.

2. *In Vitro*

Austin and Morgan[192] were the first to demonstrate the capacity of synaptosomes derived from incubated cerebral cortex slices to synthesize proteins. Subsequent studies in completely cell-free systems have facilitated information on the participation of other particulates in this process and the effects of change in the incubation system—the role of ions, Krebs cycle intermediates, energy supply, etc.[184,197–199] Aside from the usual advantages of a controlled *in vitro* environment, such studies enable some assessment of intrinsic synthesis *vs.* the contribution from axoplasmic phenomena. Attempts to determine the contribution by intrasynaptosomal mitochondria utilizing antibiotic inhibitors has led to conflicting and contradictory results, as summarized in Table V.[197–198] In some studies, cycloheximide and puromycin inhibited incorporation by 25–80% and chloramphenicol by only 25%.[197–199] Still other studies suggest that brain mitochondria and synaptosomes are insensitive to addition of chloramphenicol, but sensitive to cycloheximide.[167,184] Differences in sensitivity to antibiotics could be related to the preparative techniques (presence of cytoplasmic or mitochondrial contamination), the conditions of incubation, or to heterogeneity of the organelles, as reported in other tissues and in some cases in microorganisms[188,200] (see also Section IV, A). Very little information is available concerning heterogeneity of turnover rates for individual components of synaptosomes *in vitro*. There is some indication *in vitro* that the vesicle fraction is subject to considerably lower turnover than the soluble and membrane fractions.[194,195] Fragmentation prior to incubation *in vitro* leads to total loss of synthetic activity as noted in Table VI, showing that structural integrity is an essential requirement. In studies on optimal activity, several groups[197,198,201] have observed maximum incorporation with Na–K ratios of 5:1 (Na concentration 60–100 mM, K concentration 10–50 mM). Inhibition by ouabain (Table V) could imply that Na–K-activated ATPase is a limiting factor in the transport of amino acids. Inhibition by the addition of KCN, DNP, azide, iodoacetate, or oligomycin points to the participation of energy derived from mitochondria and the mediated transport of amino acids in turnover processes.[197,198] The absence of any effect with added ATP or energy-generating systems could be related to questions of penetrability.

The nerve-ending particle preparation presents a unique model for exploring the role of protein turnover in functional processes. Several groups have concluded that intact synaptosomes possess the intrinsic capacity to

TABLE VI

Effect of Added Materials on Protein Turnover in Synaptosomes[a]

	Concentration (mm or μg/ml)	Percent of control	Comments	Reference
(1) Addition showing no effect				
ATP, ADP, AMP glucose, α-ketoglutarate	1 mM	100	No requirement for added source of energy	197–199
RNase, DNase	50–200 μg	100	Absence of microsomal synthesis	
(2) Additions showing some inhibition				
Oligomycin	100 μg	50	Some indication for energy	
DNP, KCN	1 mM	50	from mitochondria	197, 198
Glutamate	1–5 mM	80		
Cycloheximide, chloramphenicol	0.1 mM	25	Inhibition dependent on Na–K–Mg ratios (5:1:0.8 optimal for incorporation)	167, 169, 197, 198
Puromycin, tetracycline	0.1 mM	25		201
Linomycin	0.1 mM	25		
Actinomycin D	0.1 mM	25		
Oubain	1.0 mM	80		
Osmotic shock Triton X-100, SDS Saponin Sonication	0.1–1.0 %	0	Intact structure required for synthesis	197–199

[a] Intact synaptosomes are not dependent on an exogenous source of energy but synthesis is inhibited by materials interfering with oxidative metabolism of mitochondria. Results demonstrate that disruption of synaptosomes leads to cessation of turnover; radioactivity is chiefly associated with the soluble and membrane fractions (specific activity 274 compared to 330 cpm/mg protein for the intact particles, and the synaptosomal mitochondria[170]).

synthesize some of their own protein in addition to that derived from other cytoplasmic components, or adjacent mitochondria.[197,198,204,205] As in all other cases, synthesis presupposes an adequate supply of essential precursors (amino acids, RNA's, and energy supply), and these could serve as rate-limiting factors and indirectly affect function. Marchbanks and Whittaker[206] have likened the synaptosome to an anucleate cell with considerable capacity to carry out complex metabolic functions *in vitro*. If they are capable of independent protein turnover, this may serve in part for the formation or release of enzymes directly involved in the quantal release of neurohormones. Studies from our laboratory show that synaptosomal fractions contain a sufficient content of hydrolytic enzymes to ensure complete breakdown of most proteins to polypeptides or amino acids.[2e] Among the enzymes shown to be present were acid and neutral proteinases and dipaptidyl arylamidases. Some peptide hydrolases from brain rapidly inactivate neurohypophyseal hormones, and angiotensin II, suggesting that their presence at nerve endings could serve a function for the release (or inactivation) of physiologically active peptides or amino acids.

C. Nuclei

The central role played by the nuclear apparatus in protein turnover is especially pertinent to brain cells that lack the ability for cellular division (mitosis) or regeneration.[207] In early studies, doubts were expressed about an independent protein synthesis in nuclei, although this is now reported to occur *in vivo* and *in vitro*.[1c,s,20c] Allfrey and Mirsky[208] linked protein synthesis within nuclei to that of cytoplasmic components. The Na^+-dependence for turnover could indicate a transport-dependent process for the supply of amino acids to nuclear components. Such concepts are supported by the reported sodium dependence for the concentrative uptake of various amino acids by crude nuclear fractions.[209] Navon and Lajtha[209] showed that uptake was inhibited by low concentrations of sodium, or by the presence of cyanide or DNP, further illustrating a requirement for energy in transport processes.

Special attention has been directed to the role of histones in differential gene activity, particularly in nondividing tissues characterized by a constant level of nuclear DNA. Histones are known to be involved in regenerative processes in liver and to undergo methylation prior to RNA synthesis in isolated calf-thymus nuclei.[16a] Histones appear to act as unspecific repressors of chromosomal DNA, limiting the number of genes available for RNA synthesis. Although specificity has been claimed for some lysine-rich histones, fine control of the genome is thought to reside in the acidic or the nonhistone proteins present in chromatin[4e]; in cerebellar and pituitary tissue the nonhistone proteins account for 22% of total chromatin protein.[210] The constant composition of the histones derived from different tissues in a number of species has attracted considerable attention.[211] Brain tissue is characterized by a specific acidic protein known as S-100; although present largely in glial cells, it has been detected in nuclei of neurons by immunofluorimetric methods.[1b,213] The possible existence of other brain-specific proteins in nuclei could be important to genome expression and regulation in turnover. Induction of enzymes by hormones or related materials (theophylline, glucagon, dibutyryl cyclic AMP), in liver and other tissues could result from: (a) control at the level of transcription (histones or acidic proteins); (b) translation of nuclear *m*RNA; or (c) transport of *m*RNA and associated accessory factors to the cytoplasm.[16a,212] Effects of histones on gene expression are altered by acetylation, methylation, or phosphorylation by alteration of the binding charges; the role of such materials in hormone induction may be mediated by cyclic AMP.[16a,12g,322]

The nucleolus is frequently overlooked, yet this component is a known site for formation of ribosomal species and undergoes reversible changes in size during starvation or regenerative processes.[4e]

1. *In Vivo*

Autoradiographic and other studies provide ample evidence for turnover of nuclear proteins in several different tissues, including brain.[1k,74,75,214–216] Studies on the decay rates of labeled nuclei prepared by feeding animals from conception to 26 days with tritium precursors indicate a mean half-life of 24 days for the bulk of brain nuclear protein.[71] The absence of change in residual radioactivity on restoration of label-free diet for 100 days suggests the presence of stable nuclear components accounting for 21 % of total proteins. The percentage of "stable" nuclear proteins observed in these studies may be on the high side, since they are difficult to reconcile with the mean half-life, 15–20 days, for most CNS proteins. Relatively stable acid-extractable histones were observed by Piha *et al.*[216] in pulse-labeling experiments with Lys-^{14}C administered intraperitoneally or subcutaneously; depending on the duration of exposure to label, the half-lives varied between 54 and 104 days (Fig. 9, Table IB). Similar studies by Burdman *et al.*[215] with Leu-^{14}C given intraperitoneally showed higher incorporation into the acidic (chromatin) proteins compared to histone fractions, with some age-related differences for selected neuronal components. Incorporation is reported higher in nuclei from neurons compared to glial enriched fractions[215,217] (Fig. 10). Sellinger and Johnson[155] in studies of subcellular fractions of neuronal perikarya reported a lower incorporation in microsomes and mitochondria, but higher than in the supernatant fraction (Fig. 11).

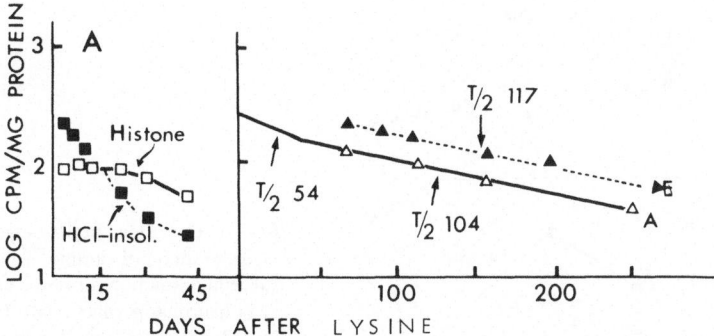

Fig. 9. Change in specific activity and turnover of histones in adult or embryo mouse brain. The specific activity of nuclear proteins was recorded after systemic injection of Lys-^{14}C in adult mice (A). Turnover as indicated in the figure was determined after injection of Lys-^{14}C in three equal portions to adult mice or pregnant mothers; adults (△———△) and embryos (▲-------▲).[216]

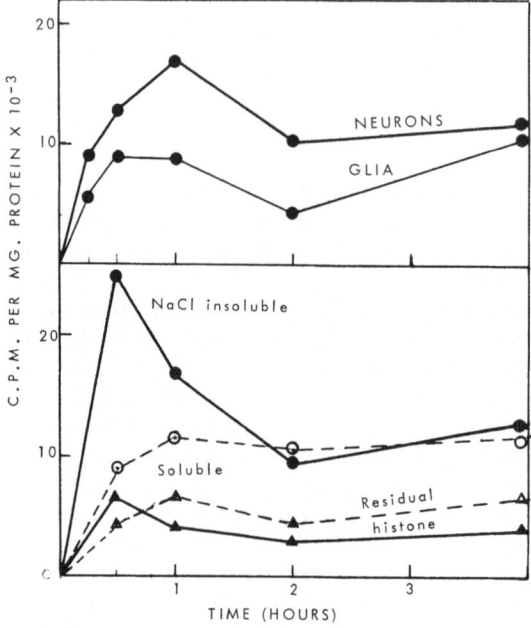

Fig. 10. Incorporation *in vivo* in glial or neuronal enriched nuclei of adult rats following an i.p. pulse of Leu-^{3}H for periods noted in the upper figure and into different protein fractions of cell nuclei in the lower figure. Data based on those of Burdman *et al.*[215,218] The results demonstrate higher incorporation into nuclei of neuronal enriched fractions at short time periods.

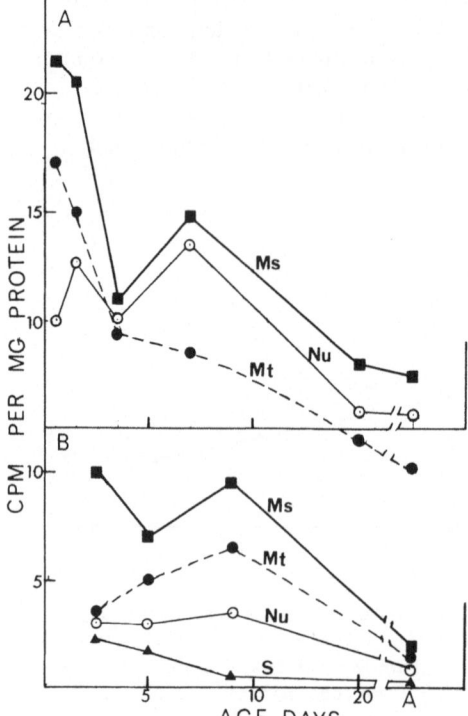

Fig. 11. Developmental change in the incorporation of amino acids into subcellular fractions of whole brain (A), or of neuronal perikarya (B). The data in (A) are from Oja[5] and are for Tyr-^{3}H administered intraperitoneally 2 hr prior to sacrifice. Data in (B) are from Johnson and Sellinger[155] for a 10-min pulse of Phe-^{14}C administered intrathecally.

2. In Vitro

Studies with isolated preparations are handicapped by the considerable damage resulting from isolation procedures; consequently the optimal conditions for study of turnover are not well established.[1c,20c] Nuclei prepared in the presence of Triton X-100 display low rates of incorporation, which could result from damage to the nuclear apparatus or removal of cytoplasmic factors essential for synthesis.[20c] Optimal incorporation is related to the ratio of sodium to potassium, with inhibition by high concentrations of K^+.[217] Lovtrup–Rein[217] observed a twofold higher incorporation into the nuclei of the neuropil compared to neurons, optimal at pH 7.4 in the presence of 13 mM sodium and 5 mM magnesium. In such studies, brain nuclei displayed greater capability for incorporation than comparable liver preparations.[218] In a comparison of nuclear proteins Burdman et al.[219] observed a higher incorporation into chromatin and residual material than into histone fractions, confirming the heterogeneity seen in vivo.

V. CHANGES DURING DEVELOPMENT

Ontogenesis is characterized by a greater intensity of metabolism for a large range of cerebral components in young animals than in adults.[1s,2b,d,13,24] The discontinuities apparent in growth for different organs, and problems associated with sampling tissue subject to a fluctuating environment, add to difficulties in obtaining quantitative measurements for turnover. The appearance of new protein species along with decay of some preexisting components necessitates acquiring qualitative knowledge of the proteins concerned and also data on their extractability and solubility.[2e] Ideally, measurement of incorporation during development is based on the specific activity of the precursor pool and expressed as amino residues incorporated into protein of known composition. Some proteins are virtually absent at first (basic myelin or S-100 acidic proteins) and appear in high concentration during myelinization or during other active phases of development.[1b,y] Few definitions exist on what constitutes an "adult" on the basis of biochemical criteria; behavioral studies suggest that development extends well beyond myelinization or deposition of other components. There is little agreement on the optimal ages for sampling tissue, and conclusions are frequently expressed on the basis of time points that fail to cover all the critical phases of cytodifferentiation. Little is understood about the remarkable mechanisms initiating the major ontogenetic events, or the adaptive changes characterizing maturation of the nervous system. Many of the morphological and biochemical changes are described in this series[1] or elsewhere[2e,d,13] and emphasis is placed on those features considered critical to protein turnover. These include altered influence during development by: (a) supply and transport of essential nutrients; (b) biogenesis of organelles involved in synthesis and breakdown; (d) pathological states (inborn metabolic errors); (e) environmental factors (states of excitation and exhaustion, sensory deprivation.[1u] In addition to the known alterations with age of the permeability barriers to amino acids,[3,5] there are variations in intracellular metabolism.[2e] Alterations in the extra- and intracellular spaces[20d] and associated amino acid pools would result in changes in the level and distribution of precursors. The intensive anabolic phase during development may require immediate reutilization of any amino acid formed by breakdown of unwanted protein species. The orderly change of components and their integration during development imply a highly sophisticated mechanism for regulating the level and appearance of different cellular components.

A. Studies *In Vivo*

Many different approaches are available for seeking correlative morphological and biochemical changes that occur during the maturational process. Representative of such studies are (a) the changed enzyme patterns characteristic of growth in the brain, (b) distribution of synthetic or catabolic systems within the cell or at different brain regions during development, and (c) the neurochemical correlates of behavior. The time course of developmental changes is dependent on the species employed. In our own studies related to turnover and breakdown in rat brain, we measured representative hydrolases in relation to four well-defined ontogenetic periods: (a) the phase of active cell proliferation prior to birth, (b) the period of axonal growth and increased dendritic complexity following birth (0–10 days in the rat), (c) myelination (10–25 days), and (d) the slow deposition of solids such as lipids and proteins in the adult. The rat is particularly suited to such studies since it is born with nearly its full complement of neuronal cells in an unmyelinated state. There is evidence that some glial cells show a limited ability to divide in adult brain,[220] as confirmed by postnatal changes in DNA in some mammalian species. In seeking the characteristic growth periods of brain, we measured the net weight gain per day and the relative rate of increase. The maximum deposition of solids based on net increase occurred at the tenth day (coincident with myelination), but the percent increase was highest at birth, followed by a fall in rate until the fifteenth day (Fig. 12). Data based on the appearance of nucleic acids, or on the pool of free amino acids as related to change in brain weight or in the level of DNA itself, exhibit a number of discontinuities in growth (Figs. 12, 13) that coincide in some cases with myelination. The change in the level of cerebral components is frequently nonlinear with age and different spectra for half-lives may depend on the age selected for analysis. It might be concluded that the slow but steady increase in brain weight observed in older or senescent animals represents a departure from the ideal steady-state conditions; this may be related to the materials that resist degradation and accumulate in vacuoles in the form of residual bodies (lipofuscin).[1q] Discontinuities in growth may be related to anatomical differentation; for example, many enzymes and brain components exhibit caudal-rostal gradients (cholinesterase, cholineacetylase, phospholipids,

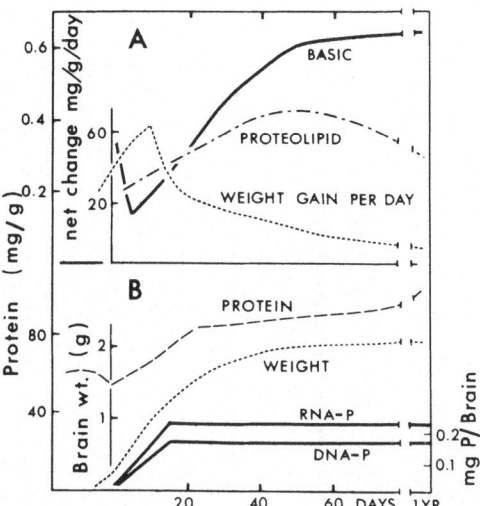

Fig. 12. Developmental change in protein and nucleic acid components of rat brain. Data in A are from Gaitonde and Martenson[275] and show the change in basic protein extracted from whole rat brain with 0.03 N-HCl as compared to the proteolipid fraction extracted with chloroform–methanol. Data on the change of nucleic acid phosphate are from Mandel et al.[309] Other data in the figure are derived from our own studies and show change in brain weight or protein expressed in several different ways.[2b]

several proteins, etc.[6]). Protein levels in rat cortex were reported by Pasquini[221] to be relatively stable at 14 days compared to a continued synthesis in the cerebrum for a period up to 30 days. Study of gross overall change *in vivo* will not reveal finer temporal and anatomical differences in an organ as diverse as the CNS. The observed change in protein components, among others the increasing lipid content with age, could reflect change in the relative solubility and extractability.

In studies with rats, aged 15–150 days, Gaitonde and Richter[63] observed a decline with age in rates of incorporation in brain, but not in liver on administration of Met-^{35}S 3 hr prior to sacrifice. A further indication of age-related differences was provided by Lajtha et al.[3,222] in a comparison of

Fig. 13. Developmental change in rat brain constituents expressed in a manner to illustrate the discontinuities during growth. Incorporation (A) and the free pool of amino acids (B) expressed in units per mole of DNA exhibit the largest maxima during the known periods of active myelinization. Data based on that of Miller[4d] and for changes in Glu and Ala on Agrawal and Himwich[1a,13b] and our own studies.[1r,2b]

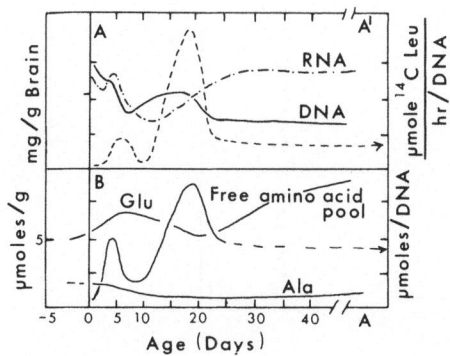

TABLE VII

Half-Life Time (In Days) of Proteins of Organs of Immature and Adult Mice as Measured with Lysine-^{14}C and Leucine-^{14}C[a]

Time after injection (min)	Brain					Liver					Muscle				
	Adult		8 days		1 day	Adult		8 days		1 day	Adult		8 days		1 day
	Lysine	Leucine	Lysine	Leucine	Leucine	Lysine	Leucine	Lysine	Leucine	Leucine	Lysine	Leucine	Lysine	Leucine	Leucine
2	2.8					0.9					3.5				
3		3.8			0.7		0.4			0.2		5.1			0.9
5	3.5	8.1	3.4		1.2	1.3	0.9	1.7		0.4	7.6	7.1	2.4		1.5
10	5.5	11	2.2		1.7	2.6	1.3	1.1		0.7	7.6	14	1.7		2.0
20	6.2	15	3.0		2.4		2.4	2.7		1.4	10	24	2.5		2.9
30	6.9	20			3.0		3.9			2.0	13	28			3.2
45	10	24	2.4		3.3		5.7			4.0	17	30	2.3		4.1
60	15	27			4.1					4.7	24	34			5.0

[a] Data based on that of Lajtha.[324] Adult mice were 100–120 days of age and compared to 8-day- and 1-day-old animals.

young mice (1–8 days) with adults, both sacrificed 2–60 min after i.v. or i.p. pulse of Lys-^{14}C or Leu-^{14}C. Taking into account the change of specific activity of the pool of free amino acids with time (Figs. 2, 13), the half-lives calculated on the basis of the specific activity of protein-bound Lys or Leu in brain showed a highly significant decrease in turnover with age (Table VII). The half-life of mice aged 1 day ranged from 1 to 4 days, depending on the precursor and time of exposure to label, compared to 3 to 27 days in adults. Exposure to label for short periods would measure rapidly turning over proteins, which are more evident in the younger animals. In other tissues, the half-lives calculated for total brain proteins are longer than for liver proteins but shorter than for muscle proteins.

Comparison of incorporation in the fetus with that at later ages shows a decrease in protein synthesis at birth followed by a subsequent decline with age. The results of Shain et al.[76] show that premature delivery is accompanied by a decline in protein synthesis, suggesting unusual intrauterine influences on mechanisms of turnover. Malnutrition, irradiation, and anoxia during the prenatal period can retard development; there are alterations in protein turnover at birth as judged by incorporation of labeled amino acids into tissues.[13,76] In a study of incorporation of intraperitoneally injected Tyr-^{3}H into subcellular fractions of rat, Oja[5] observed the highest labeling pattern in microsomal supernatant, followed by the crude mitochondrial and nuclear fractions at 1 day; in older animals incorporation in nuclei exceeded that of mitochondria. The half-lives, estimated by exposing rats to pulses of Tyr-^{3}H for periods of 5 min to 6 days, fell from 4 to 9 days in newborns to 1 to 3 days at 2 weeks of age, and rose to 4 to 6 days in the adult. This unusual labeling pattern could reflect a changing composition of Tyr content of brain proteins with age, or alternatively, factors affecting the metabolism and pool size of the precursors. Studies on uptake of label indicated no significant change in protein-bound Tyr, but the concentration of free Tyr was higher in young animals and declined more slowly in older animals.[5,223] Similar observations have been made for other amino acids when the concentration in plasma exceeded the normal level, i.e., a greater influx in young animals compared to efflux.[224] Alteration in transport of precursors during development could serve as a control mechanism influencing anabolic rates. Pool size of the precursor amino acids may be influenced by compartmentation, which is absent or at a minimum level of complexity in newborns[1j] (see Section VI, D). Increase in compartmentation could reflect an increase in a number of intracellular organelles, in dendritic processes, and in the neuron–glia ratios. Recent studies by Johnson and Sellinger,[155] and Norton and Poduslo,[148] indicate an increase in glial cell population with age, with the highest neuron–glia ratios at 5 days. Johnson and Sellinger[155] observed a higher incorporation into isolated cells at 5, 10, and 18 days than at 43 days after an intrathecal pulse of Phe-^{14}C (see Section III, C). In subcellular fractions of isolated cells, incorporation was highest at all ages in the microsomal fraction, with very little in the cell sap; it was maximal at 5 days and absent at 43 days (Fig. 11) Murthy and others[1s,2d] have proposed that alterations in protein turnover by microsomal fractions are related to an alteration in transcriptional and translational factors (see Section VII). The increased incorporation of mitochondrial fractions at 18 days is consistent with the increase in the number of organelles observed by Samson and Jacobs.[225]

B. Studies *In Vitro*

Greenberg *et al.*[62] directed attention to age-related differences in isolated systems with the observation that incorporation of Gly-^{14}C was higher in homogenates of young chicken than in those of older animals. This observation stimulated reinvestigation of incorporation *in vivo*, since previous indications were that proteins in brains are more stable than those of other organs (Section II, A). As well as in homogenates, incorporation *in vitro* can be demonstrated to occur in slices, cell suspensions, and some subcellular fractions. Thus, Shepartz and Turczyn[226] found a higher incorporation of neutral amino acids in brain homogenates of mice aged 1 day compared to 7 weeks. Gelber *et al.*[227] found a 3–4 times higher incorporation of Leu-^{14}C in homogenates of 14–16 day-old rat brain compared to 4–5-week-old tissue, with no comparable changes in liver preparations. Their finding that incorporation was stimulated by L-thyroxine when β-hydroxybutyrate was the oxidizable substrate was attributed in later studies by Klee and Sokoloff[228] to differences in rates of energy production of young versus adult mitochondria. This interpretation is consistent with the known heterogeneity of brain mitochondria with respect to their content of citric acid cycle enzymes (see Section IV, A). In whole cell suspensions of mouse brain aged 4–30 days, Johnson and Luttges[229] observed a rapid decrement in the ability to incorporate a variety of labeled amino acids, with activity at 12 days only 10% of that at 4 days. Oja[5] observed bigger age-related differences *in vitro* than *in vivo*, particularly in the presence of ATP and oxygen; incorporation of Tyr-^{3}H in brain homogenates from rats 1 day old was 5 times higher than that at 4 days of age, and 18 times higher than adults. Oja[5] attributed the decreased synthesis to an increase in catabolism as measured by the endogenous release of protein-bound Tyr *in vitro*, maximal at 14 days. This observation is consistent with our own studies on change in proteolytic enzymes with age as reviewed in Chapter 2 of this volume. Although an increased catabolism was observed *in vitro*, this may not reflect the true situation *in vivo* since a decreased rate of synthesis during the critical period of myelination would not favor brain growth. Studies with isolated systems, particularly homogenates, may lack the anatomical relationships or other factors critical to turnover. These include an energy supply, the nature of the precursor pools, or damage as the result of enzyme release from inactive or organelle-bound forms (lysosomes, etc). Isolated preparations containing intact or isolated whole cells, such as slices, cell suspensions, or neurons and glia appear to be more favorable for studies on turnover.

1. Cell-Containing Systems

Age-related differences in the incorporation of labeled amino acids have been reported in slices of cerebral cortex[101,107,230,233] and spinal cord[277] and peripheral nerve preparations (Section III). Mokrash and Manner[233] were the first to observe lower rates of incorporation in slices of rat brain from neonatal and newborn than from adults (Fig. 4). Incorporation of a mixture of unspecifically labeled amino acids into protein fell at birth, with markedly lower levels at 15 and 90 weeks of age. In contrast, incorporation into proteolipid fractions soluble in chloroform–methanol exhibited a maximum between 8–16 days, coinciding with onset of myelination. Essentially identical results were obtained by Ito and Arimatsu[233] in studies of mouse brain, with maximal incorporation into protein at birth, and with a small peak of incorporation in proteolipid at the period of myelin deposition. De Vellis et al.[236] observed a decreased incorporation of Leu-^{14}C into protein and proteolipid-protein of rat brain stem slices with age, with a further decrease following prenatal irradiation. Orrego and Lipmann[230] in rat cortex slices observed maximal incorporation of Val-^{14}C at 4 days in contrast to fetal or adult brain slices. They concluded that protein synthesis was directed by a labile RNA species that had a half-life of 2.6 hr and was inhibited by actinomycin D. Age-related differences were observed for a ribonuclease-sensitive incorporation of uridine-^{14}C in cortex slices; the inhibition was about 3 times as much in adults as in newborns.[237] Dunlop et al.[107] in our laboratory have observed some age-related differences in terms of sensitivity to antibiotic inhibition. In rat cortex slices, incorporation into protein of various amino acids was inhibited 90% by cycloheximide but less in the presence of high exogenous concentrations of amino acid (Lys, 35%; Arg, 60%; Val and Leu, 85%). Incubating slices in the presence of media containing a high concentration of amino acid ensures maximum saturation of tissue pools. The cycloheximide-sensitive component observed at different medium levels of amino acids represented a higher percentage in newborns than adults. This is good evidence for an age-related difference since transport factors are less likely to be rate limiting at high medium concentration of precursor.

2. Subcellular Fractions

A considerable number of studies demonstrate age-related differences in isolated subcellular fractions. The ribosomal complex is the major site of protein formation and has been the focus of study in relation to development as described in detail by Roberts[1q] and Rappoport et al.[1s] in the Handbook. Bondy and Perry[238] were among the first to observe a higher incorporation of Val-^{14}C in fetal microsomes of rabbit compared to adults as confirmed in later studies by Murthy and Rappoport,[241] Adams and Lim,[239] and Yamagami et al.[240] Despite such observations, the evidence for change in the components affecting synthesis and ultimately turnover is subject to dispute.[1q,241,242] The likely candidates include the protein factors involved in the initiation or termination of peptide chain elongation (Fig. 1); the supply of translational and transcriptional factors, including mRNA and tRNA; the level and specificity of tRNA synthetases; and translocation factors. A full discussion of this area is beyond the scope of this review except to note several studies on change in template activity with age (as measured by changes in incorporation with addition of RNA or poly-uridine), in polysome profiles, tRNA synthetases, and in ribonuclease activities.[1c,q,2d,242,334] Rappoport et al.[15] reported some reduction in tRNA synthetase level with age with minimal levels at 55 days, corresponding with changes in ribosomal profiles. Other groups have been unable to detect age-related differences in template activity, in ribosomal profiles, in tRNA synthetase, or in levels of ribonuclease.[1q,242] Most studies were conducted with ribosomal preparations, since interpretation of results with microsomal membranes is more complex owing to the presence of endoplasmic membranes with variable binding properties. Studies in liver show that lipid and protein components of endoplasmic reticulum membrane are assembled asynchronously.[244,245,247,338] With Leu-^{14}C as label given intravenously, the half-life for microsomal NADPH-cytochrome c reductase, and cytochrome b_5 varied from 3 to 5 days; with double-labeling procedures accompanied by enzyme isolation and purification, NAD glycohydrolase was determined to be about 18 days.[247] NAD glycohydrolase also occurs on other membranes, notably the plasma membrane; Bock et al.,[247] have emphasized that turnover in this case is dependent not only on rates of synthesis and decay, but also on rate of transport from one membrane to another, particularly if the structures are biogenetically related. In shorter periods of exposure to label, specific activities of plasma membranes increased 1.5-fold, in contrast to a decrease with rough and smooth microsomal proteins. These and other studies[248] are compatible with product-precursor relationships, for example, discharge of secretory proteins followed by binding. Problems in the preparation of pure fractions have limited studies on age-related differences for other subcellular organelles. The reported stimulation of microsomal synthesis with the addition of crude mitochondria from immature brain was ascribed by Klee and Sokoloff[228] to differences in ATP or ATP-generating systems. Ito and Arimatsu[233] and others[95] were unable to detect any differences in rates of labeled amino acid incorporation into mitochondria from rats of different ages purified on a discontinuous sucrose gradient. Incorporation into protein but not the chloroform–methanol proteolipid fractions in adults actually exceeded that of younger animals. In the case of partially purified synaptosomes, however, incorporation into protein and proteolipid fractions exhibited a marked age-related difference, with a two-to three-fold fall at 6 days of age. Qualitative differences in the composition of immature mitochondria, but not of mitochondria of adults, occur in liver; they have not been explored in brain. Liver of newborns contains an extremely high concentration of Cu^{2+} and an extraordinarily high concentration of cystine.[249] "Mitochondrocuprein" is localized on the inner membrane and may serve as a storage for Cu^{2+} during development, particularly for cytochrome oxidase that rapidly increases during the neonatal period.[249]

VI. GENERAL CONSIDERATIONS

The reason for the higher rate of turnover in the immature brain has not been found. There are several possibilities: (a) immature brain may have essentially the same type of proteins as the mature one except for the relatively greater proportion of the metabolically active species; (b) immature brain may have metabolically active proteins that are absent in the mature organ; and (c) the same protein may have a higher rate of metabolism in newborn than in adult. In (a) and (b) a change in the protein composition, in (c) a change in the metabolic machinery would be primarily responsible for developmental changes.

A. Turnover of Components Specific to Brain

Thudichum may well have been the first to emphasize specific brain components, but recent interest stems from the isolation of S-100 by Moore and McGregor.[250] Study of well-defined protein species, particularly if specific antisera can be prepared, conform to the criteria cited earlier for study of turnover. The presence of large amounts of these materials and their changes during development suggest important functional roles. A partial listing of proteins unique to the nervous system includes a number of acidic proteins such as S-100, 14-3-2,[1b] and antigen-α,[251] soluble brain proteins described by Warecka and Bauer[252] or others,[253] proteolipids and basic protein of myelin.[20g,254] Other proteins that deserve interest include neurofibrillar and tubular proteins, neurostenin (actinlike), neurophysin, and some glycoproteins.[18,28,255–257] Only limited studies are available on qualitative and quantitative changes in proteins with development. Immunological and electrophoretic studies suggest some changes in composition and extractibility with age.[28,231] Wender and Waligora[258] reported a higher concentration of His, Ser, and Val in fetal white matter but lower His, Ser, and Val in proteins of gray matter in comparison to postnatal periods. In our own studies (Toth and Lajtha, unpublished) we observed a remarkably similar composition in brain proteins of newborns and adults for different species (mouse, rat, hen, and guinea pig).

Clues to the possible function of a brain component may be provided by: (a) appearance during development; (b) determination of its turnover; (c) alteration in its metabolism as a result of specific inhibitors, or the result of training. Some specific components increase during development, notably S-100 and basic protein of myelin (Fig. 12); others are reported to decrease, notably neurotubule protein.[259–262] Enzymes themselves, although not unique to brain, are proteins frequently available in highly purified form and conveniently form antienzymes by immunological techniques. Although most enzymes increase during development, some are known to decrease, notably glutamotransferase and phenylphosphatase.[17] Finer anatomical differences in enzyme patterns may be masked by determinations on whole brain material. Different rates of appearance of S-100 were observed in brain regions dependent on the species studied. In rats, this protein reaches maximal values at 10–20 days; complement fixation measurements show a caudo-cranial distribution in guinea pig and in the human fetus. Earlier studies suggested a biphasic turnover with two components, one with a half-life of hours and the other, several days.[1b,213] Cicero and Moore[263] observed a half-life of 16 days in adult rats for S-100 (isolated on polyacrylamide gels) following an intraventricular injection of Leu-^{3}H; interestingly, incorporation only reached a peak after 48 hr (Fig. 14). Herschman[264] recently studied S-100 turnover in several clonal lines of rat and human astrocytoma. The human material accumulated S-100 (determined quantitatively by complement fixation) continuously while in the growing phase, whereas rat cultures produced protein only during the stationary phase. The decay of radioactivity of cells grown in the presence of Leu-^{3}H indicated a half-life of S-100 in human cells of 4–5 days compared to 2 to 3 days for total protein and in rat cells of 4 days compared to 60 hr for total protein. Regulation of synthesis and breakdown in cultured tumor cells may be pertinent to mechanisms *in vivo*; cessation of cell division and growth by "contact-initiation" is a characteristic of tumor cells and could be involved in the appearance of S-100 in the cultured human cells (type CHB_4) either by (a) initiation of synthesis of (b) decreased degradation. The first explanation would imply that cell density plays a role in genome expression in malignancy. In relation to S-100 turnover, Hydén and Lange,[253] (Section VI, E) have reported significant increase in specific brain areas of an S-100 type of protein.

Relatively few detailed studies are available on turnover of neurotubule (tubulin) and neurofibrillar proteins. As in the case of S-100, tubulin probably consists of more than one component, present in all brain fractions and comprising some 35% of the soluble proteins. It is characterized by its ability to bind to colchicine, or guanine nucleotides, and it interacts with *Vinca* alkaloids (vinblastin) to form crystalloid precipitates.[257] These agents block axoplasmic flow but not chemical activity of the nerve, and a functional role for the axonal components has yet to be demonstrated. The ability of tubulin subunits to aggregate into linear forms suggests that they may be involved in transport of material in the axon, possibly in association with actin like (neurostenin) materials.[255,265] Feit[259] has reported the half-life of tubulin from 5-day-old and adult mice to be about 4 days in both cases (Fig. 14) following intracerebral injection of radioactive leucine. Purified tubulin and neurostenin proteins are susceptible to breakdown by purified acid proteinase, which could therefore play an important role in structural changes seen in the axon (reversible polymerization of tubulin). Neurotubule formation is inhibited in cloned neuroblastoma cells grown in the presence of protein synthesis inhibitors (cycloheximide), particularly in the neurite outgrowths.[20f] Subunits for neurotubule formation may be present in the nerve cell body and assembled in the axon.

Glucosamine appears to be incorporated directly into protein at nerve endings, although the protein acceptor may be of perikaryal origin.[7e,266] Holian et al.[266] recently reported two groups of brain glycoproteins with

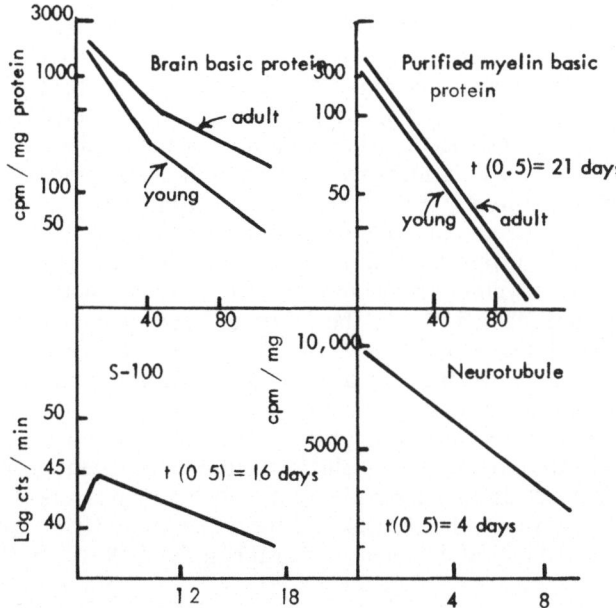

Fig. 14. Comparison of half-life determinations for selected components of the nervous system following the administration of Leu-^{3}H. Results for basic protein from Wood and King[276] using subcutaneous injection in adult rats; that for S-100 from Cicero and Moore[263] using intraventricular injection in adult rats; and for neurotubule from Feit[259] using intracerebral injection in 4-day-old mice. The crude brain protein shows two populations of proteins in young and adults with different half-lives; S-100 shows maximum incorporation at 48 hr.

half-lives of 18 hr and 4 days and 12.5 days in adults injected intraperitoneally with glucosamine-^{14}C. By comparison, incorporation into gangliosides was considerably slower. Turnover of fucose glycoproteins has not been studied in detail, but these (unlike the glucosamine materials) are transported primarily with the rapid component of axoplasmic flow. Incorporation of sulfate-^{35}S into brain mucopolysaccarides (predominantly chondroitin sulfate) decreases with age although incorporation with other proteins is relatively constant.[267] The half-life values of mucopolysaccharides or sulfur-containing proteins are unavailable. Since Met-^{35}S can form other thio-compounds, the use of labeled S may be unspecific in relation to brain proteins except possibly in the case of chondroitin sulfate.[63,267]

B. Myelin

Myelin represents an important component of the nervous system, and alteration in its rates of deposition during growth or subsequently in composition is accompanied by severe malfunction.[1w,19,254] In adults, the proteins of myelin represent a significant proportion of total brain proteins, and earlier concepts for its relative inertness[133] are now difficult to reconcile with the general observation that 90% or more of cerebral proteins are subject to great lability, with half-lives of under 20 days. Revaluation of this earlier data (taking into account the variation) would indicate considerably shorter half-lives.[133] Considerably less variation is encountered if rates of incorporation are directly related to specific activity of precursor pool as advocated in the introduction. Considerations pertinent to turnover studies in myelin include: (a) purity and structural nature of the isolated fraction particularly since myelin lamellae tend to form vesicle structures containing cytoplasmic components,[2e] (b) the relative metabolism of the different layers of the myelin sheath, (c) penetrability of label to all areas of the sheath, (d) the role of nodes of Ranvier and Schmitt–Lanternmann clefts, (e) the nature of the precursor pools, (f) the possible presence of proteolytic enzymes and the question of reutilization, (g) the role of Schwann or satelite cells in turnover. Several models for myelin structure have been proposed that would permit penetration of label longitudinally between layers of the sheath, or in the space between the sheaths and the axolemma[254,270]; penetration of precursor materials could occur also at the nodes of Ranvier.[269]

1. *In Vivo*

Incorporation of labeled amino acid or glucose into myelin proteins or proteolipids in the whole animal demonstrates considerable lability of myelin components, comparable to that of other brain structures.[43,89,154,271,272,339] Smith[271] estimated the half-life of myelin proteins and proteolipids to be about 35 days using intraperitoneal glucose-^{14}C in rats. Different rates of incorporation into proteins and lipids emphasize the heterogeneity of turnover in the myelin sheath; some lipids appear to be stable (cholesterol, cerebrosides, sphingomyelin, serine phosphatide) and others subject to turnover (inositol, phosphatide, lecithin, and proteolipids). Studies in our laboratory by D'Monte et al.[2e,43] showed a surprisingly high incorporation of Lys-^{14}C into the myelin fraction following intracisternal injection into rats and separation on a continuous sucrose gradient (Fig. 15). At first it was concluded that incorporation was attributable to cytoplasmic contamination by mitochondria, but 60% of the radioactivity was retained following treatment with hypertonic medium, and repurification on a series of sucrose and cesium-chloride gradients by the methods of Spohn and Davison.[20e] Highly purified myelin, fractionated into its proteolipid components by the methods of Einstein et al.[273] or Eng et al.[274] showed some 50% of the radioactivity to be associated with the basic protein fraction, 30% in the structural (Wolfgram) proteolipids, but only 6% in the Folch-Lees proteolipid material (Fig. 15). The measurement of the different components, based on the decay of radioactivity with time, indicated an exceptionally short half-life of 10–20 days; even shorter half-lives were reported following administration of tritiated leucine to rabbits by Blomstrand and Hamberger[154] (Fig. 4). Gaitonde and Martenson[275] studied turnover of basic

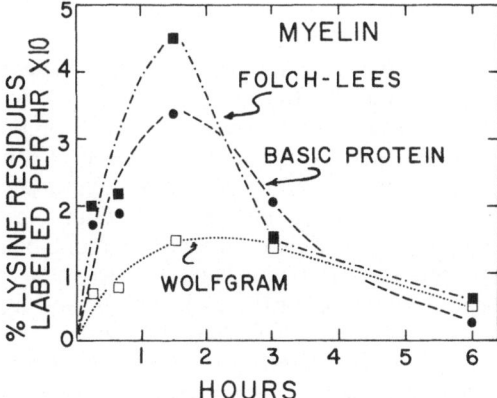

Fig. 15. Turnover of protein and proteolipid components of purified myelin[2e] extracted by the Triton X-100 procedure of Eng.[20] For other details see Fig. 6(a). Results from D'Monte, Mela, and Marks.[43] Results are expressed as incorporation × 10.

proteins of rat brain during postnatal development and observed a higher relative specific radioactivity during the period 1–10 days after birth for basic protein of myelin compared to other proteolipids. More recently, turnover of basic protein in brain was studied in detail by Wood and King[276] who reported a half-life of 21 days for 10-day-old and adult rats injected subcutaneously with Leu-^{3}H (Fig. 14). All studies on turnover in the whole animal must take into account the structural peculiarities characteristic for myelin, particularly the possibility of the trapping of precursors and reutilization of amino acids formed by breakdown within the sheath. As discussed in the companion review (Chapter 2, this volume), crude myelin fractions exhibit a variety of hydrolase activities, notably acid and neutral proteinase, aminopeptidase, and arylamidases. In our studies, purification by procedures indicated above led to a loss of all proteinase activities except the hydrolysis of small peptides such as leucyl-glycyl-glycine (aminopeptidase) and 2',3'-AMP-diesterase. Myelin purified by alternative procedures (sucrose gradient) is reported to contain native neutral proteinase.[1q]

2. In Vitro

Recent studies demonstrate the incorporation into myelin of labeled precursors in slices of spinal cord[277–279] or in cerebral cortex[154] and in crude homogenates.[2e] The relative merits of slices were previously discussed and may be especially relevant in the case of myelin. Glucose is a good precursor since it gives rise to acetyl-CoA that can be channeled into glutamate and thence other amino acids or lipids. In studies by Smith,[277] incorporation was linear with time for periods up to 3 hr, with different rates of labeling for the components subject to turnover (stable versus unstable lipids, see above). Additional studies by Smith[278] on the relative rates for glycine and leucine incorporation into the different categories of proteolipids and proteins separated by the method of Eng[20g] further illustrate the heterogeneity of turnover for the different myelin components. In contrast to studies *in vivo*, the incorporation into the basic protein fractions was at a lower rate than the acid-insoluble material (Wolfgram) or the interface materials following extraction with chloroform–methanol in the presence of Triton X-100. Cycloheximide inhibited incorporation of labeled amino acids without affecting the incorporation of acetate into lipids.[278] Such data suggest that under these conditions, turnover of proteins and lipids can be uncoupled, pointing to the operation of different synthetic mechanisms. As noted in the case of mitochondrial membranes, the integrity of structure is dependent on the integration of the mechanisms controlling turnover of both lipid and protein moieties; in myelin formation, proteolipid complexes can be seen as vital to continued function. The incorporation of labeled amino acids observed by D'Monte *et al.*[2e] in both crude mitochondrial fractions containing myelin or in the myelin-enriched material, would indicate the presence of sufficient cytoplasmic factors in juxtaposition to myelin lamellae to ensure protein synthesis. Highly purified myelin lamellae themselves clearly do not contain the appropriate enzymatic machinery for endogenous synthesis, implying that *in vivo* proteins are supplied by adjacent structures. Various structures within myelin could act as reservoirs for cytoplasmic factors necessary for synthesis; these include the Schmitt–Lanternmann clefts, nodes of Ranvier, etc. The heterogeneous rates of turnover of the various components of the myelin sheath pose an intriguing question about the mechanisms involved and point out the possible alterations this structure may undergo.

C. Hormonal Influences

Ontogenesis is suited to the study of hormonal influences because of their intimate link to maturational processes. In insects, results with the maturational hormone ecdysone[279] indicate control of turnover at a transcriptional level by gene repression, but definitive experiments showing a comparable relationship in mammalian systems are unavailable. The adverse effects of glandular deficit (thyroidal, hypothalamic, gonadal) have been reviewed extensively[1t,2,280]; thyroidectomy is accompanied by a striking neuronal hypoplasia with altered rates of myelin metabolism, and with a large but unspecific decrease in succinic dehydrogenase, glutamate decarboxylase, activated ATPase, ChE, AChE, GABA transaminase, etc. Administration of thyroxine to deficient animals prior to the critical phases of growth can ameliorate some adverse effects on turnover or on cytodifferentiation; when administered to immature animals this hormone accelerates development to an adult type of behavior.[281] The exact site of hormonal action is still an enigma; although it has been speculated that some hormones directly alter genome expression, their role in turnover could also be secondary to their diverse effects on rates of transport, size of amino acid pools, carbohydrate metabolism, and electrolyte balance.[1t,282] As noted, thyroxine can uncouple energy derived from mitochondria and thus affect the level of ATP required for production of activated acyl-tRNA's.[1t] Neonatal hypothyroidism resulted in a lower level of nucleic acids in brain,[280] but during the postnatal period an administration of thyroxine led to an increase in their level in combination with other hormones. The synergism frequently noted on administration of a mixture of hormones (thyroxine and growth hormone) implies a complex endocrine interaction on mechanisms of turnover. Newborns treated with growth hormone manifest a neuronal hyperplasia with an increased level of DNA.[280] Thyroxine and its analogues cause mitochondrial swelling *in vitro* accompanied by an increased incorporation of amino acid precursors.[235] Steroid hormones (progesterone and deoxycorticosterone) and several cations (Ca^{2+}, Zn^{2+}, Cu^{2+}, Ag^{2+}) markedly inhibit incorporation associated with variable degrees of swelling. The unique role of hormones and effects on mitochondrial proteins are unknown, but several groups have suggested the involvement of unique protein components that subsequently initiate change in turnover.[235] The role of anabolic steroids and ACTH on turnover has attracted considerable interest, although the effects are highly variable, depending on the tissue and the stage of development. Corticosterone is reported to inhibit brain growth and to depress the production of nucleic acid.[1t] Takahashi et al.[2m] did not observe any change in brain size in hypophysectomized rats aged 6 weeks, but noted some increase in Phe-^{14}C incorporation when administrated subcutaneously to animals in contrast to sham-operated littermates.

The stimulation of protein synthesis, most specifically ovalbumin, by administration of steroids to the chick provides a system for analysis of hormonal affects on macromolecular synthesis, where large amounts of specific protein are synthesized. Studies by Palmiter et al.[283] show that administration of the extradiol-17β to newborn chicks for 10 days results in the formation of a new cell type known as tubular gland cells, which specializes in the synthesis of ovalbumin and lysozyme. Secondary administration of estrogen or progesterone stimulates synthesis *in vivo* or synthesis in cultured explants. During a 30-hr culture period the rate of ovalbumin synthesis declined relative to total protein synthesis with a half-life of about 14 hr. Addition of actinomycin D inhibits uridine-^{3}H incorporation into RNA by 95%, but has little effect on ovalbumin synthesis; such findings suggest that there is no effect by this steroid on rRNA synthesis required for the induction of secretory protein synthesis.

1. Turnover of Hormones

In many cases peptidyl hormones are derived from precursor proteins that are subject to the normal patterns of synthesis and breakdown (turnover) and thus can ultimately affect rates of body growth and development.

Sachs[1p] has shown a considerable lag in the appearance of labeled vasopressin in granules of the hypothalamic tract following pulse-labeling experiments with Cys-^{35}S placed directly in the third ventricle. These results are consistent with the formation of an unidentified precursor macromolecule in the nerve cell body, followed by cleavage to yield the active octopeptide and its associated transport protein, neurophysin. Inhibition by puromycin points to a ribosomal site for synthesis; administration of labeled puromycin prevented formation of the hormone, but radioactivity appeared in the hypothalamic granules when the inhibitor was given 90 min after the pulse dose. The mechanisms by which releasing or inhibitory factors control the production and turnover of peptidyl hormones is unclear, except that such factors are themselves small peptides and are probably formed and inactivated by proteolytic enzymes. Physiologically active octopeptides once released into the plasma have a comparatively short half-life; it was estimated by Ginsberg[284] to be in the range of 1–10 min, demonstrating the potency of such hydrolases (Chapter 2, this volume).

D. Compartmentation

The evidence is convincing that in adult brain the free amino acids are present in several compartments. These compartments probably have an important influence on the turnover of proteins (the level and distribution of precursor influencing product formation) and also on the measurement of turnover (the true specific activity of the precursor in the compartment of synthesis may be different from the average specific activity of all compartments). The change in the free amino acid compartment with development similarly may have important effects on the developmental changes in turnover and the measurement of such changes. The appearance of label in glutamate compared to glutamine (from glucose or other precursors) demonstrated the existence of a small and a large compartment,[1j] and others have proposed models with additional intermediate sized compartments.[285] The absence of compartmentation phenomena in young cats[1j] would support the concept of Patel and Balazs[285] that compartments have anatomical as well as functional significance. They attributed the changed glutamate-glutamine ratios during development to the formation of neuronal processes associated with the intermediate compartments (nerve terminals) and large compartments (terminal dendrites); the small compartment was considered to represent glial and nerve cell bodies. Maximum heterogeneity of glutamate-glutamine ratios occurred at 9–21 days in the rat concomitantly with the proliferation of neuronal dendrites and deposition of myelin. Berl[286] proposed that developmental changes were associated more with change in the glutamate relative to the glutamine, and this has been subsequently confirmed in the studies of van den Berg et al.[287] Autoradiographic findings substantiate the lack of a well-differentiated structure in young animals, whose patterns are diffuse compared to those of adults.[10,75] In the studies of Patel and Balazs,[285] leucine was used as precursor since this is rapidly metabolized to yield tricarboxylic intermediates, which can be channeled into glutamate formation; only 30% of the radioactivity is recovered in leucine within 20 min of a subcutaneous injection, demonstrating the problems associated with rapid conversion of precursors used for turnover studies. Compartmentation may vary in different regions of brain and is subject to alteration during development under the influence of hormones.[281] Friede[6] has noted in his monograph a considerable phylogenetic difference for a number of enzymes and nerve tissue components, often in the direction of a caudal-rostal gradient.

E. Alterations in Protein Metabolism

Several chapters in the Handbook attest to the great significance placed on alterations in cerebral proteins caused by excitation,[1u] inhibition,[1v] drugs,[1x] hormones, undernutrition during development,[1t] or pathological conditions.[1q] The literature up to 1964 was summarized and reviewed by Lajtha[3] and is also the subject of several symposia and monographs.[15,18] Despite this wealth of information, our knowledge of the mechanisms involved, especially in relation to turnover, is rudimentary. All alterations in protein metabolism must conform to the limitations imposed by the known synthetic and degradative capacities of the nervous system.

Alterations as a result of training appear to have short, intermediate, and long-term consequences. The suggestion that those alterations involve formation and deposition of macromolecules (RNA or protein) must be compatible with the known lability of such components as demonstrated in turnover studies. This paradox could be resolved if lability took the form of a sequential replacement of multicomponent complexes such that their three-dimensional structure was unaffected.

The experiments of Lajtha and Toth,[314] and more recently Vrba and Cannon[2f] provide little evidence for the presence of any significant quantity of "metabolically inert proteins." Since no significant change in brain mass occurs over relatively long periods, the net deposition of material (superimposed on normal turnover) requires careful evaluation. During short periods of training Hydén and Lange,[253] and Ungar[15] have reported increase in S-100 acidic protein, or polypeptides. The number of cells, or macromolecules within a cell, that manifests long-lasting changes is unknown, but based on turnover studies this is very small. Brain contains 10^9 cells (10% neurons) and this in relation to the number of macromolecules could leave scope for changes beyond our current methods of resolution. The gross change in a specific component in brain as a result of training, if sustained over a long period, is clearly untenable in relation to known limitations of turnover, even more so if such changes were restricted to only a few cell types, since this would correspond to an enormous change in cell mass. Kety[29] has proposed a more dynamic relationship for "memory" involving a complex feedback mechanism mediated by neurotransmitters.

The ability of organ cells to replicate identical daughter cells, or the induction of new forms of enzymatic activity, or the production of antibodies, has led Roberts and Flexner[288] to propose the existence of a "genetic-type memory." If applied to alterations in protein metabolism, this postulate implies an induced transcriptional change resulting in a permanent metabolic format. The absence of marked changes in nuclear DNA, or its associated polymerases in adult brain (Fig. 12), would seem to indicate an important role for the RNA and ribosomal entities.[289] It is of interest to turnover that studies in bacteria show that alteration of breakdown processes is frequently accompanied by alterations in tRNA (Section VI, A). Information to date on change in genome components or levels of protein components themselves with altered functional states has failed to yield data sufficient to formulate an acceptable hypothesis. Inborn errors affecting protein metabolism with consequences to development and behavior are in most cases genetically determined. The nature of the enzyme defects, or other changes in protein metabolism were considered in a previous review[3] and are summarized elsewhere in the Handbook.[1]

Methods for investigating alterations are frequently similar to those for evaluating turnover (Section II). The two principal approaches are methods *III, a, b*, the measurement of a net change (*de novo* synthesis or protein loss) following induction or physiological stress and methods *III, c, d*, the behavioral deficits following the use of inhibitors. The same assumptions must be involved in a critical evaluation of such results; namely, nature of the component(s) involved, the effects of inhibitors on degradative as well as synthetic changes, etc. (Section II, A). This area is covered in greater depth in a number of excellent reviews dealing with memory formation, its consolidation or disruption.[288,289] Formation of memory and its definition must still be regarded as an enigma in terms of biochemical change; it is nevertheless an expression of the great plasticity of the nervous system to recognize, or recall, or respond to different afferent sensory inputs.

The pattern of RNA synthesis in rabbit brain,[290] incorporation of Lys-^{14}C in guinea pig auditory system,[291] change in the visual system, changes in nuclear RNA of catfish brain induced by olfaction,[292] incorporation into synoptic nuclei as a result of osmotic stress with change in synthesis of hypothalamic hormones[1p,293] illustrate the association of protein alteration with function. The visual system and its response to light has received intensive study stemming from the original observations of Nissl in 1892[294] showing protein changes (Nissl substance) in retinal ganglion cells on prolonged stimulation.[295,296] Subsequent studies by several groups with enucleated fetal animals, or removal of an eye, showed transneural atrophy in the contralateral cortical regions receiving afferent pulses. Such preparations provide excellent controls since avian brain is without lateral connections between hemispheres.[297–299] Recent studies by Bondy and Margolis[300] with similar avian preparations or with unilateral deprivation by suturing one eyelid show a significant fall in the incorporation of protein and nucleic acid precursors in the contralateral cerebrums and optic lobes. The rapid change in the incorporation of uracil into RNA might indicate that effects of Leu incorporation are secondary to an altered level of transcriptional factors.

Visual deprivation in newborns can result in severe atrophy of retinal cells and optic nerves and alterations in the incorporation of nucleic acid and protein precursors.[299,305] Studies of visual deprivation in adults have led to highly variable results: Altman and Das[75] failed to detect autoradiographic differences and Metzger et al.[203] failed to detect differences in incorporation of tritium into subcellular fractions (microsomes, nuclei, mitochondria, etc.) in using "split-brain monkeys" conditioned to shock or light stimuli. In similar preparations, Talwar et al.[301] reported increased incorporation on exposure to flashing lights. Rose[302] observed an increased incorporation into protein of chicken brain subjected to "imprinting" or dark adaptation and then exposed to light.

The studies mentioned above imply that afferent impulses can influence protein metabolism, as further illustrated by the many studies on denervation of muscle or incorporation studies in vitro using the lateral geniculate nucleus (monkeys) following optic lobe section.[306–308] Electrical pulses may have diverse effects, depending on the preparation studied in cerebral slices. Orrego and Lipmann[100] observed an inhibition of incorporation, but in squid axons Fisher et al.[141,142] observed an increase when incubated in the presence of a mixture of amino acids. In the case of slices the effect may be secondary to exhaustion of labile phosphates and alterations in RNA. Recent studies by Burnel et al.[303] show some remarkable light-dark adaptations in Limulus with incorporation decreased as a result of light exposure in the primary visual cells (but not adjacent areas); the authors postulate that the heavy demands made by visual responses resulting in competition for available ATP energy by the Na^+ pump mechanism at the expense of synthetic mechanism. An important consideration to the light-dark cycle is the question of diurnal variation. Remarkable changes in brain function occur during sleep but the accompanying metabolic patterns are not well defined. In other organs, notably liver, tryptophan pyrrolase, tyrosine, α-ketoglutarate transaminase, and albumin are subject to marked circadian rhythm based on enzyme activity measurements or incorporation of Leu-^{14}C into protein.[1,304]

VII. LINKAGE OF SYNTHESIS TO DEGRADATION

Schoenheimer and others[8] advanced the concept of protein turnover some 30 years ago, but despite considerable knowledge concerning protein synthesis only primitive schemes can be advanced for mechanisms linking protein formation and its degradation. Any such scheme pertaining to turnover in brain must satisfy a number of criteria: (a) substrate specificity—materials formed more rapidly (short half-life) must be broken down at equal rates to preserve tissue composition; (b) alteration—a parallel change in rate of degradation when synthesis is altered and vice versa; (c) mechanisms of turnover must conform to the abrupt cessation of cellular proliferation during development (i.e., stable level of DNA in adults). The levels of nucleic acid of DNA and RNA increase only during the first and second ontogenetic studies as observed in *in vivo*[309-310] and *in vitro*[1c] (Fig. 12). Several recent studies have attempted to define some of the factors linking synthesis to degradation in terms of change in ribosomes or ribosomal RNA, or in the level of aminoacyl-*t*RNA's as noted in Fig. 1.[34,311-313] Formation of protein requires a very large number of accessory factors and enzymatic steps, each one being a potential rate-limiting factor to turnover. The scheme depicted in Fig. 1 is based largely on work in bacterial cells and for reasons already cited, extrapolation to mammalian cells could be misleading.

A. Ribosomal RNA and Aminoacyl-*t*RNA

Bacteria in exponential growth do not appreciably degrade protein[26,30-34] but when deprived of nitrogen, a carbon source, or a required amino acid, degradation is increased several-fold.[33] Conditions of starvation or "stepdown" are known to drastically reduce net RNA synthesis, whereas conditions favoring exponential growth reduce breakdown.[34] A similar correlation was noted by A. Goldberg[306] for RNA synthesis and protein breakdown in muscle: rapid growth (as a result of induced work or administration of growth hormone) gave increased RNA synthesis and decreased protein breakdown, while atrophy resulted in opposite changes. In cultured liver cells Bolcsfoldi *et al.*[311] also observed a decreased rate of RNA synthesis on deprivation of Leu and Glu. About 80% of total cellular RNA is ribosomal and overall synthesis in a cell depends on the content of ribosomes and their polymeric state.[4e] Henshaw *et al.*[312] in a recent study observed a marked fall in ribosomal activity in liver and muscle of fasted rats. They concluded on the basis of determining rates of synthesis based on a unit of ribosomal RNA that the depression during fasting was due to a reduction in ribosomes and polymeric forms, and the ability to initiate peptide attachment. Brain and testis, and some tumors, were distinguished by the ability to sustain synthesis (based on unit *r*RNA) during starvation, but no explanation was presented. The inference could be that regulation in certain nonregenerative tissues is determined by transport factors, and intracellular pools of amino acids or aminoacyl-*t*RNA's required for synthesis. This consideration is particularly valid since a number of studies clearly show that levels of aminoacyl-*t*RNA play a role in regulating synthesis of *r*RNA in bacteria, and in protein breakdown.[34,315] In microorganisms and possibly in mammalian cells, the increased breakdown during starvation may represent a physiological response needed to prime the intracellular pools of amino acids required for synthesis of proteins essential for survival. If this is the case, there are a number of unexplained phenomena: (a) Which proteins are recognized for degradation? (b) How are proteases activated? (c) Does interference with synthesis interfere with formation of degradative or synthetic enzymes themselves required for turnover? (d) The questions related to energy dependence. It is implausible that during starvation cells sustain themselves for indefinite periods by destroying their proteins in order to create new protein species. This could occur only to a limited extent for unwanted proteins, or less essential structural materials. Although proteolytic enzymes can rapidly degrade denatured materials, or abnormal proteins prepared with amino acid analogues or synthesized in the presence of low concentrations of puromycin,[34] the manner in which normal cell constituents are selected for degradation under conditions of starvation or "step-down" in bacteria is unclear. Proteins may be sacrificed to supply precursors for biosynthesis of essential enzymes, as suggested by depression of enzymes promoting synthesis of Val and Ile in bacteria subject to "step-down" conditions.[34] Depression of a similar nature has been shown to be regulated by the level of corresponding charged tRNA's. A. Goldberg[34] has suggested that the level of aminoacyl-tRNA's may serve to regulate the rate of breakdown. When the level of charged tRNA's is low, this will stimulate breakdown of proteolytic enzymes present in the cell to form the relevant amino acid precursor. This is supported by the finding that agents blocking transfer of amino acids from charged tRNA to the growing polypeptide chain (chloramphenicol) inhibit both synthesis and breakdown. In conditions of exponential growth when charged tRNA's are not limiting, chloramphenicol is without effect. Studies with bacteria demonstrate that degradation can be uncoupled from mechanisms of synthesis, but comparable studies in mammalian cells are unavailable. Piha et al.[315] reported that most aminoacyl-tRNA transferases decreased during the postnatal period up to 14 days of age by about 30%. The level then increased to that seen at birth and remained unchanged at later ages. It is premature to correlate altered levels of transferases with altered rates of degradation in mammalian tissues but studies may provide some clues on the rapid increase in protein synthesis seen during myelination.

Altered catabolic rates on addition of DNP or CN^- seen in bacterial and mammalian systems imply the requirement of energy (see Chapter 2, this volume) although this may be linked to a secondary effect such as transport of protein to sites of destruction (see Fig. 1).

B. Polymerases

Enzymes responsible for the synthesis of DNA and RNA have attracted considerable interest in relation to the mechanisms controlling synthesis of proteins and their breakdown. DNA polymerase (E.C.1.7.7.7., DNA-dependent) exhibits an age-related change in rats, reaching a maximum at 6–10 days in postnatal periods.[1c,316,317] The normal turnover of DNA in animal tissue is a subject of some dispute and is of considerable importance to turnover, since the quantity of mRNA available from nuclear sources (as transcribed by DNA) is disturbingly small—1% in liver and probably about the same in other nucleated cells.[4e] Recent studies by Bell[318] in embryonic muscle tissue indicate, contrary to previous views, the presence of a small 7S "informational" DNA closely associated with protein, which could act as a template for mRNA synthesis in the cytoplasm. Church and Consigli[319] have reported fragmentation of DNA in cloned tumor cells of rat pituitary, which could account for the presence of this *l*-DNA species. The smaller species may be derived from a precursor material in the nucleus known as *N*-some. The presence of a fragmented DNA in the cytoplasm of brain cells capable of acting as a template for *m*RNA synthesis could be important to protein turnover in cells lacking the ability for division. It could be speculated that cytoplasmic DNA would serve to amplify genome expression during the exceptionally high demands of ontogenesis during development.

In contrast to the constant level of DNA in adult cells, the RNA species are subject to constant turnover throughout life, but reach maximal levels in the developing animal coincidental with DNA.[320] RNA polymerase (DNA-dependent) is maximal at birth but decreases with age up to about 30 days in the rat brain and is minimal in the adult.[1c] In bacteria, RNA polymerase is associated with a factor termed "sigma," which is essential for turnover since removal from the enzyme core results in inactivation.[321] Animal cells and some viruses contain a second DNA polymerase which can use RNA or/and DNA as the template.[324,325] This enzyme appears to be characteristic of leukemic white blood cells and is absent in normal white blood cells. DNA polymerase (RNA or DNA-dependent) occurs in normal proliferating cells from human, chicken, mouse, and rat embryos. The activity decreases as the embryos grow older; brain and muscle tissue lose the activity early, and kidney and liver tissue lose the activity last. The presence of this enzyme in young but not adult brain may be correlated with the necessary changes associated with protein turnover during development.

C. Initiation and Release of Peptidyl Chains

Alteration of factors affecting initiation and release of proteins from ribosomes may themselves act as rate-limiting steps to turnover. Multiple functions are ascribed to ribosomes; in addition to recognition of the initiation site on *m*RNA, they provide sites for binding of amino-acyl *t*RNA's, initiation factors, and transfer factors, and they interact with other ribosomal subunits in a precise and ordered fashion.[326,336] Current concepts involve movement of the ribosomes along the *m*RNA during translocation in association with a peptidyl transferase reaction. In bacteria, the major ribosomal species are 30*S* and 50*S* particles, which combine in an unknown manner to form a 70*S* unit. In mammalian cells, the equivalent units are 40*S* and 60*S* which combine to form an 80*S* unit.[327] Among the possible sites available for the control of synthesis in bacteria are the initiation factors termed F 1-3[328] involved in the attachment of fMet-*t*RNA in the presence of an *m*RNA having at or near its 5 end the codons AUG or GUG.[328] Mechanisms of initiation in mammalian cells are more complex than in bacterial systems. Studies with washed reticulocyte ribosomes (AUG-dependent) show attachment of both fMet-*t*RNA and the nonformylated derivative requiring two different initiation factors.[329] The next step in synthesis involves a combination of the two ribosomal units in the presence of GTP and two soluble protein factors, Ts and Tu. Translocation or movement of the growing peptide chain from the initiation (P) site to the acceptor (A) site requires an additional soluble G factor; chain termination requires the participation of release factors R-1 and R-2 in association with the hydrolytic enzymes, its properties similar to aminopeptidases of acylases.[2e,323,332,333] Finally, the maintenance of turnover requires dissociation of the ribosomal complex with recyclization of the smaller ribosomal species for purposes of synthesis; in bacterial systems dissociation is facilitated by a factor S.[330,331] Among the many unexplained facts in mechanisms of synthesis is the sudden sensitivity of the ester bond linking the peptidyl chain with the ribosome to hydrolytic cleavage and regeneration of the terminal codon CCA.[334] In the presence of low concentrations of puromycin, abnormal proteins are formed by early chain termination, since this inhibitor has properties akin to acyl-*t*RNA. Chain length is determined by the *m*RNA, and on completion of the genetic information, in the presence of the appropriate release factors, hydrolytic enzymes are activated for purposes of cleaving the ester bonds. Involvement of degradative enzymes at this critical stage suggests an important role in mechanisms of turnover.

The complex mechanism of protein synthesis so brilliantly elucidated in recent times thus presents many control points, some of which may take part in a feedback control mechanism connecting rates of synthesis to rates of breakdown. Our knowledge of the mechanisms of breakdown is too meager to recognize such control steps, but the likelihood exists that feedback control from breakdown to synthesis and from synthesis to breakdown operates in a highly specific way (substrate, region, stimulus, development, etc.) and at several points, to regulate the turnover of protein in a precise manner.

VIII. REFERENCES

Reviews, Monographs and Symposia

1. *Handbook of Neurochemistry* (A. Lajtha, ed.), Plenum Press, New York (1969–1971).
   Vol. 1. *Chemical Architecture of the Nervous System*
   (a) Chap. 3, W. A. Himwich and H. Agrawal (amino acid pools). (b) Chap. 5, S. Bogoch (proteins) and Chap. 11, E. G. Brunngraber (glycoproteins). Chap. 6, B. W. Moore (S-100). (c) Chap. 7, D. A. Rappoport, R. R. Fritz, and J. L. Myers (nucleic acids).
   Vol. 2. *Structural Neurochemistry*
   (d) Chap. 3, H. Davson (cerebrospinal fluid). (e) Chap. 5, G. Levi (spinal cord). (f) Chap. 6, K. A. C. Elliott and Chap. 7, J. A. Harvey and H. McIlwain (use of brain slices). (g) Chap. 8, J. Altman (DNA metabolism). (h) Chap. 9, H. Koenig, (lysosomes). (i) Chap. 16, G. Porcellati (peripheral nerve) and Chap. 17, R. Koenig (peripheral nerve-RNA). (j) Chap. 19, S. Berl and D. D. Clarke (compartmentation). (k) Chap. 21, B. Droz (autoradiography). (l) Chap. 22, C. W. M. Adams (histochemistry).
   Vol. 3. *Metabolic Reactions in the Nervous System*
   (m) Chap. 5, N. Marks (peptide hydrolases). (n) Chap. 7, G. Guroff and W. Lovenberg; Chap. 8, M. K. Gaitonde; Chap. 9, C. F. Baxter and Chap. 10, C. J. Van de Berg (amino acids).

Vol. 4. *Control Mechanisms in the Nervous System*
(o) Chap. 3, E. Costa and N. H. Neff (amine turnover). (p) Chap. 17, H. Sachs and Chap. 25, R. J. Wurtman (neurosecretion).
Vol. 5. *Metabolic Turnover in the Nervous System*
(q) Chap. 1, S. Roberts (protein synthesis), and Chap. 2, N. Marks and A. Lajtha (protein breakdown). (r) Chap. 9, S.-C. Cheng (tricarboxylic cycle). (s) Chap. 13, D. A. Rappoport, R. R. Fritz, and S. Yamagani (development). (t) Chap. 17, L. Sokoloff (thyroid).
Vol. 6. *Alterations of Chemical Equilibrium in the Nervous System*
(u) Chap. 2, G. P. Talwar and U. B. Singh (excitation). (v) Chap. 6, B. W. Agranoff (memory). (w) Chap. 19, G. Porcellati (demyelination). (x) Chap. 20, D. H. Clouet (narcotics).
Vol. 7. *Pathological Chemistry of the Nervous System*
(y) Chap. 6, E. R. Einstein (basic proteins).

2. *Protein Metabolism of the Nervous System* (A. Lajtha, ed.), Plenum Press, New York (1970). (a) Chap. 1, S. Roberts, C. E. Zomzely, and S. C. Bondy (protein synthesis). (b) Chap. 2, N. Marks and A. Lajtha (breakdown). (c) Chap. 4, B. Droz and H. L. Koenig (localization). (d) Chap. 5, M. R. V. Murthy (development). (e) Chap. 8, B. D'Monte, N. Marks, R. K. Datta and A. Lajtha (development). (f) Chap. 9, R. Vrba and W. Cannon (turnover). (g) Chap. 10, D. Richter (turnover and function). (h) Chap. 11, E. Koenig (peripheral nerve). (i) Chap. 12, L. Austin, I. G. Morgan and J. J. Bray (peripheral nerve). (j) Chap. 13, S. Ochs (axoplasmic transport). (k) Chap. 14, F. Lipmann (slices). (l) Chap. 16, S. E. Geel and P. S. Timaras (hormones). (m) Chap. 17, S. Takahashi, N. W. Penn, A. Lajtha and M. Reiss (hormones). (n) Chap. 18, L. Sokoloff (hormones). (o) Chap. 31, G. Porcellati (Wallerian degeneration). (p) Chap. 34, E. R. Einstein and L-P Chao (EAE).

3. A. Lajtha, Protein metabolism of the nervous system, *Int. Rev. Neurobiol.* **6**:1–98 (1964) and Alterations and pathology of cerebral protein metabolism, *Int. Rev. Neurobiol.* **7**:1–40 (1964).

4. *Mammalian Protein Metabolism* (H. N. Munro and J. B. Allison, eds.), Academic Press, New York (1964–1970).
Vol. 1
(a) Chap. 1, H. N. Munro (historical concepts of turnover). (b) Chap. 7, A. Neuberger and F. F. Richards (turnover in whole animals). (c) Chap. 11, J. B. Allison and J. W. C. Bird (nitrogen excretion).
Vol. 3
(d) Chap. 26, S. A. Miller (development).
Vol. 4
(e) H. N. Munro (mechanisms of synthesis). (f) Chap. 31, F. T. Kenney (hormonal regulation). (g) Chap. 32, R. T. Schimke (regulation of breakdown). (h) Chap. 34, H. N. Munro (pools). (i) Chap. 36, R. J. Wurtman (diurnal rhythms).

5. S. S. Oja, Studies on protein metabolism in developing rat brain. *Ann. Acad. Sci. Fenn.* **131** (5A) 7–78 (1967).

6. R. Friede, *Topographic Brain Chemistry.* Academic Press, New York (1966).

7. *Cellular Dynamics of the Neuron* (S. H. Barondes, ed.), Academic Press, New York (1969). (a) B. Droz and H. L. Koenig (turnover in axons). (c) A. Edström (peripheral nerve-RNA). (d) H. Hydén and P. W. Lange (S-100 behavior). (e) S. H. Barondes (sites of synthesis).

8. R. Schoenheimer, *The Dynamic State of Body Constituents.* Harvard University Press, Cambridge, Mass.

9. R. T. Schimke and D. Doyle, Control of enzyme levels in animal tissues, *Ann. Rev. Biochem.* **39**:929–976 (1970).

10. D. Richter, Factors influencing the protein metabolism of the brain, *Brit. Med. Bull.* **21**:76–80 (1965) and in *Neurochemistry*, 2nd ed., pp. 276–287, C. C. Thomas, Springfield, Ill. (1962).

11. *Macromolecules and the Function of the Neuron*, Z. Lodin and S. P. R. Rose, eds. (a) B. Schultze and P. Kleihaus (turnover *in vivo*) p. 42. (b) A. Edstrom and J.-E. Edstrom (spinal cord RNA) p. 103. (c) E. Koenig (peripheral nerve), p. 121. (d) B. Jakoubek and E. Gutmann (turnover in spinal cord). (e) S. P. R. Rose (visual changes).

12. *Advances in Biochemical Psychopharmacology* (E. Costa and P. Greengard, eds), 1969–1970. *Vols. 1, 2*, H. R. Mahler (turnover in synapses), p. 27. (b) M. W. Gordon and G. C. Deanin (mitochondria), p. 165. *Vol. 2*, (c) E. Costa (turnover models), p. 169. (d) P. F. Davison (neurofilaments). (e) E. Koenig (peripheral nerve). (f) H. Hydén and P. W. Lange (behavior), *Vol. 3*. (g) P. Greengard and J. F. Kuo (phosphorylation of histones-cyclic AMP).

13. *Developmental Neurobiology* (W. A. Himwich, ed.). (a) J. Dobbing (nutrition), p. 241. (b) H. C. Agrawal and W. A. Himwich (amino acids).

14. G. Ungar, *Excitation*. C. C. Thomas, Springfield, Ill. (1963).

15. Symposia. *Molecular Mechanisms in Memory and Learning.* (G. Ungar, ed.) and *Biochemistry of Brain and Behavior* (R. E. Bowman and S. P. Datta, eds.), Plenum Press, New York (1970).

16. Symposia. (a) Biochemical mechanisms involved in regeneration, *Fed. Proc.* **29**:1429–1460, P. J. Fitzgerald (mechanisms), V. G. Allfrey (genes); (b) Nutrition and cell development, *Fed. Proc.* **29**:1489–1521. (a) S. A. Miller (neonatal aspects), M. Winnick (nerve-cell growth), and (c) *Regeneration and Related Problems* (V. Kiortsis and H. A. L. Thampusch, ed.), North-Holland, Amsterdam (1964).

17. H. McIlwain, *Biochemistry of the Central Nervous System.* Little Brown, Boston (1966).

18. S. Bogoch, *The Biochemistry of Memory*, Oxford University Press (1968).

19. C. W. M. Adams, ed., *Neurohistochemistry*, Elsevier, Amsterdam (1965).

20. *Methods in the Neurosciences*, Vol. 1 (N. Marks and R. Rodnight, eds.), Plenum Press, New York (1972). (a) B. I. Roots and P. V. Johnston (neurons and glia). (b) W. Norton (neurons and glia-bulk separation). (c) B. S. McEwen and R. E. Zigmond (nuclei). (d) S. R. Cohen (tissue spaces). (e) M. Spohn and A. N. Davison (myelin). (f) M. Shelanski (neurotubules). (g) L. Eng (proteolipids).

21. D. B. Roodyn and D. Wilke, *The Biogenesis of Mitochondria.* Methuen, London (1968).

22. H. Hydén, Protein metabolism in the nerve cell during growth and function, *Acta Physiol. Scand. Suppl.* **17**:5–136 (1943).

23. H. Waelsch and H. Weil-Malherbe, Neurochemistry and psychiatry, *in Psychiatrie der Gegenwart, Forschung und Praxis*, **1B**:1–96 (1964).

24. H. Waelsch and A. Lajtha, Protein metabolism in the nervous system, *Physiol. Revs.* **41**:709–736 (1961).

25. *Problems of the Biochemistry of the Nervous System* (A. V. Palladin, ed.), Pergamon Press, 1964, pp. 3–26 (turnover).

26. Symposia, Endogenous metabolism with special reference to bacteria, *Ann. N.Y. Acad. Sci.* **102**:515–793. (a) R. T. Schimke, M. B. Brown, and E. T. Smallman (arginase), p. 587. (b) J. Mandelstam (turnover), p. 621.

27. Biomathematics: (a) J. M. Reiner, *The Organism as an Adaptive Control System.* Prentice-Hall, New York (1968). (b) D. S. Riggs, *Control Theory and Physics of Feedback Mechanisms.* Williams and Wilkins, Baltimore (1970). (c) E. Wolstenholme and J. Knight, *Homeostatic Regulators.* J. A. Churchill, London (1969). (d) J. M. Smith, *Mathematical Ideas in Biology.* Cambridge University Press, 1968. (e) Symposium, Protein levels and turnover, *in Protides of the Biological Fluids* (E. H. Peeters, ed.), pp. 211–245, Elsevier, Amsterdam (1967).

28. *Neuroscience Research Bulletin*, Symposia on axoplasmic flow **5**: 309–416 (1967), **6**: 115–216
    (1968), Synaptic function **8**: 327–450 (1970). See also Neurosciences Research Symposium
    Summaries, Vols. 1–3, M.I.T. Press, Cambridge, Mass. (1969–1970).
29. *The Neurosciences: First and Second Study Program* (F. O. Schmitt, ed.). The Rockefeller
    Univ. Press, New York (1967, 1970).

General References

30. J. Mandelstam, The intracellular turnover of protein and nucleic acids and its role in bio-
    chemical differentiation, *Bacteriol. Rev.* **24**:284–308 (1960).
31. N. S. Willets, Intracellular protein breakdown in non-growing cells of *E. coli, Biochem. J.*
    103:453–466 (1967).
32. M. J. Pine, Response of intracellular proteolysis to alteration of bacterial protein and
    implications in metabolic regulation, *Bacteriol.* **93**:1527–1533 (1967); and Intracellular
    breakdown in the LI2IO ascites leukemia, *Cancer Res.* **27**:522–525 (1967).
33. K. Nath and A. L. Koch, Protein degradation in *E. coli, J. Biol. Chem.* **245**:2889–2900
    (1970).
34. A. L. Goldberg, A role of amino-*t*RNA in the regulation of protein breakdown in *E. coli,*
    Proc. Nat. Acad. Sci. **68**:362–366 (1971).
35. D. L. Buchanan, Total carbon turnover measured by feeding a uniformly labeled diet,
    Arch. Biochem. Biophys. **94**:500–511 (1961).
36. R. Swick, Measurement of protein turnover in rat liver, *J. Biol. Chem.* **231**:751–764 (1958).
37. A. L. Koch, Evaluation of the rates of biological processes from tracer kinetic data, *J.*
    Theoret. Biol. **3**:283–303 (1962).
38. R. T. Schimke, The importance of both synthesis and degradation in the control of
    arginase levels in rat liver, *J. Biol. Chem.* **239**; 3808–3817 (1964).
39. F. T. Kenny, Induction of tyrosine-α-ketoglutarate transaminase in rat liver, *J. Biol. Chem.*
    237:3495–3498 (1962), and F. T. Kenny, Turnover of rat liver tyrosine transaminase:
    Stabilization after inhibition of protein synthesis, *Science* **156**:525–527 (1967).
40. R. T. Schimke, E. W. Sweeney, and C. M. Berlin, The roles of synthesis and degradation in
    the control of rat liver tryptophan pyrrolase, *J. Biol. Chem.* **240**:322–331; and Studies of the
    stability *in vivo* and *in vitro* of rat liver tryptophan pyrrolase, *J. Biol. Chem.* **240**:4609–4620
    (1965).
41. B. Poole, F. Leighton, and C. deDuve, The synthesis and turnover of rat liver peroxisomes,
    J. Cell. Biol. **41**:536–546 (1969).
42. J. P. Jost, E. A. Khairallah, and H. C. Pitot, Studies on the induction and repression of
    enzymes, *J. Biol. Chem.* **243**:3057–3066 (1968).
43. B. D'Monte, P. Mela, and N. Marks, Turnover of myelin, *Eur. J. Biochem.* (in preparation)
    (1971).
44. J. M. Phang, G. A. M. Finerman, B. Singh, L. E. Rosenberg, and M. Berman, Compart-
    mental analysis of collagen synthesis in fetal rat calvaria, *Biochim. Biophys. Acta* **230**:
    146–159 (1971).
45. D. H. Clouet and A. Neidle, The effect of morphine on the transport and metabolism
    of intracisternally injected Leu in the rat, *J. Neurochem.* **17**:1069–1074 (1970).
46. B. Sadasivudu and A. Lajtha, Metabolism of amino acids in incubated slices of mouse brain,
    J. Neurochem. **17**:1299–1311 (1970).
47. Y. V. Belik and L. S. Krachko, Intensity of methionine ^{35}S incorporation into the nuclear
    and cytoplasmic proteins of cat brain tissue, *UKr. J. Biochem.* **31**:322–329 (1959).
48. R. T. Schimke, R. Gauschow, D. Doyle, and I. M. Arias, Regulation of protein turnover in
    mammalian tissues, *Fed. Proc.* **27**:1223–1230 (1968).

49. I. M. Arias, D. Doyle, and R. T. Schimke, Studies on the synthesis and degradation of proteins of the endoplasmic reticulum of rat liver, *J. Biol. Chem.* **244**:3303–3315 (1969).
50. J. C. Warterlow and J. M. L. Stephen, Diet and turnover of liver and muscle protein, *Clin. Sci.* **35**:287–305 (1968).
51. P. J. Garlick, Turnover rate of muscle protein measured by constant intravenous infusion of ^{14}C-Gly, *Nature* **223**:61–62 (1969).
52. R. W. Swick, A. K. Rexroth, J. L. Strange, The metabolism of mitochondrial proteins: III. The dynamic state of rat liver mitochondria, *J. Biol. Chem.* **243**:3581–3587 (1968).
53. J. B. Balinsky, G. E. Shambaugh, and P. P. Cohen, Glutamate, dehydrogenase biosynthesis in amphibian liver preparations, *J. Biol. Chem.* **245**:128–137 (1970).
54. N. Marks, R. K. Datta, and A. Lajtha, Partial purification of brain arylamidases and aminopeptidases, *J. Biol. Chem.* **243**:2882–2889 (1968).
55. C. Voit, *Z. Biol.* **2**:307 (1866) (see ref 4a).
56. O. Folin, A theory of protein metabolism, *Am. J. Physiol.* **13**:117–138 (1905).
57. H. Borsook and G. L. Keighley, The "continuing" metabolism of nitrogen in animals, *Proc. Roy. Soc. (Lond.). B.* **118**:488–521 (1935).
58. F. Friedberg and D. M. Greenberg, Partition of intravenously administered amino acids in blood and tissues, *J. Biol. Chem.* **168**:411–413 (1947).
59. H. Tarver and L. M. Morse, The release of the sulfur from tissues of rats fed labeled methionone, *J. Biol. Chem.* **173**:53–61 (1948).
60. H. Borsook and C. L. Deasy, The metabolism of proteins and amino acids, *Ann. Rev. Biochim.* **20**:209–226 (1951).
61. F. Friedberg, H. Tarver, and D. M. Greenberg, The distribution pattern of sulfur-labeled methionine in the protein and the free amino acid fraction of tissues after intravenous administration, *J. Biol. Chem.* **190**:39–53 (1948).
62. D. M. Greenberg, F. Friedberg, M. P. Schulman, and T. Winnick, Studies on the mechanism of protein synthesis with radioactive carbon-labeled compounds, *Cold. Spr. Harb. Symp. Quant. Biol.* **13**:113–117 (1948).
63. M. K. Gaitonde and D. Richter, The metabolic activity of the proteins of the brain, *Proc. Roy. Soc.* **145**:83–99 (1956).
64. A. Lajtha, S. Furst, A. Gerstein, and H. Waelsch, Amino acid and protein metabolism of the brain. I. Turnover of free and protein bound lysine in brain and other organs, *J. Neurochem.* **1**:289–300 (1957).
65. A. Niklas, E. Quincke, W. Maurer, and H. Neyen, Messung der Neubildungsraten und biologischen Halbwertzeiten des Eiweisses einzelner Organe und Zellgruppen bei der Ratte, *Biochem. Z.* **330**:1–20 (1958).
66. B. Schultze, W. Oehlert, and W. Maurer, Vergleichende autoradiographische Untersuchung mit ^{3}H, ^{14}C und ^{35}S markierten Aminosäuren zur Grösse des Eiweisstoffwechsels einzelner Gewebe und Zellarten bei Maus, Ratte und Kaninchen, *Beitr. Path. Anat.* **120**:58–84 (1959) and **122**:406–431 (1960).
67. R. S. Piha, R. M. Bergström, A. J. Uusitalo, and S. S. Oja, Studies in the metabolism of brain proteins I and II, *Ann. Med. Exp. Fenn.* **41**:485–497 and 498–515 (1963).
68. K. von Hungen, H. R. Mahler, and W. J. Moore, Turnover of protein and ribonucleic acid in synaptic subcellular fractions from rat brain, *J. Biol. Chem.* **243**:1415–1423 (1968).
69. R. Lim and B. W. Agranoff, Protein metabolism in goldfish brain, *J. Neurochem.* **16**:431–445 (1969).
70. R. C. Thompson and J. E. Ballou, Studies of metabolic turnover with tritium as a tracer. V. The predominately non-dynamic state of body constituents in the rat, *J. Biol. Chem.* **223**:795–809 (1956).
71. A. A. Khan and J. E. Wilson, Studies of turnover in mammalian subcellular particles: brain nuclei, mitochondria and microsomes, *J. Neurochem.* **12**:81–86 (1965).

72. E. Klika and Z. Lodin, The incorporation of ^{35}Met in the meninges and CNS of cats, *Acta Histochem.* **10**:198–209 (1960).
73. D. H. Ford, Changes in brain accumulation of amino acids and adenine associated with changes in the physiological state, in *Brain-Barrier Systems* (A. Lajtha and D. H. Ford, eds.), *Prog. in Brain Res.* **29**:401–415 (1968).
74. F. T. Mérei and F. Gallyas, Quantitative determination of the uptake of Met-^{35}S in different regions of the normal rat brain, *J. Neurochem.* **11**:251–256 and 265–270 (1964).
75. J. Altman and G. D. Das, Autoradiographic and histological studies, *Anat. Rec.* **148**:535–546 (1964), and *J. Comp. Neuro.* **124**:319–336 (1965), **126**:337–390 (1966).
76. R. J. Schain, M. J. Carver, J. H. Cophenhaver, Postnatal changes in protein metabolism of brain, *Pediat Res.* **3**:135–139 (1965) and with N. R. Underdahl, *Science* **156**:984–985 (1967).
77. J. M. Reiner, The study of metabolic turnover rates by means of isotopic tracers. II. Turnover in a simple reaction system, *Arch. Biochem.* **46**:80–90 (1953).
78. S. Furst, A. Lajtha, and H. Waelsch, Amino acid and protein metabolism of the brain—III, *J. Neurochem.* **2**:216–225 (1958).
79. H. Rahmann, Zum Stofftransport im Zentralnervensystem der Vertebraten, *Z. f. Zellforsch.* **66**:878–890 (1965).
80. L. F. Pantchenko, Protein turnover at various levels of the CNS and in the liver in growing and adult animals, *J. Physiol. (U.R.S.S.)* **44**:243–248 (1958).
81. N. Marks, Some neurochemical correlates of axoplasmic flow, *Dis. Nerv. System* **31**:1–13 (1970).
82. N. Marks, R. K. Datta, and A. Lajtha, Distribution of amino acids and exo- and endopeptidases along vertebrate and invertebrate nerves, *J. Neurochem.* **17**:53–63 (1970).
83. I. M. Korr, P. N. Wilkinson, and F. W. Chomock, Axonal delivery of neuroplasmic components to muscle cells, *Science* **155**:343–345 (1967).
84. G. Koya and R. L. Friede, Segmental incorporation of Leu-^{3}H in rat spinal cord, *J. Anat.* **105**:47–57 (1969).
85. G. S. L. Appeltauer and E. E. A. Saa, Incorporation of Lys-^{14}C into spinal roots, spinal ganglia and peripheral nerves of the rat, *Exper. Neurol.* **14**:484–495 (1966).
86. A. Globus, H. D. Lux, and P. Shubert, Somadendritic spread of intracellulary injected tritiated glycine in cat spinal motoneurons, *Brain Res.* **11**:440–445 (1968).
87. E. Edström and J. Sjöstrand, Protein synthesis in the isolated Mauthner or nerve fibre of goldfish, *J. Neurochem.* **16**:67–81 (1969).
88. A. Edström, Amino acid incorporation in isolated Mauthner nerve fibre components, *J. Neurochem.* **13**:315–321 (1966).
89. R. Sammeck and R. O. Brady, Differential metabolism of myelin basic proteins in various regions of the CNS, *Trans. Amer. Soc. Neurochem.* **2**:104 (1971).
90. W. Tourtellotte, On cerebrospinal fluid immunoglobulin-G (IgG) quotients in multiple sclerosis and other diseases (A review and a new formula to estimate the amount of IgG synthesized per day by the CNS), *J. Neurol. Sci.* **10**:279–304 (1970).
91. J. Clausen, J. Matzke, and W. Gerhardt, Agar-gel micro-electrophoresis of proteins in the CSF: normal and pathological findings, *Acta. Neurol. Scand.* **40**:Suppl. **10**:49–56 (1964).
92. C. E. Lumsden, The proteins of the c.s.f. in MS in *Multiple Sclerosis: A Reappraisal* (D. McAlpine, C. E. Lumsden, and E. D. Acheson, eds.), pp. 352–399, and 345–380. E. and S. Livingstone, Edinburgh (1965).
93. G. M. Hochwald, Influx of serum proteins and their concentration in spinal fluid along the neuraxis, *J. Neurol. Sci.* **10**:269–278 (1970).
94. R. W. P. Cutler, G. V. Walters, and J. P. Hammerstad, The origin and turnover rates of c.s.f. albumin and γ-globulin in man, *J. Neurol. Sci.* **10**:259–268 (1970).

95. R. A. Menzies and P. H. Gold, The turnover of mitochondria in a variety of tissues of young adult and aged rats, *J. Biol. Chem.* **246**:2425–2429 (1971).

96. A. Lajtha and N. Marks, Dynamics of protein metabolism of the nervous system, *in The Future of the Brain Sciences* (S. Bogoch, ed.), Plenum Press, New York (1969).

97. H. Koenig and B. Rich, An autoradiographic study of nucleic acid and protein turnover in the mammalian neuraxis, *J. Biophys. Biochem. Cytol.* **4**:785–792 (1958).

98. I. D. Frantz, R. B. Lotfield, and W. W. Miller, Incorporation of ^{14}C from carboxyl-labeled DL-Ala into the proteins of liver slices, *Science* **106**:544–545 (1947).

99. O. Lindan, J. H. Quastel, and S. Sved, Biochemical studies on chlorpromazine. 2. Effects on incorporation into proteins and breakdown of Gly-^{14}C by isolated rat brain cortex, *Canad. J. Biochem.* **35**:1145–1150 (1957).

100. F. Orrego and F. Lipmann, Protein synthesis in brain slices. Effects of electrical stimulation and acidic amino acids, *J. Biol. Chem.* **242**:665–671 (1967).

101. L. C. Mokrasch and P. Manner, Incorporation of ^{14}C-amino acid and ^{14}C-palmitate into proteolipids of rat brains *in vitro*, *J. Neurochem.* **10**:541–547 (1963).

102. J. Folbergrova, Incorporation of labeled amino acids into the proteins of brain cortex slices *in vitro* in the presence of other non-radioactive amino acids, *J. Neurochem.* **13**:553–562 (1966).

103. T. Tursky, J. Krizko, L. Halcak, and M. Brechtlova, Effects of psychopharmacological agents on brain metabolism—I. Effect of imiprimaine and prothiadene upon incorporation of L-Phe into protein and lipids of brain slices, *Biochem. Pharmacol.* **14**:1645–1649 (1965).

104. J. Järnefelt and M. O. Huttunen, Synthesis of protein and nucleic acids in brain cortex slices, *in Regulatory Functions of Biological Membranes* (J. Järnefelt, ed.), *B.B.A. Library* **11**:208–215, Elsevier, Amsterdam (1968).

105. C. Prives and J. H. Quastel, Effect of stimulation in biosynthesis of nucleotides and RNA in brain slices *in vitro*, *Biochim. Biophys. Acta* **182**:285–294 (1969).

106. M. O. Huttonen, Protein and ribonuclei acid metabolism in rat brain cortex slices, Thesis, University of Helsinki, Finland (1969).

107. D. Dunlop, W. Van Elden, and A. Lajtha, Protein synthesis in rat brain slices, *J. Neurochem.* (in preparation).

108. F. Lipmann, Effect of electrical and chemical stimulation on protein synthesis in brain slices, *in Protein Metabolism of the Nervous System* (A. Lajtha, ed.), pp. 305–312, Plenum Press, New York (1970).

109. C. T. Jones and P. Banks, The effect of electrical stimulation on the incorporation of C-[U-^{14}C] valine into protein of chopped tissue from guinea-pig cerebral cortex, *Biochem. J.* **118**:791–800 and 801–812 (1970).

110. Y. Takahashi and Y. Akabane, Protein metabolism of rat brain slices, *Canad. J. Biochem.* **38**:1149–1157 (1960).

111. K. Mase, Y. Takahashi, and K. Ogata, The incorporation of [^{14}C] Glycine into the protein of guinea-pig slices, *J. Neurochem.* **9**:281–288 (1962).

112. C. Blomstrand, Effect of hypoxia on protein metabolism in neuron- and neuroglia cell-enriched fractions from rabbit brain, *Exper. Neurol.* **29**:175–188 (1970).

113. M. Chvapil, J. Hurych, and E. Mirejovská, Effect of long term hypoxia on protein synthesis in granuloma and in some organs in rats, *Proc. Exper. Biol. Med.* **135**:613–617 (1970).

114. A. W. Brown and J. B. Brierley, The nature, distribution and earliest stages of anoxic-ischemic nerve cell damage in rat brain as defined by the optical microscope, *Brit. J. Exp. Pathol.* **159**:87–106 (1968).

115. B. Jakoubek, B. Semiginovsky, M. Kraus, and R. Erdossova, The alteration of protein metabolism of brain cortex induced by anticipation stress and ACTH, *Life Sci.* **9**:1169–1179 (1970).

116. D. M. Kipnis, E. Reiss, and E. Helmreich, Functional heterogeneity of the intracellular amino acid pool in mammalian cells, *Biochim. Biophys. Acta* **51**:519–524 (1961).

117. M. J. Clemens and A. Korner, Amino acid requirement for the growth hormone stimulation of incorporation of precursors into protein and nucleic acid of liver slices, *Biochem. J.* **119**:629–634 (1970).

118. S. Roberts and B. S. Morelos, Regulation of cerebral metabolism of amino acids—IV. *J. Neurochem.* **12**:373–387 (1965).

119. R. C. Hider, E. B. Fern, and D. R. London, Relationship between intracellular amino acids and protein synthesis in the extensor digitorum longus muscle of rats, *Biochem. J.* **114**:171–178 (1969); **121**:817–827 (1971).

120. F. H. Portujal, D. H. Elowyn, and H. Jeffay, Free lysine compartments in rat liver cells, *Biochim. Biophys. Acta* **215**:339–347 (1970).

121. N. A. Peterson and C. M. McKean, The effects of individual amino acids on the incorporation of labeled amino acids into protein by brain homogenates, *J. Neurochem.* **16**:1211–1217 (1969).

122. B. M. Hanking and S. Roberts, Stimulation of protein synthesis *in vitro* by elevated levels of amino acids, *Biochim. Biophys. Acta* **104**:427–438 (1965).

123. J. P. Roscoe, M. D. Eaton, and G. Chin Choy, Inhibition of protein synthesis in Krebs 2 Ascites cells and cell-free systems by Phe and its effect on Leu and Lys in the amino acid pool, *Biochem. J.* **109**:507–514 (1968).

124. L. S. Jefferson and A. Korner, Influence of amino acid supply on ribosomes and protein synthesis of prefused rat liver, *Biochem. J.* **104**:826–832 (1967) and **111**:703–712 (1969).

125. H. C. Agrawal, A. H. Bone, and A. N. Davison, Effect of Phe on protein in the developing rat brain, *Biochem. J.* **117**:325–331 (1970).

126. E. Raghupathy, N. A. Peterson, and C. M. McKean, Effects of phenothiazines on *in vitro* cerebral protein synthesis, **19**:993–1000 (1970).

127. N. Abadom, K. Ahmed, and P. G. Scholefield, Biochemical studies of tofranil, *Canad. J. Biochem. Physiol.* **39**:551 (1961).

128. S. Navon and A. Lajtha, Uptake of morphine in particular fractions from rat brain, *Brain Res.*, **24**:534–536 (1970).

129. J. Wells, Effect of water deprivation on uptake of DL-^{35}Cystine in the hypothalamo-hypophysial system, *Exper. Neurol.* **8**:470–481 (1963).

130. D. H. Clouet and H. Waelsch, Amino acid and protein metabolism of the brain—IX. The effect of an organophosphorous inhibitor on the incorporation of Lys-^{14}C into the proteins of rat brain, *J. Neurochem.* **10**:51–63 (1963).

131. E. Koenig and G. B. Koelle, Mode of regeneration of AChE in cholinergic neurons following irreversible inactivation, *J. Neurochem.* **8**:169–188 (1961).

132. A. Lajtha, Protein metabolism in nerve, *in Chemical Pathology of the Nervous System* (D. Richter, ed.), pp. 268–369, Pergamon Press, Oxford (1961).

133. A. N. Davison, Metabolically inert proteins of the central and peripheral nervous system, muscle and tendon, *Biochem. J.* **78**:272 (1961).

134. B. Grafstein, B. McEwen, and M. L. Shelanski, Axonal transport of neurotubule protein, *Nature* **227**:289–290 (1970).

135. M. Singer and M. M. Salpeter, The transport of His-^{3}H through the Schwann and myelin sheath into the axon, including a re-evaluation of myelin function, *J. Morph.* **120**:281–316 (1966).

136. J. D. Caston and M. Singer, Amino acid uptake and incorporation into macromolecules of peripheral nerve, *J. Neurochem.* **16**:1309–1318 (1969).

137. Y. Takahashi, M. Nomura, and S. Furwawa, *In vitro* incorporation of amino-^{14}C acids into proteins of peripheral nerve during Wallerian degeneration, *J. Neurochem.* **7**:97–102 (1961).

138. D. F. Matheson, Incorporation of Glycine-^{14}C into protein of the adult rat peripheral nerve: effect of inhibition, *J. Neurochem.* **15**:179–185 (1968).

139. D. F. Matheson, Influence of age in the incorporation of Gly-^{14}C into isolated rat nerve segments, *J. Neurochem.* **15**:187–194 (1968), and into chicken nerve, *J. Neurochem.* **16**:215–223 (1969).

140. D. F. Matheson, Some aspects of lipid and protein metabolism in developing rat optic nerves, *Brain Res.* **24**:271–283 (1970).

141. S. Fisher and S. Litvak, The incorporation of microinjected ^{14}C-amino acids into TCA insoluble fractions of the giant axon of the squid, *J. Cell. Physiol.* **70**:69–74 (1967).

142. S. Fischer, M. Cellino, P. Gariglio, I. Tellez-Nagel, Protein and RNA metabolism of squid axons (*Dosidicus gigas*), *J. Gen. Physiol.* **51**:72–80s (1968).

143. M. Luxoro, Incorporation of amino acids labeled with ^{14}C in nerve proteins during activity and recovery, *Nature* **88**:1119–1120 (1960).

144. D. D. Wheeler and L. L. Boyarsky, Influx of glutamic acid in peripheral nerve-characteristics of influx, *J. Neurochem.* **15**:1019–1031 (1968).

145. M. Yamaguchi, T. Yanos, T. Yamaguchi, and A. Lajtha, Amino acid uptake in the peripheral nerve of the rat, *J. Neurobiol.* **1**:419–433 (1970).

146. C. Blomstrand, Thesis. Studies on protein metabolism in neuronal glial-cell enriched fractions from brain tissue, University of Göteborg, 1971.

147. M. Satake and S. Abe, Preparation and characteristics of nerve cell perikarya from rat cerebral cortex, *J. Biochem.* (*Tokyo*) **59**:72–75 (1966).

148. W. T. Norton and S. E. Poduslo, Neuronal soma and whole neuroglia of rat brain. A new isolation technique, *Science* **167**:1144–1145 (1970).

149. O. Z. Sellinger, J. B. Azcurra, D. E. Johnson, W. G. Ohlsson, and Z. Lodin, Independence of protein synthesis and drug uptake in nerve cell bodies and glial cells isolated by a new technique, *Nature* (in press).

150. A. Hamberger, H. A. Hansson, and J. Sjostrand, Surface structure of isolated neuron, *J. Cell. Biol.* **47**:319–331 (1970).

151. D. H. Ford and R. K. Rhines, Accumulation of ^{3}H-lysine in various types of neurons in male rats, *J. Neurol. Sci.* **10**:179–183 (1970).

152. L. Hertz, Neurological localization of K and Na effects on respiration in brain, *J. Neurochem.* **13**:1373–1387 (1966).

153. B. Jakoubek, E. Gutmann, J. Fisher, and A. Babicky, Rate of protein renewal in spinal montoneurons of adolescent and old rats, *J. Neurochem.* **15**:633–641 (1968).

154. C. Blomstrand and A. Hamberger, Protein turnover in cell-enriched fractions from rabbit brain, *J. Neurochem.* **16**:1401–1407 (1969) and *in vitro*, **17**:1187–1197 (1970).

155. D. E. Johnson and O. Z. Sellinger, Protein synthesis in neurons and glial cells of the developing rat brain: an *in vivo* study, *J. Neurochem.* 1971 (in press).

156. A. L. Flanagas and R. E. Bowman, Differential metabolism of RNA in neuronal enriched and glial-enriched fractions of rat cerebrum, *J. Neurochem.* **17**:1237–1245 (1970).

157. C. Blomstrand, A. Hamberger, and T. Yamagihara, Subcellular distribution of radioactivity in neuronal and glial-enriched fractions after incorporation of Leu-^{3}H *in vivo* and *in vitro*, *J. Neurochem.* (in press, 1971).

158. G. R. Dutton and S. Barondes, Microtubular protein: synthesis and metabolism in developing brain, *Science* **166**:1637–1638 (1969).

159. J. A. Burdman, Incorporation *in vivo* of radioactive leucine into neuronal and glial nuclear proteins of rat brain, *J. Neurochem.* **17**:1555–1562 (1970).

160. Y. Takahashi, C. S. Hsü, and S. Honura, Potassium and glutamate effects on protein synthesis in isolated neuroglial cells, Brain Res. 23:284–287 (1970).
161. H. Lotrup-Rein, Protein synthesis in isolated nuclei of nerve and glial cells, J. Neurochem. 19:433–444 (1970).
162. J. A. Burdman and L. I. Journey, Protein synthesis in isolated nuclei from adult rat brain, J. Neurochem. 16:493–500 (1969).
163. K. Hemminki, M. O. Huttonen, and J. Järnefelt, Some properties of brain cell suspensions prepared by a mechanical-enzymic method, Brain Res. 23:23–34 (1970).
164. B. Tiplady and S. P. R. Rose, Amino acid incorporation into protein of neuronal and neuropil fractions in vitro, Biochem. J., 117:65P (1969); J. Neurochem. 18, 549–558 (1971).
165. T. S. Work, J. L. Coote, and M. Ashwell, Biogenesis of mitochondria, Fed. Proc. 27:1174–1179 (1968) and Ann. Rev. Biochem. 39:251–290 (1970).
166. M. Rabinowitz and H. Swift, Mitochondrial nucleic acids and their relation to the biogenesis of mitochondria, Physiol. Revs. 50:376–427 (1970).
167. M. K. Campbell, H. R. Mahler, W. J. Moore, and S. Tewari, Protein synthesis systems from rat brain, Biochemistry 5:1174–1184 (1966).
168. A. Hamberger, N. Gregson, and A. L. Lehninger, The effect of acute exercise on amino acid incorporation into mitochondria of rabbit tissues, Biochim. Biophys. Acta 186:373–383 (1969).
169. M. A. Goldberg, Protein synthesis in isolated rat brain mitochondria and nerve ending, Brain Res., 27:319–328 (1971).
170. R. P. Wagner, Genetics and phenogenetics of mitochondria, Science 163:1026–1031 (1969).
171. M. M. K. Nass, Mitochondrial DNA: advances problems and goals, Science 165:25–35 (1969).
172. D. J. L. Luck, Formation of mitochondria in Neurospora crassa, J. Biol. Chem. 16:483–499 (1963) and J. Cell. Biol. 24:461–470 (1965).
173. M. J. Fletcher and D. R. Sanadi, Turnover of rat liver mitochondria, Biochim. Biophys. Acta 51:356–360 (1961).
174. E. Bailey, C. B. Taylor, and W. Bartley, Turnover of mitochondrial components of normal and essential fatty-acid deficient rats, Biochem. J. 104:1026–1032 (1967).
175. R. W. Swick, A. K. Rexrot, and J. L. Stange, The metabolism of rat liver mitochondria, J. Biol. Chem. 243:3581–3587 (1968).
176. D. S. Beattie, R. E. Basford, and S. B. Kontz, The turnover of the protein components of mitochondria from rat liver, kidney and brain, J. Biol. Chem. 242:4584–4586 (1967).
177. A. Neidle, C. J. van Den Berg, and A. Grynbaum, The heterogeneity of rat brain mitochondria isolated on continuous sucrose gradients, J. Neurochem. 16:225–234 (1969).
178. N. Marks, B. D'Monte, C. Bellman, and A. Lajtha, Protein metabolism in cerebral mitochondria. I. Hydrolytic enzymes and amino acid incorporation into mitochondrial membranes, Brain Res. 18:309–324 (1970).
179. L. Salganicoff and R. E. Koeppe, Subcellular distribution of pyruvate carboxylase, diphosphate pyridine nucleotides and trisphosphopyridine nucleotide isocitrate dehydrogenases, and malate enzyme in rat brain J. Biol. Chem. 243:3416–3420 (1968).
180. F. Hajos and S. Kerpel-Fronius, Electron histochemical observation of succinic dehydrogenase activity in various parts of neurons, Exp. Brain Res. 8:66–78 (1969).
181. J. B. Clark and W. J. Nicklas, The metabolism of rat brain mitochondria, J. Biol. Chem. 245:4724–4731 (1971).
182. S. H. Barondes, On the site of synthesis of the mitochondrial protein of nerve endings, J. Neurochem. 13:721–727 (1966).

183. H. S. Bachelard, Amino acid incorporation into the protein of mitochondrial preparations from cerebral cortex and spinal cord, *Biochem. J.* **100**:131–137 (1966).

184. M. W. Gordon and G. G. Deanin, Protein synthesis by isolated rat brain mitochondria and synaptosomes, *J. Biol. Chem.* **243**:4222–4226 (1968).

185. R. D. Cunningham and W. F. Bridges, Brain and liver mitochondrial protein synthesis: potassium dependent chloramphenicol inhibition, *Biochem. Biophys. Res. Comm.* **38**:99–105 (1970).

186. H. B. Bosmann and B. A. Hemsworth, Intraneural mitochondria incorporation of amino acid, and monosaccharides into macromolecules by isolated synaptosomes and synaptosomal mitochondria, *J. Biol. Chem.* **245**:363–371 (1970).

187. D. Halder, Protein synthesis in mammalian brain mitochondria, *Biochem. Biophys. Res. Comm.* **38**:129–134 (1970), and **42**:899 (1971).

188. M. A. Ashwell and T. S. Work, Contrasting effects of cycloheximide on mitochondrial protein synthesis *in vivo* and *in vitro*, *Biochem. Biophys. Res. Comm.* **32**:1006–1011 (1968).

189. K. B. Freeman, Inhibition of mitochondrial and bacterial protein synthesis by chloramphenicol, *Canad. J. Biochem.* **48**:479–485 (1970).

190. H. R. Mahler, L. R. Jones, and W. J. Moore, Mitochondrial contribution to protein synthesis in cerebral cortex, *Biochem. Biophys. Res. Comm.* **42**:384–389 (1971).

191. G. Brunner and W. Neupert, Turnover of outer and inner membrane proteins of rat liver mitochondria, *FEBS Letters* **1**:153–155 (1968).

192. L. Austin and I. G. Morgan, Incorporation of ^{14}C-labelled leucine into synaptosomes from rat cerebral cortex *in vitro*, *J. Neurochem.* **14**:377–387 (1967).

193. A. Lajtha, S. Furst, and H. Waelsch, The metabolism of the proteins of the brain, *Experienta* **23**:168–172 (1957).

194. E. de Robertis, Molecular biology of synaptic receptors, *Science* **171**:963–971 (1971).

195. S. J. Morris, H. J. Ralston, and E. M. Shooter, Studies on the turnover of mouse brain synaptosomal proteins, *J. Neurochem.* (in press), 1971.

196. H. Cramer, Zur Inkorporation von Phe-^{3}H in Proteine der Circumventrikulären Organe bei Katzen und Meerschweinchen Autoradiographische Untersuchung, *Exp. Brain Res.* **11**:343–359 (1970).

197. I. G. Morgan and L. Austin, Synaptosomal protein synthesis in a cell-free system. (a) *J. Neurochem.* **15**:41–51 (1968): Ion effects and protein synthesis in synaptosomal fraction. (b) *J. Neurobiol.* **2**:155–167 (1969).

198. S. H. Appel, B. W. Festoff, L. Autilio, and A. V. Escueta, Biochemical approaches to the study of synaptic function, *Biological Psych.* **2**:219–233 (1970).

199. L. A. Autilio, S. H. Appel, P. Pettis and Pier-Luigi Gambeti, Biochemical studies of synapses *in vitro*. I. Protein synthesis, *Biochemistry* **7**:2615–2622 (1968).

200. W. Sebald, A. J. Schwab, and Th. Bücher, Cycloheximide resistant amino acid incorporation into mitochondrial protein from *Neurospora crassa*, *FEBS letters* **4**:243–246 (1969).

201. A. V. Escueta and S. H. Appel, Biochemical studies of synapases *in vitro*. II. Potassium transport, *Biochemistry* **8**:725–733 (1969).

202. A. A. Abdel-Latif and C. G. Abood, *In vivo* incorporation of Ser-^{14}C into phospholipids and proteins of the subcellular fractions of developing rat brain, *J. Neurochem.* **13**:1189–1196 (1966).

203. H. P. Metzger, M. Cuenod, A. Grynbaum, and H. Waelsch, The use of tritium oxide as a biosynthetic precursor of macromolecules in brain and liver, *J. Neurochem.* **14**:99–105 (1966), and *Life Sci.* **5**:1115–1120 (1966).

204. H. B. Bosmann and B. A. Hemsworth, Intraneural mitochondria. Incorporation of amino acid and monosaccharides into macromolecules by isolated synaptosomes and synaptosomal mitochondria, *J. Biol. Chem.* **245**:363–371 (1970).

205. M. K. Gordon, K. G. Bench, G. G. Deanin, and M. W. Gordon, Histochemical and biochemical study of synaptic lysosomes, *Nature* **217**:523–527 (1968).

206. R. M. Marchbanks and V. P. Whittaker, The biochemistry of synapses, in *The Biological basis of Medicine* (E. E. Bittar, ed.) **5**:39–76 (1969).

207. L. Guth and W. F. Windle, The enigma of central nervous regeneration, *Exper. Neurol.* (Supp.) **5**:1–43 (1970).

208. V. G. Allfrey, Changes in chromosomal proteins at times of gene regulation, *Fed. Proc.* **29**:1447–1460 (1970 and Biosynthetic reactions in the cell nucleus, in *Aspects of Protein Synthesis* (C. B. Anfinsen, ed.), pp. 247–345, Academic Press, New York (1970).

209. S. Navon and A. Lajtha, The uptake of amino acids by particulate fractions from brain, *Biochim. Biophys. Acta* **173**:518–531 (1969).

210. L. M. J. Shaw and R. C. C. Huang, A description of two procedures which avoid the use of extreme pH conditions for the resolution of components isolated from chromatins prepared from pig cerebellar and pituitary nuclei, *Biochemistry* **9**:4530–4542 (1970).

211. A. Neidle and H. Waelsch, Histones: Species and tissue specificity, Science **145**:1059–1061 (1964).

212. D. F. Scott, R. D. Reynolds, H. C. Pitot and V. R. Potter, Co-induction of the hepatic amino acid transport system and lysosome aminotransferase by theophylline glucagon and dibutyryl-cyclic AMP *in vivo*, *Life Sci.* **9II**:1133–1140 (1970).

213. H. Hydén and B. S. McEwen. A glial protein for the nervous system, *Proc. Nat. Acad. Sci.* **55**:354–358 (1966).

214. D. H. Clouet and D. Richter, The incorporation of Met-^{35}S into proteins of the rat brains, *J. Neurochem.* **3**:219–229 (1959).

215. J. A. Burdman, Incorporation *in vivo* of radioactive Leu into neuronal and glial nuclear proteins of rat brain, *J. Neurochem.* **17**:1555–1562 (1970).

216. R. S. Piha, M. Cuenod, and H. Waelsch, Metabolism of histones of brain and liver, *J. Biol. Chem.* **241**:2397–2404 (1966).

217. H. Lovtrup-Rein, Protein synthesis in isolated nuclei of nerve and glial cells from rat brain, *Brain Res.* **19**:433–444 (1970).

218. J. A. Burdman, K. Haglid, and A. R. David, Protein synthesis in fractions from isolated brain cell nuclei, *J. Neurochem.* **17**:669–676 (1970).

219. J. A. Burdman and C. J. Journey, Protein synthesis in isolated nuclei from adult rat brain, *J. Neurochem.* **16**:493–500 (1969).

220. I. Smart and C. P. Leblond, Evidence for division and transformation of neuroglial cells in the mouse brain as derived from autoradiography after injection of thymidine-^{3}H, *J. Comp. Neurol.* **116**:349–368 (1961).

221. J. M. Pasquini, B. Kaplun, C. A. Garcia-Argiz, and C. J. Gomez, Hormonal regulation of brain development, *Brain Res.* **6**:621–634 (1967).

222. A. Lajtha, Amino acid and protein metabolism of the brain—V. Turnover of Leu in mouse tissues, *J. Neurochem.* **3**:358–365 (1959).

223. G. Guroff and S. Udenfriend, Uptake of aromatic amino acids by the brain of mature and newborn rats, in *Progress in Brain Research* **9**:187–197, Elsevier, Amsterdam (1964).

224. A. Lajtha and E. Toth, The brain barrier system. II. Uptake and transport of amino acids by the brain, *J. Neurochem.* **8**:216–225 (1961).

225. F. E. Sampson and R. J. Jacobs, Mitochondrial changes in developing rat brain, *Amer. J. Physiol.* **199**:693–696 (1960).

226. B. Shepartz and M. Turczyn, Oxidation of L-amino acids and incorporation into protein in the homogenates of brain at two stages of development, *J. Neurochem.* **10**:825–829 (1963).

227. S. Gelber, P. L. Campbell, G. E. Deibler, and L. Sokoloff, Effects of L-thyroxine on amino acid incorporation into protein in mature and immature rat brain, *J. Neurochem.* **11**:221–229 (1964).

228. C. B. Klee and L. Sokoloff, Mitochondrial differences in mature and immature brain, *J. Neurochem.* **11**:709–716 (1964).

229. T. C. Johnson and W. W. Luttges, The effects of maturation on *in vitro* protein synthesis by mouse brain cells, *J. Neurochem.* **13**:545–552 (1966).

230. F. Orrego and F. Lipmann, Protein synthesis in brain slices. Effects of electrical stimulation and amino acids, *J. Biol. Chem.* **242**:665–671 (1967).

231. P. C. Rajam, C. J. Gaundreau, A. Grady, and S. T. Rundletl, Preparation, derivation and partial characterization of organ-specific antigens from human brain, *Immunology*, **17**:367–385 (1969).

232. K. F. Swaiman and C. E. Nelson, Soluble protein nitrogen and total protein nitrogen in developing rabbit brain, *J. Neurochem.* **14**:905–910 (1967).

233. K. Ito and Y. Arimatsu, A prominent site of immature protein synthesis, *Scientific Papers*, College of Gen. Education, University of Tokyo, **18**:41–53 (1968).

234. M.-L. Vahvelainen and S. S. Oja, The uptake and incorporation into protein of Tyr-^{3}H by slices prepared from developing rat brain cortex, *Brain Res.* **13**:227–233 (1969).

235. J. Buchanan, M. P. Primack, D. F. Tapley, Relation of mitochondrial swelling to thyroxine-stimulated protein synthesis, *Endocrinology* **87**:993–999 (1970).

236. J. DeVellis, O. A. Schjeide, and C. D. Clemente, Protein synthesis and enzymic patterns in the developing brain following head x-irradiation of newborn rats, *J. Neurochem.* **14**:499–511 (1967).

237. F. Orrego, Synthesis of RNA in normal and electrically stimulated brain cortex slices *in vitro*, *J. Neurochem.* **14**:851–858 (1967).

238. S. C. Bondy and S. V. Perry, Incorporation of labeled amino acids in the soluble protein fraction of rabbit brain, *J. Neurochem.* **10**:603–609 (1963).

239. D. H. Adams and L. Lim, Amino acid incorporation by preparations from the developing rat brain, *Biochem. J.* **99**:261–265 (1966).

240. S. Yamagami, R. Fritz, and D. A. Rappoport, Biochemistry of the developing rat brain. VII. Changes in the ribosomal system and nuclear RNA's *Biochim. Biophys. Acta* **129**:532–547 (1966).

241. M. R. V. Murthy and D. A. Rappoport, Biochemistry of the developing rat brain. VI. Preparation and properties of ribosomes, *Biochim. Biophys. Acta* **95**:121–132–145 (1965).

242. M. P. Lerner and T. C. Johnson, Regulation of protein synthesis in developing mouse brain tissue, *J. Biol. Chem.* **245**:1388–1393 (1970).

243. T. C. Johnson and G. Belytschko, Alteration in microsomal protein synthesis during early development of mouse brain, *Proc. Nat. Acad. Sci.* **62**:849–851 (1969).

244. G. Dallner, P. Sickevitz, and G. E. Palade, Biogenesis of endoplasmic reticulum membranes. I, II, *J. Cell Biol.* **30**:73–96, 97–117 (1966).

245. Y. Kuriyama, T. Omura, P. Siekevitz, and G. E. Palade, Effects of phenobarbital on the synthesis and degradation of the protein components of rat liver microsomal membranes, *J. Biol. Chem.* **244**:2017–2026 (1969).

246. E. D. Kiehn and J. J. Holland, Membrane and nonmembrane proteins of mammalian cells. Synthesis, turnover, and size distribution, *Biochemistry* **9**:1716–1728 (1970).

247. K. W. Bock, P. Siekevitz, and G. E. Palade, Localization and turnover studies of membrane nicotinamide adenine dinucleotide glycohydrolase in rat liver, *J. Biol. Chem.* **246**:188–195 (1971).

248. T. Kawasahi and I. Yamashina, Metabolic studies of rat liver plasma membrane using ^{14}C-glucosamine, *Biochim. Biophys. Acta* **225**:234–238 (1971).

249. H. Porter, Neonatal hepatic mitochodrocuprein, *Biochim. Biophys. Acta* **229**:143–154 (1971).

250. B. W. Moore and D. McGregor, Chromatographic and electrophoretic fractionation of soluble proteins of brain and liver, *J. Biol. Chem.* **240**:1647–1653 (1965).

251. G. S. Bennett and G. M. Edelman, Isolation of an acidic protein from rat brain, *J. Biol. Chem.* **243**:6234–6241 (1968).

252. K. Warecka and H. Bauer, Studies on "brain-specific" proteins in aqueous extracts of brain tissue, *J. Neurochem.* **14**:783–787 (1967).

253. H. Hydén and P. W. Lange, Correlation of the S-100 brain protein with behavior, *Exper. Cell. Res.* **62**:125–132 (1970).

254. W. T. Norton, The myelin sheath, *in Cellular and Molecular Basis of Neurology and Disease* (E. Goldstein and S. Appel, eds.), Lee and Febiger, Philadelphia (in press) (1971).

255. S. Berl and S. Puszkin, Mg^{2+}-Ca^{2+}-activated ATPase system isolated from mammalian brain, *Biochemistry* **9**:2058–2067 (1970).

256. J. B. Kirkpatrick, L. Hyams, V. L. Thomas, and P. M. Howley, Purification of intact microtubules from brain, *J. Cell Biol.* **47**:384–394 (1970).

257. P. F. Davison and F. C. Huneeus, Fibrillar proteins from squid axon. I. Neurofilament protein. II. Microtubule protein, *J. Mol. Biol.* **52**:429–434 (1970).

258. M. Wender and Z. Waligora, The content of amino acids in the proteins of the developing nervous system of the guinea-pig I–II, *J. Neurochem.* **7**:259–263 (1961); **9**:115–118 (1962).

259. H. Feit, Synthesis and turnover of tubulin, Thesis. Dept. of Mol. Biol. Albert Einstein Medical School, Yeshiva, University, 1971.

260. H. Feit, G. Dutton, S. Barondes, and M. L. Shelanski, Metabolism of microtubule protein in mouse brain, *J. Cell. Biol.* **47**:60a (1970).

261. T. J. Cicero, M. W. Cowan, and B. W. Moore, Changes in the concentrations of the brain specific proteins, S-100 and 14-3-2 during the development of the avian optic fectum, *Brain Res.* **29**:1–10 (1970).

262. H. R. Herschman, L. Levine, and D. Vellis, Appearance of a brain specific-specific antigen (S-100 protein) in the developing rat brain, *J. Neurochem.* **18**, 629–633 (1971).

263. T. J. Cicero and B. W. Moore, Turnover of the brain specific protein, S-100, *Science* **169**:1333–1334 (1970).

264. H. Herschman (personal communication), UCLA Brain Information Service BIS report No. 8.

265. M. R. Adelman, G. G. Borisy, M. L. Shelanski, R. C. Weisenberg, and E. W. Taylor, Cytoplasmic filaments and tubules, *Fed. Proc.* **27**:1186–1193 (1968).

266. O. Holian, D. Dill, and E. G. Brunngraber, Incorporation of radioactivity of D-glucosamine-1-^{14}C into heteropolysaccharide chains of glycoproteins in adult and developing rat brain, *Arch. Biochem. Biophys.* **142**:111–121 (1971).

267. K. Kuriyama and T. A. Okada, Incorporation of sulfate-^{35}S into developing mouse brain: subcellular fractionation and electron microscopic studies, *Exper. Neurol.* **30**:18–29 (1971).

268. G. Banker and C. W. Cotman, Characteristics of different amino acids as protein precursors in mouse brain: advantages of certain carboxyl-labeled amino acids, *Arch. Biochem. Biophys.* **142**:565–573 (1971).

269. A. Hirano and H. M. Dembitzer, A structural analysis of the myelin sheath in the CNS, *J. Cell. Biol.* **34**:555–567 (1967).

270. B. G. Uzman and E. T. Hedley-Whyte, Myelin: Dynamic or stable, *J. Gen. Physiol.* **51**:8–18 (1968).

271. M. E. Smith, The turnover of myelin in the adult rat. *Biochim. Biophys. Acta* **164**:285–293 (1968); *J. Neurochem.* **18**:739–747 (1971).

272. M. E. Smith and L. F. Eng, The turnover of the lipid contents of myelin, *J. Am. Oil Chem. Soc.* **42**:1013–1015 (1965).

273. E. R. Einstein, J. Csejtey, and N. Marks, Degradation of encephalitogen by purified brain acid proteinase, *FEBS Letters* **1**:191–195 (1968).

274. L. F. Eng, F. C. Chao, B. Gerstl, D. Pratt, and M. G. Tavastsjerna, The maturation of human white matter. Fractionation of the myelin membrane proteins, *Biochemistry* **7**: 4755–4765 (1968).

275. M. K. Gaitonde and R. E. Martenson, Metabolism of highly basic proteins of rat during post-natal development, *J. Neurochem.* **17**:551–563 (1970).

276. J. G. Wood and N. King, Turnover of basic protein of rat brain, *Nature* **229**:56–57 (1971).

277. M. E. Smith, An *in vitro* system for the study of myelin synthesis, *J. Neurochem.* **16**:83–92 (1969).

278. F. Rawlins and M. Smith, Myelin synthesis *in vitro*: a comparative synthesis between CNS and PNS, *Trans. Amer. Soc. Neurochem.* **2**:102C (1971), and M. E. Smith, Biosynthesis of myelin proteins *in vitro*, *Fed. Proc.* **29**:471 (1970).

279. V. B. Wigglesworth, *Insect Hormones*. Oliver and Boyd, Edinburgh (1970).

280. S. Zamenhof, J. Mosley, and E. Schullar, Stimulation of the proliferation of cortical neurons by prenatal treatment with growth hormone, *Science* **152**:1396–1397 (1966).

281. R. Balazs, S. Kovacs, W. A. Cocks, A. L. Johnson, and J. T. Eayrs, Effect of thyroid hormone on the biochemical maturation of rat brain: postnatal cell formation, *Brain Res.* **25**:555–570 (1971).

282. D. B. Hudson, A. Vernadakis, and P. S. Timiras, Regional changes in amino acid concentration in the developing brain and the effects of neonatal administration of estradiol, *Brain Res.* **23**:213–222 (1970).

283. R. D. Palmiter, T. Oka, and R. T. Schimke, Modulation of ovalbumin synthesis by estradiol-17β and actinomycin-D as studied in explants of chick oviduct in culture, *J. Biol. Chem.* **246**:724–737 (1971).

284. M. Ginsburg, *Handbook of Experimental Pharmacology* Vol. 23, pp. 286–371 (1968), Springer-Verlag, Berlin.

285. A. J. Patel and R. Balazs, Manifestation of metabolism compartmentation during the maturation of rat brain, *J. Neurochem.* **17**:955–971 (1970).

286. S. Berl, Compartmentation of glutamic acid metabolism in developing cerebral cortex, *J. Biol. Chem.* **240**:2047–2054 (1965).

287. C. J. van den Berg, Compartmentation of glutamate metabolism in the developing brain, *J. Neurochem.* **17**:973–983 (1970).

288. R. B. Roberts and L. B. Flexner, The biochemical basis of long term memory, *Quart. Rev. Biophys.* **2**:135–173 (1969).

289. E. Glassman, The biochemistry of learning: an evaluation of the role of RNA and protein, *Ann. Rev. Biochem.* **38**:605–646 (1969).

290. A. Vitale-Neugebauer, A. Giuditta, B. Vitale, and S. Giaguinto, Pattern of RNA synthesis in rabbit cortex during sleep, *J. Neurochem.* **17**:1263–1273 (1970).

291. W. E. Davies, The incorporation of Lys-^{14}C into the protein of the guinea pig central auditory system, *J. Neurochem.* **17**:1319–1326 (1970).

292. D. A. Rappoport and H. F. Daginawala, Changes in nuclear RNA of brain induced by olfaction in catfish, *J. Neurochem.* **15**:991–1006 (1968).

293. A. Norström, S. Enestrom, and A. Hamberger, Amino acid incorporation into proteins of the supraoptic nucleus of the rat after osmotic stress, *Brain Res.* **26**:95–103 (1971).

294. F. Nissl, Uber die Ausreissung der Ganglienzellen am Facialiskern des Kaninchens Nach Ausreissung des Neuron, *Allg. Z. Psychiat.* **48**:197–198 (1892).

295. S. H. Kung, Incorporation of tritiated precursors in the cytoplasm of normal and chromatolytic neurons as shown by autoradiography, *Brain Res.* **25**:656–660 (1971).

296. A. J. Carlson, Changes in the Nissl's substance of the ganglion and the bipolar cells of the retina of the Brandt Comorant during prolonged normal stimulation, *Am. J. Anat.* **2**:341–347 (1902/03).

297. A. Hess, Optic centers and pathways after eye removal in fetal guinea pigs, *J. Comp. Neurol.* **199**:91–115 (1958).

298. I. Lindner and K. Umrath, Veränderungen der Sehsphäre I und II in ihrem monokularen and binokularen Teil nach Extirpation eines Auges beim Kaninchen, *Deut. Z. Nervenheilk.* **172**:495–525 (1955).

299. A. H. Riesen, Effects of stimulus deprivation on the development and atrophy of the visual sensory system, *Amer. J. Orthopsychiatry* **30**:23 (1960).

300. F. L. Margolis and S. C. Bondy, Effect of unilateral enucleation on protein and ribonucleic metabolism of avian brain (unilateral suturing), *Exper. Neurol.* **27**:344–352 (1970), and **27**:353–358 (1970).

301. G. P. Talwar, S. P. Chopra, B. K. Goel, and B. D'Monte, Correlation of the functional activity of the brain with metabolic parameters III. Protein metabolism of the occipital cortex in relation to light stimulus, *J. Neurochem.* **13**:109–116 (1966).

302. S. P. R. Rose, Changes in visual cortex on first exposure of rats to light, *Nature* **215**:253–255 (1967).

303. M. Burnel, H. R. Mahler, and W. J. Moore, Protein synthesis in visual cells of *Limulus, J. Neurochem.* **17**:1493–1499 (1970).

304. A. V. LeBouton and S. D. Handler, Diurnal incorporation of Leu-^{3}H into liver protein, *FEBS Letters* **10**:78–80 (1970).

305. H. Rahmann, Uber den Einfluss adaquater Lichtreizung auf die biochemische und morphologische Ausprägung der Sehninde der Maus, *Z. f. Zellforsch.* **67**:561–574 (1965).

306. A. J. Goldberg, Protein turnover in skeletal muscle II. Effects of denervation and cortisone on protein catabolism in skeletal muscle, *J. Biol. Chem.* **244**:3223–3229 (1969).

307. M. H. Dresden, Denervation effects on new T limb regeneration: DNA and RNA and protein synthesis, *Develop. Biol.* **19**:311–320 (1969).

308. C. Kupfer and J. L. Downer, Ribonucleic acid content and metabolic activity of lateral geniculate nucleus of monkey following afferent derivation, *J. Neurochem.* **14**:257–263 (1967).

309. P. Mandel, H. Rein, S. Harth-Edel, and R. Mandell, Distribution and metabolism of ribonucleic acid in the nervous system, *in Comparative Neurochemistry* (E. Richter, ed.), pp. 149–163, Pergamon Press, 1964.

310. J. Dobbing and J. Sands, Timing of neuroblast multiplication in developing human brain, *Nature* **226**:639–640 (1970).

311. G. Bolcsfoldi, L. Poels, and E. Eliasson, RNA metabolism in human cells during amino acid deprivation, *Biochim. Biophys. Acta* **228**:664–675 (1971).

312. E. C. Henshaw, C. A. Hirsch, B. E. Morton, and H. H. Hiatt, Control of protein synthesis in mammalian tissues through changes in ribosome activity, *J. Biol. Chem.* **246**:436–446 (1971).

313. A. J. Sussman and C. Gilvarg, Protein turnover in amino acid-starved strains of E. Coli, K-12 differing in their RNA content, *J. Biol. Chem.* **244**:6304–6306 (1969).

314. A. Lajtha and E. Toth, Instability of cerebral proteins, *Biochem. Biophys. Res. Comm.* **23**:294–298 (1966).

315. R. S. Piha, R. K. Airas, and L. I. Aäri, Changes in the activity of amino acid: *t*RNA-ligases in the developing brain of the mouse, *Suomen Kemistilehti* **39**:204–208 (1966).

316. J-Fu Chiu and S.-C. Sung, DNA nucleotidyl transferase action of the developing rat brain, *Biochim. Biophys. Acta* **209**:34–42 (1970).

317. R. A. Ehrenkranz and M. Winick, DNA polymerase in normal rat brain during ontogeny, *J. Cell. Biol.* **47**:249 (1970).

318. E. Bell, *I*-DNA: Its packaging into *I*-somes and its relation to protein synthesis during differentiation, *Nature* **224**:326–328 (1969).

319. R. L. Church and R. A. Consigli, DNA fragmentation in a clonal line of rat pituitary tumor, *Biochem. Biophys. Res. Comm*, **42**:31 (1971).

320. S.-C. Sung, DNA synthesis in the developing rat brain, *Canad. J. Biochem.* **47**:47–50 (1969).

321. R. R. Burgess, A. A. Travers, J. J. Dunn, and E. K. F. Bantz, Factor stimulating transcription by RNA polymerase, *Nature* **221**:43–46 (1969), and **228**:748–751 (1970).

322. S. C. Bondy, S. Roberts, and B. S. Morelos, Histone-acetylating enzyme of brain, *Biochem. J.* **119**:665–672 (1970).

323. A. L. Beaudet and C. T. Caskey, Mammalian peptide chain termination, I and II. *Proc. Nat. Acad. Sci.* **67**:99–106; and 619–24 (1971).

324. H. M. Temin, *Biology of Large RNA Viruses* (R. D. Barry and B. W. J. Mahy, eds.), Academic Press, London (1970).

325. E. M. Scolnick, S. A. Aaronson, G. J. Todaro, and W. P. Parks, RNA dependent DNA polymerase activity in mammalian cells, *Nature* **229**:318–321 (1971); and tumor viruses, *Proc. Nat. Acad. Sci.* **67**:1034–1041 (1970).

326. P. Lengyel and D. Söll, Mechanism of protein biosynthesis, *Bact. Rev.* **33**:264–301 (1969).

327. A. K. Falvey and T. Staehelin, Structure and function of mammalian ribosomes. I. Isolation and characteristics of active liver ribosomal subunits. II. Exchange of ribosomal subunits at various stages *in vitro* polypeptide synthesis, *J. Mol. Biol.* **53**:21–34 (1970).

328. J. S. Dubnoff and U. Maitra, Isolation and properties of polypeptide chain initiation factor F II from *E. coli*: evidence for a dual function, *Proc. Nat. Acad. Sci.* **68**:318–323 (1971).

329. D. A. Shafritz, D. G. Laycock, and W. French Anderson, Puromycin-peptide bond formation with reticulocyte initiation factors M, and M$_2$, *Proc. Nat. Acad. Sci.* **68**:496–499 (1971).

330. R. Kaempfer, Dissociation of ribosomes on polypeptide chain termination and origin of single ribosomes, *Nature* **228**:534–537 (1970).

331. S. H. Miall, T. Kato, and T. Tamaohi, A factor promoting dissociation of *E. coli* ribosomes, *Nature* **226**:1050–1052 (1970).

332. H. A. Klein and M. R. Capecchi, Polypeptide chain termination, *J. Biol. Chem.* **246**:1055–1061 (1971).

333. S. Raeburn, J. F. Collins, H. M. Moon, and E. S. Maxwell, Aminoacyltransferase II from rat liver, *J. Biol. Chem.* **246**:1041–1048; 1049–1054 (1971).

334. T. C. Johnson, Regulatory mechanisms responsible for alterations in protein and nuclei and synthesis in developing brain tissue, *in Cellular Aspects of Growth and Differentiation in Nervous Tissue* (D. Pease, ed.), UCLA Forum in Med. Sciences (1971).

335. H. Aurich, Die Rolle des Aminosaüre-Pools bie der Regulation des Proteins-und Nuklein-saüurestoff-wechsels, *Math.-Naturwiss. Reihe* **4**:727–738 (1968) Karl-Marx Univ., Leipzig.

336. M. Nomura, Bacterial ribosome, *Bact. Rev.* **34**:228–277 (1970).

337. D. A. Jones and H. McIlwain, Amino acid distribution and incorporation into proteins isolated electrically stimulated cerebral tissues, *J. Neurochem.* **18**:41–58 (1971).

338. P. J. Dehlinger and R. T. Schimke, Size distribution of membrane proteins of rat liver and their relative rates of degradation, *J. Biol. Chem.* **246**:2574–2583 (1971).

339. J. A. Benjamins, N. Herschkowitz, J. Robinson, and G. M. McKhann, The effects of inhibitors of protein synthesis on incorporation of lipids into myelin, *J. Neurochem.* **18**: 729–738 (1971).

Chapter 19

RECEPTORS: POSSIBLE MOLECULAR COMPLEXES INVOLVED IN RECEPTOR SITES AND TRANSMITTER STORAGE MECHANISMS

J. R. Smythies

Department of Psychiatry
University of Edinburgh
and
Neuroscience Program
University of Alabama

I. A POSSIBLE ROLE FOR RNA

In considering the molecular mechanisms underlying receptor site activity and ionic and other channels in excitable membrane, it is usually assumed that the macromolecules concerned must be protein, proteolipid, or glycoprotein. This material has been recently reviewed by Ehrenpreis *et al.*[1] However, now that RNA has been reported in membrane,[2–6] it is necessary to add ribonucleoprotein to this list of macromolecules, and this review will enquire into the possible role or roles of RNA in membrane. It might be concerned with local protein synthesis or it may have some other functions. It is very likely that the ionic transport mechanisms and the receptor sites controlling them are very complex and may be composed of any of these macromolecules in combination. However, in any enquiry into the possible nature of the receptor site, or of the ion transport mechanisms, RNA possesses one considerable advantage. Its molecular specification is set within very narrow limits unlike the other three candidates, where enormous variation is possible. Thus it is possible to determine by molecular models exactly how a segment of RNA could bind the putative transmitters and their agonists and antagonists, and it is possible to suggest how RNA could participate in ion transport mechanisms. Namba and Grob[7] have isolated a ribonucleoprotein from muscle that binds tubocurarine. This binding is inhibited by ACh. They therefore suggest that RNA may be concerned in the cholinergic receptor. Passow *et al.*[8] have suggested that nucleic acids, located in yeast cell membrane, may be involved in Ca^{2+} binding. The sites binding Ca^{2+} in this location have been identified as phosphoryl and carboxyl. The phosphoryl groups appear to belong to RNA, the evidence being the quantitatively similar affinity for a number of different basic dyes, and the reduction by ribonuclease of both the binding of Sr^{2+} and Ca^{2+} by plant cells, as well as the staining of the cell surface of yeast cells and sea urchin eggs by toluidine blue and methylene blue.[9–11]

A segment of double-stranded helical RNA running across a membrane can form a tube capable of transporting ions by two mechanisms. In the first, the two grooves in the side of the molecule could be converted into tubes by polyamines, such as spermine, which are known to be closely associated with RNA, attaching their N^+ to the phosphate O^-. Molecular models indicate that the spermine molecule is the right length, so that when bound across the narrow groove of a fully contracted double helix, it converts it into a tube approximately 8 Å in diameter, whereas in the major groove it can only bind from phosphate O^- to base pair $O\delta-$, forming a narrower tube (6 Å in diameter). Molecular models indicate that the former tube is large enough to admit the hydrated Na^+ ion (6.4 Å in minimum diameter) and the latter tube, the smaller hydrated K^+ ion (4 Å in minimum diameter). Furthermore, as Fig. 1 shows, ouabain, which inhibits the Na^+ pump, has the precise molecular shape required to block such as "minor groove" polyamine-RNA channel, and *all* its hydrogen bonding hydroxyl groups are complementary to hydrogen bonding sites on the inner surface of the spermine-RNA complex, as shown in Fig. 2. Likewise, the smaller "major groove" channel is similarly blocked by the tetraethylammonium ion both by steric squeeze and an ion-dipole bond to purine N:, guanine NH, or uridine O. Thus the minor groove channel could be the Na^+ channel part of the N^+ pump for active transport out of the cell, and the major groove

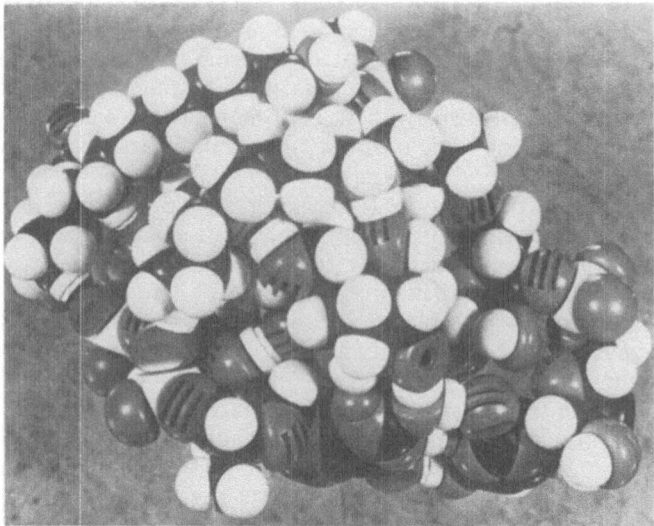

Fig. 1. A molecular model of ouabain blocking a channel composed of two spermine molecules and double helical RNA.

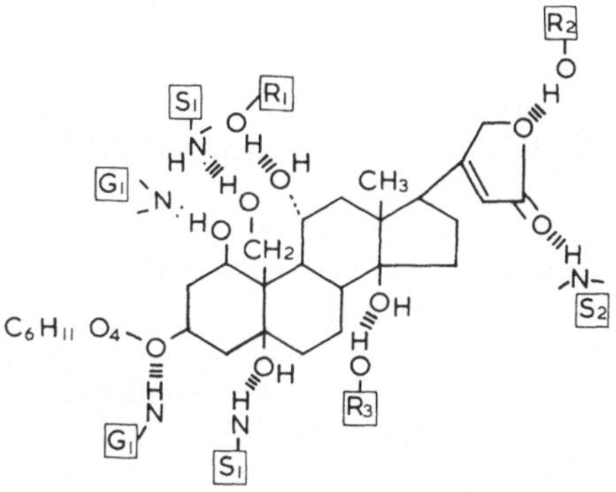

Fig. 2. A line drawing of the ouabain–spermine–RNA complex. Hydrogen bonds are shown by narrow stripes.

channel could be the K^+ channel for entry of K^+ into the cell. However, all such theories must take into account that water in membrane may be ordered water which would mean that both these postulated channels would normally be blocked by water hydrogen bonded to the many charged atoms in the walls. In this context the difference between the hydration shells of $Na^+ 3(H_2O)$ and $K^+ 2(H_2O)$ may be relevant. In Na^+ the hydration shell forms an A region, that is, the three water molecules are tightly bound and they will tend to increase the ordering of surrounding water molecules. Thus $Na^+ 3(H_2O)$ would be unable to pass through a narrow channel through membrane blocked by hydrogen-bonded water molecules. In $K^+ 2(H_2O)$, on the other hand, the hydration shell forms a B region, that is, the two water molecules are loosely bound and can decrease the ordering of surrounding water molecules and even dislodge water molecules hydrogen bonded to macromolecules. Thus hydrated K^+ ions could pass through a narrow channel through membrane because it would not be prevented from doing so by hydrogen-bonded water. The demonstration that membrane water is ordered has led to claims that the Na^+ pump cannot exist.[12] However, one possible mechanism might be as follows. The Na^+ pump depends on a Na^+, K^+ activated ATPase located on the inside of the membrane. This may be located at the inner entry to the Na^+ channel and it may provide the energy for dehydrating hydrated Na^+: in other words, the basic function of ATP may be to remove the water of hydration from the hydrated Na^+ ion and feed naked Na^+ ions into the spermine-RNA Na^+ channel. Here the first Na^+ ions fed in would immediately become rehydrated at the expense of the bonded water molecules blocking the tube. Later Na^+ ions would push past the hydrated Na^+ ions and eventually they would be discharged at the surface where they would be hydrated by ordinary water molecules. Hydrated K^+ ions could not pass down the Na^+ channel which would be almost filled by hydrated Na^+ ions.

The second main problem in excitable membrane is the nature of the channel through which Na^+ enters during the action potential. The molecule of RNA can also provide this. Double-stranded helical RNA crossing the membrane provides a potential tube larger than those we have considered so far. The hydrogen-bonded base pairs can be regarded an internal "shutter." If these are disrupted and, the RNA denatured, this "shutter" will open and the RNA and associated polyamines and protein will form a large tube through which Na$^+$ could pass down its electrochemical gradient. Thus a transmitter could act by binding to the external segment of RNA in such a way as to disrupt the hydrogen bonds. Small transmitter molecules could bind to double helical RNA in one of three loci (1) the spare NH group and unused O orbital on the hydrogen-bonded base pairs (2) by intercalation between base pairs and (3) to the phosphate groups. However, it is difficult to see how the last attachment could disrupt the Watson–Crick hydrogen bonds, and the phosphates provide no source of variability. Specificity as to which transmitter would act would be determined by the sources of variability in the RNA molecule. These sources are as follows: (1) the particular base pairs involved (2) whether the major or minor groove comes to the surface and (3) the degree of torsion of the helix.

Of these the most critical is (3) since the interatomic distances and the bonding angles are profoundly altered by altering the degree of torsion on the helix. Fortunately, this variable may be fixed if we consider the significance

Fig. 3. A molecular model of the complex formed by two prostaglandin molecules and a four base-pair segment of double helical RNA. The base pair configuration shown is CG:GC ("nicotinic").

of the stereochemical relationship between prostaglandins and helical RNA illustrated in Fig. 3. Two PG molecules, in energetically favorable conformation, are complementary to a section of double helical RNA with four base pairs if this is placed in a particular extended configuration maintained by protein. Each PG molecule can bind to the RNA by four hydrogen bonds as shown, and in the case of 19-OH PG's by five. Their functions in this position would appear to include the following. They convert the normal hydrophilic and negatively charged surrounds of the postulated primary binding site for transmitters such as ACh (that is, the NH groups of guanine and the spare O orbitals of cytosine) into a deep lipophilic pit with high lipophilic walls on all sides except the two "acute" corners, as may be observed in the figure. The PG molecules provide extensive lipophilic binding sites for transmitters such as ACh. The AU base pair is unlikely to be involved in this location as adenine lacks an NH_2 group in the minor groove and there is no hydrogen bond at risk. With two PG molecules in place (bound to base pairs 1 and 4) the binding of different transmitters (to base pairs 2 and 3) can be shown by molecular models to be determined by the particular base pairs and the particular PG's involved.

II. SPECIFICATION OF RECEPTOR SITES FOR TRANSMITTERS

A receptor that fills the stereochemical requirements for binding acetylcholine in the nicotinic site is provided by a CG:GC base pair combination. The upper PG must be F and ACh binds by its carbonyl O to the guanine NH, its N^+ to cytosine O plus extensive lipophilic contacts along one side of the ACh molecule and the PG hydrocarbon chain.

In the muscarinic site the middle base pairs must be GC:GC and the upper PGE. In this case ACh binds by its "ether" oxygen to guanine NH of base pair 2; and N^+ to cytosine O of base pair 3 plus extensive lipophilic contacts. The strong positive charge on the ACh quaternary N^+ could disrupt the hydrogen bond on base pair 3 and thus initiate channel opening. Likewise, molecular models indicate a glutamate receptor as GC:CG.

In the case of glycine the molecule is too short to span across two base pairs like the other transmitters, but it can span one (N^+ to cytosine O; O^- to guanine NH). In this case, the prostaglandin concerned could be PG dinor F2α. The dinor PG's have only 18 carbon atoms instead of the usual 20 and when two are bound as described above they need only a three base pair segment of RNA and delineate a site with only one base pair available. Such a site provides a precise stereochemical fit for strychnine with two hydrogen bonds, and the larger ACh site provides a precise fit for toxiferine 1 with two hydrogen bonds and two ionic bonds, both with extensive lipophilic contacts.

In the case of 5-HT and the catecholamines, these could bind to different locations on the RNA molecule than those we have so far considered.

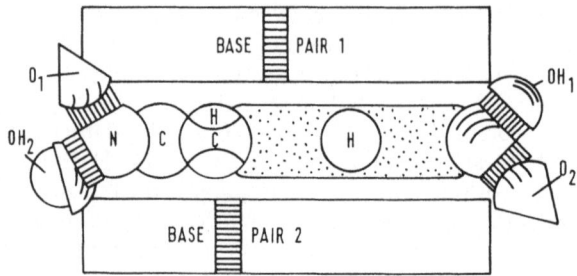

Fig. 4. Binding of 5-HT by intercalation.

Experimental data[13-15] indicate that 5-HT and its antagonists, such as LSD, bind to nucleic acids by intercalation between base pairs and π cloud overlap. Molecular models show that when the 5-HT molecule is thus intercalated in double-stranded helical RNA, it can form four hydrogen bonds to two ribose ring oxygens and two ribose hydroxyls (Fig. 4). Each PG-RNA complex could bind three molecules of 5-HT intercalated between the four base pairs. Each "end" site (between base pairs 1 and 2, and 3 and 4) could bind d-LSD (but not its three inactive stereoisomers) as well as Δ^{9-} tetrahydrocannabinol. In the latter, the benzene ring intercalates between the base pairs, the ring hydroxyl hydrogen bonds to the ribose ring oxygen, and the complex B and C ring system forms a tilted delta-shaped clump that fits precisely into the acute angle between the two prostaglandin (E) side chains with extensive lipophilic contact. The D ring and diethylamide groupings perform the same function for d-LSD (PG in this case must be F).

Serotonin could have two functions in this location. Three molecules bound in the three available sites would stabilize the double helix by restoring the π cloud stacking energy lost by the extended configuration of the RNA and by the hydrogen bonding cross-linkage between the two strands of the RNA. However, if the 5-HT enters the site with its electron-donating 2 position touching the π cloud of cytosine (an electron receptor), its function could be connected with a charge transfer reaction which might tend to disrupt the helix. Thus its action might depend on its orientation on entering the site.

On the other hand, NE, and in particular its potent antagonist perphenazine (and other phenothiazines), could bind in the major groove to a fully contracted helix with no prostaglandin on the other (minor groove) side. The ring system of the phenothiazine is known to intercalate into nucleic acids[16] and the 2-chloro group ensures that it does so from the major groove. This locates the long phenothiazine tail running down the major groove bonding to NH or O groups available. NE binds across four base pairs: 3 OH to base 4 O, 4 OH to base 4 NH, 3 OH to base 2 O, N^+ to base 1 O. Thus 5-HT could stabilize the double helix in its extended configuration and might thus be expected to inhibit ACh action, whereas NE could stabilize the double helix in its contracted configuration and if this latter is associated with an open Na^+ channel, it should potentiate ACh (or other primary transmitter) action. NE action and prostaglandin action should clearly be mutually exclusive. Table I summarizes the specification of some of the receptor sites based on this hypothesis.

Additional support for this hypothesis comes from a consideration of molecular models of veratridine and tetrodotoxin. I postulated above that the open channel for the inward flow of Na^+ was the inside of a segment of double-stranded helical RNA with the hydrogen bonds disrupted: thus in the walls there will be two segments of single-stranded RNA (as well as polyamine and protein). The remarkable molecule of veratridine (Fig. 5) has a line of oxygen atoms so arranged on one side that it can bind along its length, by no less than eleven hydrogen bonds, to an eight base-pair segment of RNA as illustrated in Fig. 6. The other side of the veratridine molecule facing the inside of the Na^+ channel is purely lipophilic. Tetrodotoxin has a similar band of oxygen atoms forming a band around the molecule, only this is arranged quite differently. The tetrodotoxin molecule is shaped remarkably like a cricket ball with the band of oxygens and hydroxyls arranged like the seam around its equator, plus a tongue coming out from one side with three nitrogen atoms. This can insert itself between two strands of RNA which have just separated and prevent further separation by binding to the base pairs on each side. This model has been fully developed elsewhere[17] and it can account for a wide range of structure activity data on the agonists and antagonists of transmitters and other membrane-active drugs.

TABLE I

Specifications of the Receptor Sites

		ACh(m)	ACh(n)	Glut.[b]	Gly.[b]	5-HT (excit)
Base pair	1	CG	GC	+	+	GC
	2	GC[a]	CG	GC	Either	GC
	3	GC	GC	CG	+	CG
	4	Either	Either	+	−	CG
Groove		Mi	Mi	Mi	Mi	Intercal.
PG		2Es or 1E, 1F	2Fs or 1F, 1E	20c	18c	2Fs

[a] Counting left to right.
[b] + Depends on particular PG; e.g., for F, 1 is GC and 4 is CG; for E, 1 is CG and 4 is GC; for A an AU base pair can be used.

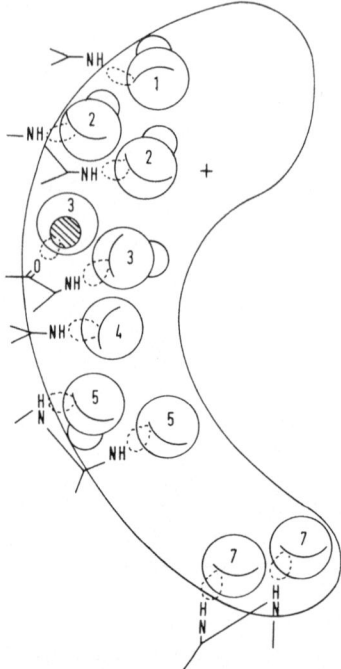

Fig. 5. Molecular model of veratridine (drawing).

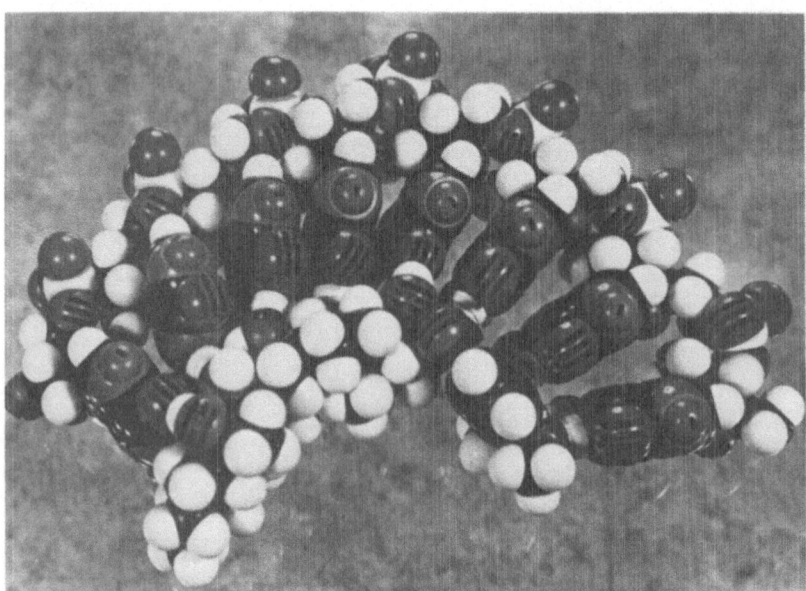

Fig. 6. Molecular model of veratridine bound to an eight base segment of poly G.

III. AN ALTERNATIVE MODEL INVOLVING NUCLEOSIDES

The model can be correlated with Durell's theory,[18] which takes account of the extensive experimental data involving a phospholipid in the receptor, as follows. At one stage in Durell's mechanism, he invokes a compound of the following type: nucleotide–phospholipid (linked by covalent bonds through the phosphates). A series of such molecules could form a complex with single-stranded RNA by Watson–Crick base pairing and a hydrogen bond between the 3 OH of one ribose and the oxygen of the nucleotide phosphate of the adjacent molecule (which would take the place of the covalent bond in this position in RNA). This complex would require divalent ions to neutralize the charges on the extra phosphates and it would be further stabilized by extensive lipophilic interactions between adjacent phospholipid hydrocarbon side chains. The complex could very well form a conformation practically identical to the RNA conformation as far as the base-ribose portions, essential for this hypothesis, are involved and all the stereo-chemical relations worked out on double-stranded RNA would apply to this complex. This model does not require the postulation of any conformational change in a protein to provide the space occupied by the open Na^+ channel, as this space could be provided by the disruption of the complex by ACh and the floating away of the phospholipid (possibly attached to protein) on the hydrocarbon "pool."

IV. A MODEL INVOLVING A PROTEIN–NUCLEOSIDE COMPLEX

Recent investigations have shown that a stereochemically practically equivalent complex can be constructed in two alternative ways. In one the single strand RNA of the complex described in the last paragraph, can be replaced by a polypeptide chain of at least seven amino acids, in which amino acids 1, 3, 5, and 7 are either arginine, glutamate, or glutamine. This is because, in terms of the requirements of this model, the following are equivalent: arginine and guanine; glutamate and cytosine; glutamine and uracil. Thus the following are complementary: arginine and cytosine; glutamate and guanine; glutamine and adenine. This means that we can remove the single strand RNA from an RNA–nucleotide complex model, and replace it with a polypeptide chain matching the nucleoside bases as described. This complex will now bind two PG molecules to form the lipophilic rhomboid with amide NH groups taking the place, on one side, of the ribose OH groups. The second complex is obtained by now replacing the complexed nucleosides with a second polypeptide chain, so that the GC base pair of an all-RNA complex is equivalent to an arginine–glutamate amino acid pair in an all-polypeptide complex.

The stereochemical relationship between a nucleic acid and a polypeptide complexed with nucleosides is such that, for the purposes of binding

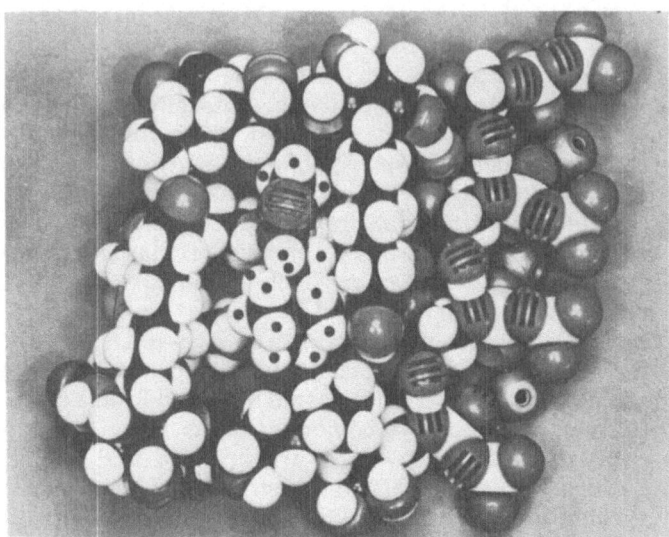

Fig. 7. ACh bound in a model 3 muscarinic receptor.

transmitters, their agonists and antagonists, there are only very slight differences between them.

Thus there are at least four possible types of complex that can form, with prostaglandins, good working models of receptors: (1) all RNA, (2) RNA–nucleoside, (3) polypeptide–nucleotide, and (4) all polypeptide. The nucleoside in turn can be phosphorylated to various degrees or linked to lipid. The translation required in Table I is arginine for "G" and glutamate for "C." Some preliminary structure-activity relationship studies indicate, for some receptors, that the most probable basis is a form of (3), which I have called a PG–PNPL complex (prostaglandin–polypeptide–nucleophospholipid). An example of model (3) (ACh muscarinic) is shown in Fig. 7.

The choice between these alternatives can be made by experiment.

V. STORAGE SITES

A. Acetylcholine

Recent experiments using fluorescent techniques[19] on the binding of morphine to nucleosides have shown a very marked binding to poly C (poly C > poly A > poly U) and to CMP and an equally good binding to poly IC. This indicates that morphine is binding to loci on cytosine unmasked by the bond to inosine. As poly C is noted for its good stacking (as opposed to poly U, for example, which does not stack and poly A which stacks poorly), we examined molecular models of CMP molecules stacked in a simple helix in an energetically preferred conformation. We observed that this provided good binding sites for ACh (N^+ to phosphate O^- : "ether" O to phosphate OH of adjacent CMP molecules and $-CH_3$ (lipophilic) to ribose 4 and 5 CH). Morphine now binds by its N^+ and the surrounding flat lipophilic surface of its B and C rings to the complementary flat lipophilic surface presented by the ring O and 4 and 5 CH of one ribose, the 2 and 3 CH's of the adjacent ribose, and the 5 and 6 CH's of adjacent cytosines. This locates the morphine A ring hydroxyl in a position to make a hydrogen bond with the carbonyl O of ACh. There are also extensive lipophilic contacts between one morphine molecule and the next and between the morphine and ACh hydrocarbons. Thus morphine in this locus would clearly prevent the release of ACh. Other morphine-like compounds such as methadone and meperidine also fitted the complex remarkably well. Thus we suggested that the ACh storage site in synaptic vesicles, or elsewhere, is simply stacked molecules of CMP, stabilized by divalent ions between phosphate O^-'s.

B. Catecholamines

This result led us to examine the data on the storage mechanism for catecholamines in the adrenal medulla.[19] These are known to contain ATP, Mg^{2+}, Ca^{2+}, protein, and prostaglandin F1α (and no other PG).[20] Molecular models indicated that there is only one way to assemble these components into a complex such that four molecules of a catecholamine or two molecules of serotonin can bind per molecule of ATP. The protein chain must have glutamine for every other amino acid. Adenine is complementary to glutamine and binds with the hydrogen bond at risk on this major side (i.e., the side equivalent to the major groove of RNA). The adenine ATP ladder (stabilized by divalent ions between phosphate O's) will now bind PGF on the major side, PGA on the minor side, but not PGE. This complex will now bind four molecules of a catecholamine (in pairs linked by chelation from their OH's to a suitable metal ion) or two molecules of 5-HT. These molecules partly intercalate between the rungs of the amino acid–base ladder and bind by their N's to phosphate and carbonyl O's. Moreover, the complex can now bind ACh on the major side (carbonyl O to adenine NH, N^+ to glutamine O (disrupting this bond) and extensive lipophilic contacts with the adjacent PGF) in such a manner as would disrupt the complex and release the bound amines. The model also has close stereochemical relations with reserpine. Thus this model explains the data and we have also shown it to have close stereochemical relations to cocaine, imipramine, amphetamine and its congenors, and angiotensin, which can all bind to the ATP–polypeptide–PG complex in such a way as to block the uptake of catecholamines into the intercalation site, which may be the basis for their biological action.[19] The chromaffin granule does not contain PG thus the actual storage complex may be simply an ATP–protein–divalent ion complex. The PGF–ATP–protein–divalent ion complex may be located in the external membrane of the chromaffin cell (and other adrenergic neurones) where it can act as an ACh (nicotinic) receptor. The binding of ACh would release calcium ions (as well as PGF 1α) which could enter the chromaffin granules and disrupt, in turn, the storage complex.

It should also be noted that the role described above for RNA in the active transport channels could be filled as well by a double helix of stacked nucleotides or of a protein-nucleoside complex. A suitable protein for this has just been identified in hen oviduct— that is, polyglutamate, which would form the required tubes in a complex with guanine nucleosides (GMP, GDP, or GTP) and spermine molecules.

C. Cyclic AMP

Many of the effects of the PG's can be explained if they could be supposed to promote the binding of cyclic AMP (and other adenine nucleosides such as ADP and ATP) in some metabolically inactive store, from which the nucleoside would be released by a catecholamine. Such a store could consist of an array of nucleoside molecules bound to their complementary amino acid (glutamine) in a locus as every alternate amino acid in a protein chain. This will bind PGA on the minor side and PGF on the major side (but not PGE) which would tie the complex together. It can also bind NE (on the major side) so as to disrupt the complex. Prostaglandins could also be involved in binding ATP onto the active site on adenylcyclase in conjunction with a catecholamine.

If the adenine–cyclic AMP pairs are interspersed with (single) glutamate–guanine pairs, this can now bind PGF(A) as before and be disrupted by epinephrine, and if the single interspersing pairs are arginine–cytosine, this will now bind any PG (E and F on major side ; A on minor) and be specifically blocked by polyphoretin phosphate (a specific antagonist of PG action[21]) which, as the dimer, can bind by no less than five good hydrogen bonds and four lipophilic contacts.

VI. REFERENCES

1. S. Ehrenpreis, J. A. Fleisch, and T. W. Mittag, Approaches to the molecular nature of pharmacological receptors, *Pharm. Rev.* **21**:131 (1969).
2. C. B. Kasper and D. M. Kashing, Isolation and characterization of a mammalian nuclear membrane, *Fed. Proc.* **28**:404 (1969).
3. H. W. S. King and W. Fitschen, Characterization of RNA from the smooth endoplasmic reticulum of rat liver, *Biochim. Biophys. Acta* **155**:32–37 (1968).
4. M. Takagi and K. Ogata, Direct evidence for albumin biosynthesis by membrane bound polysomes in rat liver, *Biochem. Biophys. Res. Comm.* **33**:55–50 (1968).
5. L. S. Tulegenova, N. P. Rodionova, and V. S. Shapot, Chemical fractionation of microsomes and metabolic activity of membrane RNA from rat liver, *Biochim. Biophys. Acta* **166**:265–267 (1968).

6. I. G. Morgan and L. Austin, Synaptosomal protein synthesis in a cell-free system, *J. Neurochem.* **15**:41–51 (1968).

7. T. Namba and D. Grob, Cholinergic receptors in skeletal muscle, *Ann. N.Y. Acad. Sci.* **144**:772–802 (1967).

8. H. Passow and B. Lowenstein, An all-or-none response in the release of potassium by yeast cells treated with methylene blue and other basic redox dyes, *J. Gen. Physiol.* **43**:97–107 (1960).

9. M. Stacey, *in The Nature of the Bacterial Surface* (A. A. Milne and N. W. Pine, eds.), p. 29, C. C. Thomas Co., Springfield, Ill. (1949).

10. A. I. Lansing and T. B. Rosenthal, The relation between ribonucleic acid and ionic transport across the cell surface, *J. Cell Comp. Physiol.* **40**:337–345 (1952).

11. J. Hiraoka and H. Takada, *J. Inst. Polytech. Osaka City Univ., Series B*, **8**:59 (1957).

12. F. W. Cope, A non-equilibrium thermodynamic theory of leakage of complexed Na^+ from muscle, with NMR evidence that the non-complexed fraction muscle Na^+ is intra-vacuolar rather than extra-cellular, *Bull. Math. Biophys.* **29**:691–704 (1967).

13. K. L. Yielding and H. Sterglanz, Lysergic acid diethylamide (LSD) binding to deoxyribonucleic acid (DNA), *Proc. Soc. Exp. Biol. Med.* **128**:1096–1098 (1968).

14. T. E. Wagner, *In vitro* interaction of LSD with purified calf thymus DNA, *Nature* **222**:1170 (1969).

15. J. R. Smythies and F. Antun, Binding of tryptamine and allied compounds to nucleic acids, *Nature* **223**:1061 (1969).

16. S. Ohnishi and H. McConnell, Interaction of the radical ion of chlorpromazine with deoxyribonucleic acid, *J. Amer. Chem. Soc.* **87**:2293 (1965).

17. J. R. Smythies, *Int. Rev. Neurobiol.* **13** (in press).

18. J. T. Garland and J. Durell, *Int. Rev. Neurobiol.* **13** (in press).

19. J. R. Smythies, F. Antun, G. Yank, and C. Yorke, Molecular mechanisms of storage of transmitters in synaptic terminals, *Nature* (in press).

20. P. W. Ramwell, J. E. Shaw, W. W. Douglas, and A. M. Poisner, Efflux of prostaglandin from adrenal glands stimulated by acetylcholine, *Nature* **210**:273–274 (1960).

21. K. E. Eakins and S. M. M. Karim, The nature of the prostaglandin-blocking activity of polyphoretin phosphate, *J. Pharm. Exp. Therap.* (in press).

Chapter 20

DISSIPATIVE TRANSPORT PROCESSES

William A. Brodsky and Adil E. Shamoo

Departments of Physiology and Biophysics
Mount Sinai Medical and Graduate Schools of the City University of New York

Irving L. Schwartz

Departments of Physiology and Biophysics
Mount Sinai Medical and Graduate Schools of the City University of New York
and
The Medical Research Center
Brookhaven National Laboratory, Upton, N.Y.

I. INTRODUCTION

A. Concepts and Evolution

The earliest concept of the plasma membrane was that of an ultrathin porous film separating two aqueous solutions in physical as well as in biological systems. In physical systems, use of the porous film model led to the development of the laws of osmotic pressure and to the laws governing the equilibrium distribution of ions across membranes. In biological systems, use of the model has accounted for some, but not for all of the data on distribution and transfer rates of materials across cell membranes. Recent developments have been concerned with the chemical architecture of biological membranes as well as the material transfers across cell membranes.

The current concept of the plasma membrane[1,2] is that of a bilayer film composed of phospholipid and protein molecules which may or may not interact with molecules which move across the membrane through its aqueous and nonaqueous interstices. In addition, the plasma membrane is apparently a metabolically responsive entity capable of regulating the material exchange between the cell and its external environment.

The evolution of cell membranes along with that of cellular life itself has been related to the origin of organic matter on the planet.[3-5] The latter has been ascribed to interactions between electrical discharges (lightning) and the primitive gaseous milieu consisting of methane, water, ammonia, and CO_2 leading to the formation of amino acids, nitriles, and hydroxy organic acids. The first biomolecular structures could have arisen in the aqueous environment from the adsorption to clay and subsequent polymerization of such amino and hydroxy acids (under the proper conditions of temperature, humidity, and electromagnetic radiation), with the formation of membranes and precursor enzyme proteins.[6] These membranes in the form of spherical sacs could have provided the original framework for the evolution of life.

B. Selective Permeability

Even the most primitive of the biological membranes holds the enzyme proteins, nucleic acids, and other macromolecules together in one package, thereby providing isolation of the cellular metabolism from the environment, and forming a barrier to the leakage of essential metabolites from the cell, as well as to that of toxic substances into the cell. Apart from its role as a barrier, the living membrane can play a role in the facilitation of the migration of some substances into or out of the cell—e.g., glucose entry into the cell or urea efflux from the cell occurs more rapidly than that which is known to occur through artificial membranes of similar dimensions. Thus, the membrane functions as a selective sieve promoting exchanges of essential substances between the cell interior and its surroundings. Finally, constituents of the membrane can bind with certain substances (sodium, glucose, and amino acids), and move them from one to the other boundary fluid in an "uphill" direction utilizing chemical energy derived from metabolic reactions.

In the case of the transport of nonelectrolyte, "uphill" means net movement across the membrane from a region of low concentration in one boundary fluid to a region of high concentration in the opposite boundary fluid; and in the case of transport of electrolytes (ions), "uphill" means movement from a region of low electrochemical potential energy in one boundary fluid to a region of high electrochemical potential energy in the opposite boundary fluid. In other words, the charged particle can be moved against the transmembrane gradient of electrochemical potential.[7]

In the general sense, "uphill" or active transport means the net movement of any substance across the membrane from a region of low to a region of high internal energy of the translocated substance. Parenthetically, this does not exclude the possibility that an active transport process can facilitate "downhill" movement of the translocated substance.

C. Integration of Membrane Properties and Cell Functions

Because it can control the flow and balance of materials into and out of cells, the membrane is involved in the regulation of the chemical composition and volume of the cell fluid and in the regulation of those metabolic activities which depend upon delivery of exogenous material—e.g., the dependence of oxygen consumption rate upon sodium transport in some cells.

1. Relation to Common Needs of All Cells

The control of transmembrane flows with the consequent homeostatic balance of material within cells plays important roles in cellular growth, reproduction, and adaptation.

During the growth of some cells, the premitotic swelling is accompanied by transient accumulation of electrolyte and water until a new steady state of material flow exists which implies that a change in the permeability properties of the membrane is a required part of the process of growth or reproduction.

During adaptation of certain microorganisms to changes in the external environment, genetically induced synthesis of new protein causes changes in the chemical composition and selective permeability of bacterial cell membranes as well as changes in the enzymatic makeup of the cell interior.[8] The biosynthetic process is presumed to be mediated by the environmentally induced removal of a repressor molecule from a specific operon system in the bacterial DNA with the consequent synthesis of three proteins—one of which, the so-called "permease," causes the appropriate change in membrane permeability.

2. Relation to Special Needs of Some Cells

The current picture of the living cell is that of a highly organized internally regulated set of enzymatically catalyzed reactions, some of the free energy of which can be funneled via membrane processes into useful functions. Examples include: (1) the transmission of excitation along the axon; (2) the excitation-contraction phenomenon in muscle; (3) the maintenance of constant chemical composition and volume of all cells; and (4) the production of secretory fluids by the kidney and by all of the exocrine glands (sweat, salivary, gastric, intestinal, mammary, bronchial epithelia, etc.).

II. CONCEPT OF THE MEMBRANE

A. Structural

The classical picture of the molecular architecture of cell membranes is that of a biomolecular leaflet of phospholipids, 50–100 Å in thickness.[9] A given pair of phospholipid molecules is oriented so that the two phosphatidyl groups (the ionizable groups) face outwardly toward each of the aqueous boundary fluids to form the membrane interfaces, while the hydrocarbon moieties face inwardly toward each other to form the inner portion of the membrane. Proteins are supposed to be distributed over the membrane surface in a sheetlike conformation. Evidence for the bilamellar array has been assembled from the patterns of X-ray diffraction[10] and electron microscopy[11,12] of the myelin sheath. The "unit-membrane" pattern of electron micrographs[11] consists of two bands, each of 25 Å in width, separated by a low density band about 25 Å wide. The spacing of the dense bands in the electron micrographs corresponds to the pattern of X-ray repeats; and the dark band appearance has been attributed by some workers to the reaction of the ionizable phosphatidyl groups with the osmium stain.

Revision of the classical picture has been prompted by work showing: (1) that osmium tetroxide reacts with the unsaturated bonds of the fatty acyl chains, rather than with the phosphatidyl head groups and consequently that the "unit membrane" thickness varies from 50 to 200 Å[13,14]; (2) that the main chemical forces between lipid and protein are hydrophobic and not electrostatic in nature[15]; (3) that the conformation of isolated membrane proteins is mostly α-helical and random coil with a small amount in the sheetlike or β-conformation—as deduced from spectra of optical rotatory dispersion and circular dichroism, and that the helical and coiled molecules penetrate the entire thickness of the membrane; and (4) that there exist several stable arrays of model lipid-water mixtures including bilamellar, hexagonal, rhombohydric and polyhydric, as can be deduced from patterns of X-ray diffraction.[16,17]

According to a current picture, reviewed by Korn,[18] the cell membrane is composed of a bilayer of phospholipid fitted into aggregates of α-helical and randomly coiled molecules. The picture is that of a dynamic membrane, with transitions between the various arrays (bilamellar, hexagonal, and globular micelles, etc.) of the phospholipids, and with conformational transitions of the proteins. Such a protein-rich dynamic structure can interact with specific transported molecules as well as with specific cellular metabolites for its own biosynthesis and required energy transformations.

B. Biochemical

Cell membranes are supposed to contribute a major component to the insoluble "microsomal" pellet, separated from disrupted cells by conventional methods of ultracentrifugation. Chemical analyses have revealed that such pellets consist mainly of phospholipids, proteins, cholesterol, and small amounts of polysaccharides.[19]

The protein-containing fraction has been implicated in the selectivity of transport across cell membranes; and specific binding between isolated membrane proteins and transportable substrates has been demonstrated in several types of cells.[20] The phospholipid fraction of the membrane has recently been implicated as a nonspecific translocator of sugars through nonpolar solvents and across the liquid–liquid interfaces between such solvents and water.[21-23]

Binding of solutes which traverse cell membranes (penetrant molecules) has been demonstrated in membrane proteins isolated from cells which transport sodium, galactosides, sulfate, and calcium.[20] The exact connection between substrate-specific binding to isolated proteins and substrate transport across intact cell membranes remains uncertain; but the nature of this connection has been approached closely in a few cases. For example, the parallelism between kinetic parameters of Na–K-sensitive ATPase in microsomal pellets and those of sodium transport by the intact system has been demonstrated.[24-27] In addition, substrate-specific binding can be demonstrated in proteins eluted from bacterial membranes which transport the substrate (e.g., substrates such as sulfate or galactosides), but not from mutant strains which do not transport the substrate.

Translocation of the carrier-substrate complex within the membrane has been approached directly by LeFevre who has studied the movement of glucose-^{14}C phospholipid extracts from red cells across an artificial liquid system of water–chloroform–water. Indirect approaches to the translocation process have been made by studying the acceleration of ion transport across intact mitochondria[28,29] or across synthetically prepared phospholipid membranes of about 70 Å thickness (black lipid membranes) after addition of certain antibiotics (valinomycin, nigericin, monactin–dinactin, etc.)[30–33] and of synthetic polycylic ethers which are known to form positively charged clathyrate complexes with the cations.[34]

Another important biochemical concept of the cell membrane is that of a coupling transducer between metabolism and transport. The free energy released from cellular metabolism must be funneled into the active transport process in some manner. However, in some cases, the coupling may operate in the reverse direction—i.e., the free energy released from the downhill ion transport may be funneled into metabolic reactions—as has been inferred from the sodium-dependent rate of oxygen consumption in secretory cells which transport sodium[35]; and from the dependency of oxidative phosphorylation in mitochondria upon the transfer of hydrogen ions and electrons across the mitochondrial membranes.[36] Apart from the question of whether transport or metabolism provides the primary driving force, the coupling between scalar processes of metabolism and vectorial processes of transport is an example of the Curie–Prigogine principle. This physical principle holds that coupling between scalar and vectorial processes is feasible only when the two processes operate in series in a nonisotropic medium—e.g., such as that formed by the interfaces between a cell membrane and its aqueous boundary fluids.[37]

Whereas the current biochemical approach is concerned largely with interactions between penetrant molecules and carrier substances and with bioenergetic coupling of transport and metabolism, the classical concept of permeability is based on the translocation of free ions or molecules which do not form mobile carrier complexes during their passage across the membrane, i.e., the permeability of model membranes to a given substance has been classically defined only in terms of dissipative processes, such as diffusion. Where the transported substance apparently binds with a membrane substance to form a mobile-carrier complex the permeability must be determined from the back-diffusion or leakage rate of the transported substance—a determination fraught with the difficulty of knowing how much of the leak is through a path in parallel with that of the carrier-complexing reaction.[38]

C. Physiological

Physiological studies of the cell membrane have dealt mainly with the kinetic and thermodynamic aspects of transport processes.

1. *Kinetic*

The flux of molecules across membranes has been related to transmembrane gradients through the "membrane permeability"—which is defined as flux per unit concentration gradient. This permeability of many kinds of membranes varies with concentration gradients, electrical potential gradients, temperature, molecular size of the permeant molecule, and the number of hydrogen bonding sites on it. This means that the mechanism of transfer across several cell membranes is not described by simple diffusion of uncharged particles and/or that the path of diffusion is not necessarily that of a simple aqueous pore penetrating the membrane. The kinetic approach has exploited the concept of diffusion of a molecule by random jumps across the lattice network of a membrane in a manner similar to that of diffusion of molecules in liquids or solids.

2. *Thermodynamic*

In the steady state, cells maintain transmembrane concentration gradients of most native permeant molecules. Physiologists have studied the nature of forces required to produce and maintain such gradients (or distribution ratios).[39] In some cases (e.g., Donnan ratios of ions across capillary membranes), the distribution is an inevitable physical consequence of the known forces acting across the membrane. In other cases, the steady-state distribution ratios cannot be explained by the known transmembrane forces (i.e., gradients of electrochemical potential and hydrostatic pressure); therefore chemical forces between membrane constituents and the transported material have been invoked. In any case, the interior of most cells is K-rich and Na-poor while their external bathing fluids, except for unicellular freshwater organisms, are Na-rich and K-poor. Such a distribution of ions, together with maintenance of a constant cell volume, requires that the transport of at least one ion (sodium) be coupled to an energy-yielding reaction of metabolism. The concomitant transport of other ions (e.g., potassium, chloride) could be dissipative ("passive") and/or active in nature.

III. MECHANISMS OF PENETRATION

A. Lattice *vs.* Porous Pathway

A comprehensive theory of the motion of molecules through cell membranes would include the movement of a water-soluble molecule across the aqueous-membrane interface, its subsequent passage across a hydrophobic, nonisotropic barrier, and its emergence from the barrier across the second interface into the opposite aqueous boundary fluid. The relative motion of molecules in a membrane, like that in the solid or liquid state, is dependent upon the steric configurations and water content of interstices of the membrane lattice, the spatial distribution of charged groups, the surface energy barriers of interfaces between hydrophobic and hydrophilic regions, and the frictional, electrical, and chemical interactions between molecules in motion relative to one another. Two different theories of diffusion or viscous flow through membranes have been used—the lipid lattice theory and the aqueous pore theory (Chapter 3, Ref. 35).

1. Lattice Theory

This theory holds that the membrane lipids in the bimolecular leaflet array constitute a symmetrical hydrophobic barrier to the passage of water-soluble penetrant molecules. The penetrant molecules interact with the membrane surface so that hydrogen bonds between the penetrant and the aqueous solvent are broken before entry of the penetrant into the hydrophobic phase of the membrane. If this step were rate-limiting, the rate of penetration across the cell membrane would be an inverse function of the number of hydrogen bonding groups on the penetrant molecule, which has proved to be the case in several instances.[40] After entry, the penetrant becomes temporarily trapped by forming weak bonds with some of the atoms in the lipid lattice, a step analogous to the formation of transition complexes in the Eyring theory of diffusion through a liquid lattice. The penetrant, by virtue of the environmental thermal energy, shakes loose from the lattice site, migrates in the interstices until it anchors to another lattice site by hydrogen bonding, and by a series of such random walk steps, penetrates the entire thickness of the membrane (Chapter 3, Ref. 35).

The overall picture predicts several observations on the rate of penetration of lipid-soluble, uncharged molecules. Thus, a large series of such molecules penetrate biological membranes at rates related directly to the oil-water solubility coefficients of the penetrants; and the transfer rates, less than 1/1000th of those in an equivalent thickness (50–200 Å) of water, are of the order of those found in synthetic highly cross-linked rubberlike polymers. Moreover, the transfer rates of several molecules across cell membranes are increased four to sixfold by substituting nonhydrogen bonding groups ($-CH_2$) for the hydrogen bonding groups ($-OH$ etc.) on the penetrant.[35,39]

On the other hand, the lattice picture, as presented, fails to account for the following: (1) the ability of some membranes (nerve, muscle, red cell) to discriminate between cations and anions; (2) the fact that water permeability estimated from the net flow of water during osmosis is greater than that estimated from the unidirectional diffusional flow of isotope (D_2O) during zero net flow of water in some membranes (red cell, frog skin, etc.); and (3) the fact that some penetrants (glucose, urea) move through the membrane at aberrantly high rates (compared to those expected in passive diffusion across cell membranes or across synthetic polymerized rubberlike membranes).

2. Aqueous Pore

This theory holds that tortuous water-filled channels of about 3.5 Å radius penetrate the entire thickness of the membrane. The walls of such channels (or pores) are lined with fixed charged groups—the position, density, and sign of charge which are dependent upon the conformational array of phospholipid and protein molecules of the surrounding membrane structure.

The model has been used to account for the cation-anion selectivity of such membranes (nerve, muscle, red cell, etc.) as well as for the high ratio of osmotic to diffusional permeability of these membranes to water. The physical basis for osmotic flow of solvent across semipermeable membranes has been related to the frequency of collisions of water and soluble molecules with the pore apertures on both surfaces of the membrane; and in this sense, osmotic flow through pores may be compared to the kinetic-molecular theory of the gaseous state.

Stein has reviewed the theoretical and experimental evidence for the pore theory and has noted its major defects.[35] Thus, a pore radius of 3.5 Å, as estimated for the human red cell[41] is sufficient to explain the membrane selectivity due to sieving for some, but not for all penetrants. Moreover, the high osmotic permeability relative to diffusional permeability for water (a property allegedly unique to porous membranes), has been found in presumably pore-free synthetic phospholipid films[42] and in thick films of mesityl oxide.[43]

3. Limitations of Pore and Lattice Theories

Neither the pore nor the lattice theory, without additional qualification, accounts for the following: (1) the solute-solute interaction as seen in "single file" diffusion of potassium across the axonal membrane; (2) the selectivity among different cations (Na vs. K) or among different anions (Cl vs. HCO$_3$) as is found in many cell membranes; and (3) the interaction between the permeant and a membrane constituent as inferred from the kinetic patterns of the passive and/or active transport of physiologically important substances (sugar, amino acids, and sodium) across almost all cell membranes.

B. Phenomenological Approach

The uncertainties inherent in the lattice and pore models have stimulated the use of irreversible (nonequilibrium) thermodynamics in the study of membrane processes. The thermodynamic approach, independent of any specific mechanism of transmembrane flow and of path geometry, provides a precise definition of membrane permeability with respect to the passive transfer of any permeant. The physical basis for irreversible thermodynamics and its relation to the phenomenological laws governing irreversible processes was developed by Onsager, and first applied to transport processes in biological membranes by Kedem and Katchalsky.[44,45]

The approach is to use the experimentally established laws showing that the flow rates (of matter or energy) in several irreversible processes are directly proportional to driving forces—as is the case in Fick's law of diffusion or Ohm's law of electrical flow. In a single flow system, the flow is related directly to a conjugate force by a "straight" phenomenological coefficient. In a multiflow system, each flow is related to all of the driving forces—(conjugate and nonconjugate) in the system; and the relation to the non-conjugate forces is by "coupling" coefficients. An example of a coupled process is the flow of heat induced by an electrical force, and conversely, the flow of electricity induced by a temperature gradient (i.e., the thermoelectric effects responsible for the operation of thermoelectric coolers and heaters and thermocouples).

The set of equations representing the "straight" and "coupled" dependencies of a set of flows and forces can be expressed in matrix form as:

$$J_i = \sum_j L_{ij} x_j, \qquad i = 1, 2, \ldots n \qquad (1)$$

where J_i denotes flux of the ith component in a multiflow system. L_{ij} denotes the coefficient in the ith row and the jth column of the L matrix designed so that $i = j$ for the "straight" coefficients; $i \neq j$ for the coupling coefficients and X_j denotes the jth driving force.

Expanding from matrix notation to the algebraic form, the set of fluxes and forces becomes:

$$J_1 = L_{11}X_1 + L_{12}X_2 + \cdots L_{1n}X_n$$

$$J_2 = L_{21}X_1 + L_{22}X_2 + \cdots L_{2n}X_n$$

$$\vdots \qquad\qquad \vdots$$

$$J_n = L_{n1}X_1 + L_{n2}X_2 + \cdots L_{nn}X_n \qquad (2)$$

whence L_{11}, L_{22}, L_{nn}—i.e., L $(i = j)$ are the straight phenomenological coefficients (e.g., the diffusion coefficient in Fick's law, or the electrical conductance in Ohm's law); and the L_{ij}'s $(i \neq j)$ are the coupling coefficients which relate the nonconjugate forces to flux.

In a more familiar notation, let J_v, J_d, J_e and J_q represent relative flow rates of solvent, of solute, or electricity and of heat, respectively, and let X_p, X_d, X_e, X_q represent the corresponding conjugate forces (gradients of pressure, of solute concentration, of electrical potential, and of temperature) chosen so that the "straight" phenomenological coefficients, L_p, L_d, L_e, and L_q correspond to hydraulic permeability, diffusive conductance, electrical conductance, and thermal conductance respectively. Correspondingly, the straight (or uncoupled flows) L_pX_p, L_dX_d, L_eX_e and L_qX_q become diagonal elements in the matrix of simultaneous equations representing the multiflow interacting system as follows:

$$J_v = \boxed{L_p X_p} \quad\quad + L_{pd} X_d \quad\quad + L_{pe} X_e \quad\quad + L_{pq} X_q$$

$\quad\quad$ (hydraulic flow) $\quad$ (osmosis) $\quad\quad$ (electro $\quad\quad\quad$ (thermo
$\quad\quad\quad\quad\quad\quad\quad\quad\quad\quad\quad\quad\quad\quad\quad\quad\quad$ osmosis) $\quad\quad\quad$ osmosis)

$$J_d = L_{dp} X_p \quad\quad + \boxed{L_d X_d} \quad\quad + L_{de} X_e \quad\quad + L_{dq} X_q$$

$\quad\quad$ (filtration) $\quad\quad\quad$ (diffusion) $\quad\quad$ (electro $\quad\quad\quad$ (thermo
$\quad\quad\quad\quad\quad\quad\quad\quad\quad\quad\quad\quad\quad\quad\quad\quad$ phoresis) $\quad\quad\quad$ diffusion)

$$J_e = L_{ep} X_p \quad\quad + L_{ed} X_d \quad\quad + \boxed{L_e X_e} \quad\quad + L_{eq} X_q \quad\quad (3)$$

$\quad\quad$ (streaming $\quad\quad\quad$ (diffusion $\quad\quad$ (electric $\quad\quad\quad$ (thermo-
$\quad\quad\quad$ current) $\quad\quad\quad\quad$ current) $\quad\quad\quad$ flow) $\quad\quad\quad\quad$ electric
$\quad\quad\quad\quad\quad\quad\quad\quad\quad\quad\quad\quad\quad\quad\quad\quad\quad\quad\quad\quad\quad\quad\quad\quad\quad$ current)

$$J_q = L_{qp} X_p \quad\quad + L_{qd} X_d \quad\quad + L_{qe} X_e \quad\quad + \boxed{L_q X_q}$$

$\quad\quad$ (pressure $\quad\quad\quad$ (differential $\quad\quad$ (electro $\quad\quad\quad$ (heat flow)
$\quad\quad\quad$ pyresis) $\quad\quad\quad\quad$ pyresis) $\quad\quad\quad$ pyresis)

where the straight coefficients, L_{ii}, appear with the conjugate forces in the boxed diagonal terms ($i = j$) of the matrix, and where the coupling coefficients, L_{ij} appear with the nonconjugate forces in the off-diagonal terms ($i \neq j$) of the matrix. Onsager's theorem shows that the coupling coefficients, related in a symmetrical and reciprocating manner, can be expressed in the form,

$$L_{ij} = L_{ji} \quad\quad (4)$$

which equates the cross coefficients in any pair of coupled processes. Apart from simplifying the mathematical manipulations involved in evaluating L_{ij} terms, the Onsager equality provides operational insight into the nature of interacting forces and fluxes in multiflow systems.

$\quad$ The permeability of a membrane to a solute corresponds to one of the conductance elements (L_d) in the L_{ij} matrix, and its evaluation depends upon ensuring that all other forces are vanishingly low during the experiments or upon performing enough experiments to solve for all L_{ij} terms.

$\quad$ Whereas the phenomenological equations are useful for conceptualization, the quantities defined and the form in which they are cast are not identical to what is usually measured in experiments on irreversible processes. The phenomenological form for the simple diffusion of a nonelectrolytic solute, S, through an aqueous solvent system is analogous to Fick's law which holds that the relative velocity, J_d, of a diffusing species is proportional to its gradient of chemical potential or

$$J_d = \mathscr{L}_d(-d\mu_s/dx), \quad\quad (5)$$

where J_d (usually expressed in units of volume per unit area per unit time, or moles per unit area per unit molal concentrations per unit time) is the velocity of a solute relative to that of solvent, where $\mathscr{L}_d$, the constant of proportionality, is the so-called straight phenomenological coefficient which is not a function of concentration and which can be regarded as the reciprocal of the coefficient of friction between solute and solvent molecules and where $d\mu_s/dx$, the gradient of chemical potential, is the conjugate driving force impelling directionally oriented motion to the solute particles.

The operational form for expressing this type of diffusion is:

$$J_s = -P_s \Delta C_s \tag{6}$$

where J_s is the flux of solute relative to the fixed inertial frame (e.g., the membrane) at *any* concentration of solute; where P_s is the permeability of the membrane to S, and where ΔC_s is the transmembrane difference in concentration of S.

J_d, the velocity of solute relative to that of solvent, the molal velocity of solute, may be related to J_s as follows:

$$J_d = J_s/C_s - J_v \tag{7}$$

where J_v is the velocity of solvent relative to membrane, and where J_s, the measured solute flux across the membrane at any concentration, is normalized with respect to C_s so that its units are consistent with those of J_d and J_v.

Substitution of Eq. (5), which is Fick's first law, into Eq. (7) gives

$$J_s = C_s \mathscr{L}_d(-d\mu_s/dx) + C_s J_v \tag{8}$$

which requires the restriction that $J_v = 0$ be consistent with the uncoupled type of pure diffusion described by Eqs. (5) and (6). With this restriction ($J_v = 0$) and in the steady state, since $\mu_s = \mu_s^0 + RT \ln C_s$, Eq. (8) may be integrated to yield

$$J_s = -\mathscr{L}_d RT \Delta C_s/\Delta X$$
$$= -D_s \Delta C_s/\Delta X \tag{9}$$

where $\mathscr{L}_d RT = D_s$, the self-diffusion coefficient; and where $-D_s/\Delta X = P_s$, the permeability coefficient, which leads to

$$J_s = P_s \Delta C_s \tag{10}$$

Further insight into the relation between J_d and J_s may be gained by relating the velocity of a diffusing molecule to its flux through the identity,

$$dn/dt = CAJ \tag{11}$$

where dn/dt, the total flux, is the number of moles of solute at concentration, C, moving normal to an area of solution, A, at a velocity of J and where the frame of reference for the solute is not defined. Relating Eq. (11) to the phenomenological and practical forms of Eqs. (5) and (6) respectively yields the expressions

$$J_d = \frac{dn_s/dt}{C_s A} \tag{12}$$

and

$$J_s = \frac{dn_s/dt}{A} \tag{13}$$

which imply that $J_d C_s = J_s$ when $J_v = 0$, as is evident from Eq. (7).

However, the inadvertent application of this restriction to the evaluation of permeability of membranes across which solvent flow occurs ($J_v \neq 0$) has resulted in erroneous estimates of membrane permeability. Even in a closed system of fixed volume, diffusion of solute through free solution or through membranes is inevitably accompanied by counterdiffusion of solvent.

In the case of a coupled process of solvent and solute flow across a membrane, the phenomenological equations encompassing both flows are:

$$J_v = L_p \Delta P + L_{pd} RT \Delta C_s \tag{14}$$

$$J_d = L_{dp} \Delta P + L_d RT \Delta C_s \tag{15}$$

where J_v is the velocity of solvent relative to that of solute; where J_d is the velocity of solute relative to that of solvent; where $\Delta P = X_p$ and $RT \Delta C_s = X_d$; where the solute is a nonelectrolyte and where the flow of heat and electricity are ignored.

The equations can be solved under steady-state conditions for three different cases—that of a "perfect" osmotic membrane; that of a "leaky" osmotic membrane; and that of a "nonselective" membrane.

1. Perfect Osmotic Membrane

In a perfect osmotic membrane, the separation of solvent from solute is complete, since only solvent penetrates the membrane, and the relative motion of solvent with respect to solute is expressed as

$$J_v = -J_d \tag{16}$$

which leads to

$$(L_p + L_{dp})\Delta P + (L_{pd} + L_d)RT \Delta C_s = 0 \tag{17}$$

and which requires that

$$(L_p + L_{dp}) = 0 = (L_{dp} + L_d)$$

for nonzero values of ΔP and ΔC, and consequently

$$L_p = L_d = -L_{dq} = -L_{pd} \tag{18}$$

indicating that a perfect osmotic membrane is completely described by one coefficient.

For the condition $J_v = 0$, where $-L_{pd}/L_p = 1$, where ΔP is the stopping pressure and where ΔC_s is the transmembrane difference of solute concentration, we obtain from Eq. (14)

$$\Delta P = -(L_{pd}/L_p)RT\Delta C_s$$

and, therefore,

$$\Delta P = RT\Delta C_s \tag{19}$$

which is the equation for the van't Hoff law of osmotic pressure.

2. Leaky Osmotic Membranes

The coupling relationships between the two forces and the two flows, applicable to leaky as well as to perfect osmotic membranes, can be derived separately in ratio form from Eqs. (14) and (15) under each of the following conditions: $J_v = 0$; $\Delta C_s = 0$; $J_d = 0$; and $\Delta P = 0$.

For $J_v = 0$, the pressure required (the effect) to stop the solvent flow induced by a given transmembrane concentration gradient (the cause), is

$$(\Delta P/RT\Delta C_s)_{J_v=0} = -L_{pd}/L_p = \sigma_s \tag{20}$$

where σ_s, the Staverman reflection coefficient, is 1.0 in the case of a perfect osmotic membrane and where $0 < \sigma_s < 1.0$, in the case of a leaky osmotic membrane. Thus, the stopping pressure in a real (leaky) membrane is less than that in an ideal (nonleaky) membrane for a given transmembrane gradient of concentration.

For $\Delta C_s = 0$, the relative velocity of solute per velocity of solvent is defined as

$$(J_d/J_v)_{\Delta C_s=0} = L_{dp}/L_p = \sigma_s \tag{21}$$

which illustrates the flow ratio induced by applying a hydrostatic pressure across the membrane. In a perfect osmotic membrane, the relative velocity of solute relative to that of solvent is maximal and $J_d = -J_v$, since 100% of the solute is "reflected" from the membrane. In a "leaky" membrane some solute is dragged across the membrane by solvent flow and, accordingly, the reflection of solute at the membrane is less than 100%, or $\sigma_s < 1.0$.

For the condition that $J_d = 0$, the chemical concentration gradient (the effect) required to stop the relative solute flow induced by an applied hydrostatic pressure (the cause) is

$$(RT\Delta C_s/\Delta P)_{J_d=0} = -L_{dp}/L_d \tag{22}$$

which is not the reciprocal of the stopping pressure relation of Eq. (20) where the relative flow of solvent, J_v, and not that of solute, J_d, is zero.

But $J_d = -J_v = 0$ in a perfect osmotic membrane under conditions of $J_v = 0$, suggesting that the requirement for van't Hoff law can be satisfied by the experimental maneuver of Eq. (22) under the proper conditions.

In the case of a leaky membrane, the condition $J_d = 0$ requires further analysis. In this case Eq. (7) reduces to

$$J_s = J_v C_s \tag{23}$$

indicating that there is a bulk flow of solute and solvent through the leaky membrane when $J_d = 0$.

For the condition that $\Delta P = 0$, it can be seen from Eqs. (14) and (15) that the velocity of solvent relative to that of solute is

$$(J_v/J_d)_{\Delta P = 0} = L_{pd}/L_d \tag{24}$$

which illustrates the flow ratio produced by a transmembrane concentration gradient. In a perfect (nonleaky) membrane, $|J_v/J_d| = 1$ and in a real (leaky) membrane, $|J_v/J_d| < 1.0$. This means that the velocity of solvent relative to that of solute is less in the real membrane than in the ideal osmotic membrane.

By definition,

$$J_s = (J_v + J_d)C_s \tag{25}$$

where J_s is the solute flux referred to the membrane and where $J_v C_s$ and $J_d C_s$ are the contributions of hydraulic and diffusional flows to the net solute flux. Using Eqs. (4), (13), (14), (15), (20), and (21), and solving for J_s in terms of J_v, $RT\Delta C_s$, L_p, and L_d leads to

$$J_s = (1 - \sigma_s)C_s J_v + (L_d - \sigma_s^2 L_p)C_s RT\Delta C_s \tag{26}$$

which can describe solute flux across a perfect osmotic membrane ($\sigma_s = 1$); across a leaky osmotic membrane ($0 < \sigma_s < 1$) or across a nonselective membrane ($\sigma_s = 0$).

Thus when $\sigma_s = 1.0$, $J_s = 0$, which states that net solute flux is zero across a perfect osmotic membrane; and when $\sigma_s = 0$, we have the condition required for a completely nonselective membrane.

3. Nonselective Membrane

In the case of a nonselective membrane which both solvent and solute penetrate equally well, there is no solute reflection ($\sigma_s = 0$). It follows that the osmotic pressure difference vanishes for all values of ΔC_s, because there is no velocity of solute relative to solvent other than that defined by pure diffusion in free solution.

From Eq. (26), when $\sigma_s = 0$, we obtain,

$$J_s = C_s J_v + L_d C_s RT\Delta C_s \tag{27}$$

which describes net solute flux as the algebraic sum of solute flow carried by bulk flow $C_s J_v$ and the diffusional flow ($L_d C_s RT\Delta C_s$) across a nonselective membrane. Clearly this formulation applies as well to the case of bulk flow and diffusion in free solution. In other words, if $J_v = 0$, the pure diffusional flux $J_s = L_d C_s RT\Delta C_s$ is defined; and if $\Delta C_s = 0$, the pure bulk flow, $J_s = C_s J_v$ is defined.

C. Reversal of Osmotic Flow in Biological Systems

Osmotic work is done whenever a mechanism operates to effect a transfer of water from a concentrated to a more dilute solution. Osmoregulatory processes are exemplified by the function of the gills of fish and the renal tubules, gastrointestinal and other transporting epithelia of mammals.

Inasmuch as a primary mechanism for the active transport of water molecules has never been demonstrated, there has been much interest in indirect devices for moving water up its gradient of chemical potential. Such devices depend on osmotic gradients, on differential membrane permeabilities, and/or on the geometry of the diffusion path.

An active mechanism for sodium transport operating in a system with an appropriate array of membranes (the three-compartment, two-membrane system) or with long, narrow channels (the standing gradient system) can produce uphill movement of water.

1. Three-Compartment, Two-Membrane System

Figure 1 is a schematic plot of the osmotic concentration versus distance across three compartments (LF, $CELL$, ISF) and across two membranes (M_1 and M_2) bounding a cell which can transport water from a hypertonic fluid in LF to a hypotonic fluid in ISF.

The system operates to pull water osmotically from LF to the hypertonic fluid in the cell, increasing the hydrostatic pressure and consequently forcing both solute and water into compartment ISF.

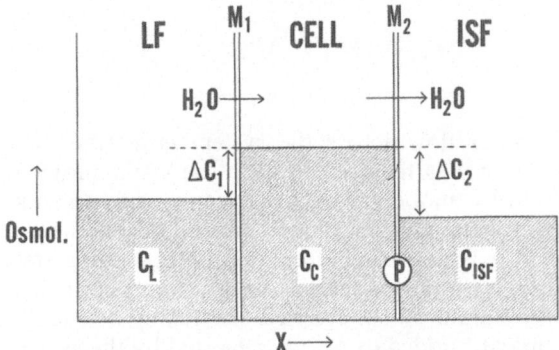

Fig. 1. Osmolality (Osmol.) versus distance (x) in a three-compartment, two-membrane system, where: LF denotes the lumen fluid compartment; CELL denotes intracellular fluid, the central compartment; ISF denotes the interstitial fluid compartment; M_1 and M_2 denotes the membranes. The osmolalities in each compartment are C_L, C_C, and C_{ISF}, and the sodium pump, if present, is denoted by P.

In the artificial system of Curran and McIntosh,[46] the mechanism operates under the conditions

$$0 < \sigma_2 < \sigma_1 < 1.0$$

indicating that M_1 is a relatively tight osmotic membrane, and M_2, a relatively leaky one. The fluid pulled osmotically across M_1 would be hypotonic to the boundary fluid bathing M_1, and more importantly, hypotonic to the fluid pushes out of the central compartment across M_2. This process would reduce the osmolality of the central compartment, ultimately causing cessation of the net solvent flow.

The figure shows two unequal and oppositely oriented osmotic gradients—ΔC_1 at M_1, and ΔC_2 at M_2. The osmotic pressure differences at each membrane are

$$\Delta P_1 = \sigma_1 RT \Delta C_1 \tag{28}$$

which is oriented to drive water from left to right, and

$$\Delta P_2 = \sigma_2 RT \Delta C_2 \tag{29}$$

which is oriented to drive water from right to left. Even though $\Delta C_2 > \Delta C_1$, the net overall flow of water will be directed from left to right if $\Delta P_1 > \Delta P_2$, which requires that

$$\sigma_1 \Delta C_1 > \sigma_2 \Delta C_2 \tag{30}$$

In other words, as long as the ratio of the two reflection coefficients, σ_1/σ_2, exceeds that of the two osmotic gradients, $\Delta C_2/\Delta C_1$, water will flow from the hypertonic compartment, LF, to the hypotonic compartment, ISF. The net driving pressure across M_1 and M_2 is

$$\Delta P_{net} = \Delta P_1 - \Delta P_2, \tag{31}$$

$$\Delta P_{net} = RT(\sigma_1 \Delta C_1 - \sigma_2 \Delta C_2) > 0 \tag{32}$$

Since the water flow across the system is a dissipative process, an extra source of power is required to keep the water moving continuously against the overall osmotic gradient between compartments LF and ISF. In the artificial system, Curran added "concentration energy" by continuously refreshing the hypertonic fluid in the central compartment.

In the biological system, such as an intestinal epithelial cell, a metabolically driven sodium pump, coupled electrically or chemically to potassium transfer, may be added to M_2 as suggested by the symbol, Ⓟ, in Fig. 1. The sodium pump removes sodium from the cell, keeping its concentration therein low, and consequently potassium is driven into the cell, keeping its concentration high. The outer solutions bathing M_1 and M_2 are kept high in sodium and low in potassium concentration. Water and sodium (with chloride) move down their transmembrane gradients at M_1, while little, if any, potassium is transferred. This transfer reduces the potassium concentration in the cell but increases the sodium concentration and the hydrostatic pressure in the cell. At M_2, sodium is pumped out of the cell, potassium is driven in, and water is driven out hydrostatically. If the parallel back-leak of sodium is low, the sodium pumping will drive potassium from the solution bathing M_2 into the cell, and offset the diluting action of the sodium efflux. This requires that the force of the electric field due to sodium pumping be greater than the osmotic force driving water into the cell.

The sodium pump added to the Curran model of a three-compartment system acts to maintain the required hypertonicity of the central compartment. Together with the two membranes whose water permeabilities are described by $\sigma_1 > \sigma_2$, the system can produce a net flow of water from LF to ISF—as long as the osmotic force across M_1 exceeds that across M_2, as shown in Eqs. (28–32).

2. Standing Osmotic Gradients

The maintenance of osmotic gradients (of the size of those between hypertonic urine and systemic plasma) across the width (10–30 μ) of single renal tubular cells by a "diffusion pump" model[47] results in the dissipation of more energy than is available to the cells.[48] This disadvantage has been offset by the use of models requiring maintenance of gradients in long, narrow channels or compartments where there exists a high length-to-area ratio of the diffusion path.[49,50] Such channels have been found between cells of secretory tissue,[51] and in the renal interstitial fluid where the gradient is "stretched" along the length of renal pyramid from the cortico-medullary junction to the papillary tip.[39]

a. Intercellular. Figure 2 is a schematic representation of the microscopic anatomy (upper panel) and the osmotic concentration profile (lower panel) of an epithelial cell (e.g., gall bladder, intestine, turtle bladder) which transfers water from the mucosal to the serosal fluid in the absence of an overall osmotic gradient.[52]

According to Diamond's scheme,[51] the salt and water diffuse across the mucosal membrane into the cytoplasm in the first step. From the cytoplasm, salt is pumped by an active process into the long, narrow channels between the cells. The salt pumping process results in the formation of a hypertonic fluid in the channel adjacent to the tight junction. Then water is transferred passively down its transmembrane gradient from cytoplasm to channel—a process which dilutes the channel fluid progressively (as seen in the figure) and accelerates the linear flow of fluid from the tight junction to the open end of the channel at the region of the basement membrane. The fluid emerging into the serosal fluid is either hypertonic to or nearly isotonic with the mucosal fluid. Diamond has suggested that such a mechanism could account for the secretion of hypertonic fluid by the salt glands of seagulls and of sea turtles, and for the secretion of isotonic fluid by the gall bladders of fish and mammals. He further points out that for secretion of hypertonic fluid, the system requires one rather than two membranes.

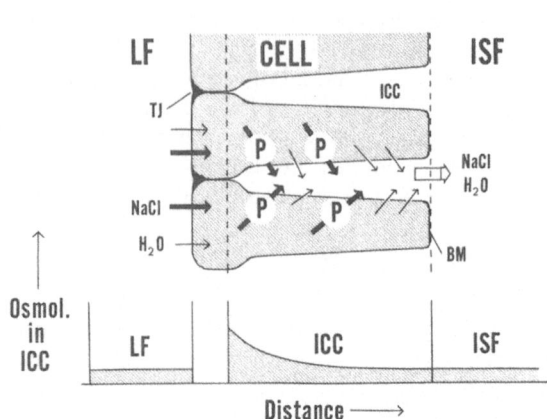

Fig. 2. *Upper*: Structure of epithelial mucosal cell where LF, CELL, and ISF denote the lumen fluid, the cell and the interstitial fluid; and where TJ denotes the tight junction, BM, the basement membrane, ICC, the intercellular channel, and Ⓟ, the sodium pump. Thick dark arrows indicate NaCl transfer; thin dark arrows indicate water transfer and the thick light arrow denotes water and NaCl transfer.

Lower: Osmolality versus distance. Profiles shaded for lumen fluid, for intracellular channel and for ISF, while the cytoplasmic osmolality is not specified.

Schilb and Brodsky have shown that inhibition or cessation of salt pumping in the turtle bladder results in a transient acceleration of the net water flux from mucosa to serosa.[53] This suggests that the turtle bladder mucosal membrane is more of an osmometer than is the membrane lining the intracellular channels, and that the reptilian bladder functions like a three-compartment, two-membrane system (described above), whereas the gall bladder functions like a one-membrane standing gradient system.

b. Renal. The standing gradient in the interstitium of the mammalian renal medulla is produced by a set of sodium pumps (countercurrent multiplier effect) in the ascending limb of Henle's loop, and is maintained by the action of the countercurrent exchange system of the vasa recta.

The action of the renal gradient, a region of increasing hypertonicity from the cortex to the medullary tip, is to remove water osmotically from the descending limb of Henle's loop and from the terminal portion of the collecting duct. As a result, a urine low in volume and hypertonic to the systemic plasma (but isoosmotic with the interstitial fluid of the renal medullary tip) can be excreted.

The overall effect of the countercurrent gradient, together with sodium pumps and a counterflow capillary network, is to move water from the hypertonic fluid of the collecting duct to the interstitium and ultimately to the isotonic systemic blood.

As in the micro type of standing gradient in epithelial cells, the renal gradient, by itself, cannot perform the task of reversing osmosis without the energy delivered by the sodium pumps of the countercurrent multiplier array in the ascending limb of Henle's loop.

c. Nitella. Osterhout produced a standing osmotic gradient in a long narrow cylindrical plant cell, *Nitella*, by immersing one end in a beaker of hypertonic fluid, and the other in a beaker of hypotonic fluid.[49,50] After a time, an intracellular standing gradient was obtained. The next step in the procedure, the interposing of *Nitella* between two new solutions (other than the ones used to produce the standing gradient), is illustrated in Fig. 3.

Figure 3 depicts the standing gradient in *Nitella* (upper sketch) and the concentration profile extending from compartment L through the *Nitella* to compartment R (the lower graph). The ends of the *Nitella*, containing the preset gradient (C_1 at M_1 and C_2 at M_2) were fixed into two compartments, L and R, separated by a cork stopper S, such that M_1 made contact with the solution in L, and M_2 with that in R. Since $C_L > C_R$, the overall osmotic pressure difference, $-RT(C_L - C_R)$, was oriented to drive water from R to L. The osmotic gradients across each membrane ($-RT\Delta C_L$ across M_1 and $-RT\Delta C_R$ across M_2) were oriented in opposite directions such that the gradient across M_1 would drive water from left to right while that across M_2 would drive water from right to left. But $\Delta C_L > \Delta C_R$, and consequently the net driving force on water was oriented to move water from the hypertonic solution in *L* to the hypotonic solution in *R*. The process requires that a hydrostatic pressure develop in the *Nitella* cell as the result of solvent flow across M_1 into the cell and that this intracellular pressure must exceed the osmotic pressure difference at M_2.

The system is essentially a three-compartment system in which the osmotic permeabilities of each of the membranes are equal to one another ($\sigma_1 = \sigma_2$). Osmotic flow against the overall gradient is occasioned by the magnitude and orientation of the two boundary gradients together with the standing gradient in the central compartment. In the three-compartment system of Curran, osmotic flow against the overall gradient is occasioned primarily because of the difference in the osmotic permeabilities ($\sigma_1 \neq \sigma_2$) of the two membranes.

Fig. 3. *Upper*: *Nitella* interposed between two solutions (L, the hypertonic side and R, the hypotonic side) by a stopper, S. *Lower*: Osmolality profile where C_L and C_R denote osmolality levels in solutions L and R; where C_1 and C_2 denote the preset boundary osmolalities in the *Nitella*; where
$$\Delta C_L = C_1 - C_L,$$
$$\Delta C_R = C_2 - C_R;$$
where $\Delta C_L > \Delta C_R$; and where water moves from L to R until $\Delta C_L = \Delta C_R$.

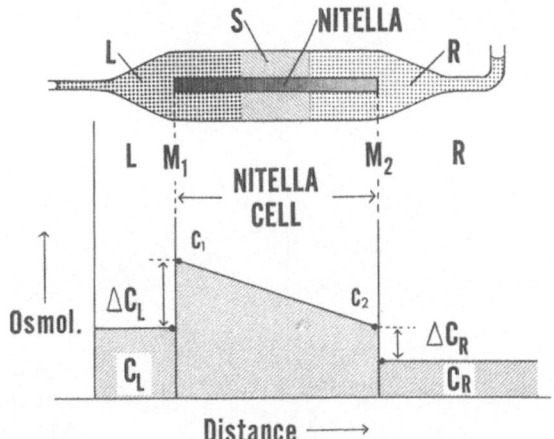

3. Two Solute

Meschia and Setnikar[54] have described an apparent reverse osmosis across a single membrane bathed by aqueous solutions containing two different solutes—one, denoted 1, which penetrates the membrane almost as readily as water ($\sigma_1 \approx 0$), and another, denoted 2, which does not penetrate ($\sigma_2 = 1.0$). Since $\Delta P = \sigma_s RT \Delta C_s$, the gradient due to the first penetrant would induce no osmotic driving pressure, whereas that due to the non-penetrant would account for all of the osmotic pressure difference generated.

Figure 4 depicts the situation schematically. As long as $C_{2L} < C_{2R}$, molecules of water (and of solute, 1) will move from left to right—as if this were a simple binary solution—even if the total osmolality in L, $(C_{1L} + C_{2L})$ exceeds that in R, $(C_{1R} + C_{2R})$.

Analytically,

$$\Delta P = \sigma_1 RT \Delta C_1 + \sigma_2 RT \Delta C_2 \tag{33}$$

but $\sigma_1 = 0$, and consequently,

$$\Delta P = \sigma_2 RT \Delta C_2 \tag{34}$$

In short, since the membrane cannot discriminate between the molecules of solute 1 and those of the aqueous solvent, the direction and magnitude of the osmotic driving force is almost wholly a function of the gradient of the nonpenetrant solute 2.

D. Ionic Flow Across Membranes

1. Nernst Equilibrium

According to Gibbs and Guggenheim, the internal energy of a charged particle in solution at constant pressure and temperature is a function of

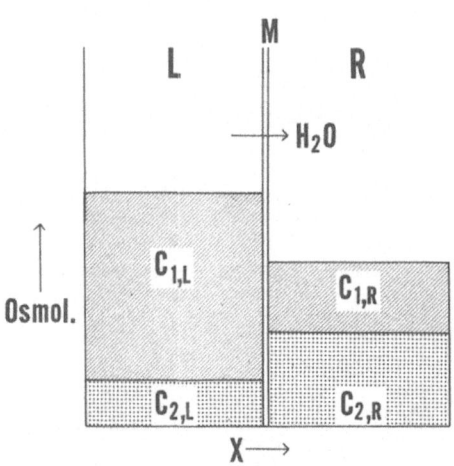

Fig. 4. Osmolality versus distance (x) in a system of two solutions, L and R bounding a membrane, M. With two solutes, 1 and 2, and with the total osmolality $(C_{1,L} + C_{2,L})$ on the left greater than that on the right $(C_{1,R} + C_{2,R})$ water will move toward the dilute solution (as indicated by the arrow) if only one of the two solutes is a nonpenetrant.

its chemical and electrical potential energy at a point in space, or

$$\bar{\mu}_{i\alpha} \equiv \mu_{i\alpha} + z_i F E_\alpha \tag{35}$$

where the subscripts denote the ion in question (i) and its location at a point in space (α), where $\bar{\mu}_{i\alpha}$ is the electrochemical potential of the ith ion at a locus α; z_i is the valence of the ith ion; F is the Faraday constant; and E_α is the electrical potential energy at the locus α; and where all terms are normalized with respect to the molal concentration of the ion. The chemical potential term, $\mu_{i\alpha}$, is defined by the equation,

$$\mu_{i\alpha} = \mu_{i\alpha}^0 + RT \ln a_{i\alpha} \tag{36}$$

where μ_i^0 is a constant; where $a_{i\alpha}$ is the chemical activity; and where $a_i = C_i f_i$. If the activity coefficient, f_i, is taken as a unity

$$a_{i\alpha} = C_{i\alpha} \tag{37}$$

where $C_{i\alpha}$ is the chemical concentration.

If a membrane selectively permeable to the ith ion, separates two aqueous solutions containing the ion, and if the electrochemical potential energy of the ion in solution 1 is $\bar{\mu}_{i,1}$ and in solution 2, $\bar{\mu}_{i,2}$, then the difference of electrochemical potential across the membrane is

$$\bar{\Delta}\mu_i = RT \ln C_{i,1}/C_{i,2} + z_i F \Delta E \tag{38}$$

which says that the difference of electrochemical potential, $\Delta\bar{\mu}_i$, and not just the difference of chemical concentration, is what determines the direction of motion of the ion across the membrane, between solutions 1 and 2. At equilibrium, with zero net motion of the ion, $\Delta\bar{\mu}_i \equiv 0$, and consequently

$$z_i F(\Delta E_i^0) = -RT \ln C_{i,1}/C_{i,2} \tag{39}$$

from which it follows that

$$\Delta E = \Delta E_i^0 = -\frac{RT}{z_i F} \ln C_{i,1}/C_{i,2} \tag{40}$$

which is the well-known Nernst relationship expressing the balance of electrical (ΔE_i^0) and diffusional forces ($RT/z_i F \ln C_{i,1}/C_{i,2}$) acting on the ith ion under conditions of transmembrane equilibrium. In terms of Briggsian logarithms, the Nernst equation can be reduced to

$$\Delta E_i^0 = -60 \log C_{i,1}/C_{i,2} \tag{41}$$

where $2.3\ RT/zF \approx 60\ \text{mV}$ at 25°C.

The Nernst equation, applicable to ion-selective membranes (permeable to cations only, or to anions only) may be used in two ways: (1) by following the transmembrane potentials generated after varying the concentration ratios of the permeant ion; or (2) by following the equilibrium concentration ratios reached after applying various electric fields across a membrane interposed initially between two identical solutions containing the permeant ion. In method (1), the equilibrium potentials, ΔE_i^0, are reached immediately after changing the transmembrane concentration ratio, because the requirement of zero net ionic flow is guaranteed by the charge selectivity of the membrane. In case (2), the equilibrium potential is reached only *after* the permeant ion has been moved across the membrane and redistributed such that its diffusional force corresponding to the new transmembrane concentration ratio is equal to the force of the externally applied voltage across the membrane.

Figure 5(a) is a schematic plot of the concentrations of potassium ($[K]_1$ and $[K]_2$) bounding a membrane (M) permeable only to cations. The state of electrochemical equilibrium for potassium is indicated by the equality of the two oppositely oriented K fluxes, $M_{K(2 \to 1)} = M_{K(1 \to 2)}$. Figure 5(b) shows the plot of values of the transmembrane electrical potential differences at equilibrium ($-\Delta E^0$) for a family of $[K]_1/[K]_2$ ratios. The slope of the solid line, 60 mV, is in keeping with the state of electrochemical equilibrium for each and every value of $[K]_1/[K]_2$, and with the perfect charge selectivity (cation permeable) of the membrane.

The slope of the dotted line, < 60 mV, represents a plot of $-\Delta E$ *vs.* $\log [K]_1/[K]_2$ for a membrane with imperfect charge selectivity, e.g., for a membrane which is permeable to cations and anions. As long as $[K]_1$ is not equal to $[K]_2$, the system cannot be in a state of transmembrane electrochemical equilibrium, whence the transmembrane diffusion potentials are denoted by ΔE rather than ΔE^0.

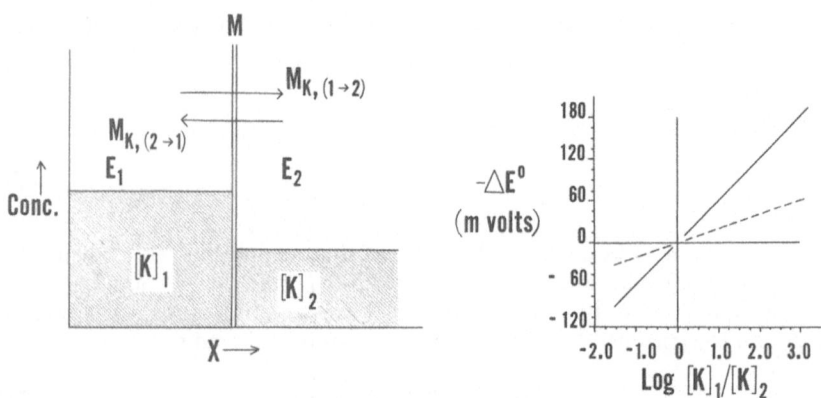

Fig. 5. (a) Concentration of potassium versus distance across a membrane, M interposed between two solutions, 1 and 2 containing potassium at concentrations, $[K]_1$ and $[K]_2$ respectively. The transmembrane flux of K from 1 to 2 is denoted by $M_{K(1 \to 2)}$, and that from 2 to 1, by $M_{K(2 \to 1)}$. (b) Transmembrane equilibrium potential difference ($-\Delta E^0$) versus $\log [K]_1/[K]_2$ in a perfect cation-selective membrane. Dotted line indicates the slope for a nonequilibrium condition in a "leaky" membrane.

2. Diffusion Potential

The presence of ion gradients across a perfect cation-permeable (or anion-permeable) membrane cannot induce a net transmembrane flow of ions. On the other hand, the presence of such gradients across an imperfect or leaky membrane does induce a net transmembrane flow of ions. This is because the leaky, charged membrane permits migration of a counterion through the leak pathway while the cationic (or anionic) permeant migrates through the ion-specific pathway. Thus, two ionic flows are induced by the concentration gradient—that of the "specific" permeant species through the appropriately charged selective "pore," and that of its counterion through a parallel pore of nonselective nature. The transmembrane potential difference generated by the gradient of the specific permeant ion is lowered by the concomitant motion of its counterion through the leak. This means that transmembrane diffusion of an ion pair is an irreversible process in which the electric field generated by the flow of one ion drags its counterion through the membrane. In other words, diffusion of an ion pair involves coupling of the flow of one ion to that of its counterion, and can be represented in the phenomenological form as,

$$\mathcal{J}_1 = \mathcal{L}_{11}\left(\frac{\Delta\bar{\mu}_1}{\Delta X}\right) + \mathcal{L}_{12}\left(\frac{\Delta\bar{\mu}_2}{\Delta X}\right) \tag{42}$$

$$\mathcal{J}_2 = \mathcal{L}_{21}\left(\frac{\Delta\bar{\mu}_1}{\Delta X}\right) + \mathcal{L}_{22}\left(\frac{\Delta\bar{\mu}_2}{\Delta X}\right) \tag{43}$$

where $\mathcal{J}_1$ and $\mathcal{J}_2$ are the fluxes of the electropositive and electronegative ion respectively; $\mathcal{L}_{11}$ and $\mathcal{L}_{22}$, the "straight" phenomenological coefficients; $\mathcal{L}_{12}$ and $\mathcal{L}_{21}$, the "coupling" coefficients; $\Delta\bar{\mu}/\Delta X$, the electrochemical gradient (i.e., the driving force); and where the volume flux and heat flux are taken as zero.

During the process of net transfer of the ions, 1 and 2, across the membrane, the ion-specific current densities are

$$I_1 = z_1 F \mathcal{J}_1, \quad \text{and} \tag{44}$$

$$I_2 = z_2 F \mathcal{J}_2 \tag{45}$$

where the units for z are equivalents/mole; for F, coulombs/equivalent; and for $\mathcal{J}$, moles/area/time; and consequently the units for I, the current density, are in coulombs/time/area or amperes/area.

Transforming Eqs. (42) and (43) from units of flux to units of current density, and substituting L_{ij} for $\mathcal{L}_{ij}/\Delta X$, leads to

$$I_1 = z_1 F L_{11}(\Delta\bar{\mu}_1) + z_1 F L_{12}(\Delta\bar{\mu}_2) \tag{46}$$

$$I_2 = z_2 F L_{21}(\Delta\bar{\mu}_1) + z_2 F L_{22}(\Delta\bar{\mu}_2) \tag{47}$$

The law of conservation of charge, applied to the diffusional transfer of the cation and anion, requires that

$$I_1 + I_2 = 0 \tag{48}$$

Now let $L_{12} = L_{21} = 0$, on the assumption that there is no frictional interaction (coupling) between the cations and anions; and substituting Eqs. (46) and (47) into Eq. (48) leads to

$$z_1 L_{11} \Delta\bar{\mu}_1 = -z_2 L_{22} \Delta\bar{\mu}_2 \tag{49}$$

Given that $\Delta\bar{\mu}_1 = RT \ln C_{1,\alpha}/C_{1,\beta} + z_1 F \Delta E$ for the electrochemical gradient of cation between solutions α and β; $\Delta\bar{\mu}_2 = RT \ln C_{2,\alpha}/C_{2,\beta} + z_2 F \Delta E$ for the corresponding anionic gradient, where ΔE, the electrical potential between α and β is $E_\alpha - E_\beta$; and given that $C_{1,\alpha}/C_{2,\alpha} = C_{1,\beta}/C_{2,\beta} = 1.0$, because for univalent electrolytes the cation concentration equals the anion concentration in solution α as well as in solution β for a single ion pair; it can be shown that

$$-\Delta E = \left(\frac{L_{11} - L_{22}}{L_{11} + L_{22}} \right) \frac{RT}{z_1 F} \ln \frac{C_{1,\alpha}}{C_{1,\beta}} \tag{50}$$

and

$$-\Delta E = (t_1 - t_2) \frac{RT}{z_1 F} \ln \frac{C_{1,\alpha}}{C_{1,\beta}} \tag{51}$$

which is the well-known equation for the electrical potential generated by the diffusion of an electrolyte from α to β; and where t_1 and t_2, the "transference numbers" of cation and anion respectively are $t_1 = L_{11}/L_{11} + L_{22}$ and $t_2 = L_{22}/L_{11} + L_{22}$.

In the case of a nonselective membrane as defined by the condition that $t_1 = t_2 = 0.5$, then ΔE vanishes, which means that there is no relative mobility between cation 1 and anion 2. The diffusion of KCl in free solution yields a near zero level of ΔE, meaning that the mobility of K is nearly equal to that of Cl in free solution. Consequently, KCl is a good electrolyte for determining the relative cationic or anionic selectivity of a membrane during diffusion of KCl through the membrane.

In the case of a perfect cationic-selective membrane, $t_1 = 1.0$ and $t_2 = 0$, and the diffusion equation reduces to the Nernst relationship,

$$\Delta E^0 = -(RT/z_1 F) \ln (C_{1,\alpha}/C_{2,\alpha})$$

In the case of a perfect anion-selective membrane, $t_2 = 1.0$ and $t_1 = 0$, and by analogous reasoning to that used above, the same Nernst relationship would hold, but the orientation of ΔE would be opposite to that across the cation-selective membrane, because $z_1 = -z_2$ and $\Delta E_1^0 = -\Delta E_2^0$.

In the case of a "relatively" cation-permeable membrane, the value of t_1 for K^+ is greater than 0.5 and less than 1.0; and in the case of a "relatively" anion-permeable membrane, the value of t_1 for K^+ is greater than 0 and less than 0.5.

The equation for diffusion potential, used conventionally to determine relative mobilities of cations and anions in free solution (Hittorf transference cells), can be used to determine the relative ion selectivity of membranes as follows.[55] If a membrane of unknown ion permeability is interposed between two solutions of univalent electrolyte such as KCl, and if the concentration of one solution (the α solution) is ten times greater than that of the other (the β solution), the diffusion potential, in terms of transference numbers, Briggsian logarithms, and molal concentrations of K is

$$\Delta E = -(t_K - t_{Cl})60 \log \frac{[K]_\alpha}{[K]_\beta} \tag{52}$$

where ΔE is the diffusion potential in millivolts; where t_K and t_{Cl} are the Hittorf transference numbers corresponding to $\mathcal{L}_1$ and $\mathcal{L}_2$, the intramembrane mobilities of K and Cl; and where $2.3\ RT/zF = 60$ (in millivolts).

Since $[K]_\alpha = 10[K]_\beta$,

$$\Delta E = -(t_K - t_{Cl})60 \tag{53}$$

an expression which permits the determination of the ion selectivity of the unknown membrane from measured values of the transmembrane potential. For perfect ion selectivity, $\Delta E = \pm 60$ mV—i.e., the α side would be electronegative for a cation-selective membrane where $t_K = 1.0$, and electropositive for an anion-selective membrane where $t_{Cl} = 1.0$. For no ion selectivity, $\Delta E = 0$, since $t_K = t_{Cl} = 0.5$. For relative cationic or anionic selectivity, ΔE takes on values between -60 mV and $+60$ mV.

In short, the degree of ion selectivity of the unknown membrane is determined from the ratio of the measured ΔE to the ideal ΔE of 60 mV; and the relative cationic or anionic preference of the membrane, from the orientation of ΔE.

3. Gibbs–Donnan Equilibria

The generation of diffusion potentials across a membrane equally permeable to all penetrating cations and anions is possible provided that the membrane is impermeable to another ion (say to proteinate) in one of the bathing solutions. If such a membrane is interposed between a solution of Na, Cl, and proteinate (side 1) and a solution of Na and Cl (side 2), the conditions for electrochemical equilibrium (no net ionic flow) and osmotic equilibrium (no net solvent flow) can be determined by the Gibbs–Donnan relationship which is derived herewith.

Under the condition of electrochemical equilibrium,

$$\Delta \bar{\mu}_{Na} = \Delta \bar{\mu}_{Cl} = 0 \tag{54}$$

and consequently

$$\frac{RT}{z_+ F} \ln [Na]_1/[Na]_2 = \frac{RT}{z_- F} \ln [Cl]_1/[Cl]_2 \tag{55}$$

which reduces to

$$[Na]_1/[Na]_2 = [Cl]_2/[Cl]_1 \tag{56}$$

the "Donnan ratios," which satisfy the condition of electrochemical equilibrium where there is zero net transfer of any ion across the membrane. However, the condition of electrochemical equilibrium, as defined above, is not consistent with the condition of osmotic equilibrium. This is because

$$[Na]_1[Cl]_1 = [Na]_2[Cl]_2 \tag{57}$$

and in units of equivalent concentration,

$$[Na]_1 = [Cl]_1 + [Prot]_1 \tag{58}$$

and

$$[Na]_2 = [Cl]_2. \tag{59}$$

Substituting in and eliminating all terms in [Cl] from Eq. (57) gives

$$[Na]_1^2 - [Na]_1[Prot]_1 = [Na]_2^2 \tag{60}$$

whence

$$[Na]_1 > [Na]_2 \tag{61}$$

and by analogous algebra,

$$[Cl]_2 > [Cl]_1 \tag{62}$$

Thus the equality of Eq. (57) can be compared to a rectangle of dimensions $[Na]_1 \times [Cl]_1$, which is equal in area to a square of dimensions $[Na]_2 \times [Cl]_2$. But the minimum perimeter about quadrilaterals of equal area is that about a square, and so, in units of osmolality,

$$[Na]_1 + [Cl]_1 > [Na]_2 + [Cl]_2 \tag{63}$$

showing that the osmolality of side 1, without even including the proteinate, is hypertonic to that of side 2, whence the requirements for osmotic equilibrium are not satisfied by the Donnan conditions of electrochemical equilibrium.

To satisfy the requirements of both osmotic and electrochemical equilibria simultaneously, an osmotic or hydrostatic force must be applied to the system. This can be done by adding the appropriate concentration, say ΔC_s, of a nonpenetrant nonelectrolyte to side 2, or by applying the osmotically equivalent hydrostatic pressure ΔP, to side 1 such that the osmolality of solution 1 equals that of solution 2.

In a first approximation such an osmotic equality may be expressed as follows:

$$[Na]_1 + [Cl]_1 + [Prot]_1 = [Na]_2 + [Cl]_2 + \Delta C_s \tag{64}$$

$$[Na]_1 + [Cl]_1 + [Prot]_1 = [Na]_2 + [Cl]_2 - \Delta P/RT \qquad (65)$$

where $\Delta C_s = -\Delta P/RT$ (van't Hoff).*

One of the outstanding biological examples of a Gibbs–Donnan system is the capillary—bounded on its luminal surface by a protein-rich plasma and bounded on its outer surface by protein-poor interstitial fluid.

At the arteriolar end of the capillary (where glucose and O_2 are carried to the cells), the hydrostatic pressure within the capillary induces an ultra-filtration of plasma water with all of the nonprotein constituents (including the electrolytes, glucose, and O_2) from the vascular to the interstitial fluid. The osmotic disequilibrium during ultrafiltration, occasioned by the high level of arteriolar transcapillary pressure, ΔP_A, may be expressed by the inequality,

$$[Na]_1 + [Cl]_1 + [Prot]_1 < [Na]_2 + [Cl]_2 - \Delta P/RT \qquad (66)$$

At the venular end (where CO_2 and urea are carried away from the cells), the hydrostatic pressure in the capillary has decreased ($\Delta P > 0$), and the conditions for electrochemical and osmotic equilibrium do not hold. This results in a reverse ultrafiltration (reabsorption) of interstitial water with all of its dissolved constituents (including the electrolytes, CO_2, and urea) from the interstitial to the vascular fluid. The osmotic disequilibrium during absorption of the interstitial fluid into the capillary, occasioned by the low level of venular transcapillary pressure, ΔP_v, and the high level of protein (consequent to the prior ultrafiltration across the arteriolar end), may be expressed by the inequality

$$[Na]_1 + [Cl]_1 + [Prot]_1 > [Na]_2 + [Cl]_2 - \Delta P_v/RT \qquad (67)$$

At some point between the arteriolar and venular end of the capillary, the state of transcapillary electrochemical equilibrium coexists with that of

---

* The transmembrane osmotic pressure at equilibrium can be expressed in terms of the gradients as follows:

$$\Delta P = RT(\Delta[Na] + \Delta[Cl] + \Delta[Prot]) \qquad (65a)$$

where $\Delta[X] = [X]_2 - [X]_1$, in the case of each solute.

Implicit to all of the osmotic equations at equilibrium is that

$$\sigma_{Na} = \sigma_{Cl} = \sigma_{Prot} = 1.0 \qquad (65b)$$

which is in full accord with all of the classical data[3,39] showing that the measured osmotic pressure is the algebraic sum of the partial osmotic pressures due to the ion gradients and to the proteinate gradient.

However, one of us (A. E. Shamoo, unpublished report) has recently observed that the equilibrium condition of $\sigma_{ion} = 1.0$ must be reconciled with the nonequilibrium condition of $\sigma_{ion} = 0$ without assuming any physical or chemical change in the membrane itself. This can be done by relating the terms $RT\Delta[Na]$ and $RT\Delta[Cl]$ to the electrical part ($zS\Delta E$) rather than to the chemical part of the electrochemical potential energy of Na and Cl, from which it can be shown that σ_{salt} (not σ_{ion}) is zero under any and all conditions.

osmotic equilibrium, whence no transcapillary flow of solute or water occurs, and the osmotic equilibrium is that expressed by Eqs. (64) and (65).

Thus, the chemical composition and volume of the vascular and interstitial fluids are related by a Gibbs–Donnan system involving blood pressure, tissue turgor, and the colloid osmotic pressure of the blood. Were it not for the level of hydrostatic pressure of the blood, interstitial fluid in excessive amounts would cross the capillary membrane, and cause progressive expansion of the plasma volume. Such an expansion would be limited by the resultant dilution of plasma proteins and by the elastic recoil of the vascular walls. Conversely, the dissipation of blood pressure at the venular end of the capillary is required to prevent accumulation of the plasma in the interstitial spaces.

E. Ussing Flux Ratio

Another approach to the study of ion permeability of biological membranes is the analysis of unidirectional ionic fluxes estimated from the radioactive isotope fluxes. The ion flux ratio, introduced by Ussing[56] and Teorell,[55] is defined as the transmembrane flux of a given ion in one direction divided by the flux of the same ion in the opposite direction. Assumptions required for the analysis and interpretation of the flux ratio are as follows: (1) the velocity of ions moving through aqueous-filled membrane channels in one direction is not affected by that of ions moving in the opposite direction, i.e., there is independence of ionic motion; and (2) the velocity of ionic motion depends upon the transmembrane gradients of chemical potential and of electrical potential.

Ussing applied the Gibbs–Guggenheim concept of the electrochemical potential of an ion to derive the force term acting on the ith ion. He related this force to the ionic velocity by a frictional coefficient, and then showed the relation between the force and the ionic flux ratio.

By definition (see Eq. 35),

$$\bar{\mu}_i = \mu_i + z_i FE$$

where $\bar{\mu}_i$ is the electrochemical potential energy per mole of the ith ion; the chemical potential energy per mole; and $z_i FE$, the electrical potential energy per mole. Differentiation of $\bar{\mu}_i$ with respect to distance, x, leads to an expression containing the diffusional and electrical forces, or

$$\frac{d\bar{\mu}_i}{dx} = \frac{RT}{C_i}\frac{dC_i}{dx} + z_i F\frac{dE}{dx} \tag{68}$$

the electrochemical force per mole of the ith ion, where C_i, the concentration term, is obtained by assuming that the activity coefficient, $f_i = 1.0$, and that the chemical activity, $a_i = f_i c_i$. Dividing both sides of Eq. (68) by Avogadro's number, N_o, gives the electrochemical force acting on a single ion,

$$-\frac{1}{N_o}\frac{d\bar{\mu}_i}{dx} \tag{69}$$

which is related by a frictional coefficient, G_i, to the average velocity of the ith ion as follows:

$$v_i = -\frac{1}{G_i N_o} \frac{d\bar{\mu}_i}{dx} \tag{70}$$

But v_i, a molal velocity, is also defined as the number of ion particles moving across a unit of area at unit concentration—i.e.,

$$v_i = \frac{1}{AC_i} \frac{dn_i}{dt} \tag{71}$$

where n_i is the number of particles moving across area A in time, t, and where $dn_i/dt = M_i$, the total flux across area A at concentration C_i in the positive direction, is related to the molal velocity, v_i, by the expression,

$$M_i = AC_i v_i = -\frac{AC_i}{G_i N_o} \frac{d\bar{\mu}_i}{dx} \tag{72}$$

After substitution of Eq. (68) into Eq. (72), the latter equation can be rearranged, and multiplied by an integrating factor, I, to obtain an integrable exact differential equation,

$$I\left(\frac{M_i G_i N_o}{ART}\right) dx = -I\,dC_i - IC_i\left(\frac{zE}{RT}\right) dE \tag{73}$$

where

$$I = \exp(zFE/RT) \tag{74}$$

from the laws of the calculus.

Then the integration of Eq. (73) between the limits, $C_i = C_{i,1}$ at $x = 0$, $C_i = 0$ at $x = \Delta x$, leads to an expression for the flux of the ith ion from $x = 0$ to $x = \Delta x$ (i.e., from compartment 1 across the membrane to compartment 2), or

$$M_{1 \to 2} \frac{G_i N_o}{ART} \int_{x=0}^{x=\Delta x} \exp\left(\frac{zFE}{RT}\right) dx = C_{i,1} \exp(zFE_1/RT) \tag{75}$$

Simultaneously, the oppositely moving flux, $-M_{2 \to 1}$, of the same ion (or of its isotope) can be obtained by integrating Eq. (73) between the limits $C_i = C_{i,2}$ at $x = \Delta x$ and $C_i = 0$ at $x = 0$. Since the ion flux from compartment 1 to 2 is through the same path, and is independent of the ion flux from compartment 2 to 1, one can divide $M_{1 \to 2}$ by $M_{2 \to 1}$ with the resultant cancellation of the complicated integral term on the left side of Eq. (75), and consequently the flux ratio is

$$\frac{M_{1 \to 2}}{M_{2 \to 1}} = \frac{C_{i,1}}{C_{i,2}} \exp[zF(E_1 - E_2)/RT] \tag{76}$$

where $C_{i,1}$ is the chemical concentration of the ith ion in compartment 1 $(x = 0)$; and $C_{i,2}$, the concentration in compartment 2 $(x = \Delta x)$.

If $C_{i,1} = C_{i,2}$

$$\frac{M_{1 \to 2}}{M_{2 \to 1}} = \exp(z_i F \Delta E / RT) \tag{77}$$

and

$$\Delta E = 60 \log \frac{M_{1 \to 2}}{M_{2 \to 1}} \tag{78}$$

where ΔE, defined as $E_1 - E_2$, is expressed in the form of the Nernst relationship, and where the oppositely oriented unidirectional fluxes correspond to the activity (or concentration) terms.

Equations (75–78) are valid expressions for both radioactive and nonradioactive isotopic fluxes and concentrations of a given ion or molecule.

Figure 6(a) is a schematic diagram of a membrane interposed between two identical solutions (1 and 2) wherein steady-state conditions of sodium flux $(M_{1 \to 2}, M_{2 \to 1})$ and transmembrane potential $(E_1 - E_2)$ prevail; and Fig. 6(b) is a plot of values of the logarithm of the flux ratio versus transmembrane potential for four different types of membrane—all under the condition that $C_{i,1} = C_{i,2}$.

Line A denotes the flux ratio of the ith ion as a function of the electrical potential across a membrane permeable only to the ith ion. The finding of such a pattern in a biological membrane may be interpreted as a consequence of passive ionic transfer because the flux pattern can be completely explained by the known driving forces (e.g., ΔE, in this case) acting on the permeant ion. The main criterion for passive rather than active transport of any ion is that the net flux vanishes $(M_{1 \to 2} = M_{2 \to 1})$ when the external driving forces are eliminated (e.g., $\Delta E = 0$ in the case of Fig. 6).

If the biological membrane is not so simple as that required in the derivation of the flux ratio function, other flux patterns (e.g., B, C, or D) will be generated under the same external conditions. Thus, line B denotes a pattern of passive ion transfer across a membrane where a significant portion of each flux is not affected by the force of electric field—as is often seen in the process known as exchange diffusion wherein a penetrant cation associates and dissociates—at the two membrane–aqueous interfaces with an intramembrane molecule functioning as a carrier. For example, the penetrant cations exchange with H ions for an association site on the carrier. The cation-carrier complex, electrically neutral and unaffected by electric fields, moves passively across the lipid membrane by free diffusion. If the mechanism of penetration were solely by exchange diffusion, a flat line of zero slope would be produced. Patterns such as line B in Fig. 6 could then result from selective ion penetration by electrochemical diffusion coupled with exchange diffusion.

Line C is interpreted as a consequence of a passive type of transfer called "single file diffusion" across a membrane with long, narrow, charged

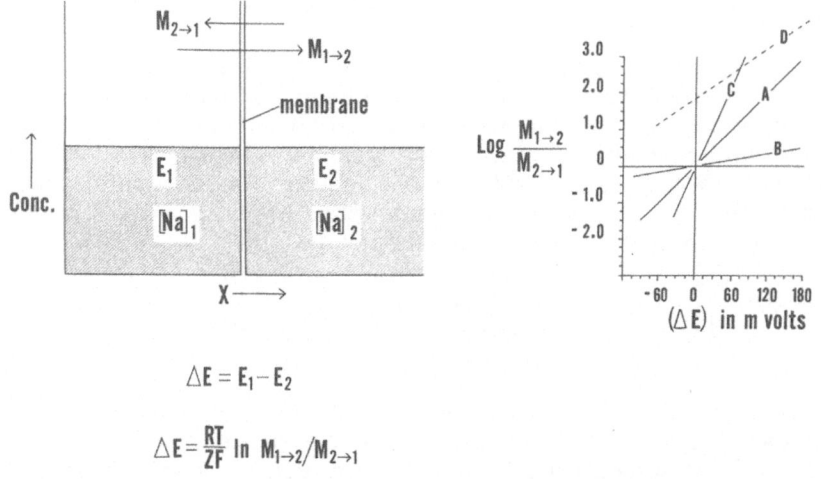

$$\Delta E = E_1 - E_2$$

$$\Delta E = \frac{RT}{ZF} \ln M_{1\to2}/M_{2\to1}$$

Fig. 6. (a) Schematic of a hypothetical living membrane interposed between two identical Na-containing solutions, 1 and 2. The membrane mechanism pumps Na from 1 to 2; thus the unidirectional flux of Na from 1 to 2, $M_{(1\to2)}$, is greater than that from 2 to 1, $M_{(2\to1)}$. (b) Logarithm of the ion flux ratio versus transmembrane potential difference (ΔE). Line A denotes the pattern for electrochemical diffusion of sodium as defined by Ussing; line B, the pattern for exchange diffusion of Na superimposed on electrochemical diffusion; line C, the pattern for "single file" diffusion of Na; and line D, a pattern for active transport of Na.

pores. This type of transfer, observed in the squid axon, will be discussed in the next section of this paper.

Line D, the flux ratio actually found in the case of sodium transport across the isolated frog skin,[56] is a consequence of "active" transport of the ith ion (sodium) across the biological membrane. In this case, a net transfer of sodium occurs in the absence of any known externally applied forces (e.g., $\Delta\bar\mu_{Na} = 0$); thus some force, metabolically generated, must be operative to explain the net ionic motion observed.

This interpretation was applied by Ussing to the fact that a finite net flux of sodium took place across the short-circuited frog skin (to be discussed further in Chapter 23) bathed on both sides by identical sodium-rich Ringer solutions; and to the fact that the net sodium flux was equal to the short-circuiting current. Short-circuiting conditions have been applied to many in vitro biological membrane preparations (frog skin, frog stomach, cornea, lens, intestine, and bladders of frogs and turtles) because of the high yield of electrophysiological information obtained under such conditions. For example, the net flux and short-circuiting current are zero for any passively transported ion, and are nonzero for any actively transported ion. In the case of a membrane which actively transports one ion, say sodium, the short-circuiting current is equal to the net sodium transport. In the case of a membrane which actively transports several ions, the short-circuiting current is equal to the algebraic sum of all of the net ion transports.

F. Single-File Diffusion

Hodgkin and Keynes[57] reasoned that the flux ratio of potassium across the membrane of DNP-poisoned squid axons ought to be governed by the forces of the transmembrane electric field and of the transmembrane gradient of potassium. In other words, K influx and efflux were considered as passive processes, independent of one another, while the poisoned membrane was considered to be potassium-selective. Applying Ussing's equation to the potassium influx gives an equation of the form

$$M_{K,in} \int_{x=0}^{x=\Delta x} \frac{G_K N_o}{ART} \exp \left(\frac{zFE}{RT} \right) dx = C_{K,out} \exp (z_K FE_{out}/RT) \qquad (79)$$

where influx ($M_{K,in}$) depends upon the electric field and the outside concentration, $C_{K,out}$. On the other hand, $M_{K,in}$ is independent of the efflux ($M_{K,out}$) and of the inside concentration of potassium ($C_{K,in}$). It follows that the flux ratio, $M_{K,in}/M_{K,out}$, should be related to the ratio of the chemical concentrations and to the transmembrane electric potential, $\Delta E = E_{in} - E_{out}$, by the expression

$$\frac{M_{K,in}}{M_{K,out}} = \frac{C_{K,out}}{C_{K,in}} \exp (-z_K F \Delta E/RT) \qquad (80)$$

which can be reduced to the forms,

$$-(\Delta E - \Delta E_K^0) = \frac{RT}{z_K F} \ln \left(\frac{M_{K,in}}{M_{K,out}} \right) \qquad (81)$$

and

$$-(\Delta E - \Delta E_K^0) = 60 \log (M_{K,in}/M_{K,out}) \qquad (82)$$

where $-\Delta E_K^0 = RT/z_K F \ln (C_{K,in}/C_{K,out})$; where ΔE is the transmembrane electrical potential difference; where $\Delta E - \Delta E_K^0$ is the net driving force acting on the potassium ions; and where the equation is similar in form to, but more general in meaning than the Nernst relationship.

Hodgkin and Keynes varied the external K concentration from 10 to 200 mM and fixed the transmembrane potential (ΔE) for each K level such that $M_{K,in} = M_{K,out}$—i.e., such that the net flux of potassium was always zero; and under these conditions they showed that $\Delta E = -60 \log C_{K,in}/C_{K,out}$, which established the potassium selectivity of the axon and which proved that transfer of K ion across the axonal membrane was indeed a passive transport process.

However, under conditions of $M_{K,in} \neq M_{K,out}$, the Ussing relationship broke down insofar as the flux ratio did not relate to $C_{K,in}/C_{K,out}$ and ΔE in the manner predicted by the equation. For example, when $\Delta E = 62 \, mV$,

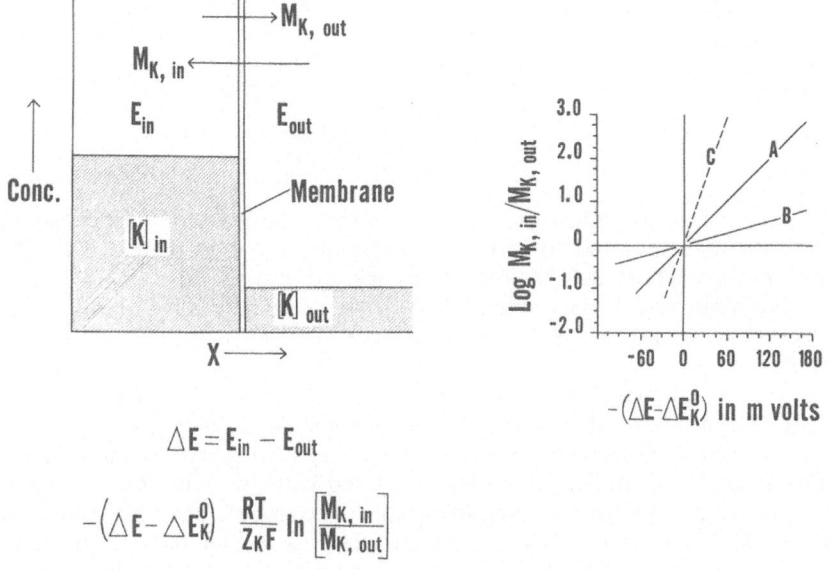

$$\Delta E = E_{in} - E_{out}$$

$$-\left(\Delta E - \Delta E_K^0\right) \quad \frac{RT}{Z_K F} \ln\left[\frac{M_{K, in}}{M_{K, out}}\right]$$

Fig. 7. (a) Concentration of K bounding the poisoned membrane of squid axon; $[K]_1$ in axoplasm, and $[K]_2$ externally. Net K flux into axon shown by arrows where $M_{K,in} > M_{K,out}$, and where $-\Delta E$ denotes the transmembrane potential. (b) Logarithm of the K flux ratio versus the transmembrane electrochemical potential, $-(\Delta E - \Delta E_K^0)$. Lines A, B, and C satisfy conditions analogous to those defined in the legend of Fig. 6.

$[K_{in}] = 274$ mM, and $[K_{out}] = 10.4$, the predicted value of $M_{K,in}/M_{K,out}$ was 0.4; but the measured value was 0.1.

Figure 7(a) presents a schematic of the DNP-poisoned membrane of a squid axon bounded by axoplasm with a K concentration of $[K]_{in}$ and by artificial sea water with a K concentration of $[K]_{out}$. The transmembrane potential is denoted by $E_{in} - E_{out}$ or ΔE; and the influx and efflux by $M_{K,in}$ and $M_{K,out}$, respectively. Figure 7(b), a plot of log $M_{K,in}/M_{K,out}$ versus $\Delta E - \Delta E_K^0$, shows the flux pattern observed across the squid axon as compared to that expected across a membrane with ideal K selectively and with bidirectionally independent K fluxes. Whereas the theoretical plot of $\Delta E - \Delta E_K^0$ requires a tenfold increase in flux ratio for each 60 mV increment of $\Delta E - \Delta E_K^0$, the actual data showed a tenfold increase of flux ratio for each 23 mV of $\Delta E - \Delta E_K^0$. Moreover, the assumption of independence of oppositely oriented fluxes did not hold up, in that increasing $[K_{out}]$ by tenfold resulted in a reduction of $M_{K,out}$ of 2.5, and an increase of $M_{K,in}$ of 30-fold. The measured flux ratio increased by a factor of 1000 instead of by the predicted factor of 10 for each 60 mV increase in the net driving force, $\Delta E - \Delta E_K^0$. This is reflected in the slope of line C, which is greater than that of the ideal case (line A) in the figure.

Such data suggested that the K ion appeared to move across the membrane as if it (K ion) had a valence of 2.5–3.0, or as if about 3 potassium ions had to engage with the membrane before one got translocated.

In other words,

$$-(\Delta E - \Delta E_K^0) = \frac{60}{n} \log \frac{M_{K,in}}{M_{K,out}} \tag{83}$$

where $n = 2.5$–3.0, which fits the notion that K ions are constrained to move through long narrow pores in single file, and that there are 1.5–2.0 ions, on the average, in a channel at any instant of time.

Hodgkin and Keynes then showed the physical reality of single file diffusion in an ingeniously contrived experiment on a simple mechanical model.

Two symmetrical cubically shaped chambers, the walls of which were made of aluminum and "Perspex," were connected by a short tube (analogous to a pore in a membrane) in one case, and by a long narrow tube in the other case. The chamber on the left was filled with 100 blue steel balls and that on the right with 50 silver steel balls. The double chamber was then shaken for a uniform length of time, and the flux of blue balls from left to right was compared with that of silver balls from right to left. In the case with the short connecting pore, the flux ratio was approximately 2.7 blue/1 silver, as expected from the concentration ratio. In the case with the long narrow pore in which there were about 3 balls at any given time, at least 4 collisions were necessary for the translocation of a single ball. The frequency with which 4 consecutive collisions occur is theoretically 2^4 times greater on the 100-ball side than on the 50-ball side. As expected for such a "long pore," the flux ratio was 18 blue/1 silver, which was close to the predicted ratio of 2^4.

The preceding discussion has described transport through membrane of solutes moving down their transmembrane gradients of electrochemical potential energy with the consequent dissipation of energy. However, many observations on ion transport across biomembranes cannot be explained under this restriction. For example, the saturation kinetics of and the effect of inhibitors on passive transport processes and, particularly, the recognition of active transport processes require interactions between the transported species and a constituent of the membrane. Such considerations have indeed dominated the modern approach to biotransport—an approach focused on the isolation, identification, and molecular characterization of ion carriers in nonaqueous phases of biological and artificial membranes.

VI. REFERENCES

1. W. Stoeckenius and D. M. Engelman, Current models for the structure of biological membranes, *J. Cell. Biol.* **42**:613–646 (1969).
2. D. Chapman, (ed.), *Biological Membranes*, Academic Press, New York (1968).

3. J. T. Edsall and J. Wyman, *Biophysical Chemistry*, Vol. 1, Academic Press (1958).

4. S. L. Miller, Production of some organic compounds under possible primitive earth conditions, *J. Am. Chem. Soc.* **77**:2351–2361 (1955).

5. J. D. Bernal, *The Physical Basis of Life*, Routledge and Kegan Paul, London (1951).

6. A. Katchalsky, Biological organization and thermodynamics Symposium IV. Third International Biophysics Congress, IUPAB, Cambridge, Mass. (1969).

7. T. Rosenberg, Accumulation and active transport in biological systems I. Thermodynamic considerations, *Acta Chem. Scand.* **2**:14–33 (1948).

8. F. Jacob and J. Monod, Genetic regulatory mechanisms in the synthesis of proteins, *J. Mol. Biol.* **3**:318–356 (1961).

9. H. Davson and J. F. Danielli, *The Permeability of Natural Membranes*, Cambridge Univ. Press, London (1952).

10. F. O. Schmitt, R. S. Bear, and K. J. Palmer, X-ray diffraction studies on the structure of the nerve myelin sheath, *J. Cell. Comp. Physiol.* **18**:31–41 (1941).

11. J. D. Robertson, New observations on the ultrastructure of the membranes of frog peripheral nerve fibers, *J. Biophys. Biochem. Cytol.* **3**:1043–1047 (1957).

12. J. B. Finean, The nature and stability of nerve myelin, *Intern. Rev. of Cytol.* **12**:303–336 (1961).

12a. H. Fernandez-Moran, New approaches in the study of biological ultrastructure by high resolution electron microscopy, *in Symposia of the International Society for Cell Biology* (R. J. C. Harris, ed.), pp. 411–428, Vol. I, Academic Press, New York (1962).

13. E. D. Korn, II. Synthesis of bis(methyl 9, 10-dihydroxy-sterate) osmate from methyl oleate and osmium tetroxide under conditions used for fixation of biological material, *Biochim. Biophys. Acta* **116**:317–324 (1966).

14. E. D. Korn, III. Modification of oleic acid during fixation of amoebae by osmium tetroxide, *Biochim. Biophys. Acta* **116**:325–335 (1966).

15. D. F. H. Wallach, Membrane lipids and the conformations of membrane proteins, *in Membrane Proteins*, Proc. Sympos. N.Y. Heart Association, pp. 3–26, Little, Brown and Co., Boston (1969).

16. V. Luzzati, X-ray diffraction studies of lipid-water systems, *in Biological Membranes* (D. Chapman, ed.), pp. 71–124, Academic Press, New York (1968).

17. D. A. Haydon and J. Taylor, The stability and properties of bimolecular lipid leaflets in aqueous solutions, *J. Theoret. Biol.* **4**:281–296 (1963).

18. E. D. Korn, Structure and function of the plasma membrane, *in Biological Interfaces: Flows and Exchanges*, Proc. Sympos. N.Y. Heart Assoc., pp. 257–278, Little Brown and Co., Boston (1968).

19. L. L. M. van Deenen and J. de Gier, Chemical composition and metabolism of lipids in red cells of various animals species, *in The Red Blood Cell* (C. Bishop and D. M. Surgenor, eds.), pp. 243–308, Academic Press, New York (1964).

20. A. B. Pardee, Membrane transport proteins, *Science* **162**:632–637 (1968).

21. P. G. LeFevre, The behavior of phospholipid-glucose complexes at hexane/aqueous interfaces, *in Currents in Modern Biology*, Vol. I, pp. 29–38, North Holland Publishing Co., Amsterdam (1967).

22. C. Y. Jung, J. E. Chaney, and P. G. LeFevre, Enhanced migration of glucose from water into chloroform in the presence of phospholipids, *Arch. Biochem. Biophys.* **126**:664–676 (1968).

23. P. G. LeFevre, C. Y. Jung, and J. E. Chaney, Glucose transfer by red cell phospholipids in $H_2O/CHCl_3/H_2O$ three layer systems, *Arch. Biochem. Biophys.* **126**:677–691 (1968).

24. R. L. Post, C. R. Merritt, C. R. Kinsolving, and C. D. Albright, Membrane adenosine triphosphatase as a participant in the active transport of sodium and potassium in the human erythrocyte, *J. Biol. Chem.* **235**:1796–1802 (1960).

25. E. T. Dunham and I. M. Glynn, Adenosinetriphosphatase activity and the active movements of alkali metal ions, *J. Physiol.* (*London*) **156**:274–293 (1961).
26. R. E. Solinger, C. F. Gonzalez, Y. E. Shamoo, H. R. Wyssbrod, and W. A. Brodsky, Effect of ouabain on ion transport mechanisms in the isolated turtle bladder, *Am. J. Physiol.* **215**:249–261 (1968).
27. Y. E. Shamoo and W. A. Brodsky, The Na + K dependent adenosine triphosphatase in the isolated mucosal cells of turtle bladder, *Biochim. Biophys. Acta* **203**:111–123 (1970).
28. H. Lardy, Influence of antibiotics and cyclic polyethers on ion transport in mitochondria, *Fed. Proc.* **27**:1278–1282 (1968).
29. B. C. Pressman, Ionophorous antibiotics as models for biological transport, *Fed. Proc.* **27**:1283–1288 (1968).
30. P. Mueller, D. O. Rudin, H. T. Tien, and W. C. Westcott, Symposium on the plasma membrane. Reconstitution of excitable cell membrane structure *in vitro*, *Circulation* **26**:1167–1177 (1962).
31. H. T. Tien and A. L. Diana, Biomolecular lipid membranes: a review and a summary of some recent studies, *Chem. Physics Lipids* **2**:55–101 (1968).
32. D. C. Tosteson, Effect of macrocyclic compounds on the ionic permeability of artificial and natural membranes, *Fed. Proc.* **27**:1269–1277 (1968).
33. G. Eisenman, S. M. Ciani, and G. Szabo, Some theoretically expected and experimentally observed properties of lipid bilayer membranes containing neutral molecular carriers of ions, *Fed. Proc.* **27**:1289–1305 (1968).
34. C. J. Pedersen, Ionic complexes of macrocyclic polyethers, *Fed. Proc.* **27**:1305–1309 (1968)
35. W. D. Stein, *The Movement of Molecules across Cell Membranes* (Chapter 6), Academic Press, New York (1967).
36. P. Mitchell, Chemiosmotic coupling in oxidative and photosynthetic phosphorylation, *Biol. Rev.* **41**:445–502 (1966).
37. I. Prigogine, *Introduction to Thermodynamics of Irreversible Processes* (Chapter 4), Interscience, John Wiley and Sons, New York (1961).
38. O. Kedem, and A. Essig, Isotope flows and flux ratios in biological membranes, *J. Gen. Physiol.* **48**:1047–1070 (1965).
39. H. Davson, *A Textbook of General Physiology*, J. & A. Churchill Ltd., London (1964).
40. J. Dainty and B. Z. Ginzburg, The permeability of the protoplasts of *Chara australis* and *Nitella translucens* to methanol, ethanol and isopropanol, *Biochim. Biophys. Acta* **79**:122–128 (1964).
41. D. A. Goldstein and A. K. Solomon, Determination of equivalent pore radius for human red cells by osmotic pressure measurement, *J. Gen. Physiol.* **44**:11–17 (1960).
42. T. E. Thompson, The properties of bimolecular phospholipid membranes, *in Cellular Membranes in Development* (M. Locke, ed.), pp. 83–96, Academic Press, New York (1964).
43. V. W. Sidel and J. F. Hoffman, Water transport across membrane analogues, *Fed. Proc.* **20**:137 (1962).
44. O. Kedem and A. Katchalsky, Thermodynamic analysis of the permeability of biological membranes to non-electrolytes, *Biochim. Biophys. Acta* **27**:229–246 (1958).
45. O. Kedem and A. Katchalsky, A physical interpretation of the phenomenological coefficients of membrane permeability, *J. Gen. Physiol.* **45**:143–179 (1961).
46. P. F. Curran and J. R. McIntosh, A model system for biological water transport, *Nature* **193**:347–348 (1962).
47. J. E. Franck and J. E. Mayer, An osmotic diffusion pump, *Arch. Biochem.* **14**:297–313 (1947).
48. W. A. Brodsky, W. S. Rehm, W. H. Dennis, and D. G. Miller, Thermodynamic analysis of the intracellular osmotic gradient hypothesis of active water transport, *Science* **121**:302–303 (1955).

49. W. J. V. Osterhout, Movements of water in cells of *Nitella*, *J. Gen. Physiol.* **32**:553–557 (1949).

50. W. J. V. Osterhout, Transport of water from concentrated to dilute solutions in cells of *Nitella*, *J. Gen. Physiol.* **32**:559–566 (1949).

51. J. M. Diamond and W. H. Bossert, Standing gradient osmotic flow. A mechanism for coupling of water and solute transport, *J. Gen. Physiol.* **50**:2061–2081 (1967).

52. J. McD. Tormey and J. Diamond, The ultrastructural route of fluid transport rabbit gall bladder, *J. Gen. Physiol.* **50**:2031–2059 (1967).

53. T. P. Schilb and W. A. Brodsky, Transient acceleration of transmural water flow by inhibition of sodium transport in turtle bladders, *Amer. J. Physiol.* **219**:590–596 (1970).

54. G. Meschia and I. Setnikar, Experimental study of osmosis through a collodion membrane, *J. Gen. Physiol.* **42**:429–444 (1958).

55. T. Teorell, Transport phenomena in membranes, *Discussions Faraday Soc.* **21**:9–26 (1956).

56. H. H. Ussing, Distinction by means of tracers between active transport and diffusion. The transfer of iodide across the isolated frog skin, *Acta Physiol. Scand.* **19**:43–56 (1949).

57. A. L. Hodgkin and R. D. Keynes, The potassium permeability of a giant nerve fiber, *J. Physiol. (London)* **128**:61–88 (1955).

Chapter 21

CARRIER-MEDIATED TRANSPORT PROCESSES

H. R. Wyssbrod, W. N. Scott, W. A. Brodsky, and I. L. Schwartz

The Departments of Physiology, Biophysics, and Ophthalmology
Mount Sinai Medical and Graduate Schools of
the City University of New York
New York, N.Y.

The Institute for Medical Research and Studies
New York, N.Y.

and

The Medical Research Center
Brookhaven National Laboratory
Upton, N.Y.

I. INTRODUCTION

In Chapter 20[1] the movement of solutes across biological membranes is treated in relation to processes which *do not*, in general, involve a chemical interaction between the permeant and the membrane, and which are thermodynamically dissipative in character—i.e., the free energy of the matter under observation decreases during transport.

The present chapter considers processes which *do* involve a chemical interaction between the permeant and the membrane and which may be dissipative (downhill) or nondissipative (uphill) in character.

A. Flux (J)

The net transmembrane flux of substance S from side 1 to 2 is denoted by $J_S^{1 \to 2(net)}$, for which typical physical units are the number of moles of S which cross the membrane per unit of membrane area per unit of time. Whenever $J_S^{1 \to 2(net)} > 0$, net movement is from side 1 to 2; whenever $J_S^{1 \to 2(net)} < 0$, net movement is from side 2 to 1.

In biological systems, solution, which consists primarily of water, moves across membranes in response to transmembrane osmotic and pressure gradients. Volume flow of solution can result in appreciable amounts of transmembrane movement of solute. This moiety of solute flux is usually attributed to "solvent drag," since it is associated with the frictional inter-action between solute and solvent. *In this chapter, only the moiety of solute flux exclusive of the "solvent drag" moiety is considered.* Solutions on each side of the membrane under consideration are assumed to be adjusted osmotically so that volume flow is eliminated. Otherwise, all measured fluxes would have to be corrected for the moiety resulting from the "solvent drag" effect before they could be related to any of the transport models presented in this chapter. In practice, it is not always easy to make this correction. In biological systems such as kidney and intestine in which trans-membrane volume flow is normally appreciable, the "solvent drag" moiety of flux for some solutes may be of the same magnitude as the "independent" moiety (the moiety of concern in this chapter). Therefore, in these cases, the effect of "solvent drag" on solute movement must be considered if the transport properties of these systems are to be understood fully.

B. Electrochemical Potential ($\tilde{\mu}$)

On each side of the membrane, S can be characterized by its electro-chemical potential ($\tilde{\mu}_S$), which denotes the availability of Gibbs free energy per mole of S. Typical physical units of $\tilde{\mu}_S$ are joules per mole of S. Gibbs free energy is a measure of the work capacity of the system when temperature and pressure remain constant, e.g., the synthesis of macromolecules, the work of muscular contraction, or osmotic work to concentrate solutions.

The electrochemical potential of S on either side i ($\tilde{\mu}_{S(i)}$) may be expressed in terms of parameters which are accessible to measurement[2]

$$\tilde{\mu}_{S(i)} = \tilde{\mu}^0_{S(i)}(T_i, P_i, a_S = 1 \text{ unit}, \psi = 0) + RT_i \ln a_{S(i)} + z_S F \psi_i \qquad (1)$$

where R is the gas constant, T_i is the absolute temperature on side i, $a_{S(i)}$ is the chemical activity in arbitrary units, z_S is the valence of S if S is an electro-lyte, F is the Faraday constant, ψ_i is the electrical potential on side i with respect to any reference potential, and $\tilde{\mu}^0_{S(i)}$ is the reference electrochemical potential, which is a function of temperature and pressure on side i, and which is defined when the electrical potential is at the reference level ($\psi = 0$) and when the chemical activity is one arbitrary unit. The *difference* in $\tilde{\mu}_S$, $\Delta\tilde{\mu}_S$, across the membrane, rather than either $\tilde{\mu}_{S(1)}$ or $\tilde{\mu}_{S(2)}$ is the important parameter with which we are concerned.

$$\Delta\tilde{\mu}_S = \tilde{\mu}_{S(2)} - \tilde{\mu}_{S(1)} \qquad (2a)$$

$$= \tilde{\mu}^0_{S(2)} - \tilde{\mu}^0_{S(1)} + RT \ln (a_{S(2)}/a_{S(1)}) + z_S F \Delta\psi \qquad (2b)$$

where

$$\Delta\psi = \psi_2 - \psi_1 \qquad (3)$$

Note that the orientation of $\Delta\tilde{\mu}_S$ is arbitrarily chosen so that the electro-chemical potential on side 2 is measured with respect to that on side 1. We will consider only the case where there is a gradient neither in temperature nor in pressure across the membrane. Thus, the reference electrochemical potentials $\tilde{\mu}^0_{S(1)}$ and $\tilde{\mu}^0_{S(2)}$ are equal, and Eq. (2b) can be simplified to

$$\Delta\tilde{\mu}_S = RT \ln\left(a_{S(2)}/a_{S(1)}\right) + z_S F \Delta\psi \tag{4}$$

The chemical activity $(a_{S(i)})$ and concentration $([S]_i)$ of S on side i are related by

$$a_{S(i)} = \gamma_{S(i)} \times [S]_i \tag{5}$$

where the activity coefficient $\gamma_{S(i)}$ is defined in such a way that it approaches unity as the solution becomes infinitely dilute $([S]_i \to 0)$. *In this chapter all activity coefficients are taken to be equal to unity, and concentration is used in place of chemical activity.* Thus, all formulas in this chapter represent a first approximation to the corresponding formulas which could be written to include activity coefficients. With the condition that $\gamma_{S(i)} = 1$, Eq. (4) may be simplified to

$$\Delta\tilde{\mu}_S = RT \ln\left([S]_2/[S]_1\right) + z_S F \Delta\psi \tag{6}$$

If S is a nonelectrolyte $(z_S = 0)$, Eq. (6) may be simplified to

$$\Delta\mu_S = RT \ln\left([S]_2/[S]_1\right) \tag{7}$$

where the chemical potential $(\Delta\mu_S)$ is used in place of the electrochemical potential $(\Delta\tilde{\mu}_S)$. Note that for *nonelectrolytes*, the side of higher concentration is *always* the side of higher chemical potential.

In this chapter, we assume that the solutions in the aqueous phases are perfectly mixed so that there is no change in electrochemical potential across the interface between aqueous and membrane phases. In other words, there is no unstirred layer at the interface. Consequently, the entire electrochemical potential difference $(\Delta\tilde{\mu}_S)$ between aqueous phases is dissipated within the membrane, and a negligible part of it is dissipated across either interface. For nonelectrolytes, $\Delta\mu_S$ is composed only of a chemical or concentration term [see Eq. (7*)], and consequently, there can be no difference between the solute concentration in the bulk of the aqueous phase and that in the interface at the membrane. For electrolytes, $\Delta\tilde{\mu}_S$ is composed of a chemical and an electrical term [see Eq. (6)], and consequently, there can be a change in solute concentration across the interface if the electrical potential difference across the interface is such that the electrical term negates the chemical term [see Eq. (150)]. In some biological systems, unstirred layers play an important role in dissipating part of the difference in electrochemical potential between two aqueous phases separated by a membrane and, consequently, the transport models of this chapter would not apply to these systems unless modified to take the unstirred layer into account.

---

* Equations (6) and (7) express the electrochemical potential difference of an electrolyte and the chemical potential difference of a nonelectrolyte, respectively, between aqueous phases 1 and 2. Each of these equations can also be applied to the difference across an interface if the subscript 1 denotes parameters evaluated in the bulk solution, and the subscript 2 denotes parameters in the interface at the membrane.

C. Downhill, Level, and Uphill Flow

In *downhill* (dissipative) flow, the net flux of a substance across a membrane is from the side of higher electrochemical potential to that of lower potential, and hence, the substance becomes energetically degraded or dissipated with time—i.e., its movement is associated with a decrease in its ability to do work.

In *level* flow, there is no change in electrochemical potential as the substance crosses the membrane.

In *uphill* (nondissipative) flow, the net flux of a substance is toward a region of higher electrochemical potential, and hence, the substance becomes energetically upgraded with time—i.e., its movement is associated with an increase in its ability to do work.

D. Active and Passive Transport

Rosenberg[3] originally proposed that active transport be defined as a process which results in the net transmembrane flow of a substance from a region of lower to one of higher electrochemical potential—i.e., an active process is one which results in uphill flow. On the other hand, Kedem,[4] following a suggestion by Rosenberg,[5] proposed that active transport be defined as a process in which the transmembrane flow of a substance is coupled to an off-equilibrium metabolic reaction; if no coupling is present, the transport process is passive. In this chapter, Kedem's, rather than Rosenberg's, definition of active and passive transport is adopted. The terms uphill and downhill flow correspond to Rosenberg's earlier definitions of active and passive transport, respectively.

In a metabolic reaction, there is flow of reactants, but this flow has no direction associated with it, for it occurs isotropically in one of the aqueous solutions bathing the membrane. In biological systems, there are many off-equilibrium metabolic processes which can be coupled to transport processes. The hydrolysis of adenosine triphosphate (ATP) to adenosine diphosphate (ADP) and inorganic phosphate (P_i) is an example of an off-equilibrium process which has been implicated in active transport.

A metabolic reaction can be measured in terms of flow. Consider an active transport system in which a metabolic reaction is coupled, at least in part, to the transport of a particular substance. This coupled component of the reaction can be measured in terms of moles of reactant that undergo chemical transformation per unit area of membrane per unit time. Note that the units of measurement of the metabolic reaction are the same as those for the transmembrane flux of a substance (see Section I, A), even though neither reactants nor products need cross the membrane. Hence, a metabolic reaction is often described in terms of its flow (J_M), even though this flow is isotropic.

Figure 1 shows active processes for downhill, level, and uphill flows. An off-equilibrium process on side 1 is coupled with a transport process so that the former process aids the movement of substance from side 1 to 2 and opposes the movement from side 2 to 1. The decrease in free energy of the

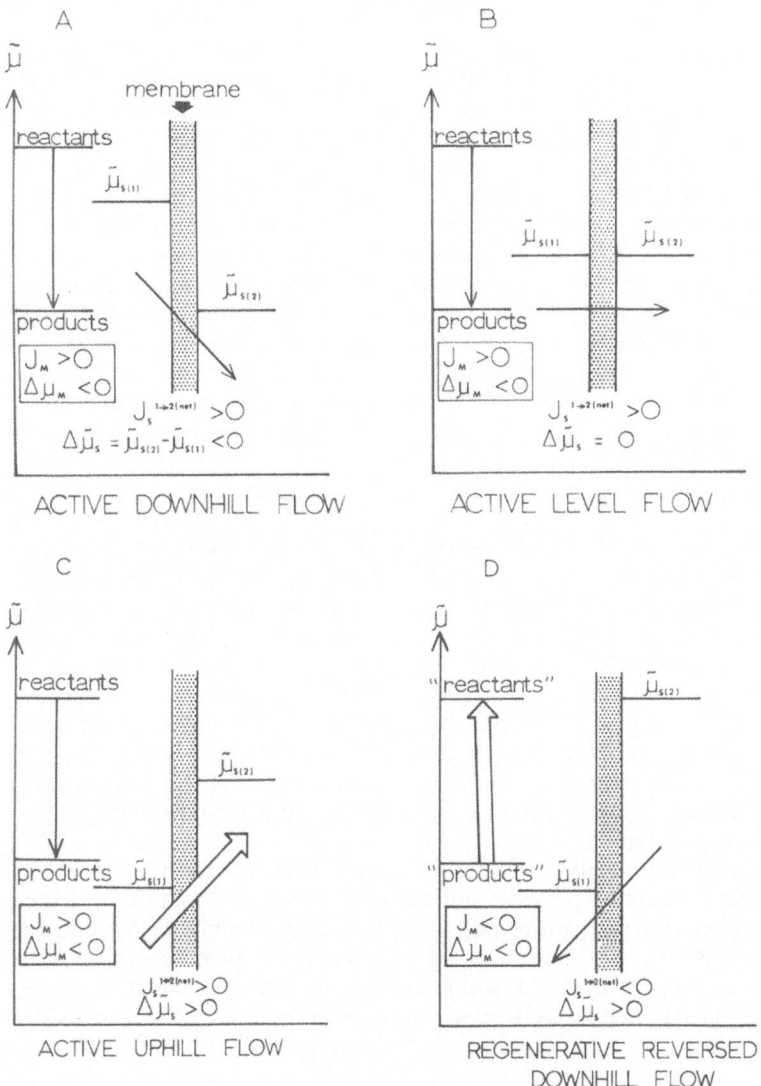

Fig. 1. Modes of active and regenerative transport.

metabolic reaction ($\Delta\mu_M < 0$) indicates that the sum of the chemical potentials for the products of the reaction is less than the sum for the reactants and hence, thermodynamically, the result of the reaction is the energetic degradation of the reactants into the products. Figure 1(a) shows active downhill flow, in which both the metabolic reaction and the difference in electrochemical potential ($\Delta\tilde{\mu}_S$) of transported substance contribute to the net movement of S from side 1 to 2. Figure 1(b) shows an example of active level flow in which net movement of S occurs even though $\Delta\tilde{\mu}_S$ has been eliminated. Figure 1(c) shows active uphill flow, in which the magnitude of $\Delta\mu_M$ for the metabolic reaction (which is oriented to move S from side 1 to 2) is sufficient to overcome the magnitude of $\Delta\tilde{\mu}_S$ (which is oriented to move S from side 2 to 1). Figure 1(d) shows reversed downhill flow, in which the magnitude of $\Delta\tilde{\mu}_S$ overwhelms $\Delta\mu_M$ and forces the transport system to operate in reverse. Thus, net transport of S is from side 2 to 1, and, if the metabolic reaction is tightly coupled to the transport system, the metabolic reaction may be reversed ($J_M < 0$)—i.e., the substances which are normally the products of the reaction combine to form the substances which are normally the reactants. If metabolic energy is expended, the system is active; if metabolic energy is stored, the system is regenerative. For example, in the erythrocyte, reversed flow of sodium has been used to regenerate ATP from ADP and P_i.[6,7]

The free energy of the entire coupled system must decrease with time. Therefore,

$$J_M \times \Delta\mu_M + J_S^{1 \to 2(net)} \times \Delta\tilde{\mu}_S < 0 \qquad (8)$$

where $\Delta\tilde{\mu}_S = \tilde{\mu}_{S(2)} - \tilde{\mu}_{S(1)}$, where substances are designated as reactants or products so as to make $\Delta\mu_M < 0$, where J_M represents the portion of the metabolic reaction directly coupled to the transport of S, and where the orientation of J_M is chosen so that in either a completely uncoupled system or during level flow, $J_M > 0$. Note that $J \times \Delta\mu$ has units of energy per unit area of membrane per unit time. A negative value of $J \times \Delta\mu$ means that free energy is dissipated with time. Although the sum of the metabolic and transport terms of Eq. (8) is negative, either—but not both—of the terms may be positive. For example, during active uphill flow (Fig. 1c) the transport term is positive; during regenerative reversed downhill flow (Fig. 1d) the metabolic term is positive.

Figure 1(c) shows that active transport results in uphill flow when the metabolic reaction provides sufficient free energy to S to move it to the region of higher electrochemical potential. It is also possible for S to move uphill passively—i.e., without coupling to a metabolic process—if the movement of S is coupled to the downhill transmembrane movement of some other substance T, and if the rate of decrease in free energy of T is sufficient to provide for the rate of increase in free energy of S—i.e.,

$$J_S^{1 \to 2(net)} \times \Delta\tilde{\mu}_S + J_T^{1 \to 2(net)} \times \Delta\tilde{\mu}_T < 0 \qquad (9)$$

The term involving S in Eq. (9) may be positive (i.e., S may move uphill) if the term involving T is sufficiently negative.

E. Synopsis of Approach

Section II. Passive Transport

A. Experimental observations are reviewed to show that simple diffusion does not account for transmembrane movement of many substances and, consequently, most solute translocations across membranes must be mediated by interaction either with some component of the membrane or with the membrane per se.

B. Models for mediation of passive transport of nonelectrolytes are presented.

C. Interaction phenomena among transported substances and their chemical analogues are considered in some detail.

D. Current concepts of conformational enzymology are applied to a model of transport.

E. A model for the passive transport of electrolytes is presented in order to demonstrate the role of the transmembrane difference in electrical potential in the overall transport process.

Section III. Active Transport

A. A model for passive transport is modified to include coupling to a metabolic process.

B. The passive transport of one substance is coupled to the transport of another substance which is actively transported. In this co-transport scheme, uphill movement of the co-transported species is indirectly active inasmuch as it depends on the active transport of the second species.

C. A metabolic process can serve as the source of or sink for a substrate. If different processes occur on different sides of a membrane, a gradient of substrate can be created, and this gradient can drive substrate across the membrane. The movement is active inasmuch as it depends upon metabolic processes to create the gradient. The redox scheme for acidification is a special case of transport resulting from metabolic sources of and sinks for H^+.

Section IV. Pseudo-Uphill Transport

"Trapping" schemes are presented to show how a substance may appear to move uphill when actually the movement is entirely downhill and may even be via simple diffusion.

Section V. Molecular Description of Carrier-Mediated Transport Systems

Some molecular characteristics of components of carrier systems are discussed and techniques for identifying subcellular components that participate in carrier-mediated transports are reviewed.

Soluble proteins that bind transport substrate with a high degree of specificity have been prepared from bacterial cells. Reconstitution experiments and genetic studies provide strong evidence that these proteins are required for the carrier-mediated transport of the bound substrate. A soluble calcium-binding protein has also been implicated in the transport of calcium by the vertebrate intestine.

A phosphotransferase system responsible for the accumulation of glucose (or glucose-6-phosphate) by bacteria has been shown to consist of membrane-bound components in addition to "soluble" protein components. Membrane sodium- and potassium-dependent adenosine triphosphatase (ATPase) has been implicated in the transport of sodium and potassium and the enzyme has been partially purified.

Section VI. Conclusion

We conclude that a molecular description of a transport system cannot be based solely upon transport kinetics, but must be founded, in addition, upon the physiochemical characteristics of the isolated, purified, and reconstituted system.

II. PASSIVE TRANSPORT

A. Properties of Passive Mediated Transport

The flux of a nonelectrolyte moving across a membrane via simple diffusion is given by the integrated form of Fick's first equation[8,9] which states that

$$J_S^{1 \to 2(net)} = (D_S'/L) \times ([S]_1' - [S]_2') \tag{10}$$

$J_S^{1 \to 2(net)}$ is net flux per unit of membrane area from side 1 to 2, L is the thickness of the membrane, $[S]_i'$ is the concentration in the membrane at the interface with the aqueous solution on side i (where side i represents either side 1 or side 2) and D_S' is the diffusion constant, which is inversely related to the amount of bonding between penetrating substance and membrane constituents (primarily lipids). If the transmembrane diffusion is the rate-limiting step, the concentration in the membrane, $[S]_i'$, is related to the concentration in the adjacent aqueous phase, $[S]_i$, by

$$[S]_i' = d_S \times [S]_i \tag{11}$$

where d_S is the partition coefficient, which serves as a measure of lipoid solubility. Substituting $d_S \times [S]_i$ for $[S]_i'$ in Eq. (10) yields

$$J_S^{1 \to 2(net)} = (d_S \times D_S'/L) \times ([S]_1 - [S]_2) \tag{12a}$$

$$= (D_S/L) \times ([S]_1 - [S]_2) \tag{12b}$$

where D_S, the apparent diffusion constant, is equal to $d_S \times D_S'$.

The *net* flux may be separated into two *unidirectional* fluxes, each of which may be determined by use of isotopic tracers (see Section II, B, 3). In general, the net flux from side 1 to 2 is the unidirectional flux from side 1 to 2 minus the unidirectional flux from side 2 to 1, i.e.

$$J_S^{1 \to 2(net)} = J_S^{1 \to 2} - J_S^{2 \to 1} \tag{13}$$

where the superscripts $1 \to 2$ and $2 \to 1$ without the notation (net) indicate unidirectional fluxes. Fick's equations for unidirectional fluxes are

$$J_S^{1 \to 2} = (D_S/L) \times [S]_1 \qquad (14a)$$

$$J_S^{2 \to 1} = (D_S/L) \times [S]_2 \qquad (14b)$$

Note that each unidirectional flux in a simple diffusion system is a function of substrate concentration on the side on which it originates, but is independent of the concentration on the side to which it moves.

Some substances (e.g., sugars and amino acids) move across biological membranes at a much greater rate than expected from simple diffusion estimated from lipoid solubility (related to d_S) and amount of bonding to membrane constituents (related to D_S'). The passive movement of these substances not only occurs at an unexpectedly high rate, but deviates in the following characteristic ways from the case of simple diffusion:

Characteristic 1. Abnormal Flux Ratios

Under circumstances in which Fick's equations for unidirectional fluxes [Eqs. (14a) and (14b)] are applicable, the ratio of unidirectional fluxes for nonelectrolytes is equal to the ratio of cis-concentrations* :

$$(J_S^{1 \to 2}/J_S^{2 \to 1})_{\text{Fick}} = [S]_1/[S]_2 \qquad (15)$$

In many biological systems, the ratio of the unidirectional fluxes for nonelectrolytes is different from that predicted by Fick's equation. In these systems the ratio of unidirectional fluxes for nonelectrolytes, here defined simply as $(J_S^{1 \to 2}/J_S^{2 \to 1})$, is usually less than the ratio of cis-concentrations, i.e.

$$J_S^{1 \to 2}/J_S^{2 \to 1} < (J_S^{1 \to 2}/J_S^{2 \to 1})_{\text{Fick}} = [S]_1/[S]_2 \qquad (16)$$

Sides 1 and 2 have been chosen so that $[S]_1 > [S]_2$. Consequently, $J_S^{1 \to 2} > J_S^{2 \to 1}$, and net flux is from side 1 to 2. Several other arrangements of the flux ratio equations are possible—e.g. from Eqs. (12b), (14a), and (14b),

$$(J_S^{1 \to 2}/J_S^{1 \to 2(\text{net})})_{\text{Fick}} = [S]_1/([S]_1 - [S]_2) \qquad (17a)$$

$$(J_S^{2 \to 1}/J_S^{1 \to 2(\text{net})})_{\text{Fick}} = [S]_2/([S]_1 - [S]_2) \qquad (17b)$$

It follows from Eqs. (13), (16), (17a), and (17b) that in many biological systems

$$J_S^{1 \to 2}/J_S^{1 \to 2(\text{net})} > (J_S^{1 \to 2}/J_S^{1 \to 2(\text{net})})_{\text{Fick}} = [S]_1/([S]_1 - [S]_2) \qquad (18a)$$

$$J_S^{2 \to 1}/J_S^{1 \to 2(\text{net})} > (J_S^{2 \to 1}/J_S^{1 \to 2(\text{net})})_{\text{Fick}} = [S]_2/([S]_1 - [S]_2) \qquad (18b)$$

When isotopic tracers are used to determine the unidirectional fluxes of a substance across biological membranes and a chemical measurement is used to determine the net flux of this substance, the equilibration of tracer

---

* Cis refers to the side on which the unidirectional flux originates. For example, side 1 is the cis-side for $J_S^{1 \to 2}$ and side 2 is the cis-side for $J_S^{2 \to 1}$.

occurs more rapidly than the equilibration of total penetrant. This finding is consistent with Eqs. (18a) and (18b). The discrepancy between the rates at which tracer and total penetrant approach equilibrium was first demonstrated by LeFevre and McGinniss[10] who studied the movement of both tracer and total glucose across human erythrocytes. The finding of a difference of several orders of magnitude in the time constants characterizing the approach to equilibrium of tracer and of total glucose indicated that transport could not have occurred by simple diffusion.

In some cases, the inequality of Eq. (16) is reversed—i.e., the flux ratio is greater than that predicted by Fick's equation. Systems which exhibit a characteristic called *cis*-stimulation demonstrate this reversal in the inequality of the flux ratio [see Eq. (120a) in Section II, C, 7].

Characteristic 2. Substrate Specificity

A substrate may penetrate at a much faster rate than its isomers or optical enantiomers—e.g., in erythrocytes, transmembrane movement of D-glucose is favored over that of its isomer, D-fructose, and over that of its optical enantiomer, L-glucose; in most biological systems, the movement of L-amino acids in usually favored over that of the corresponding D-amino acids (see Section V, B, 2). It is doubtful that solubility of substrate in the membrane or amount of bonding of substrate to membrane would differ drastically among similar substrates; hence, specific mechanisms, rather than simple diffusion, undoubtedly account for transmembrane movement of many substances. The possession of specific transport mechanisms permits biological organisms to control the composition of cellular and extracellular compartments, to select nutrients, and to eliminate waste products.

Characteristic 3. Specific Inhibitors

Some chemical agents which are known to inhibit enzymatic reactions also inhibit transport processes. These inhibitory agents are often specific for particular transport processes and, in some cases, the amount required for inhibition is too small to effect a general change in the physical properties of the membrane (see Section V, B, 2). Since Fick's equation predicts interference with transport only by such a general change, it is clear that at least some chemical inhibitors must interact with specific membrane sites concerned with transport mechanisms. Indeed, such interaction with transport sites provides one means by which certain drugs influence biological function.

Characteristic 4. Homo-cis-inhibition

Fick's equations for unidirectional fluxes [Eqs. (14a) and (14b)] predict that unidirectional flux is proportional to the concentration in the side from which the flux originates (*cis*-side) and increases without limit as the *cis*-concentration, $[S]_{cis}$, is increased. In many biological systems, however, a maximal or saturation value of unidirectional flux is reached as *cis*-concentration is increased. Certainly, Fick's equation does not predict this behavior. A limiting value of flux suggests that molecules of S compete for a limited number of reactive sites related to transmembrane transport. Interference on the *cis*-side between identical molecules is termed homo-*cis*-inhibition, a phenomenon which provides cells with a means of prevention of excessive uptake of nutrients.

Characteristic 5. Hetero-cis-inhibition

Unidirectional flux of a substance S (e.g., glucose) may be reduced by the presence of a structural analogue T (e.g., galactose) on the *cis*-side. Reciprocally, the flux of T is reduced by the presence of S. Fick's equation does not predict this interaction between substances. The interference between S and T on the *cis*-side is termed hetero-*cis*-inhibition, and suggests that both substances compete for a limited number of sites related to transport (see Section II, C, 2). In some biological systems, any one of a whole class of similar substances may provide a cell with an essential nutrient. Hetero-*cis*-inhibition provides the cell with a means of limiting the excessive uptake of similar substrates.

Characteristic 6. Counterflow

The passive uphill movement of a substance can sometimes be related to the passive downhill movement of a structural analogue. Fick's equation, which states that net movement is always downhill [see Eq. (12b)], can never account for net uphill movement. When the net movement of the substance moving uphill is in the opposite direction to that of the analogue moving downhill, the phenomenon is termed counterflow and is a consequence of hetero-*cis*-inhibition (see Section II, C, 3). Counterflow provides a cell with a means of trading a substance which it contains in abundance for one in which it is deficient, even when the latter substance is in limited supply in its extra-cellular environment (see Section V, B, 5, b).

Characteristic 7. trans-Effects

In some biological systems a unidirectional flux can be modified by the presence of the substance on the *trans*-side,* i.e., the side to which that flux is moving. In other words, a change in $[S]_2$ can change $J_S^{1 \to 2}$ (see Section II, C, 6). Fick's equations [Eqs. (14a) and (14b)] state that unidirectional fluxes are a function of concentration on the *cis*-side only (i.e., the side from which the flux originates) and not of concentration on the *trans*-side. *Trans*-effects provide a cell with a mechanism by which its content of a given substance can regulate the uptake of similar substances. The first *trans*-effect was

---

* The prefixes *cis*- and *trans*- were borrowed from the vocabulary of organic chemistry and introduced into the vocabulary of the transport field by Rosenberg and Wilbrandt.[11] In general, side i is the *cis*-side and side j is the *trans*-side for the unidirectional flux $J_S^{i \to j}$. In this chapter, we shall often use $J_S^{cis \to trans}$ as a general designation for a unidirectional flux of S.

reported by Heinz,[12] who observed a stimulatory effect of intracellular glycine upon the movement of glycine into Ehrlich mouse ascites carcinoma cells.

A passive downhill system which differs in characteristics from a Fick simple diffusion system is often termed a facilitated diffusion system.[13]

Other characteristics which distinguish passive movement across biological membranes from simple diffusion are discussed by Rosenberg and Wilbrandt,[11] Stein,[14,15] and Dowben.[16]

In the following parts of Section II, transport mechanisms which can account for some of the distinguishing characteristics of passive movement in biological systems are presented.

B. Basic Carrier Systems

1. Introduction to Carrier-Mediated Transport

The system under consideration consists of two aqueous phases separated by a lipoprotein or lipid phase. In biological systems cellular and subcellular membranes serve as the lipoprotein or lipid phase. In the previous section it was shown that simple diffusion cannot account for the movement of all substrates across biological membranes. One scheme which can explain movement across biological membranes involves the combination of the penetrant with some membrane constituent called a carrier.

Höber[17–19] proposed a carried mechanism to explain movement of glucose across the intestine. This mechanism and a similar one proposed later by Verzàr,[20,21] Shannon and Fisher,[22] and Shannon[23,24] were postulated to exist within the cytoplasm of the cell rather than within the cell membrane. The active osmotic diffusion pump, proposed by Franck and Mayer[25] to explain the movement of solvent and solute across biological tissues, can be classified as a cytoplasmic-carrier mechanism. Current attention is focused mainly on carrier systems which are operative within the membrane and not on those involving the cytoplasm.* Membrane-carrier systems have been suggested by, among others, Osterhout and Stanley,[33] Osterhout,[34,35] Lundergårdh,[36,37] Guensberg,[38] Hodgkin,[39] Ussing,[40–42] Rosenberg,[3,5] LeFevre,[43,44] LeFevre and Davies,[45] LeFevre and LeFevre,[46] Rosenberg and Wilbrandt,[11] Widdas,[47–49] and Wilbrandt.[50]

In the most general sense, the carrier may be either a major structural component of the membrane or a minor component consisting of some substance which partitions itself in the membrane. Conceptually, a carrier mechanism may take various forms—e.g., the carrier may be fixed in place in the membrane [Figs. 2(a)–2(c), 2(h) and 2(i)] or mobile [Figs. 2(d)–2(i)]. Although many workers in the field of transport restrict the term carrier to refer only to a mobile mechanism, we shall use the term carrier† to refer

---

* One notable case of a cytoplasmic carrier is the role of cytoplasmic myoglobin as the carrier of oxygen between the sarcolemma and mitochondria in muscle cells.[26–32]

† Perhaps the term portor or translocator, as suggested by Mitchell,[31,32] or mediator should be used instead of carrier. We hesitate, however, to add to the jargon of the transport field, preferring instead to expand the meaning of the term carrier.

A   Fixed Carrier: Single Site (Simple Reaction)

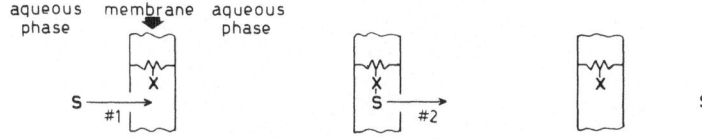

B   Fixed Carrier: Single Site (Patlak's Gate)

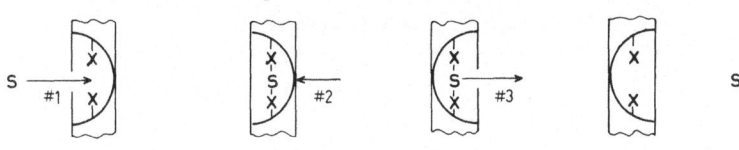

C   Fixed Carrier: Multisite (Danielli's "Bucket Brigade")

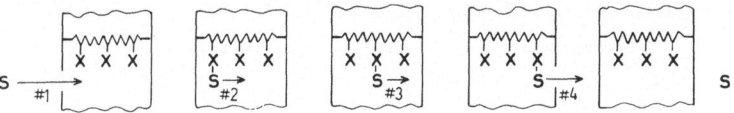

D   Mobile Carrier: Parallel Motion (Stein's Hemiport)

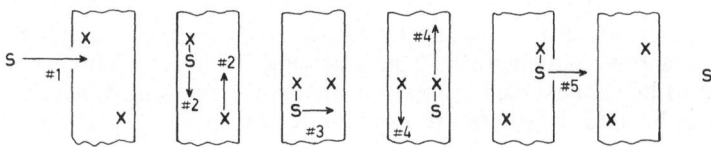

E   Mobile Carrier: Rotational Motion

F   Mobile Carrier: Transverse Motion

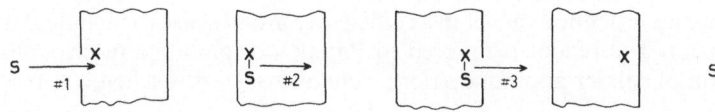

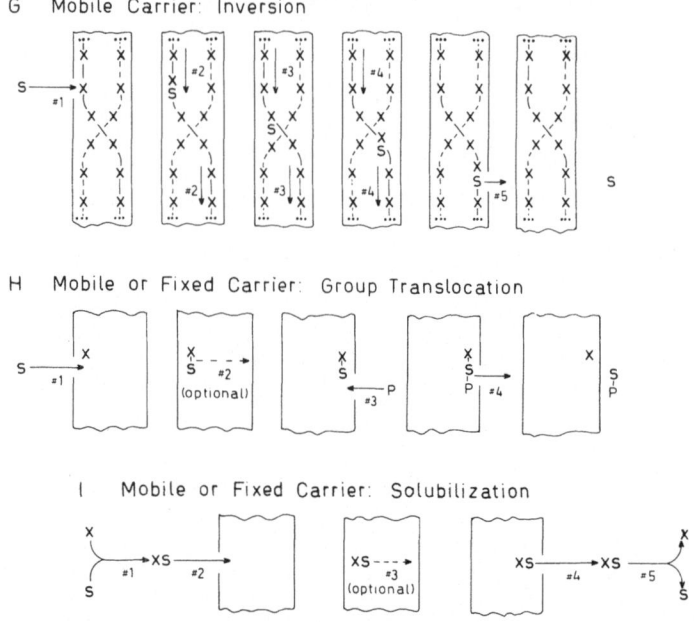

Fig. 2. Modes of carrier translocation.

to any molecule or reactive group which mediates the movement of a substance across a membrane. The following list of mechanisms is not intended to be exhaustive, but to show the diverse ways in which the term carrier can be used to specify a transport mechanism.

a. Fixed Carrier: Single Site (Simple Reaction) In the scheme in Fig. 2(a) which was proposed by Shannon[23,24] LeFevre and Davies,[45] and LeFevre and LeFevre,[46] carrier X is a fixed, nonmobile component within the membrane. Substance S is not permitted to cross the membrane as free (noncomplexed) S, but may react with X in step 1 to form XS, and then be released in step 2 into the aqueous phase on the right [(Fig. 2(a)].

b. Fixed Carrier: Single Site (Patlak's Gate). The scheme in Fig. 2(b) was proposed by Patlak,[51,52] who specifically called it a gate type noncarrier mechanism, since it could involve a modification of the potential barrier which limits the transmembrane movement of S, rather than involve the formation of a chemical substrate-carrier complex. To avoid semantic confusion, we have defined carrier in a sufficiently broad sense to include Patlak's mechanism. The reader is referred to Patlak's original manuscripts for his definition of carrier and noncarrier mechanisms. In this scheme, S reacts in step 1 with the carrier, which is oriented to receive S from the aqueous phase on the left; in step 2 the carrier undergoes a transition which orients the carrier to release S into the aqueous phase on the right in step 3 [Fig. 2(b)].

c. Fixed Carrier: Multisite (Danielli's "Bucket Brigade"). In the scheme in Fig. 2(c), which was proposed by Danielli,[13] there are a series of sites which can combine with S to form XS. In step 1, S combines with the first site to form XS; in steps 2 and 3, S hops toward the right from site to site (as if being handled by a "bucket brigade"), until it is finally released into the aqueous phase on the right [Fig. 2(c)]. The binding sites could comprise the lining of a pore which transverses the membrane, in which case the terms specific pore and multisite fixed carrier become synonymous.

d. Mobile Carrier: Parallel Motion (Stein's Hemiport). In the scheme in Fig. 2(d), which was proposed by Stein,[53] carrier sites exist on both sides of the membrane and are restricted to move in any direction parallel to the plane of the membrane. In step 1, S reacts with X on the left to form XS; in step 2, movement of XS on the left and X on the right eventually results in the alignment of XS and X so that S may be transferred from X on the left to X on the right in step 3; in step 4, movement of X on the left and XS on the right results in misalignment of X and XS; in step 5, S is released from XS into the aqueous phase on the right [Fig. 2(d)]. Stein called his mechanisms hemiports; we have expanded the definition of mobile carrier to include his hemiports.

e. Mobile Carrier: Rotational Motion. In the scheme, in Fig. 2(e), which is similar to that of LeFevre,[44] S reacts with X in step 1 to form XS, which undergoes a 180° rotation in steps 2 and 3 so that S can be released into the aqueous phase on the right in step 4 (Fig. 2e). This mechanism is similar to the single-site fixed carrier [Fig. 2(a)] inasmuch as both schemes require a parameter which characterizes the formation of XS, but it differs inasmuch as an additional parameter is needed to describe the rotational step.

f. Mobile Carrier: Transverse Motion. We have chosen this scheme, which was originally proposed by Widdas,[47,49] [Fig. 2(f)] to explain the interaction phenomena observed in biological systems (see Section II, C). These phenomena are not, however, uniquely explained by this scheme (which was chosen only because of the ease with which it could be developed mathematically), for they can also be explained by some of the other schemes shown in Figs. 2(a)–(i). In this scheme, S reacts with X in step 1 to form a mobile complex XS, which moves across the membrane in step 2 so that S can be released into the aqueous phase on the right in step 3 [Fig. 2(f)].

g. Mobile Carrier: Inversion. In this scheme in Fig. 2(g) there is no topographical difference between the inside and the outside of a membrane —i.e., the surface of a membrane resembles that of a Klein bottle or a Möbius strip. In the "unit membrane" model of J. D. Robertson,[54–57] who based his ideas on those of Danielli[58,59] (and Davson), the carrier sites (X's) are on the protein layer which is thought to coat lipid membranes. In the "repeating subunit" model of Green,[60–63] each corresponds to one of the types of subunits, not all of which need serve as carriers. If the membrane topographically resembles a Klein bottle, this coat or subunit eventually makes a transition from contact with the aqueous phase on the left to contact with the aqueous phase on the right. In step 1, S combines with one of the X's to form XS; in steps 2–4, continuous movement of the surface layers eventually results in the transition of XS from contact with the aqueous phase on the left to contact with that on the right so that S can be released into the latter phase in step 5 [Fig. 2(g)].

h. Mobile or Fixed Carrier: Group Translocation. In the scheme in Fig. 2(h), which was proposed by Roseman,[64] who based his ideas on those of Mitchell and Moyle,[65] and Mitchell,[66] S reacts with X in step 1 to form XS, which moves across the membrane in step 2 if the carrier is mobile rather than fixed; before the release of S from the carrier, S is modified by the attachment of P to S in step 3; in step 4, SP is released into the aqueous phase on the right [Fig. 2(h)]. In this scheme, S is translocated from the aqueous phase on the left, where it is free S, to that on the right, where it is "trapped" as the group S in SP, in which form the S cannot move back to the left. Although a carrier need not be an integral part of a "trapping" scheme (see Section IV), it may nevertheless be a desirable part of the scheme, since the specificity of a carrier for a substrate can determine which substances are "trapped."

i. Mobile or Fixed Carrier: Solubilization. In the scheme in Fig. 2(i), which was inspired by studies of transport-inducing antibiotics,[67–73] S reacts with X in step 1 to form XS in the aqueous phase on the left; whereas the solubility of free S in the membrane is negligible, both X and XS are highly soluble; hence, XS may enter the membrane in step 2; in step 3, XS moves across the membrane if the carrier is mobile rather than fixed; in step 4, XS enters the aqueous phase on the right; in step 5, S is released from the carrier [Fig. 2(i)]. Note that in this scheme, the carrier need not reside within the membrane at all times, as it must in the other schemes [Figs. 2(a–h)]. Thus, this carrier is conceived as a soluble constituent of both aqueous and lipid phases.

2. Generalized Model

Figure 3 shows one of many possible schemes for a carrier-mediated transport system. In this scheme substrate S is transferred from aqueous phase 1 to aqueous phase 2 via a series of reversible reactions. S can move reversibly from phase 2 to phase 1 by undergoing the series of reactions in the direction opposite to that shown by the arrows. The net direction of movement of S is determined by the difference in electrochemical potential of S across the membrane. Consider the following sequence of reactions:

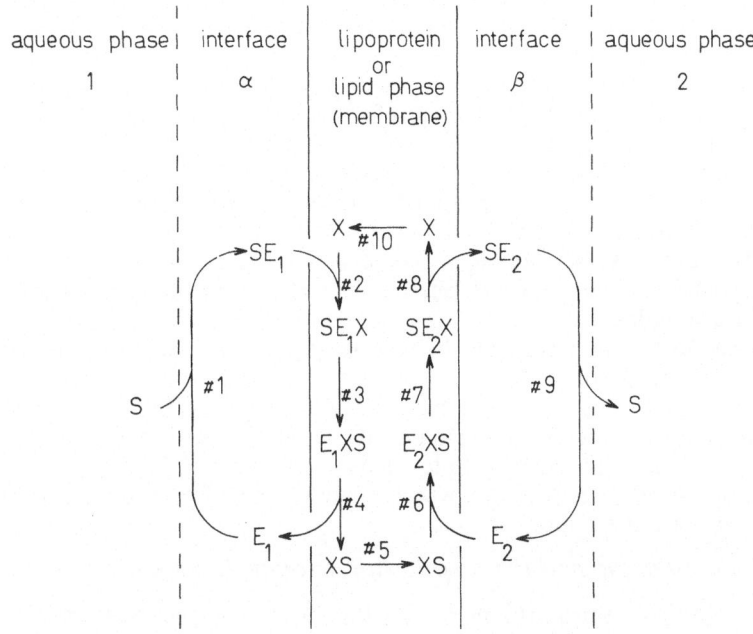

Fig. 3. A generalized transport scheme for movement of a substrate across a membrane.

Reaction 1. S combines reversibly with an enzyme E_1 located on side 1 of the membrane. E_1 provides the carrier-mediated transport system with the ability to recognize the substrate S among the various dissolved constituents of aqueous phase 1. Similar substrates can compete to form their respective enzyme-substrate complexes. The domain of E_1 is designated by the region α, which is a part of either aqueous phase 1 or the membrane. If α is a part of aqueous phase 1, E_1 is a soluble enzyme; if α is a part of the membrane, E_1 may be a major or minor structural constituent of side 1 of the membrane, or may exist at the interface between the phases by being loosely bound (e.g., hydrogen-bonded) or tightly bound (e.g., covalently bonded) to a constituent of the membrane.

Reaction 2. The complex SE_1 reacts reversibly with carrier X to form another complex SE_1X. The partition coefficient is such that the amount of free (noncomplexed) S in the membrane is negligible; thus, the amount of free S crossing the lipid phase via simple diffusion of S is negligible. The carrier X, either a major or minor constituent of the membrane, may be specific for a particular SE_1 complex, or may serve as a generalized carrier for different complexes in which the substrates are completely different in chemical nature.

Reaction 3. The enzyme-substrate complex undergoes a transition in which the bond between S and E_1 is broken as a new bond between S and X is formed. In effect, S is transferred from the enzyme to the carrier.

Reaction 4. E_1 is released, thereby leaving S bound to X as the carrier-substrate complex XS.

Reaction 5. The transition of S from side 1 to side 2 occurs only while S is attached to X. Possible mechanisms for the transition are discussed in the previous section and are shown in Figs. 2(a)–2(i).

Reaction 6. This reaction is the reverse of reaction 4. An enzyme E_2, which may be—but is not necessarily—identical to enzyme E_1, combines with XS to form a new complex E_2XS.

Reaction 7. This reaction is the reverse of reaction 3. In effect, S is transferred from the carrier to the enzyme moiety of the enzyme-carrier-substrate complex.

Reaction 8. This reaction is the reverse of reaction 2. SE_2 is released and free carrier X is re-formed.

Reaction 9. This reaction is the reverse of reaction 1. S is released into aqueous phase 2, thereby resulting in the transport of S from phase 1 to phase 2, and free enzyme E_2 is re-formed.

Reaction 10. This reaction, which is the reverse of reaction 5, serves to return the carrier to side 1.

3. *Model Building and the Kinetic Approach to Transport*

In a study of transport kinetics, the flux of substrate across a membrane is determined as a function of concentrations of substrate in aqueous phase 1 and in phase 2. If the substrate is an ion, the flux is also a function of its valence and of the electrical potential difference between the two aqueous phases. Typical physical units of flux are μmole/hour or (if the flux is normalized with respect to the membrane area) μmole/hour-cm². *In this chapter, all fluxes are normalized with respect to membrane area.*

Net flux can be determined by a chemical measurement of either the rate of disappearance of substrate in one aqueous phase or the rate of appearance in the other. The unidirectional flux from phase 1 to phase 2 can be measured by introducing a trace amount of an isotopically labeled substrate (viz., $S^\dagger$) into phase 1 and measuring its rate of appearance in phase 2.[42,74–76] As long as the concentration of $S^\dagger$ in phase 2 can be kept negligible, so that very little isotope moves back to phase 1, the rate of appearance of $S^\dagger$ in phase 2 gives a good measure of a unidirectional movement from phase 1 to 2. Introduction of a different isotope of S (viz., $S^\ddagger$) into phase 2 provides the means of determining the unidirectional flux from phase 2 to phase 1. It is assumed that there is no isotope effect—i.e., that the kinetic constants for all reactions between enzyme, carrier, and substrate are the same for S, $S^\dagger$ and $S^\ddagger$. Thus, the unidirectional flux of S from phase 1 to phase 2 may be defined as the flux of isotope $S^\dagger$ (a measurable quantity) multiplied by the ratio of [S] to [$S^\dagger$] in phase 1, and the unidirectional flux of S from phase 2 to phase 1 may be defined in a similar manner. The value found for net flux by chemical measurement of the rate of appearance of S in one phase should be the same as that found by calculating the difference between the two unidirectional fluxes which are determined isotopically.

The object of building a mathematical model is to arrive at a set of equations which relate experimentally measurable quantities. In the kinetic models developed in this chapter, the flux of substrate is expressed as a function of the concentration (which we treat as equal to the chemical activity) in aqueous phases 1 and 2 and of the electrical potential difference across the membrane (when the substrate is an ion). In the model shown in Fig. 3, ten reactions, described by twenty rate constants, are involved in the transport process. Since not all of the rate constants are independent, some can be eliminated by replacement with functions of the others. The number of independent rate constants is, nevertheless, still large, and consequently an attempt is made to reduce their number further by assuming that all but one or two of the reactions are at equilibrium—a condition in which one equilibrium constant replaces two rate constants. The remaining off-equilibrium reactions are the rate-limiting steps for the transport process. In their classic work, Rosenberg and Wilbrandt[77] considered a number of models with different rate-limiting steps in each scheme, and subsequently evaluated the application of various kinetic models to particular transport processes in biological systems.[78–81]

In the following section, a simplified version of the model shown in Fig. 3 is presented. This simplified model predicts transport phenomena which resemble Michaelis–Menten kinetics for enzymatic reactions and can account for most of the experimental findings in various transport systems. In reviewing the transport schemes presented in this chapter and elsewhere, it is important always to remember that more than one model can be constructed to fit kinetic data, and therefore, that any given model can be refuted but not conclusively established solely by such data.

4. A Simplified Model for Passive Downhill Movement of a Nonelectrolyte

Figure 4 shows a simplified version of the model shown in Fig. 3. In this model, the function of specificity for (or recognition of) substrate is performed by the carrier X rather than by an enzyme E_1 or E_2. Although the simplified model may predict all the experimental kinetic data, it is more likely that the actual mechanism of transport in a biological system more closely resembles the generalized model shown in Fig. 3 than the simplified version described in this section.

The derivation of the kinetic equation from this model is based on the following assumptions:

1. *S is translocated in the form of XS.* The nonelectrolytic substrate S can cross the membrane only in the form of carrier-substrate complex XS and not as free substrate. In an actual biological system, however, part of the S may be translocated across the membrane in the form of free S via simple diffusion.

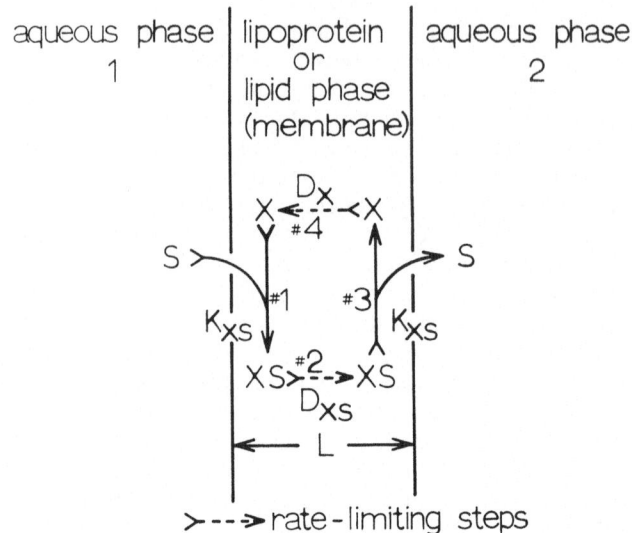

>---> rate-limiting steps

Fig. 4. A simplified scheme for passive downhill transport of a non-electrolyte across a membrane.

2. *Carrier is conserved.* Total quantity of carrier $Q_{X(total)}$ in all forms remains constant within the membrane. In other words,

$$Q_{X(total)} = Q_{X(free\ carrier)} + Q_{XS(complex)} = constant \qquad (19)$$

where Q is amount of carrier per unit area. In an actual biological system, carrier is constantly being synthesized and degraded. Nevertheless, during the course of any given short-term experiment, the amount of carrier remains approximately constant.

3. *The system is operating in the steady state.* The rate of the degree of advancement (or velocity) of each of the four reactions is equal, and, therefore, the net rate of movement of carrier-substrate complex XS from side 1 to 2 is exactly balanced by an equal and opposite movement of free carrier X from side 2 to 1. If the system were not operating in the steady state, carrier would pile up at one side of the membrane and become depleted at the other side. The time required to establish a new steady state after changing [S] in one of the aqueous phases can be estimated as follows:

$$\tau \approx \frac{amount\ of\ total\ carrier\ in\ the\ membrane\ (pmole/cm^2)}{change\ in\ steady\text{-}state\ level\ of\ flux\ (\mu mole/hr\text{-}cm^2)} \qquad (20)$$

For biological systems, the time constant, τ, is probably of no greater magnitude than milliseconds.

Epithelial tissues consist of at least three aqueous phases (one of which is intracellular) separated by two lipoid phases in series. When [S] is changed in one of the outer aqueous phases, the transport systems in the lipoid phases probably adjust to a new steady state within milliseconds, although the time required for the center aqueous phase to come to a new steady state may be on the order of minutes. Thus, the time required for the entire system to approach a new steady state is dependent upon the relatively long time constant for the center aqueous phase, rather than upon the short time constants of the transport systems.

4. *The rate-limiting reactions are the transitions of X and XS across the membrane (see reactions 2 and 4 of Fig. 4).* These transitions are described by Fick's equation for simple diffusion—i.e., the net rate of movement of X or XS across the membrane is proportional to the difference in [X] or [XS], respectively, across the membrane and is inversely proportional to membrane thickness L; the proportionality constant is called the diffusion constant and is designated by D_X or D_{XS}, respectively. In other words,

$$J_{XS}^{1 \to 2(net)} = (D_{XS}/L) \times ([XS]_1 - [XS]_2) \tag{21}$$

and

$$J_X^{2 \to 1(net)} = (D_X/L) \times ([X]_2 - [X]_1) \tag{22}$$

where J is the flux per unit area of membrane. Note that the steady-state assumption requires that

$$J_S^{1 \to 2(net)} = J_{XS}^{1 \to 2(net)} = J_X^{2 \to 1(net)} \tag{23}$$

Applying Fick's equation to the movement of X and XS across the membrane requires that X and XS be nonionic, for if they were ionic in character, the electrical potential difference across the membrane would also affect the transition. In many biological systems, X and XS are ionic in character (i.e., each carries a net charge), and, therefore, Fick's equation cannot be applied. When S is ionic in character, the charge on XS and X differs by the amount of charge on S, and the movement of XS and X across the membrane is affected differently by the electrical potential difference. In addition, the electrical potential difference which may exist at each of the two interfaces between aqueous and lipoid phases results in a difference in [S] in the bulk of the aqueous phase and [S] at the site of reaction between X and S in the lipoid phase. In Section II, E a carrier model for the passive downhill movement of an electrolyte is considered.

In considering cell membranes in which L is of the order of 70 Å, it can be argued that Fick's equation does not apply to such a small phase. Therefore, in an attempt to avoid the use of Fick's equation, Vidaver[82] described the transitions of X and XS across the membrane by kinetic rate constants. For example, the transitions of X from side 1 to 2 and from side 2 to 1 are described, respectively, by

$$J_X^{1 \to 2} = k_X^{1 \to 2} \times Q_{X(1)} \tag{24a}$$

and

$$J_X^{2 \to 1} = k_X^{2 \to 1} \times Q_{X(2)} \tag{24b}$$

where $Q_{X(i)}$ is the quantity of carrier on side i ($i = 1$ or 2) and is given by

$$Q_{X(i)} = [\overline{X}]_i \times L \tag{25}$$

where $[\overline{X}]_i$ is the average concentration that would result if the free carrier on side i were uniformly distributed throughout the membrane. In this particular treatment, we assume that the kinetic constants are equal and may be both written as k_X. Therefore, the net movement of X from side 1 to 2 is given by

$$J_X^{1 \to 2(net)} = J_X^{1 \to 2} - J_X^{2 \to 1} = k_X \times (Q_{X(1)} - Q_{X(2)}) \qquad (26a)$$

$$= (k_X \times L) \times ([\overline{X}]_1 - [\overline{X}]_2) \qquad (26b)$$

Note that this equation is equivalent to Fick's equation if $k_X = D_X/L^2$. Thus, even if X and XS do not move by simple diffusion, the equations describing their transitions across the membrane may be identical to those given by Fick's equation. Throughout this chapter, the lumped parameter (D/L^2) appears in all equations for maximal flux [e.g., see Eqs. (31, 52, 84, 88, 106, 197) and (261)]. In all cases, (D/L^2) may be replaced by a first-order rate constant (k) that has units of reciprocal time. The rate constant k may be interpreted as the probability per unit time that the carrier will make a transition from one side of the membrane to the other.

5. *The formation and dissociation of XS is so rapid that X, S and XS may be considered to be in equilibrium (see reactions 1 and 3 of Fig. 4).* One equilibrium constant rather than four kinetic constants—three of which are independent—is adequate to describe reactions 1 and 3. The equilibrium or dissociation constant K_{XS} is defined to have units of concentration and to include the partition coefficient of S between aqueous and lipoid phases:

$$K_{XS} = [X]_i \times [S]_i/[XS]_i \qquad (27)$$

where i refers to compartment 1 or 2. K_{XS} is the concentration of substrate required to convert half of the free carrier into carrier-substrate complex.

Rosenberg and Wilbrandt[77] also consider the model in which reactions 1 and 3 are off-equilibrium (and, hence, rate-limiting) and reactions 2 and 4 are so rapid that $[X]_1 \approx [X]_2$ and $[XS]_1 \approx [XS]_2$.

The remainder of this section is concerned with a kinetic model based on the five assumptions listed above and on the auxiliary assumption that diffusion constants for both the unloaded (or free) carrier and the loaded carrier (or carrier-substrate complex) are equal—i.e.,

$$D = D_X = D_{XS} \qquad (28)$$

where the D without subscript suffices to represent both D_X and D_{XS}. In Section II, C, 6, the restriction that $D_X = D_{XS}$ is removed, thereby leading to more complex phenomena, viz. to *trans*-effects.

A kinetic model to which Eq. (28) applies leads to the following Michaelis–Menten kinetics:

$$J_S^{cis \to trans} = J_{S,max} \times [S]_{cis}/([S]_{cis} + K_{S,0.5}) \qquad (29)$$

where $J_S^{cis \to trans}$ represents the unidirectional flux per unit area either from side 1 to 2 or from side 2 to 1, $J_{S,max}$ is the maximal or saturation value of unidirectional flux which is approached as $[S]_{cis} \to \infty$, and the apparent dissociation constant, $K_{S,0.5}$, is the concentration of S required for half-maximal unidirectional flux. For details of the derivation of Eq. (29), refer to Rosenberg and Wilbrandt.[77]

In this particular model, the *cis*-concentration of substrate required to achieve half-maximal unidirectional flux is identical with the equilibrium constant which links [X], [XS], and [S], i.e.,

$$K_{S,0.5} = K_{XS} \tag{30}$$

If $Q_{X(total)}$ is the total amount of carrier (X plus XS) per unit area and L is the thickness of the membrane (which is assumed to be planar), then

$$J_{S,max} = Q_{X(total)} \times D/L^2 \tag{31}$$

In addition to flux, another useful parameter is the clearance, which is defined as the unidirectional flux per unit *cis*-concentration, i.e.,

$$C_S^{cis \rightarrow trans} \equiv J_S^{cis}/[S]_{cis} \tag{32}$$

where $C_S^{cis \rightarrow trans}$ is the unidirectional clearance per unit area. For the kinetic model described by Eq. (29),

$$C_S^{cis \rightarrow trans} = J_{S,max}/([S]_{cis} + K_{S,0.5}) \tag{33}$$

The unidirectional *clearance* represents the rate at which the *cis*-side is *cleared* of substrate S by the transport process operative in the membrane. $C_S^{cis \rightarrow trans}$ does not, however, account for any unlabeled substrate which may be replenished on the *cis*-side by back movement from the *trans*-side. The development of the concept of net clearance, which is not simply the difference between the unidirectional clearances, is beyond the scope of this chapter.

Although clearance is expressed in units of volume per unit time per unit area (e.g., in $\mu l/sec/cm^2$ or $ml/hr/cm^2$), this parameter provides no information on the actual volume flow rate across the membrane. For example, if $C_S^{cis \rightarrow trans} = 1\ \mu l/sec/cm^2$, then an amount of S equivalent to that contained in 1 μl of solution on the *cis*-side moves to the *trans*-side across each square centimeter of membrane every second—i.e., 1 μl of solution is being cleared of all its S by each square centimeter of membrane every second, but this clearance provides no information on transmembrane volume flow. Alternately, the clearance per unit area of membrane is expressed as a velocity (e.g., in cm/hr), in which case it may be interpreted as the average velocity at which the substrate moves across the membrane.

It can be seen from Eq. (33) that clearance approaches a maximum as *cis*-concentration approaches zero, i.e.,

$$C_{S,max} = \lim_{[S]_{cis} \rightarrow 0} C_S^{cis \rightarrow trans} \tag{34a}$$

$$= J_{S,max}/K_{S,0.5} \tag{34b}$$

$$= J_{S,max}/K_{XS} \tag{34c}$$

$$= Q_{X(total)} \times D/(L^2 \times K_{XS}) \tag{34d}$$

Note that the greater the affinity of the substrate for the carrier (i.e., the lower the value of K_{XS}), the greater the maximal clearance.

Figures 5(a) and 5(b) show unidirectional flux and clearance as functions of *cis*-concentration. It is apparent that flux continuously increases—and clearance continuously decreases—as *cis*-concentration increases. In this model, half-maximal clearance and half-maximal flux are achieved at the same *cis*-concentration. The curves of Figs. 5(a) and 5(b) are replotted respectively as straight lines in Figs. 5(c) and 5(d), which correspond respectively to the Lineweaver–Burk[83] and the Hanes[84] plots used to display enzyme kinetics. It is also significant [Fig. 5(a)] that in the region of low *cis*-concentration ($[S]_{cis} \ll K_{S,0.5}$), flux is approximately proportional to *cis*-concentration. Hence, in this region the carrier system cannot be distinguished from a simple diffusion system by kinetic characteristics alone.

From Eq. (29), the net flux from side 1 to 2 may be found as follows:

$$J_S^{1 \to 2(net)} = J_S^{1 \to 2} - J_S^{2 \to 1} \tag{35a}$$

$$= J_{S,max} \times \left(\frac{[S]_1}{[S]_1 + K_{S,0.5}} - \frac{[S]_2}{[S]_2 + K_{S,0.5}} \right) \tag{35b}$$

$$= J_{S,max} \times K_{S,0.5} \times \frac{([S]_1 - [S]_2)}{([S]_1 + K_{S,0.5}) \times ([S]_2 + K_{S,0.5})} \tag{35c}$$

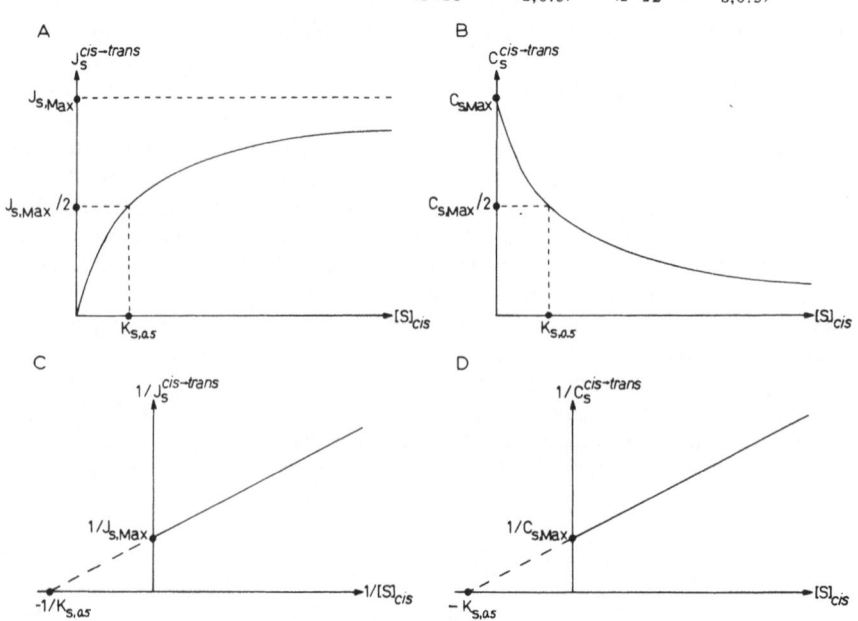

Fig. 5. Transport kinetics which resemble Michaelis–Menten enzyme kinetics. These kinetics apply to the transport scheme shown in Fig. 4 for passive downhill transport of a nonelectrolyte when all diffusion constants are equal.

From the term $([S]_1 - [S]_2)$ in the numerator of Eq. (35c), it follows that net movement of S is from side 1 to 2, i.e. $J_S^{1 \to 2(net)} > 0$ when $[S]_1 > [S]_2$ and from side 2 to 1, i.e. $J_S^{1 \to 2(net)} < 0$ when $[S]_1 < [S]_2$. Hence, this particular model is an example of a passive downhill carrier system.

We will now demonstrate how the model presented in this section can account for the following characteristics, listed in Section II, A, which distinguish biological transport from simple diffusion:

a. Abnormal flux ratios. From Eq. (29), it follows that

$$J_S^{1 \to 2}/J_S^{2 \to 1} = ([S]_1/[S]_2) \times ([S]_2 + K_{S,0.5})/([S]_1 + K_{S,0.5}) \quad \cdot \quad (36)$$

If sides 1 and 2 are chosen so that $[S]_1 > [S]_2$, then

$$([S]_2 + K_{S,0.5})/([S]_1 + K_{S,0.5}) < 1 \tag{37}$$

Substitution of Eqs. (15) and (37) into Eq. (36) yields

$$J_S^{1 \to 2}/J_S^{2 \to 1} < (J_S^{1 \to 2}/J_S^{2 \to 1})_{Fick} \tag{38}$$

If, however, concentrations of S well below saturation are used on both sides (i.e., $[S]_1 \ll K_{S,0.5}$ and $[S]_2 \ll K_{S,0.5}$), then

$$([S]_2 + K_{S,0.5})/([S]_1 + K_{S,0.5}) \approx 1 \tag{39}$$

Substitution of Eqs. (15) and (39) into Eq. (36) yields

$$J_S^{1 \to 2}/J_S^{2 \to 1} \approx (J_S^{1 \to 2}/J_S^{2 \to 1})_{Fick} \tag{40}$$

An "abnormal" flux ratio can be detected experimentally only when a saturating concentration of S is used on at least one of the sides (i.e., $[S]_1 \gg K_{S,0.5}$ and/or $[S]_2 \gg K_{S,0.5}$).

The generalized expression for "abnormal" flux ratios may be written in terms of the clearance concept:

$$J_S^{1 \to 2}/J_S^{1 \to 2} \neq (J_S^{1 \to 2}/J_S^{2 \to 1})_{Fick} \tag{41a}$$

$$\neq [S]_1/[S]_2; \tag{41b}$$

$$(J_S^{1 \to 2}/[S]_1)/(J_S^{2 \to 1}/[S]_2) \neq 1; \tag{41c}$$

$$C_S^{1 \to 2}/C_S^{2 \to 1} \neq 1; \tag{41d}$$

$$C_S^{1 \to 2} \neq C_S^{2 \to 1} \tag{41e}$$

In other words, an "abnormal" flux ratio signifies an inequality between the clearances $C_S^{1 \to 2}$ and $C_S^{2 \to 1}$.

b. Substrate specificity. The formation of a mobile carrier-substrate complex depends upon the specificity of a carrier for the substrate. The dissociation constant serves as a measure of relative specificity. Equations

(29) and (30) show that for a given *cis*-concentration of substrate, unidirectional flux increases as the dissociation constant K_{XS} decreases (hence, as the affinity between X and S increases). Similar substrates (e.g. S, T, and U) could be arranged according to the values of their respective dissociation constants:

low specificity	intermediate specificity	high specificity
low affinity	intermediate affinity	high affinity
high dissociation constant	intermediate dissociation constant	low dissociation constant
K_{XS}	$> K_{XT}$	$> K_{XU}$ (42)

In Sections II, C, 2–5, the consequences of competition between two substrates for a single carrier are examined.

c. *Specific Inhibitors.* Equations (29) and (31) show that flux is proportional to $Q_{X(total)}$ (the total amount of carrier per unit area). If a chemical agent can, however, form an immobile complex with either carrier or carrier-substrate complex, then the amount of $Q_{X(total)}$ is effectively reduced, since Eq. (27) shows that $Q_{X(total)}$ includes only the forms which participate in transport, namely, X and XS, not X-inhibitor and XS-inhibitor.

A chemical agent is an inhibitor for a given carrier if it forms an immobile complex with that carrier. The formation of such complexes may be the basis for the pharmacological action of many drugs.

d. *Homo-cis-inhibition.* Figure 5(a) clearly shows that flux reaches a limiting value as *cis*-concentration is increased.

C. Interaction Phenomena

1. *Introduction*

In Section II, B, 3 an operational definition of unidirectional flux is given, namely

$$J_S^{cis \to trans} = ([S]_{cis}/[S^\dagger]_{cis}) \times J_{S^\dagger}^{cis \to trans} \tag{43}$$

where $J_S^{cis \to trans}$ is the unidirectional flux of all forms of substrate S (including $S^\dagger$) and $J_{S^\dagger}^{cis \to trans}$ is the measurable quantity, namely, the unidirectional flux of isotopic tracer $S^\dagger$. Equation (43) can be rearranged to

$$J_S^{cis \to trans}/[S]_{cis} = J_{S^\dagger}^{cis \to trans}/[S^\dagger]_{cis} \tag{44}$$

Clearance is defined in the previous section by Eq. (32) as the unidirectional flux per unit *cis*-concentration. Thus, the terms in Eq. (44) can be replaced by the appropriate clearances:

$$C_S^{cis \to trans} = C_{S^\dagger}^{cis \to trans} \tag{45}$$

Note that although the fluxes of S and $S^\dagger$ may be quite different [see Eq. (43)], the clearances of S and $S^\dagger$ are identical if there is no isotope effect (see Section II, B, 3). It should not be surprising that clearances for S and $S^\dagger$ are identical, because it is reasonable to expect that if a given volume is cleared of $S^\dagger$ in a given period of time, then that same volume is cleared of S during the same period of time.

Consider a membrane in which nonelectrolytic substrate is translocated via simple diffusion. Fick's equations for the two unidirectional fluxes are given by Eqs. (14a) and (14b) and generalized by the following single equation :

$$J_S^{cis \to trans} = (D_S/L) \times [S]_{cis} \qquad (46)$$

where D_S is the apparent diffusion constant of S, and L is the thickness of the membrane. Note that flux is directly proportional to cis-concentration and does not approach a limiting value as $[S]_{cis} \to \infty$. It follows from the definition of clearance that

$$C_S^{cis \to trans} \equiv J_S^{cis \to trans}/[S]_{cis} = D_S/L \qquad (47)$$

Note that the two unidirectional clearances ($1 \to 2$ and $2 \to 1$) are equal and constant (i.e., depend on neither $[S]_1$ nor $[S]_2$). If the substrate is an electrolyte, the two unidirectional clearances are still independent of concentration. They are also, however, functions of transmembrane electrical potential difference. The two clearances are equal, therefore, only when the potential difference is zero.

If the movement of S across a membrane can be characterized by a constant clearance, then the process of simple diffusion is sufficient to explain the translocation process. In biological systems, however, the clearance of a given substrate often depends upon the concentrations of that substrate on both sides of the membrane; furthermore, the two unidirectional clearances are often not equal. Carrier mechanisms are usually invoked to explain such deviations from a simple diffusion system.

If addition of substrate or analogue to the cis-side results in a decrease in clearance, the phenomenon is called cis-inhibition or cis-depression; if the addition results in an increase in clearance, it is called cis-stimulation. Similarly, addition to the trans-side may result in trans-inhibition or trans-stimulation. Note that cis- and trans-effects are better determined by changes in clearance than by changes in flux. An inhibitory or stimulatory effect on clearance indicates that some interaction is occurring among the substrate molecules as they are translocated, since simple diffusion (in which there is no interaction) can account only for constant clearance. In the remainder of this section (Section II, C) a number of cis- and trans-effects and some consequences of these effects are discussed in terms of various carrier models.

2. Hetero-cis-inhibition

The simple model introduced in Section II, B, 4 and shown in Fig. 4 can account for cis-inhibition. Note that $C_S^{cis \to trans}$ in Fig. 5(b) decreases as cis-concentration increases. This kind of cis-inhibition is called homo-cis-inhibition because S interferes with its own translocation. In biological systems, the presence of a chemical analogue T (e.g., galactose) may interfere with the movement of a substrate S (e.g., glucose) and vice versa. When an increase in $[\mathrm{T}]_{cis}$ results in a decrease in $C_\mathrm{S}^{cis \to trans}$, the effect is called hetero-*cis*-inhibition.

In Fig. 6, a carrier model similar to that shown in Fig. 4 allows two substrates, S and T, to compete to form their respective carrier-substrate complexes, XS and XT. The five assumptions listed in Section II, B, 4 apply to this model, but an additional equilibrium constant, K_XT, is introduced to relate [X], [T], and [XT], and the conservation law for carrier [see Eq. (19)] must include the amount of carrier in the form XT. As in Section II, B, 4, the auxiliary assumption is made that all diffusion constants are equal, i.e.

$$D = D_\mathrm{X} = D_\mathrm{XS} = D_\mathrm{XT} \tag{48}$$

Without this auxiliary assumption, the unidirectional clearance is a function of *trans*-concentration. When diffusion constants are not equal, the expressions shown in Sections II, C, 2–5 must be modified to include *trans*-effects. The discussion of *trans*-effects is given in Section II, C, 6 rather than in this section, since it would only serve to obscure the basic properties of hetero-*cis*-inhibition.

A kinetic model based on the assumptions listed above leads to Michaelis–Menten kinetics with competitive inhibition, i.e., to

$$J_\mathrm{S}^{cis \to trans} = J_{\mathrm{S,max}} \times [\mathrm{S}]_{cis}/([\mathrm{S}]_{cis} + K_{\mathrm{S},0.5}^{cis \to trans}) \tag{49}$$

and

$$C_\mathrm{S}^{cis \to trans} = C_{\mathrm{S,max}}^{cis \to trans} \times K_{\mathrm{S},0.5}^{cis \to trans}/([\mathrm{S}]_{cis} + K_{\mathrm{S},0.5}^{cis \to trans}) \tag{50}$$

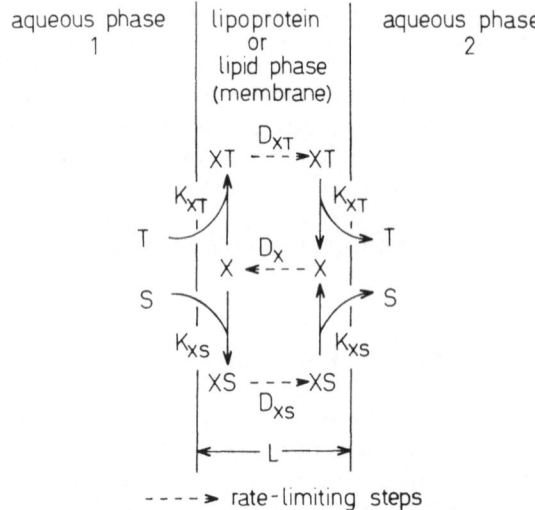

Fig. 6. A simplified scheme in which two different substrates compete for transmembrane transport on a passive carrier.

where

$$K_{S,0.5}^{cis \to trans} = K_{XS} \times (1 + [T]_{cis}/K_{XT}), \tag{51}$$

$$J_{S,max} = Q_{X(total)} \times D/L^2, \tag{52}$$

and

$$C_{S,max}^{cis \to trans} = J_{S,max}/K_{S,0.5}^{cis \to trans} \tag{53a}$$

$$= Q_{X(total)} \times D/\{K_{XS} \times (1 + [T]_{cis}/K_{XT})\} \tag{53b}$$

An analogous set of equations for the movement of T may be formed from Eqs. (49–53b) by interchanging the letters S and T in all equations.

Note that $K_{S,0.5}^{cis \to trans}$ is neither constant nor independent of direction since it is a function of the *cis*-concentration of analogue T. Equation (51) shows that the presence of an analogue increases the concentration necessary to achieve half-maximal flux and gives the appearance of a decrease in carrier affinity for the primary substrate. This phenomenon is similar to competitive inhibition observed in enzyme kinetics, since K_{XT} in Eq. (51) is analogous to an inhibitor constant. This particular model predicts an important relation among concentrations necessary to achieve half-maximal flux, namely, that the dissociation constant for carrier and substrate, T, is identical to the inhibitor constant for T[K_{XT} in Eq. (51)]. The following set of experiments can be performed to test this prediction :

Step 1. Unidirectional flux of S in the absence of T is determined as a function of $[S]_{cis}$. If Michaelis–Menten kinetics are observed, this carrier model can be used to analyze the data. From such an analysis, $K_{S,0.5}^{cis \to trans}$ is determined on a plot such as that shown in Fig. 5(c) or 5(d) and is related to K_{XS} by Eq. (51) as follows :

$$K_{S,0.5(step\ 1)}^{cis \to trans} = K_{XS} \tag{54}$$

Thus, in this step, the dissociation constant for carrier and substrate, S, is evaluated.

Step 2. Unidirectional flux of T in the absence of S is determined as a function of $[T]_{cis}$. From the equation obtained from Eq. (51) by interchanging S and T,

$$K_{T,0.5(step\ 2)}^{cis \to trans} = K_{XT} \tag{55}$$

Thus, in this step, the dissociation constant for carrier and substrate, T, is evaluated. In the next step, a test is made to see if K_{XT} is equal to the inhibitor constant.

Step 3. Unidirectional flux of S is determined as a function of $[S]_{cis}$ when $[T]_{cis}$ is set at a constant level of $K_{T,0.5(step\ 2)}^{cis \to trans}$ which is equal to K_{XT}. If K_{XT} is equal to the inhibitor constant, then $K_{S,0.5}^{cis \to trans}$ in this step should be double that in step 1 since Eqs. (51, 54) and (55) lead to the following prediction :

$$K_{S,0.5(\text{step } 3)}^{cis \to trans} = K_{XS} \times (1 + [T]_{cis}/K_{XT}) \tag{56a}$$

$$= K_{XS} \times (1 + K_{T,0.5(\text{step } 2)}^{cis \to trans}/K_{XT}) \tag{56b}$$

$$= K_{XS} \times (1 + K_{XT}/K_{XT}) \tag{56c}$$

$$= 2 \times K_{XS} \tag{56d}$$

$$= 2 \times K_{S,0.5(\text{step } 1)}^{cis \to trans} \tag{56e}$$

A similar experiment can be performed with the roles of S and T interchanged. If $K_{S,0.5}^{cis \to trans}$ in step 3 is not twice that in step 1, this model cannot account for the data. In such a case, two carrier systems with a different set of dissociation constants could be operative in parallel. The previous test was proposed by Ahmed and Scholefield.[85] The graphical method proposed by Dixon[86] to determine enzyme inhibitor constants can be used to evaluate K_{XT} in Eq. (51).

Note that $J_{S,max}$ is not a function of the *cis*-concentration of analogue T. Furthermore, according to Eq. (52), the maximal unidirectional fluxes for S and T are identical, i.e.

$$J_{S,max} = J_{T,max} = Q_{X(total)} \times D/L^2 \tag{57}$$

This equality, however, depends upon the auxiliary assumption that all diffusion constants are equal [see Eq. (48)]. If the diffusion constants are not equal, then maximal fluxes are no longer equal, and the substrate with the larger diffusion constant has the larger maximal flux; in this case, the expression for maximal flux becomes more complex than the simple form of Eq. (57), because D must be replaced by a function of D_X, $D_{X\text{-substrate}}$ and *trans*-concentrations of all substrates [e.g., see Eqs. (82–85b)].

Note that $C_{S,max}^{cis \to trans}$ in Eq. (53b) is neither constant nor independent of direction since it is a function of the *cis*-concentration of analogue T. Equation (53b) shows that the presence of an analogue decreases the maximal clearance. Even when S and T are not present together, maximal clearances for S and T are not equal unless their dissociation constants are equal. In general, the substrate with the higher affinity for the carrier (i.e., with the lower dissociation constant) has the higher maximal clearance. In addition, if the diffusion constants are not equal, the relatively simple form of Eq. (53b) must be replaced by a more complex function; in this case, even if the dissociation constants for two substrates are equal, maximal clearances for them may not be equal.

Figure 7 shows $J_S^{cis \to trans}$ and $C_S^{cis \to trans}$ as functions of $[S]_{cis}$ for various values of $[T]_{cis}$. In Fig. 7(a), with an increase in $[T]_{cis}$, an increase in $[S]_{cis}$ is necessary to achieve half-maximal flux. If the abscissa were extended to the right, it would be seen that the presence of T does not affect $J_{S,max}$. In Fig. 7(b),

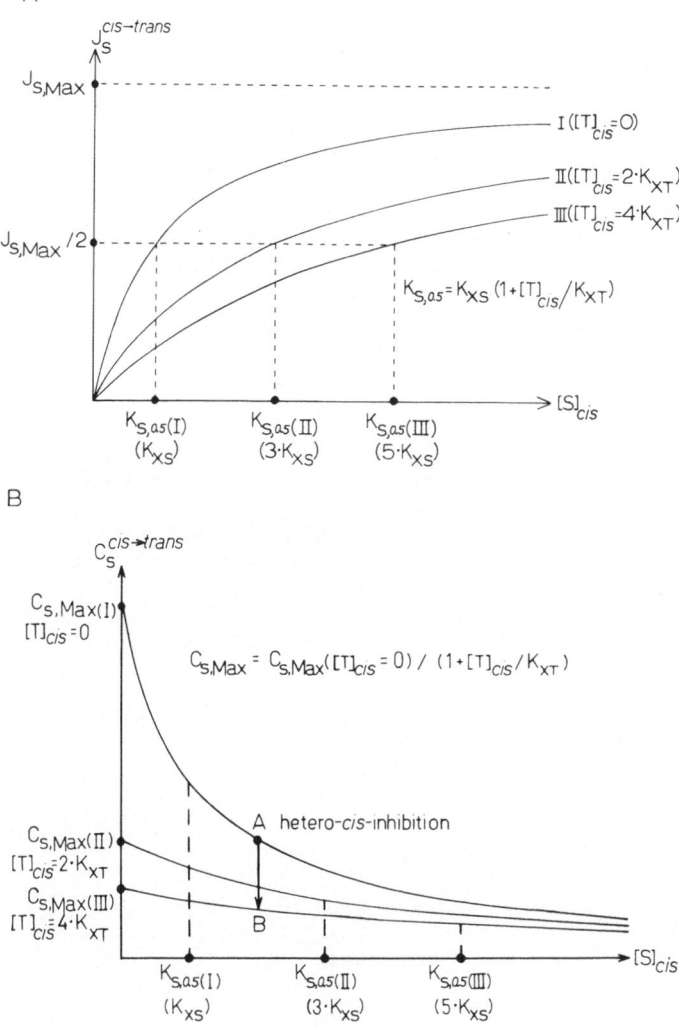

Fig. 7. Effect of the presence of substrate T on the transport of substrate S (hetero-*cis*-inhibition). These kinetics apply to the transport scheme shown in Fig. 6 for competition between two nonelectrolytic substrates for a common passive carrier when all diffusion constants are equal.

it is clear that the presence of T reduces $C_{S,max}$; note that an increase in $[T]_{cis}$ from 0 (point A) to $4 \times K_{XT}$ (point B), when $[S]_{cis}$ remains constant, results in a decrease in $C_S^{cis \to trans}$. This is an example of hetero-*cis*-inhibition.

Kinetic models which exhibit hetero-*cis*-inhibition have been presented by Widdas,[47,49] Jacquez,[87] Rosenberg and Wilbrandt,[88] and Miller.[89,90]

3. Counterflow

Addition of analogue T to one of the aqueous phases can result in the reversal of the direction of net movement of substrate S. Such reversal of flow is called counterflow and can be explained by the model of hetero-*cis*-inhibition presented in the previous section.

Consider the following example:

Condition 1. S moves across the membrane by the mechanism shown in Fig. 6 with the auxiliary assumption that $D_X = D_{XS} = D_{XT}$. The unidirectional flux of S is given by Eq. (49), which in the absence of T reduces to

$$J^{cis \to trans}_{S(\text{condition 1})} = J_{S,\max} \times [S]_{cis}/([S]_{cis} + K_{XS}) \tag{58}$$

The net flux is found as follows from Eq. (58):

$$J^{1 \to 2(net)}_{S(\text{condition 1})} = J^{1 \to 2}_{S(\text{condition 1})} - J^{2 \to 1}_{S(\text{condition 1})} \tag{59a}$$

$$= J_{S,\max} \times [S]_1/([S]_1 + K_{XS}) - [S]_2/([S]_2 + K_{XS}) \tag{59b}$$

$$= J_{S,\max} \times K_{XS} \times ([S]_1 - [S]_2)/\{([S]_1 + K_{XS}) \times ([S]_2 + K_{XS})\} \tag{59c}$$

Condition 1 is set so that $[S]_1 \geq [S]_2$. Equation (59c) shows that there is no net flow when $[S]_1 = [S]_2$, and there is downhill movement of S from side 1 to 2 when $[S]_1 > [S]_2$.

Condition 2. T is added to side 1. $J^{1 \to 2}_S$ is hetero-*cis*-inhibited by $[T]_1$, but $J^{2 \to 1}_S$ is not affected. From Eqs. (49) and (51),

$$J^{1 \to 2}_{S(\text{condition 2})} = J_{S,\max} \times [S]_1/\{[S]_1 + K_{XS}(1 + [T]_1/K_{XT})\}; \tag{60}$$

$$J^{2 \to 1}_{S(\text{condition 2})} = J^{2 \to 1}_{S(\text{condition 1})} = J_{S,\max} \times [S]_2/([S]_2 + K_{XS}) \tag{61}$$

Therefore,

$$J^{1 \to 2(net)}_{S(\text{condition 2})} = J^{1 \to 2}_{S(\text{condition 2})} - J^{2 \to 1}_{S(\text{condition 2})} \tag{62a}$$

$$= J_{S,\max} \times \frac{[S]_1}{[S]_1 + K_{XS} \times (1 + [T]_1/K_{XT})} - \frac{[S]_2}{[S]_2 + K_{XS}} \tag{62b}$$

$$= J_{S,\max} \times K_{XS} \times \frac{[S]_1 - [S]_2 - [S]_2 \times [T]_1/K_{XT}}{\{[S]_1 + K_{XS} \times (1 + [T]_1/K_{XT})\} \times ([S]_2 + K_{XS})} \tag{62c}$$

If the numerator of Eq. (62c) becomes negative despite the condition that $[S]_1 > [S]_2$, net movement of S is uphill from side 2 to 1, i.e., uphill counterflow of S results from the downhill movement of T. Therefore, the condition for counterflow is that

$$[S]_1 - [S]_2 - [S]_2 \times [T]_1/K_{XT} < 0 \tag{63}$$

Solution of Eq. (63) for $[T]_1$ yields:

$$[T]_1 > K_{XT} \times ([S]_1/[S]_2 - 1) \tag{64}$$

Thus, for the mechanism shown in Fig. 6, when $D_X = D_{XS} = D_{XT}$, knowledge of the concentration ratio of substrate S and of the dissociation constant between carrier and analogue T is sufficient to calculate the concentration of T required to induce counterflow. In the special case when there is no concentration difference of S across the membrane, Eq. (64) reduces to $[T]_1 > 0$, i.e., the addition of any amount of T on side 1 induces counterflow.

Emphasis should be placed on the fact that the uphill movement of S observed during counterflow can be accomplished by means of a passive transport system which does not involve the utilization of metabolic energy. Hence, the terms uphill transport and active transport are not synonymous.

4. Competitive Exchange Diffusion

In the previous section, it is shown that addition of analogue T results in a decrease in, or even a reversal of, the net flux of S. Simultaneous with the change in flux of S is the appearance of a net flux of T. Under certain conditions the magnitude of the flux of T is approximately equal to the change in the flux of S. Thus, a part of the flux of S appears to have been exchanged for the flux of T. Such an exchange is called competitive exchange diffusion and can be explained by the model of hetero-*cis*-inhibition presented in Fig. 6.

Consider the following example, which is taken from the previous section:

Condition 1. Before the addition of T to the system,

$$J^{1 \to 2(\text{net})}_{T(\text{condition 1})} = 0 \tag{65}$$

since no T is present.

Condition 2. After the addition of T only to side 1,

$$J^{1 \to 2}_{T(\text{condition 2})} = J^{1 \to 2}_{T(\text{condition 2})} - J^{2 \to 1}_{T(\text{condition 2})} \tag{66a}$$

$$= J_{T,\max} \times [T]_1/\{[T]_1 + K_{XT} \times (1 + [S]_1/K_{XS})\} - 0 \tag{66b}$$

Equations (66a) and (66b) are analogous to Eqs. (62a–62c). The change in net flux of T is given by

$$\Delta J^{1 \to 2(\text{net})}_T = J^{1 \to 2(\text{net})}_{T(\text{condition 2})} - J^{1 \to 2(\text{net})}_{T(\text{condition 1})} \tag{67a}$$

$$= J_{T,\max} \times [T]_1/\{[T]_1 + K_{XT} \times (1 + [S]_1/K_{XS})\} \tag{67b}$$

$$= J_{T,\max} \times ([T]_1/K_{XT})/(1 + [S]_1/K_{XS} + [T]_1/K_{XT}) \tag{67c}$$

The change in net flux of S is given by

$$\Delta J_S^{1 \to 2(net)} = J_{S(condition\ 2)}^{1 \to 2(net)} - J_{S(condition\ 1)}^{1 \to 2(net)} \tag{68a}$$

$$= J_{S(condition\ 2)}^{1 \to 2} - J_{S(condition\ 2)}^{2 \to 1}$$

$$- (J_{S(condition\ 1)}^{1 \to 2} - J_{S(condition\ 1)}^{1 \to 2}) \tag{68b}$$

$J_S^{2 \to 1}$ does not change since it does not depend upon $[T]_1$, i.e., there is no *trans*-effect. Thus, Eq. (68b) reduces to

$$\Delta J_S^{1 \to 2(net)} = J_{S(condition\ 2)}^{1 \to 2} - J_{S(condition\ 1)}^{1 \to 2} \tag{69}$$

$J_S^{1 \to 2}$ is given by Eq. (58) for condition 1 and by Eq. (60) for condition 2. Hence,

$$\Delta J_S^{1 \to 2(net)} = J_{S,max} \times [S]_1 / \{[S]_1 + K_{XS} \times (1 + [T]_1/K_{XT})\}$$
$$- J_{S,max} \times [S]_1 / ([S]_1 + K_{XS}) \tag{70a}$$

$$= -\frac{[S]_1/K_{XS}}{1 + [S]_1/K_{XS}} \times J_{S,max} \times \frac{[T]_1/K_{XT}}{1 + [S]_1/K_{XS} + [T]_1/K_{XT}} \tag{70b}$$

Note that $\Delta J_S^{1 \to 2(net)}$ is negative, i.e., the net movement of S from side 1 to 2 is decreased by the addition of T to side 1.

Remember that $J_{S,max} = J_{T,max}$ [(see Eq. (57)]. Thus, a part of the right hand side of Eq. (70b) can be replaced by Eq. (67c) as follows:

$$\Delta J_S^{1 \to 2(net)} = -\frac{[S]_1/K_{XS}}{1 + [S]_1/K_{XS}} \times \Delta J_T^{1 \to 2(net)} \tag{71}$$

Thus, the changes in fluxes of S and T are related by

$$\frac{\Delta J_S^{1 \to 2(net)}}{\Delta J_T^{1 \to 2(net)}} = -\frac{[S]_1/K_{XS}}{1 + [S]_1/K_{XS}} \tag{72}$$

Competitive exchange diffusion is observed when the ratio of fluxes is approximately minus one, i.e. when a decrease in flux of S is approximately matched by an increase in flux of T. The right-hand side of Eq. (72) is approximately minus one when $[S]_1 \gg K_{XS}$. Thus, competitive exchange diffusion is observed only when the carrier is initially saturated with S on the side of addition.

Competitive exchange diffusion and counterflow are not mutually exclusive phenomena. Consider the following example: a carrier is initially saturated with S on side 1, and the net flux of S is from side 1 to 2. Since the carrier is saturated with S, addition of T to side 1 results in competitive exchange diffusion. If a sufficient amount of T is added so that the inequality of Eq. (64) is obeyed, the direction of net flux of S reverses and is from side 2 to 1, i.e., counterflow is observed, as well as competitive exchange diffusion.

5. *Maintenance of a Concentration Gradient*

With the model of hetero-*cis*-inhibition presented in Fig. 6, it is possible to establish and maintain a concentration gradient of S across a membrane with a gradient of T. Consider a membrane separating two aqueous phases which contain two different concentrations of S, namely, $[S]_1$ and $[S]_2$. S moves across the membrane along its concentration gradient. The net movement of S can, however, be brought to zero by the appropriate degree of hetero-*cis*-inhibition of S transport by analogue T. Thus, with no net transfer of S across the membrane, the gradient of S is maintained.

In the previous two sections only the addition of T to side 1 is considered. In the following discussion, the case is considered in which T is added to both sides to bring about zero net movement of S.

The desired condition is expressed by

$$J_S^{1 \to 2(\text{net})} = 0 \tag{73}$$

which is equivalent to

$$J_S^{1 \to 2} = J_S^{2 \to 1} \tag{74}$$

Substitution of unidirectional fluxes [evaluated from Eqs. (49) and (51)] into Eq. (74) yields

$$J_{S,\text{max}} \times [S]_1 / \{[S]_1 + K_{XS}(1 + [T]_1/K_{XT})\}$$
$$= J_{S,\text{max}} \times [S]_2 / \{[S]_2 + K_{XS}(1 + [T]_2/K_{XT})\} ; \tag{75a}$$

$$([S]_1/K_{XS})/(1 + [S]_1/K_{XS} + [T]_1/K_{XT})$$
$$= ([S]_2/K_{XS})/(1 + [S]_2/K_{XS} + [T]_2/K_{XT}) \tag{75b}$$

Solving Eq. (75b) for $[S]_1/[S]_2$ yields

$$[S]_1/[S]_2 = (1 + [T]_1/K_{XT})/(1 + [T]_2/K_{XT}) \tag{76a}$$

$$= ([T]_1 + K_{XT})/([T]_2 + K_{XT}) \tag{76b}$$

Thus, with a knowledge of the dissociation constant for XT and of the concentrations of T, it is possible to calculate the concentration ratio of S that may be established. Solving Eq. (76b) for $[T]_1$ yields

$$[T]_1 = ([S]_1/[S]_2) \times [T]_2 + ([S]_1/[S]_2 - 1) \times K_{XT} \tag{77}$$

Note that the right-hand side of Eq. (77) is composed to two terms: the first term is a function of $[T]_2$ and the second is independent of $[T]_2$. The first term indicates that with increasing concentrations of T on side 2, more T is required on side 1 to maintain the gradient of S. If T is absent on side 2, the first term drops out and the second term indicates that T on side 1 alone can maintain a gradient of S only if $[S]_1 > [S]_2$. In general, to maintain a gradient of S, [T] and [S] must be greater on the same side.

Another problem of interest is to compare the concentration ratio of S, the "maintained" species, with the ratio of T, the "maintaining" species.

From Eq. (77),

$$([T]_1/[T]_2)/([S]_1/[S]_2) = 1 + (1 - [S]_2/[S]_1) \times K_{XT}/[T]_2 \qquad (78)$$

When $[S]_1 > [S]_2$, the second term on the right-hand side of Eq. (78) is positive; hence the right-hand side is greater than unity. In this case,

$$[T]_1/[T]_2 > [S]_1/[S]_2 > 1 \qquad (79)$$

Note that the "maintaining" species, T, cannot maintain a concentration ratio as large as that of the "maintained" species, S. Equation (76b) or Eq. (78) shows, however, that when $[T] \gg K_{XT}$ on both sides, then

$$[S]_1/[S]_2 \approx [T]_1/[T]_2 \qquad (80)$$

In this case, the "maintaining" species can maintain almost the same concentration ratio as that of the "maintained" species.

Details relating to the kinetic expressions in Sections II, C, 2–5 are found in Rosenberg and Wilbrandt.[91]

6. Trans-effects

In this section we consider the situation in which addition of substrate to a given side of a membrane results in the modification of the unidirectional flux moving toward that side, i.e., trans-stimulation or trans-inhibition.* Trans-stimulation is also termed accelerative exchange diffusion. The model shown in Fig. 4 and presented in Section II, B, 4 cannot explain trans-effects when $D_X = D_{XS}$, since the expressions for unidirectional flux and clearance are independent of trans-concentration [see Eq. (29)]. If the diffusion constants of free carrier X and carrier-substrate complex XS are no longer equal, i.e., if $D_X \neq D_{XS}$ the model presented in Fig. 4 explains both trans-stimulation and trans-inhibition. This model, developed by Britton,[92] Regen and Morgan,[93] Levine, Oxender, and Stein,[94] Levine and Stein,[95] Stein,[96] and Wyssbrod,[97] is characterized by Michaelis–Menton kinetics of the form

$$J_S^{cis \to trans} = J_{S,max}^{cis \to trans} \times [S]_{cis}/([S]_{cis} + K_{S,0.5}^{cis \to trans}) \qquad (81)$$

$J_{S,max}^{cis \to trans}$ and $K_{S,0.5}^{cis \to trans}$ are no longer constants but are functions of the trans-concentration, $[S]_{trans}$. $J_{S,max}^{cis \to trans}$ (the maximal unidirectional flux) and $K_{S,0.5}^{cis \to trans}$ (the cis-concentration necessary to achieve half-maximal unidirectional flux) may be different for the directions $1 \to 2$ and $2 \to 1$ and, therefore, must carry the superscript, cis → trans.

First, consider the expression for maximal unidirectional flux.

$$J_{S,max}^{cis \to trans} = J_{S,max}^0 \times \frac{1 + r_S \times ([S]_{trans}/K_{XS})}{1 + \{2r_s/(r_s + 1)\} \times ([S]_{trans}/K_{XS})} \qquad (82)$$

---

* In Section II, C, 1, a modification of clearance rather than flux is used as the criterion for stimulation or inhibition. For trans-effects, modification of the flux is equivalent to modification of the clearance inasmuch as flux and clearance are related by the cis-concentration, which is maintained constant. During trans-effects studies only trans-concentration is varied.

where r_S is the ratio of D_{XS} to D_X and $J^0_{S,max}$ is the value of maximal unidirectional flux when no substrate is present on the *trans*-side. Note that $J^{1\to2}_{S,max}$ and $J^{2\to1}_{S,max}$ are not equal unless $[S]_1 = [S]_2$ and that $J^{cis\to trans}_{S,max}$ is independent of *trans*-concentration only when $r_S = 1$, which is the special case considered in Section II, B, 4.

It is shown later in this section that when $r_S > 1$, *trans*-stimulation is observed, but when $r_S < 1$, *trans*-inhibition is observed.

Figure 8(a) derived from Eq. (82), show that $J^{cis\to trans}_{S,max}$ varies as a function of $[S]_{trans}$ for different values of r_S. Note that when $r_S > 1$, i.e., when the loaded carrier XS is "more mobile" than the unloaded carrier X, an increase in $[S]_{trans}$ results in an increase in maximal unidirectional flux, but when $r_S < 1$, i.e., when XS is "less mobile" than X, an increase in $[S]_{trans}$ results in a decrease in $J^{cis\to trans}_{S,max}$. Figure 8(b) shows an expansion of Fig. 8(a) for negative values of r_S.

In the previous discussion the variation of $J^{cis\to trans}_{S,max}$ as a function of *trans*-concentration for various values of r_S is examined. Now let us consider the limiting values of $J^{cis\to trans}_{S,max}$, i.e., values of $J^{cis\to trans}_{S,max}$ when $[S]_{trans} = 0$ and when $[S]_{trans} \to \infty$.

When $[S]_{trans} = 0$, it can be shown that

$$J^0_{S,max} = \{2/(r_S + 1)\} \times (Q_{X(total)} \times D_{XS}/L^2) \tag{83}$$

Equation (83) may be written in the form of Eq. (31), namely

$$J^0_{S,max} = Q_{X(total)} \times D_{(S)}/L^2 \tag{84}$$

if, by definition,

$$D_{(S)} \equiv \{2/(r_S + 1)\} \times D_{XS} \tag{85a}$$

$$\equiv 2 \times D_{XS} \times D_X/(D_{XS} + D_X) \tag{85b}$$

since $r_S \equiv D_{XS}/D_X$.

Except for the factor 2, Eq. (85b) is similar to the equation for an equivalent conductance which may be used to replace two electrical conductances in series. If *two* equal series conductances $D_{(S)}$ are used to replace two unequal series conductances (D_{XS} and D_X), then the factor 2 appears in the equation. Schematically,

$$\underset{\text{—}\Lambda\Lambda\Lambda\text{—}\Lambda\Lambda\Lambda\text{—}}{D_{(S)} \qquad D_{(S)}} \quad = \quad \underset{\text{—}\Lambda\Lambda\Lambda\text{—}\Lambda\Lambda\Lambda\text{—}}{D_{XS} \qquad D_X} \tag{86}$$

Now let us examine the other limiting case, namely $[S]_{trans} \to \infty$. From Eq. (82)

$$J^\infty_{S,max} \equiv \lim_{[S]_{trans} \to \infty} J^{cis\to trans}_{S,max} \tag{87a}$$

$$= \{(r_S + 1)/2\} \times J^0_{S,max} \tag{87b}$$

Substitution of Eq. (83) into Eq. (87b) yields

$$J^\infty_{S,max} = Q_{X(total)} \times D_{XS}/L^2 \tag{88}$$

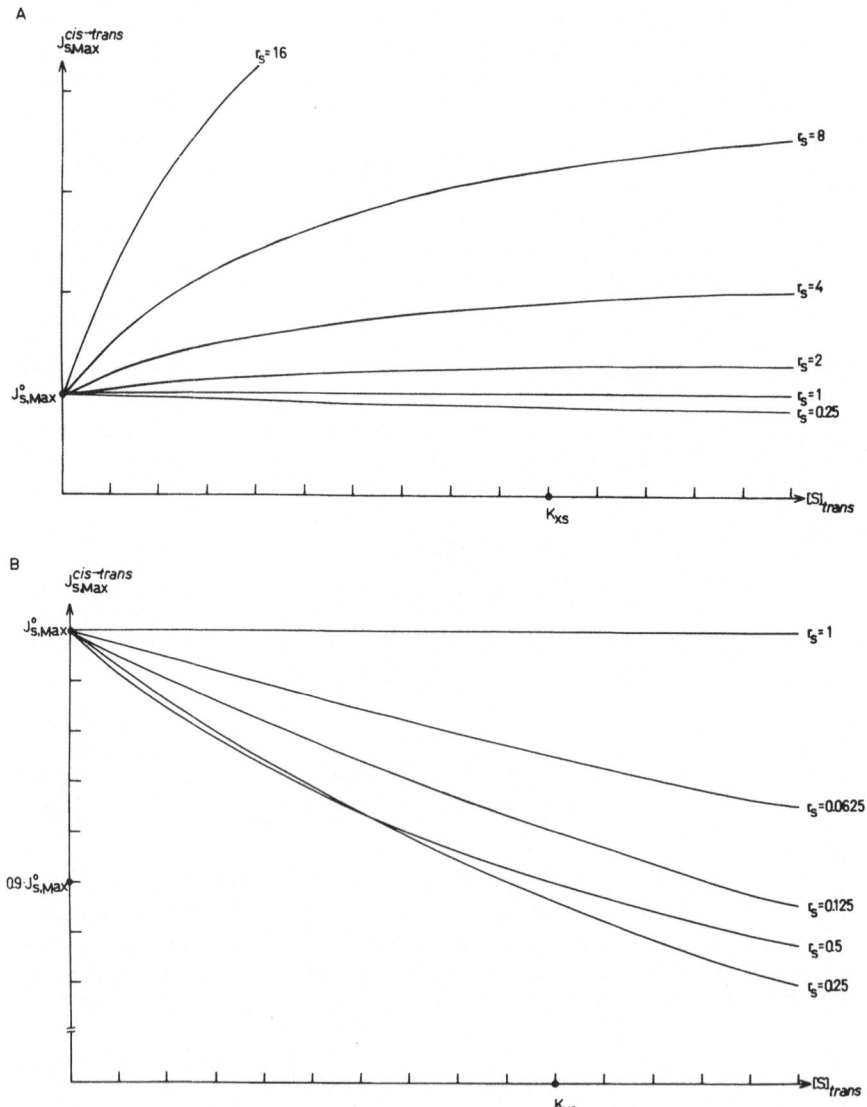

Fig. 8. Effect of *trans*-concentration on maximal undirectional flux.[97] This figure applies to the transport scheme shown in Fig. 4 for passive downhill transport of a nonelectrolyte when diffusion constants (D_{XS} and D_X) are not necessarily equal. The ratio of D_{XS} to D_X is denoted by r_S.

Note that $J^{\infty}_{S,max}$ depends upon D_{XS}, the diffusion constant of the loaded carrier, whereas $J^0_{S,max}$ depends upon both D_{XS} and D_X [see Eqs. (84–85b) and Eq. (88), respectively]. In the case of $J^0_{S,max}$, the smaller diffusion constant is the determining factor [see Eq. (85b)]. When $r_S > 1$, $D_{XS} > D_X$; hence, D_X is the determining factor. Since $J^\infty_{S,max}$ depends upon D_{XS} and $J^0_{S,max}$ depends primarily upon D_X, these two limiting values may be quite different. On the other hand, when $r_S < 1$, $D_{XS} < D_X$; hence, D_{XS} is the determining factor. Since both $J^\infty_{S,max}$ and $J^0_{S,max}$ depend primarily on D_{XS}, $J^{cis \to trans}_{S,max}$ is not as dependent on *trans*-concentration in the case of *trans*-inhibition ($r_S < 1$) as it is in the case of *trans*-stimulation ($r_S > 1$). To emphasize this point, Fig. 9 shows a plot, taken from Eq. (87b), of $J^\infty_{S,max}$ as a function of r_S. In the case of *trans*-stimulation ($r_S > 1$), *trans*-concentration may increase the maximal unidirectional flux many times. In the case of *trans*-inhibition ($r_S < 1$), however, *trans*-concentration does not have a great effect upon $J^{cis \to trans}_{S,max}$; even when r_S equals the limiting value of zero, maximal unidirectional flux varies only over a twofold range. It is apparent that any transport system which exhibits a $J^\infty_{S,max}$ that is less than one-half of $J^0_{S,max}$ cannot be explained by this model for *trans*-effects. If $J^\infty_{S,max}$ and $J^0_{S,max}$ are known from experimental data, r_S can be found from either Eq. (87b) or Fig. 9.

Let us now shift our attention from $J^{cis \to trans}_{S,max}$ to $K^{cis \to trans}_{S,0.5}$, the value of *cis*-concentration necessary to achieve half-maximal flux. It can be shown that

$$K^{cis \to trans}_{S,0.5} = K^0_{S,0.5} \times \frac{1 + \{(r_S + 1)/2\} \times ([S]_{trans}/K_{XS})}{1 + \{2r_S/(r_S + 1)\} \times ([S]_{trans}/K_{XS})} \quad (89)$$

where $K^0_{S,0.5}$ is the value of $K^{cis \to trans}_{S,0.5}$ when no substrate is present on the *trans*-side. Note that $K^{1 \to 2}_{S,0.5}$ and $K^{2 \to 1}_{S,0.5}$ are not equal unless $[S]_1 = [S]_2$ and

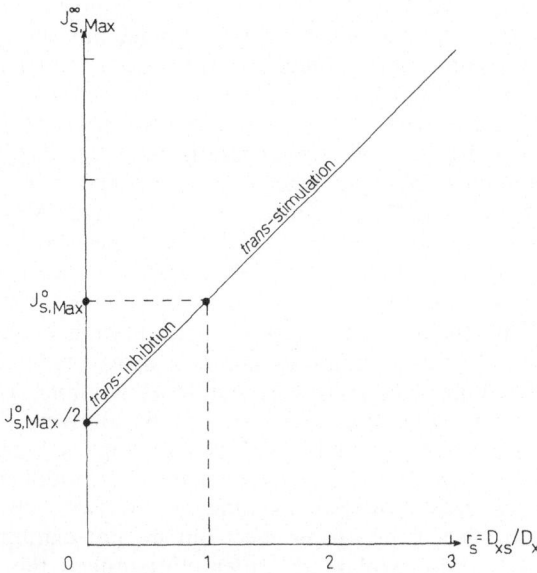

Fig. 9. The variation of maximal unidirectional flux as a function of the ratio of the diffusion constant of the loaded carrier to that of the unloaded carrier.[97] This figure applies to the transport scheme shown in Fig. 4 for passive downhill transport of a nonelectrolyte when diffusion constants are not necessarily equal. Variation is indicated by the ratio of flux when *trans*-concentration approaches infinity to flux when *trans*-concentration is zero.

that $K_{S,0.5}^{cis \to trans}$ is independent of *trans*-concentration only when $r_S = 1$, which is the special case considered in Section II, B, 4.

$K_{S,0.5}^0$ and $K_{S,0.5}^\infty$ can be expressed in terms of K_{XS}, the dissociation constant for carrier and substrate:

$$K_{S,0.5}^0 = \{2/(r_S + 1)\} \times K_{XS} \qquad (90a)$$

and

$$K_{S,0.5}^\infty = \{(r_S + 1)/(2r_S)\} \times K_{XS} \qquad (90b)$$

where $K_{S,0.5}^\infty$ is the value that $K_{S,0.5}^{cis \to trans}$ approaches as $[S]_{trans}$ approaches infinity.

In general,

$$K_{S,0.5}^{cis \to trans} < K_{XS} \text{ when } r_S > 1; \qquad (91a)$$

$$K_{S,0.5}^{cis \to trans} = K_{XS} \text{ when } r_S = 1; \qquad (91b)$$

$$K_{S,0.5}^{cis \to trans} > K_{XS} \text{ when } r_S < 1 \qquad (91c)$$

Note that K_{XS} (the dissociation constant) and $K_{S,0.5}^{cis \to trans}$ (the *cis*-concentration necessary to achieve half-maximal flux) are equal only when $r_S = 1$.

Figures 10a and 10b show $K_{S,0.5}^{cis \to trans}$ as a function of *trans*-concentration. Note that an increase in *trans*-concentration results in an increase in $K_{S,0.5}^{cis \to trans}$ both when $r_S > 1$ and when $r_S < 1$. Thus, except for the special case when $r_S = 1$, an increase in the *trans*-concentration always results in the increase in the *cis*-concentration necessary to achieve half-maximal flux. If the only effect of *trans*-concentration were on $K_{S,0.5}^{cis \to trans}$, flux would be *trans*-inhibited whenever $r_S \neq 1$ since any increase in $K_{S,0.5}^{cis \to trans}$ means that it becomes more difficult to saturate the carrier mechanism and, hence, to achieve maximal flux.

It is clear that *trans*-inhibition is manifested whenever $r_S < 1$, since an increase in *trans*-concentration decreases $J_{S,max}^{cis \to trans}$ and increases $K_{S,0.5}^{cis \to trans}$ On the other hand, when $r_S > 1$, an increase in *trans*-concentration increases both $J_{S,max}^{cis \to trans}$ (a stimulatory effect) and $K_{S,0.5}^{cis \to trans}$ (an inhibitory effect). The derivative of flux $J_S^{cis \to trans}$ with respect to *trans*-concentration is, however, always positive when $r_S > 1$ and, therefore, the stimulatory effect of *trans*-concentration on $J_{S,max}^{cis \to trans}$ must overcome the inhibitory effect on $K_{S,0.5}^{cis \to trans}$. Thus the model predicts *trans*-stimulation whenever $r_S > 1$.

In this section only homo-*trans*-effects are considered in order to keep the kinetic equations as simple as possible. The model presented in this section (Fig. 4) can be modified by allowing the carrier to combine with more than one analogue. If two analogues, S and T, exhibit a hetero-*trans*-effect which can be explained by this modified model (Fig. 6), they must also exhibit hetero-*cis*-inhibition. If two substances can be found that exhibit a hetero-*trans*-effect but do not exhibit hetero-*cis*-inhibition, then we cannot explain the interaction unless the model is further modified.

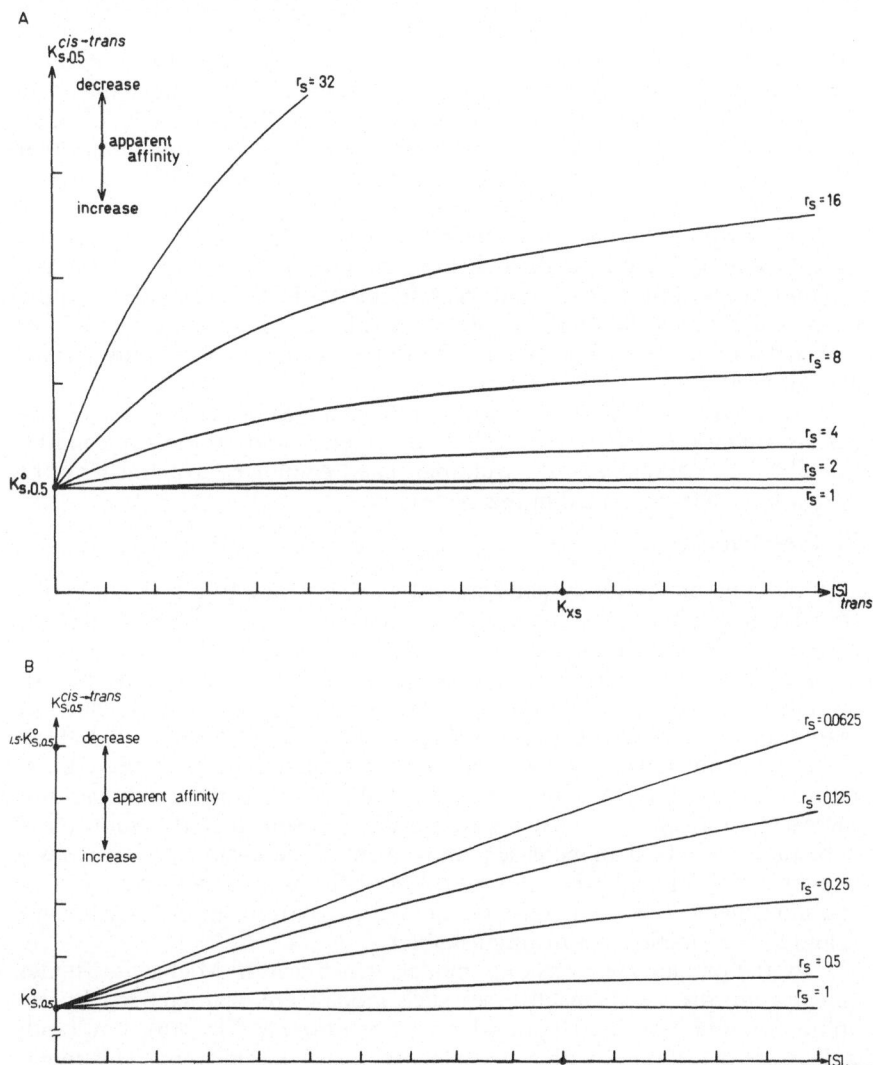

Fig. 10. Effect of *trans*-concentration on the *cis*-concentration required for half-maximal undirectional flux.[97] This figure applies to the transport scheme shown in Fig. 4 for passive downhill transport of a nonelectrolyte when diffusion constants are not necessarily equal. The ratio of D_{XS} to D_X is denoted by r_S.

In addition, substrate S must either *trans*-stimulate or *trans*-inhibit all other analogues (T, U, V, etc.). If a mixed pattern* of *trans*-stimulation and

---

* An example of a mixed pattern is a *trans*-stimulation of analogues T and V by S and a *trans*-inhibition of analogue U by S.

trans-inhibition is observed, more than one carrier system must be invoked, or the simple scheme presented above must be modified even further, because *trans*-stimulation or *trans*-inhibition is determined in the model shown in Fig. 6 only by the value of r_S, i.e., by the value of D_{XS} relative to D_X. In other words, if S *trans*-stimulates one analogue, it must *trans*-stimulate all other analogues, and cannot *trans*-inhibit any other analogue with which it interacts via the common carrier.

Hetero-*trans*-effects provide the means for regulation of the intracellular balance among various analogues. For example, an excessive intracellular amount of one amino acid could *trans*-stimulate the uptake of other amino acids which are in deficient supply, or a sufficient intracellular supply of one metabolite could *trans*-inhibit the uptake of other metabolites which are not needed.

In the literature may be found many examples of hetero- and homo-*trans*-stimulation[12,80,91-95,98-111] and hetero- and homo-*trans*-inhibition;[106,112-116] many of the examples cited cannot be explained by the particular model presented in this section.

7. Cis-stimulation

In some biological systems, addition of substrate to a given side of a membrane results in *cis*-stimulation, i.e., an increase in clearance of substrate from that side. *Cis*-stimulation is a dramatic cooperative phenomenon, as is apparent from the following example. A membrane is bathed on one side by substrate S containing isotopic tracer $S^\dagger$. The clearance from that side is found to be C so that tracer flux $J^\dagger_{initial} = C \times [S^\dagger]$. Additional S, but not $S^\dagger$, is then introduced on that side and *cis*-stimulation is observed, i.e., clearance increases to some value $(C + \Delta C)$. Consequently, tracer flux increases to $J^\dagger_{final} = (C + \Delta C) \times [S^\dagger] > J^\dagger_{initial}$. Note that the addition of substrate S results in an increase in movement of the substrate $S^\dagger$ which was present before the addition, i.e., addition of S has a cooperative effect on the movement of $S^\dagger$ (and consequently, on the movement of S itself, since S and $S^\dagger$ are kinetically indistinguishable).

In the previous sections only models which lead to Michaelis–Menten kinetics are discussed. In all cases such models are characterized by *cis*-inhibition and never by *cis*-stimulation. For example, Fig. 5(b) shows only a decrease in clearance with every increase in *cis*-concentration. This model can be modified so that *cis*-stimulation is observed, at least over part of the range of *cis*-concentration. As a possible modification, Wong[117] proposes a "polyvalent" carrier mechanism containing more than one binding site for substrate. This mechanism can explain all *cis*- and *trans*-effects, including those in which a substrate is *trans*-stimulatory over part of the range of *trans*-concentration and *trans*-inhibitory over another part of this range. In the previous section the model allows for the manifestation of either *trans*-stimulation or *trans*-inhibition—but not for a mixture of the two modes—over the entire range of *trans*-concentration.

To illustrate the ability of a polyvalent carrier model to account for *cis*-stimulation, consider the simplest of all polyvalent carriers, namely, a divalent carrier which has two binding sites. Kinetics for divalent carrier systems have been derived by Stein,[118,119] Wilbrandt and Kotyk,[120] Britton,[121] and Wyssbrod.[122] A simplified divalent carrier model is shown in Fig. 11. The derivation of the kinetic equation for this model is based on the following assumptions:

1. *S is translocated in the form of XS.* The substrate S can cross the membrane in either the form of mono-substrate complex XS or as di-substrate complex XS_2, but not as free substrate.

2. *Carrier is conserved.* In other words,

$$Q_{X(total)} = Q_X + Q_{XS} + Q_{XS_2} = constant \qquad (92)$$

where Q is the amount of carrier per unit membrane.

3. *The system is operating in the steady state.* The net amount of loaded carrier moving from side 1 to 2 must equal the net amount of unloaded carrier moving from side 2 to 1, i.e.,

$$J_{XS}^{1 \to 2(net)} + J_{XS_2}^{1 \to 2(net)} = J_X^{2 \to 1(net)} \qquad (93)$$

The net flux of substrate is given by

$$J_S^{1 \to 2(net)} = J_{XS}^{1 \to 2(net)} + 2 \times J_{XS_2}^{1 \to 2(net)} \qquad (94)$$

The factor 2 in the second term of Eq. (94) indicates that two substrate molecules are translocated for every molecule of disubstrate complex translocated.

4. *The rate-limiting reactions are the transitions of X, XS, and XS_2 across the membrane.* These transitions are described by Fick's equation

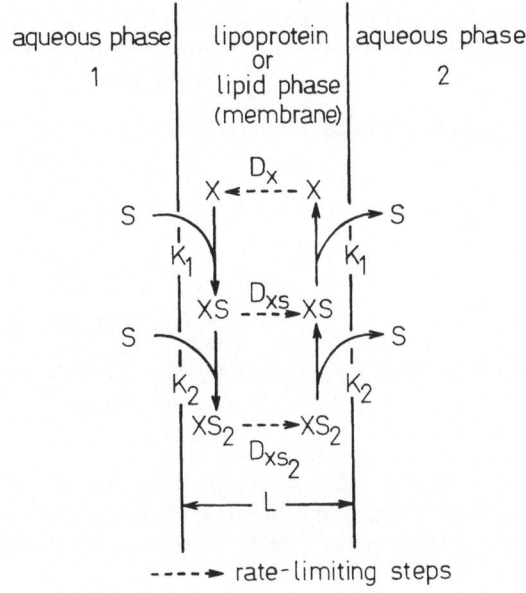

Fig. 11. A divalent carrier for passive downhill transport of a nonelectrolyte across a membrane.[122] This transport scheme can account for *cis*-stimulation.

and all diffusion constants are taken to be equal so that *trans*-effects are eliminated, i.e.,

$$D = D_X = D_{XS} = D_{XS_2} \qquad (95)$$

5. *The formation and dissociation of* XS *and* XS_2 *are so rapid that* S, X, XS, *and* XS_2 *may be considered to be in equilibrium.* Since there are two binding sites per carrier, two equilibrium constants are needed to describe the system. The divalent carrier could consist of two identical subunits, each with one binding site. The subunits could interact in such a way that the attachment of substrate to one subunit could modify the affinity of the other subunit toward the binding of a second substrate molecule. If the dissociation constant is a function only of the number of substrates bound and not of the particular sites bound, then

$$XS_{j-1} + S \overset{K_j}{\rightleftharpoons} XS_j \qquad (96)$$

where XS_j is a complex with j molecules of substrate per molecule of carrier, XS_0 is identical to free carrier X, and a different equation is needed for each possible value of $j(j = 1, 2, \ldots,$ total number of subunits per carrier). This approach is similar to that taken by Atkinson, Hathaway, and Smith[123] to explain cooperative enzyme kinetics.

K_j is the intrinsic dissociation constant that links [S], [unoccupied sites on XS_{j-1}] and [occupied sites on XS_j]. On each side i of the membrane,

$$K_j = [\text{unoccupied sites on } XS_{j-1}]_i \times [S]_i/[\text{occupied sites on } XS_j]_i. \qquad (97)$$

If n = number of subunits per carrier,

$$[\text{unoccupied sites on } XS_{j-1}]_i = \{n - (j - 1)\} \times [XS_{j-1}]_i \qquad (98)$$

since there are $(j - 1)$ occupied sites on XS_{j-1} and, consequently, $\{n - (j - 1)\}$ unoccupied sites. Since there are j occupied sites on XS_j,

$$[\text{occupied sites on } XS_j]_i = j \times [XS_j]_i \qquad (99)$$

Substitution of Eqs. (98) and (99) into Eq. (97) yields

$$K_j = \{(n + 1 - j)/j\} \times [XS_{j-1}]_i \times [S]_i/[XS_j]_i \qquad (100)$$

For the divalent mode, n = number of subunits per carrier = 2 and j can assume the values 1 and 2:

$$K_1 = 2 \times [X]_i \times [S]_i/[XS]_i; \qquad (101)$$

$$K_2 = (1/2) \times [XS]_i \times [S]_i/[XS_2]_i \qquad (102)$$

R is defined as the value of K_2 relative to K_1, i.e.,

$$R \equiv K_2/K_1 \qquad (103)$$

Consider the three following cases:

1. If $K_2 > K_1, R > 1$. In this case, the binding of the first substrate molecule hinders the binding of the second substrate molecule, i.e., since affinity is inversely related to the dissociation constant, the affinity of the

carrier for the second substrate is decreased by the binding of the first one. Here binding exhibits an antagonistic rather than a cooperative effect.

2. If $K_2 = K_1, R = 1$. Affinity of the carrier for the substrate does not depend upon the number of substrates previously bound. In this case, transport kinetics resemble Michaelis–Menten enzyme kinetics.

3. If $K_2 < K_1, R < 1$. In this case, the binding of the first substrate molecule facilitates the binding of the second, i.e., the affinity of the carrier for the second substrate is increased by the binding of the first one. It remains to be seen if this cooperative binding leads to *cis*-stimulation, i.e., to a cooperative effect on transport.

A model based on the five assumptions listed above does not lead to Michaelis–Menten kinetics (except if $K_1 = K_2$) but rather to the following formulations for unidirectional flux and clearance:

$$J_S^{cis \rightarrow trans} = J_{S,max} \times \frac{[S]_{cis}^2 + K_2 \times [S]_{cis}}{[S]_{cis}^2 + 2 \times K_2 \times [S]_{cis} + K_1 \times K_2}; \quad (104)$$

$$C_S^{cis \rightarrow trans} = J_{S,max} \times \frac{[S]_{cis} + K_2}{[S]_{cis}^2 + 2 \times K_2 \times [S]_{cis} + K_1 \times K_2} \quad (105)$$

where

$$J_{S,max} = 2 \times D \times Q_{X(total)}/L^2 \quad (106)$$

The factor 2 in Eq. (106) arises because two substrate molecules are translocated on each oligomer of carrier when transport is maximal. Note that flux and clearance are not functions of *trans*-concentration. Hence, no *trans*-effects are manifest.

$K_{S,0.5}$, the *cis*-concentration necessary to achieve half-maximal flux, is evaluated from Eq. (104) as follows:

$$(1/2) \times J_{S,max} = J_{S,max} \times \frac{K_{S,0.5}^2 + K_2 \times K_{S,0.5}}{K_{S,0.5}^2 + 2K_2 \times K_{S,0.5} + K_1 \times K_2} \quad (107)$$

Equation (107) may be solved for $K_{S,0.5}$:

$$K_{S,0.5} = \sqrt{K_1 \times K_2} \quad (108)$$

Note that $K_{S,0.5}$ is the geometric mean of the two dissociation constants. From Eqs. (103) and (108), K_1 and K_2 may be expressed as functions of R, which serves as a measure of cooperative binding, and $K_{S,0.5}$, which may be measured experimentally.

$$K_1 = K_{S,0.5}/\sqrt{R} \quad (109)$$

$$K_2 = K_{S,0.5} \times \sqrt{R} \quad (110)$$

Substitution of Eqs. (109) and (110) into Eqs. (104) and (105) yields

$$J_S^{cis \rightarrow trans} = J_{S,max} \times \frac{[S]_{cis}^2 + K_{S,0.5} \times \sqrt{R} \times [S]_{cis}}{[S]_{cis}^2 + 2 \times K_{S,0.5} \times \sqrt{R} \times [S]_{cis} + K_{S,0.5}^2} \quad (111)$$

$$C_S^{cis \to trans} = J_{S,\text{max}} \times \frac{[S]_{cis} + K_{S,0.5} \times \sqrt{R}}{[S]_{cis}^2 + 2 \times K_{S,0.5} \times \sqrt{R} \times [S]_{cis} + K_{S,0.5}^2} \qquad (112)$$

It is not clear from Eq. (112) if there is a region of *cis*-stimulation, i.e., if there is some range of *cis*-concentration over which an increase in $[S]_{cis}$ results in an increase in $C_S^{cis \to trans}$. Mathematically, the region of *cis*-stimulation is characterized by $dC_S^{cis \to trans}/d[S]_{cis} > 0$, and that of *cis*-inhibition by $dC_S^{cis \to trans}/d[S]_{cis} < 0$. The two regions are separated by a maximal clearance characterized by $dC_S^{cis \to trans}/d[S]_{cis} = 0$. The value of *cis*-concentration corresponding to maximal clearance is found by differentiating $C_S^{cis \to trans}$ in Eq. (112) with respect to $[S]_{cis}$, by setting $dC_S/d[S]_{cis}$ equal to zero, and by solving for $[S]_{cis}^{\text{max}C}$, the *cis*-concentration at which clearance is maximal. We find that when

$$dC_S^{cis \to trans}/d[S]_{cis} = 0, \qquad (113)$$

$$[S]_{cis}^{\text{max}C} = (-\sqrt{R} \pm \sqrt{1-R}) \times K_{S,0.5} \qquad (114)$$

$[S]_{cis}^{\text{max}C}$ has a real value only when $0 < R < 1$, i.e., only when binding of substrate by carrier exhibits cooperativity. Only *positive* real values of concentration, however, have physical significance. Thus, it is required that

$$[S]_{cis}^{\text{max}C} = (-\sqrt{R} \pm \sqrt{1-R}) \times K_{S,0.5} > 0; \qquad (115)$$

$$-\sqrt{R} + \sqrt{1-R} > 0; \qquad (116a)$$

$$0 < R < 0.5 \qquad (116b)$$

Since $R \equiv K_2/K_1$ by definition [see Eq. (103)], Eq. (116b) may be rewritten as

$$K_2 < 2 \times K_1 \qquad (117)$$

Equation (117) states that the affinity of XS for S must be at least twice as great as the affinity of X for S. Note that the requirement for *cis*-stimulation is stronger than that for cooperativity of binding. For example, over the range where $0.5 < R < 1$, *cis*-stimulation is not manifest even though there is cooperative binding.

Figure 12 shows $C_S^{cis \to trans}$ as a function of $[S]_{cis}$ for various values of R. Note that only when $0 < R < 0.5$ do the curves have a region of *cis*-stimulation in which clearance increases with an increase in *cis*-concentration. All regions of *cis*-stimulation are denoted by solid lines (———), and those of *cis*-inhibition by dashed lines (– – –). A maximum in clearance is indicated by an open circle (○). Note that the value of $[S]_{cis}^{\text{max}C}$, the *cis*-concentration at which clearance is maximal, increases from zero when $R = 0.5$ to $K_{S,0.5}$ when $R = 0$. In other words, for the model shown in Fig. 11, the range of *cis*-concentration corresponding to the region of *cis*-stimulation must always lie between zero and the *cis*-concentration necessary to achieve half-maximal flux ($K_{S,0.5}$).

Figure 13 shows flux $J_S^{cis \to trans}$ as a function of $[S]_{cis}$ for various values of R. As in Fig. 12, all regions of *cis*-stimulation are denoted by solid lines

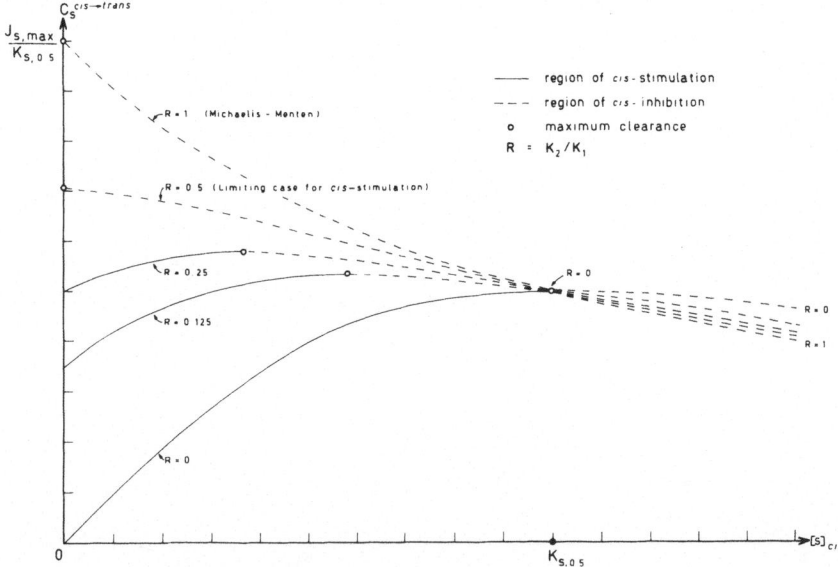

Fig. 12. Unidirectional clearance as a function of *cis*-concentration for various values of the ratio, R, of the dissociation constant of the disubstrate carrier complex to that of the monosubstrate complex.[122] This figure applies to the transport scheme shown in Fig. 11 when all diffusion constants are equal. Whereas cooperative binding is manifest when $R < 1$, cooperative transport (*cis*-stimulation) is manifest only when $R < 0.5$ and only over part of the range of *cis*-concentration.

(———), and those of *cis*-inhibition by dashed lines (– – –). The point corresponding to a maximum in clearance is indicated by an open circle (○).

All curves which exhibit a maximal clearance also exhibit a point of inflection, denoted by an open square (□). The point of inflection always occurs at a lower concentration than the point of maximal clearance. Figure 14 shows the value of $[S]_{cis}$ corresponding to these two points for all values of R associated with *cis*-stimulation. Note that it is not the point of inflection but the point of maximal clearance that denotes the end of the region of *cis*-stimulation. Thus, *cis*-stimulation is still exhibited in a region of negative curvature between the point of inflection and the point of maximal clearance.

This model may exhibit unusual flux ratios. At high concentrations ($[S]_1 > [S]_2 \gg K_{S,0.5}$), the flux ratio is less than that predicted by Fick's equation, i.e.,

$$J_S^{1\to2}/J_S^{2\to1} < (J_S^{1\to2}/J_S^{2\to1})_{\text{Fick}} \tag{118a}$$

$$< [S]_1/[S]_2 \tag{118b}$$

where sides 1 and 2 have been chosen so that $[S]_1 > [S]_2$ (consequently, $J_S^{1\to2} > J_S^{2\to1}$ and net flux is from side 1 to 2). In terms of clearance, Eq. (118b)

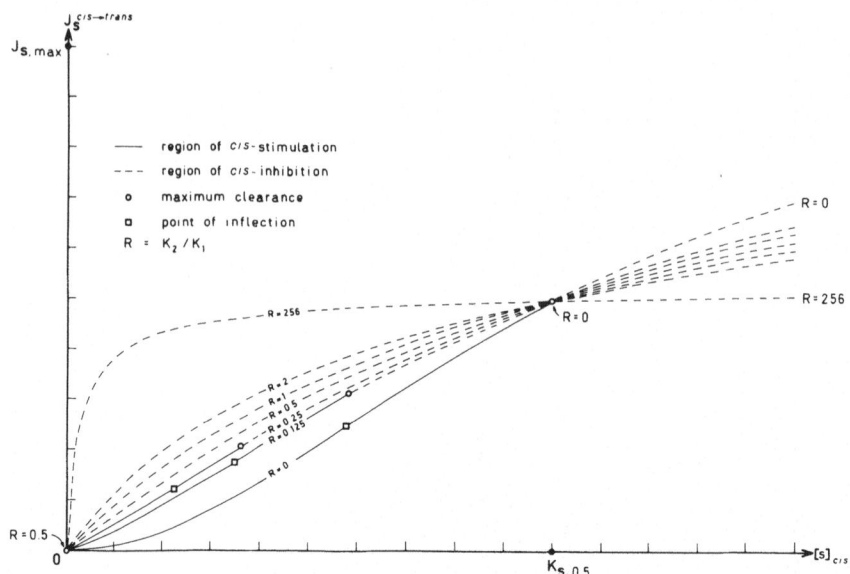

Fig. 13. Unidirectional flux as a function of *cis*-concentration for various values of the ratio, R, of the dissociation constant of the disubstrate carrier complex to that of the monosubstrate complex.[122] This figure applies to the transport scheme shown in Fig. 11 when all diffusion constants are equal. When *cis*-stimulation is manifest ($R < 0.5$), the *cis*-concentration at which maximum clearance is seen (O) is lower than the *cis*-concentration at which the curve shows a point of inflection (□). The curve denoted by $R = 1$ is similar to the Michaelis–Menten function, i.e. it is a rectangular hyperboloid.

becomes:

$$(J_S^{1 \to 2}/[S]_1)/(J_S^{2 \to 1}/[S]_2) < 1 \qquad (119a)$$

$$C_S^{1 \to 2}/C_S^{2 \to 1} < 1 \qquad (119b)$$

$$C_S^{1 \to 2} < C_S^{2 \to 1} \qquad (119c)$$

At low concentrations ($[S]_2 < [S]_1 \ll K_{S,0.5}$), the flux ratio is also "abnormal," and if the system does not exhibit *cis*-stimulation (i.e., $R > 0.5$), the flux ratio is less than that predicted by Fick's equation. If the system does exhibit *cis*-stimulation (i.e., $R < 0.5$), the flux ratio is greater than that predicted by Fick's equation, i.e.,

$$J_S^{1 \to 2}/J_S^{2 \to 1} > (J_S^{1 \to 2}/J_S^{2 \to 1})_{\text{Fick}} \qquad (120a)$$

$$> [S]_1/[S]_2 \qquad (120b)$$

$$C_S^{1 \to 2} > C_S^{2 \to 1} \qquad (120c)$$

This unusual flux ratio is characteristic of a passive nonelectrolyte system which manifests *cis*-stimulation.

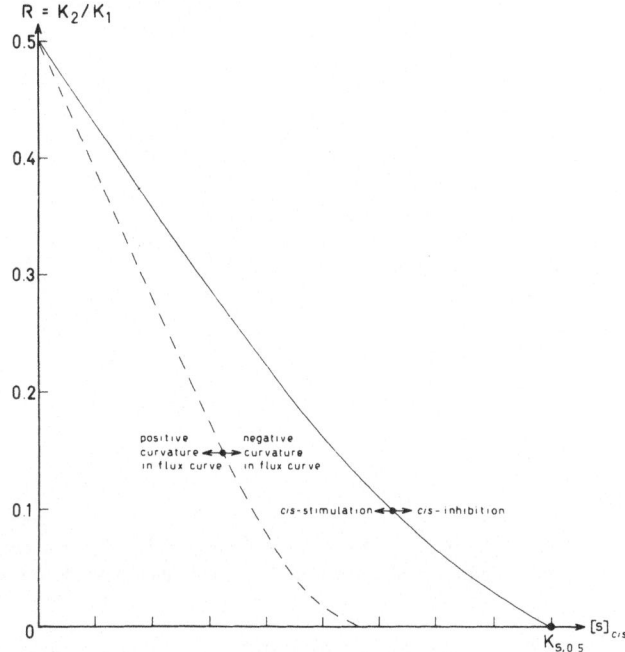

Fig. 14. The *cis*-concentration that denotes maximal clearance (——) and point of inflection in the flux curve (– – –) for all ratios (R) which lead to *cis*-stimulation.[122] This figure applies to the transport scheme shown in Fig. 11 when all diffusion constants are equal. In no case is *cis*-stimulation manifest when *cis*-concentration is greater than that required for half-maximal flux ($K_{S,0.5}$).

Figure 15 shows a geometric method for identifying the point of maximal clearance on any plot of unidirectional flux *vs. cis*-concentration. A straight line (– – –) has been superimposed on the plot so that the line both passes through the origin and lies tangent to the curve at some point other than at the origin. The point of tangency corresponds to the point of maximal clearance and is denoted by an open circle (○). Curve A exhibits a point of maximal clearance and therefore has a region of *cis*-stimulation. For curve B, the only line that both goes through the origin and lies tangent to the curve is the tangent drawn at the origin. Consequently, curve B does not have a region of *cis*-stimulation.

The simplest polyvalent carrier, namely, a divalent carrier, was used to demonstrate the ability of a carrier model to account for *cis*-stimulation. If polyvalent carriers exist in biological systems, however, they may be capable of binding more than two substrates. For example, a carrier might consist of four subunits, each of which could bind one molecule of substrate. If the kinetics for such a carrier were developed by the approach taken in

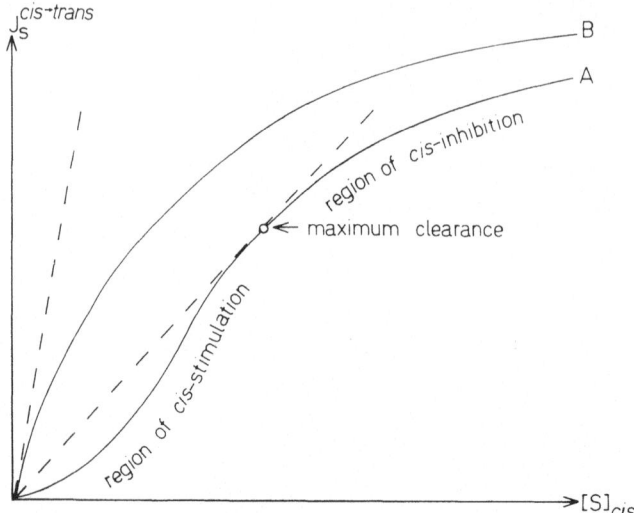

Fig. 15. A graphical method to determine if *cis*-stimulation is manifest, and if so, to find the *cis*-concentration to which maximum clearance corresponds.

this section, Eq. (96) would require four different dissociation constants to describe the system. Not only would the development be complicated by the introduction of two more unknown constants, but the simplifying assumption that each dissociation constant is a function only of the number of substrates bound and not of the particular subunits bound may not be justified.

To develop a multi-subunit model such as a four-subunit (tetrameric) model,* it is advisable to adopt one of the subunit models developed for enzymes by Monod, Wyman, and Changeux[125] or Koshland, Némethy and Filmer.[126] In these models, each oligomer consists of subunits which can exist in either of two different conformational states. In the Monod model, both conformational forms may bind substrate, but with different affinities. The model can be simplified by allowing only one conformational form to bind substrate. All subunits of a particular oligomer must be either in one conformational state or the other, i.e., all subunits must change state in concerted action. In the Koshland model, the change in conformational state in a subunit is a result of the binding of the substrate. Thus, one conformational form always has substrate bound to it; the other form never has. In the Koshland mode, all subunits of a particular oligomer need not be in the same conformational form. Only adjacent subunits interact, and this interaction, which depends on the conformational states of the two neighbors, may either stabilize or destabilize the entire multiunit oligomer. The interaction of each pair of nearest neighbors is taken into account. Since subunits

---

* The rudiments of a particular tetrameric model developed by Lieb and Stein[124] are discussed in Section II, D.

can be geometrically arranged in different ways, various arrangements of nearest neighbors are possible. Each arrangement, or geometry, leads to a slightly different kinetic pattern. Koshland's model for cooperative enzyme kinetics is similar to Pauling's model for cooperative binding of oxygen to hemoglobin.[127] Both the Monod and the Koshland model can account for either antagonistic or cooperative binding of substrate by the oligomer.

Another approach to a model for *cis*-stimulation is to assume that all intrinsic dissociation constants (K_j's) are equal (i.e., there is no cooperative binding) and that the substrate-carrier complex becomes more mobile as the carrier becomes progressively more loaded.[120] A model based on this approach is characterized by kinetics complicated by a *trans*-effect which is a consequence of the inequality among the diffusion constants for the various substrate-carrier complexes.

In biological systems, *cis*-stimulation can serve as the basis for the amplification of a chemical signal. The role of cooperative processes in amplification by biological membranes is discussed by Changeux *et al.*[128] and by Changeux and Thiéry.[129] For the model shown in Fig. 11, when $[S]_{cis}$ is doubled in the region of *cis*-stimulation, $J_S^{cis \to trans}$ is more than doubled and may be nearly quadrupled (if R approaches zero, i.e., if $K_2 \ll K_1$). A transport mechanism exhibiting *cis*-stimulation can serve as a component in the homeostatic regulation of S since flux is extremely sensitive to changes in [S] in some regions of *cis*-stimulation.

Many examples of homo- and hetero-*cis*-stimulation may be found in the literature;[118–121,130–137] many of the examples cited cannot be explained by the particular model presented in this section.

D. The Application of Current Concepts of Conformational Enzymology to Transport

Lieb and Stein[124] have developed a tetrameric model for transport of glucose across erythrocyte membranes. In this model, the tetramer is immobile and substrate is passed from one of the two subunits on one side of the membrane to one of the two subunits on the other side via a concerted conformational change in the tetramer. Lieb and Stein call their transport mechanism a "non-carrier internal transfer model." They use the word non-carrier in the sense that the carrier is not mobile.

On each side there is one subunit with high affinity for substrate and another with low affinity. The tetramer may assume either of two conformational forms (conformers). In the external conformation, the substrate binding site of each subunit is exposed to one of the external solutions. In the internal conformation, all sites are exposed to an intermediate pool which serves as the medium through which substrate is passed from one subunit to the other. Substrate molecules from the external solutions bind to the subunits when the tetramer is in the external conformation. A conformational change of the internal conformation then exposes the bound substrates to the internal pool or pools where the substrates are all released and redistributed among the subunits. The subunits exposed to a pool compete for the substrate molecules in that pool. Lieb and Stein postulated that the probability that a subunit binds a substrate molecule from the pool to which the subunit is exposed is proportional to the affinity of the subunit for the substrate. Consequently, there is a tendency for substrate molecules to be passed from a low-affinity subunit to its paired high-affinity subunit. It is possible to devise other redistribution rules in which molecular phenomena such as steric hindrance between substrate molecules interferes with the redistribution. Although various numbers and arrangements of pools are possible in this type of scheme, Lieb and Stein selected a model in which there are two distinct intermediate pools, each of which faces a high-affinity subunit on one side and a low-affinity subunit on the other side. Before the tetramer can revert from the internal conformation to the external conformation, all free substrate molecules in the internal pool must be bound to subunits, i.e., no conformational change can occur as long as free substrate remains in the internal pool.

Whereas the equilibrium distribution of conformers is of primary importance in an enzymatic system, the transition rate between conformers is of primary importance in a transport system. Each of the sixteen* different occupancy states of the tetramer is characterized by a transition rate constant. Lieb and Stein postulated that the transition between the external and internal conformers is favored by the binding of substrate to the tetramer and that transition rate is proportional to the number of substrate molecules bound. It is possible to adopt other transition rules in which the transition rate is independent of the number of bound substrate molecules or in which the transition rate depends upon the particular combination of subunits to which substrate is bound.

The transition of each of the sixteen occupancy states contributes to a unidirectional flux. Each state contributes in proportion to its concentration, transition rate constant, and the probability that *cis*-labeled substrate is passed on to a subunit on the *trans*-side and is not returned to the *cis*-side.

Lieb and Stein indicate that their tetrameric model can account for experimental data which cannot be explained by mobile carrier models such as those developed in Sections II, B, and C. The tetrameric model is a notable attempt to apply the current concepts of molecular enzymology to transport, and it would not be surprising if, in the future, many kinetic transport models will draw upon this approach.

E. A Simplified Model for Passive Downhill Movement of an Electrolyte

If the reader has no particular interest in considering such a model, he may proceed to the discussion of active transport (Section III, A) without loss of continuity.

---

* Since each of the four subunits may either be unoccupied or occupied by substrate, there are sixteen different states of occupancy possible.

In Section II, B, 4 a simple model for passive downhill movement of a nonelectrolyte is considered. This model is restricted to nonelectrolytes in order to eliminate the complexities involved in computing the effect of trans-membrane electrical potential difference on the translocation of substrate across the membrane.

Figure 16 shows a model in which neutral carrier X combines with a univalent cation S^+ to form XS^+, which then combines anion A^- to form a neutral complex XSA. Note the inclusion of the electrical potential difference $(\Delta\psi)$ which affects the movement of charged complex XS^+ across the membrane. In our development, the potential difference is oriented so that the potential on side 2 is measured with respect to that on side 1, i.e.,

$$\Delta\psi = \psi_2 - \psi_1 \tag{121}$$

This model for electrolyte transport is derived in part from the work of Ciani, Eisenman, and Szabo,[71] who developed a theory to describe the effects of neutral carriers such as the macrotetralide actin antibiotics on the electrical properties of bilayer lipid membranes and in part from the work of Britton,[121] who considered the effect of electric fields on carrier systems. This model is developed further by Wyssbrod.[138]

Most parameters in the lipoid phase (membrane) or in the aqueous interface at the membrane are denoted by a prime (e.g., $[S^+]'$), whereas

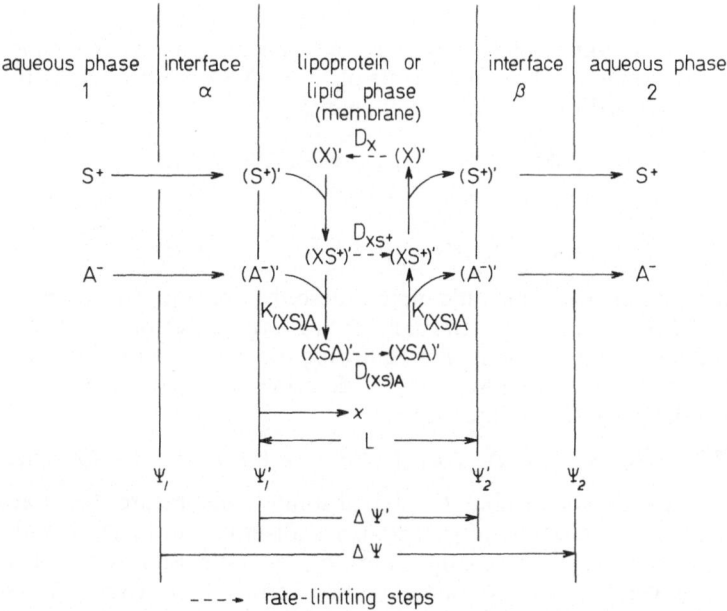

Fig. 16. A simplified model for passive downhill transport of an electrolyte across a membrane.[138]

those in the bulk aqueous phases are not so designated (e.g., $[S^+]$). The following assumptions apply to the model:

1. *Univalent cation S^+ can cross the membrane only in the forms XS^+ or XSA; univalent anion A^-, only in the form XSA; free cation and anion cannot cross the membrane.* The concentrations of S^+ and A^- within the lipoid phase can be neglected, except at each interface, where they react to form complexes and are then released from these complexes. Any free A^- which exists within the membrane serves only to neutralize the complex XS^+ and does not contribute to the flux of A^-, since the diffusion of free A^- is assumed to be low within the membrane.

2. *The system is operating in the steady state.* Several consequences of this condition are

$$J_{S^+}^{1 \to 2(net)} = J_{XS^+}^{1 \to 2(net)} + J_{XSA}^{1 \to 2(net)} \qquad (122)$$

since S^+ may cross the membrane either as XS^+ or XSA;

$$J_{A^-}^{1 \to 2(net)} = J_{XSA}^{1 \to 2(net)} \qquad (123)$$

since A^- may cross the membrane only as XSA; and

$$J_{XS^+}^{1 \to 2(net)} + J_{XSA}^{1 \to 2(net)} = J_X^{2 \to 1(net)} \qquad (124)$$

since free carrier must return to side 1 to replace the loaded forms departing from side 1.

3. *The rate-limiting reactions are the transitions of X, XS^+ and XSA across the membrane.* The movements of X and XSA are by simple diffusion and are described by Fick's equation, namely,

$$J_X^{2 \to 1(net)} = (D_X/L) \times ([X]_2' - [X]_1'), \qquad (125)$$

and

$$J_{XSA}^{1 \to 2(net)} = (D_{XSA}/L) \times ([XSA]_1' - [XSA]_2') \qquad (126)$$

Fick's equation does not adequately describe the movement of charged complex XS^+ across the membrane because the effect of the electric field must be taken into account. The following differential equation, known as the Nernst–Planck equation[139–141] describes the flux of XS^+ at every plane at distance χ from interface α:

$$J_{XS^+}^{1 \to 2(net)} = D_{XS^+} \times \{ -d[XS^+]'/d\chi + [XS^+]' \times (zF/RT) \times (-d\psi'/d\chi) \} \quad (127)$$

where R is the gas constant, T is the absolute temperature, F is Faraday's constant, ψ' is the electrical potential, χ is distance, and $[XS^+]'$ and ψ' are functions of χ. $J_{XS^+}^{1 \to 2(net)}$ remains constant, however, and does not depend upon χ during the steady state. The first term on the right-hand side of Eq. (127) is identical to Fick's equation in differential form, and the second term describes the effect of the electric field $(-d\psi'/d\chi)$ on the movement of the carrier-substrate complex which we treat as a univalent cation ($z = +1$).

4. *The electrical potential affects the movement of* XS^+. The evaluation of the electrical term in Eq. (127) is not simple and requires the introduction of further assumptions which may not be readily justified. For example, the assumption may be made that the electric field $(-d\psi'/d\chi)$ remains constant. Hence, Eq. (127) may be easily integrated between the two boundaries of the lipoid phase ($\chi = 0$ and $\chi = L$) to yield the following more useful non-differential form (see Goldman[142] for details):

$$J_{XS^+}^{1 \to 2(net)} = \frac{D_{XS^+}}{L} \times \frac{[XS^+]'_1 \times \{\exp(-\Delta\phi'/2)\} - [XS^+]'_2 \times \exp\{(\Delta\phi'/2)\}}{\{\sinh(\Delta\phi'/2)\}/(\Delta\phi'/2)}$$

(128)

where $\Delta\phi'$, a unitless normalized potential difference across the lipoid phase, is defined by

$$\Delta\phi' = \Delta\psi'/(RT/zF) = (zF/RT) \times (\psi'_2 - \psi'_1)$$ (129)

where ψ'_1 and ψ'_2 are the electrical potentials in the lipoid phase at interfaces α and β, respectively. When $T = 25°C$, the normalization factor (RT/zF) is approximately $+12.8$ mV for divalent cations, $+25.7$ mV for univalent cations, -25.7 mV for univalent anions and -12.8 mV for divalent anions. In this derivation, we treat the carrier-substrate complex as a univalent cation ($z = 1$).

Equation (128) is similar in form to Fick's equation, except that each boundary concentration is weighted by an exponential function of the potential difference and the entire expression is divided by the factor $\{\sinh(\Delta\phi'/2)\}/(\Delta\phi'/2)$, which approaches unity as the potential difference approaches zero ($\Delta\phi' \to 0$), and which is greater than unity whenever $\Delta\phi' \neq 0$. It may be shown that Fick's equation is approached as $\Delta\phi' \to 0$, i.e.,

$$\lim_{\Delta\phi' \to 0} J_{XS^+}^{1 \to 2(net)} = (D_{XS^+}/L) \times ([XS^+]'_1 - [XS^+]'_2)$$ (130)

Two other special cases are of interest, namely, when $\Delta\phi'$ is a large positive number ($\Delta\phi' \gg 1$) and when it is a large negative number ($-\Delta\phi' \gg 1$). For large positive potential differences,

$$\lim_{\Delta\phi' \gg 1} J_{XS^+}^{1 \to 2(net)} = -(D_{XS^+}/L) \times [XS^+]'_2 \times \Delta\phi'$$ (131)

The normalized potential, $\Delta\phi'$, in Eq. (131) may be replaced by the actual potential, $\Delta\psi'$, to yield

$$\lim_{\Delta\psi' \gg RT/F} J_{XS^+}^{1 \to 2(net)} = -(D_{XS^+}/L) \times (F/RT) \times [XS^+]'_2 \times \Delta\psi'$$ (132)

Equation (132) states that for large positive potential differences ($\Delta\psi' \gg RT/F$), the net flux of XS^+ is from side 2 to side 1 because the expression is negative, and the dependency of the flux upon the concentration of complex on side 2 ($[XS^+]'_2$) indicates that the net movement is determined primarily by the unidirectional movement from side 2 to 1.

Similarly, for large negative potential differences,

$$\lim_{-\Delta\psi' \gg RT/F} J_{XS^+}^{1\to2(net)} = (D_{XS^+}/L) \times (F/RT) \times [XS^+]_1' \times (-\Delta\psi') \quad (133)$$

In the case of large negative potentials $(-\Delta\psi' \gg RT/F)$, the net flux of XS^+ is from side 1 to side 2 because the expression $(-\Delta\psi')$ is positive and the net movement depends primarily upon the concentration of complex on side 1 $([XS^+]_1')$. Note that Eqs. (132) and (133) are similar in form to Ohm's law because net flux (which corresponds to electrical current in Ohm's law) is proportional to the electrical potential difference.

Equation (128) may be rewritten in the following form, which resembles Fick's equation for simple diffusion:

$$J_{XS^+}^{1\to2(net)} = (D_{XS^+}/L) \times \{f(-\Delta\phi') \times [XS^+]_1' - f(+\Delta\phi') \times [XS^+]_2'\} \quad (134)$$

where the functions $f(-\Delta\phi')$ and $f(+\Delta\phi')$ represent the correction factors which must be applied to the boundary concentrations when the equation is written in the form of Fick's equation. From Eq. (128), the correction factor is identified as

$$f(\Delta\phi') = \frac{\exp(\Delta\phi'/2)}{\{\sinh(\Delta\phi'/2)\}/(\Delta\phi'/2)} \quad (135)$$

It is easily shown that functions of plus and minus $\Delta\phi'$ are related as follows:

$$f(+\Delta\phi') - f(-\Delta\phi') = \Delta\phi' \quad (136)$$

Figure 17 shows $f(-\Delta\phi')$ and $f(+\Delta\phi')$ as functions of $\Delta\phi'$. Also shown on the abscissa is a scale which is measured in millivolts and which is evaluated for a temperature of 25°C. These factors may be applied to any ion whose transmembrane movement can be described by an equation of the form of Eq. (134) (the integrated form of the Nernst–Planck equation when the electric field is constant). The scale calibrated in millivolts, however, applies only to univalent cations and must be divided by the valence, z, before it can be generally applied to ions of any charge.

Although the assumption that the electrical field is constant makes a ready evaluation of $J_{XS^+}^{1\to2(net)}$ possible, there is actually no experimental basis on which to make this assumption. Ciani, Eisenman, and Szabo[71] do not make the constant field assumption in their treatment of this problem. Instead, they start with the Poisson equation, which relates the local deviation of the electrical field from constancy to local charge density, the latter parameter depending upon the local concentrations of XS^+, S^+, and A^-. This more basic approach, however, leads to a more complicated form for the expression for $J_{XS}^{1\to2(net)}$. Although this treatment perhaps provides a more accurate evaluation of flux, the simpler expression given by Eq. (128) provides a good first approximation which can be more readily manipulated algebraically.

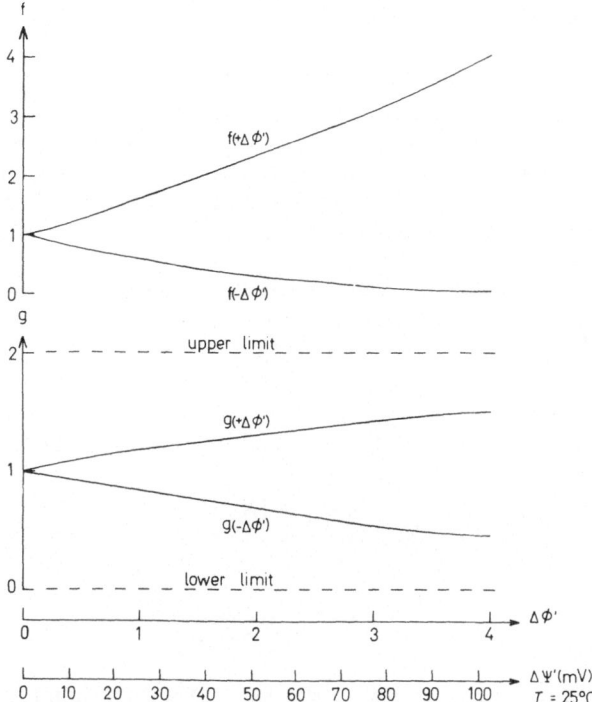

Fig. 17. Boundary concentration correction factors as a function of transmembrane potential difference.[138] Factor f is used to determine flux of an ion across a membrane and factor g is used to determine amount of an ion in the membrane when there is a constant electrical field throughout the membrane and when flux is determined by the Nernst–Planck equation. The scale calibrated in millivolts applies only to univalent cations and must be divided by the valence, z, before it can be generally applied to ions of any charge.

Unidirectional fluxes may be identified in Eq. (134):

$$J_{XS^+}^{1\to2} = (D_{XS^+}/L) \times f(-\Delta\phi') \times [XS^+]_1' ; \tag{137}$$

$$J_{XS^+}^{2\to1} = (D_{XS^+}/L) \times f(+\Delta\phi') \times [XS^+]_2' \tag{138}$$

The ratio of unidirectional fluxes may be found from Eqs. (135), (137), and (138):

$$J_{XS^+}^{1\to2}/J_{XS^+}^{2\to1} = \{f(-\Delta\phi')/f(+\Delta\phi')\} \times [XS^+]_1'/[XS^+]_2' \tag{139a}$$

$$= \{\exp(-\Delta\phi'/2)/\exp(+\Delta\phi'/2)\} \times [XS^+]_1'/[XS^+]_2' \tag{139b}$$

$$= \{\exp(-\Delta\phi')\} \times [XS^+]_1'/[XS^+]_2' \tag{139c}$$

Equation (139c) is identical to the Ussing flux ratio equation.[42,74,143]

5. *Carrier is conserved.* In other words,

$$Q_{X(total)} = Q_X + Q_{XS^+} + Q_{XSA} = \text{constant} \tag{140}$$

where Q is the amount of carrier per unit membrane. In the model of Ciani, Eisenman, and Szabo,[71] total quantity of carrier in the aqueous, but not in the lipoid, phase remains constant. In their model, macrotetralide actin antibiotics serve as carriers which solubilize electrolytes into the lipoid phase via an electrolyte-carrier complex in which form electrolytes may cross the membrane.

Q_X may be evaluated by integration of $[X]'$ across the lipoid phase, i.e.,

$$Q_X = \int_0^L [X]' \, d\chi \tag{141}$$

Since the movement of X is via simple diffusion, it is easy to show that $[X]'$ is a linear function of distance, χ, and that the evaluation of Eq. (141) leads to

$$Q_X = \frac{[X]'_1 + [X]'_2}{2} \times L \tag{142}$$

Similarly,

$$Q_{XSA} = \frac{[XSA]'_1 + [XSA]'_2}{2} \times L \tag{143}$$

Since the movement of XS^+ obeys Eq. (128) rather than Fick's equation for simple diffusion, $[XS^+]'$ is no longer a linear function of distance, χ, and consequently, Q_{XS^+} does not have the simple form of either Eq. (142) or (143). To evaluate Q_{XS^+}, Eq. (127) must be solved for $[XS^+]'$ and then $[XS^+]'$ must be integrated across the membrane in a similar fashion to Eq. (141). When this is done, Q_{XS^+} may be written in the following form:

$$Q_{XS^+} = \tfrac{1}{2}\{g(-\Delta\phi') \times [XS^+]'_1 + g(+\Delta\phi') \times [XS^+]'_2\} \times L \tag{144}$$

where the functions $g(-\Delta\phi')$ and $g(+\Delta\phi')$ represent the correction factors which must be applied to the equation, which would otherwise use the average of the two boundary concentrations to evaluate Q_{XS^+}. The correction factor assumes the rather complicated form:

$$g(\Delta\phi') = 2 \times \left(\frac{1}{1 - \exp(\Delta\phi')} - \frac{1}{\Delta\phi'} \right) \tag{145}$$

It is easily shown that functions of plus and minus $\Delta\phi'$ are related as follows:

$$g(+\Delta\phi') + g(-\Delta\phi') = 2 \tag{146}$$

Three special cases of Eq. (145) are of interest. As the potential difference approaches zero ($\Delta\phi' \to 0$), $g(\Delta\phi')$ approaches unity, and consequently Eq. (144) reduces to the following simple form which is similar to Eqs. (142) and (143):

$$\lim_{\Delta\phi' \to 0} Q_{XS^+} = \tfrac{1}{2}\{[XS^+]'_1 + [XS^+]'_2\} \times L \tag{147}$$

When $\Delta\phi' \gg 1$, $g(-\Delta\phi') \to 0$ and $g(+\Delta\phi') \to 2$. Consequently, Eq. (144) reduces to

$$\lim_{\Delta\phi' \gg 1} Q_{XS^+} = [XS^+]'_2 \times L \tag{148}$$

Similarly, when $-\Delta\phi' \gg 1$, $g(-\Delta\phi') \to 2$ and $g(+\Delta\phi') \to 0$. Consequently, Eq. (144) reduces to

$$\lim_{-\Delta\phi' \gg 1} Q_{XS^+} = [XS^+]'_1 \times L \tag{149}$$

In summary, the quantity of carrier in the form XS^+ depends upon the concentration of complex, $[XS^+]'$, on either side 1 or side 2. Whenever side 2 is highly positive with respect to side 1 ($\Delta\psi' \gg RT/F$), $[XS^+]'_2$ is the determining factor in the evaluation of Q_{XS^+}; whenever side 2 is highly negative with respect to side 1 ($-\Delta\psi' \gg RT/F$), $[XS^+]'_1$ is the determining factor.

Figure 17 shows $g(-\Delta\phi')$ and $g(+\Delta\phi')$ as functions of $\Delta\phi'$. Also shown on the abscissa is a scale which is measured in millivolts and which is evaluated for a temperature of 25°C. These factors may be applied to any ion whose transmembrane movement can be described by an equation of the form of Eq. (134) (the integrated form of the Nernst–Planck equation when the electric field is constant). The scale calibrated in millivolts, however, applies only to univalent cations and must be divided by the valence, z, before it can be generally applied to ions of any charge.

6. *The interface is sufficiently thin that a negligible amount of energy is expended when there is net movement of S^+ and A^- across it.* In other words, S^+ and A^- may be considered to be in electrochemical equilibrium across each interface between lipoid and aqueous phases. This condition of electrochemical equilibrium is expressed by the Nernst equation, namely,

$$[N]'_i = [N]_i \times \exp\{zF(\psi_i - \psi'_i)/RT\} \tag{150}$$

where N represents either S^+ or A^- with a valence z of $+1$ or -1 respectively, quantities without primes ($[N]_i$ and ψ_i) refer to values in the bulk aqueous phase, quantities with primes ($[N]'_i$ and ψ'_i) refer to boundary values in the interface at the membrane, and i may equal either 1 or 2, referring to either interface α or β, respectively. Equation (150) may be derived from the principles outlined in Section I, B.

7. *The formation and dissociation of the complexes XS^+ and XSA are so rapid that X, S^+, A^-, XS^+ and XSA may be considered to be in equilibrium.* The two equilibrium constants are chosen to have units of concentration and to include the chemical, but not electrical, partition coefficients of S^+ and A^- between aqueous and lipoid phases:

$$K_{XS^+} = [X]'_i \times [S^+]'_i/[XS^+]'_i; \tag{151}$$

$$K_{XSA} = [XS^+]'_i \times [A^-]'_i/[XSA]'_i \tag{152}$$

where i indicates the interface at which the equation is applied.

The seven assumptions listed above can serve as the basis for the solution of the unidirectional fluxes of S^+ and A^- as a function of the *cis*- and *trans*-concentrations of S^+ and A^- and as a function of the transmembrane potential difference ($\Delta\psi$), which may either be spontaneously generated or be set (clamped) at some fixed level. The spontaneously generated potential difference, often termed the open-circuited potential difference, is a function of the *cis*- and *trans*-concentrations of all ionic constituents. Therefore, in the open-circuited state, unidirectional fluxes can be expressed as functions only of *cis*- and *trans*-concentrations. Although the open-circuited state corresponds to the normal state of biological membranes, it is more useful to consider the voltage-clamped state, in which the potential difference is set and maintained at some fixed level by passing the requisite amount of electrical current (I_{ext}) through the membrane from an external source. One special case of the voltage-clamped state is the short-circuited state, in which the transmembrane potential difference is maintained at zero in order to eliminate the electrical component of the transmembrane electrochemical potential difference ($\Delta\tilde{\mu}$) of all ionic constituents [see second term of Eq. (6)]. The use of the short-circuited state to study biological membranes was introduced by Ussing and Zerahn.[143]

Several auxiliary equations describing electroneutrality must be used to solve for the kinetic equations. For example, the concentration of all positive charges must be equal to that of all negative charges at all points in all phases, i.e., the net charge at all points must be zero. In other words,

$$\sum_{j=1}^{n} (z_j \times [N_j]_i) = 0 \tag{153}$$

where i specifies any aqueous phase 1 or 2, j specifies the ionic constituent, the summation extends over n ionic components, and z_j is the valence of species j. For the simple two-constituent system presented in the list of assumptions,

$$(+1) \times [S^+]_i + (-1) \times [A^-]_i = 0; \tag{154a}$$

$$[S^+]_i = [A^-]_i \tag{154b}$$

where i may indicate either aqueous phase 1 or 2. In the interface at the membrane, Eq. (153) must be replaced by

$$[S^+]_i' + [\text{fixed positive charges in equivalents}]_i' $$
$$= [A^-]_i' + [\text{fixed negative charges in equivalents}]_i' \tag{155}$$

The contribution of the charge on the carrier-substrate complex is not included in Eq. (155) because the concentration of carrier in the membrane is negligible in comparison with the other concentration terms in the equation.

Charged groups on phospholipids in biological membranes, however, may contribute appreciably to the fixed charges in Eq. (155). In practice, quantitation of net charge could be difficult. The influence of fixed positive and negative charges located on the membrane at the interface is an example of a Gibbs–Donnan effect.[144]

Since biological membranes are often bathed by complex mixtures of ionic components, Eqs. (153) and (155) would have to contain more terms to describe a biological system.

The assumption of electroneutrality may not be valid for a region as thin as a cellular membrane. When electroneutrality is not assumed, Poisson's equation must be used to relate deviation from electroneutrality to nonconstancy of the electric field (see Ciani, Eisenman, and Szabo[71]).

Another auxiliary equation related to the maintenance of electroneutrality states that the amount of externally sent electrical current (I_{ext}) is equal to the sum of net ionic currents through the membrane—i.e.,

$$I_{ext} = \sum_{j=1}^{n} (I_j^{1 \to 2(net)}) \qquad (156)$$

Since current is related to flux by

$$I_j^{1 \to 2(net)} = z_j \times F \times J_j^{1 \to 2(net)} \qquad (157)$$

where F is the Faraday constant, Eq. (156) may be rewritten as

$$I_{ext} = F \times \sum_{j=1} (z_j \times J_j^{1 \to 2(net)}) \qquad (158)$$

In the open-circuited state (which corresponds to the normal state of a biological membrane), $I_{ext} = 0$, and therefore Eq. (158) reduces to

$$\sum_{j=1}^{n} (z_j \times J_j^{1 \to 2(net)}) = 0 \qquad (159)$$

For the simple two-constituent system presented in the list of assumptions,

$$(+1) \times J_{S^+}^{1 \to 2(net)} + (-1) \times J_{A^-}^{1 \to 2(net)} = 0; \qquad (160a)$$

$$J_{S^+}^{1 \to 2(net)} = J_{A^-}^{1 \to 2(net)} \qquad (160b)$$

Equation (160b) states that the net fluxes of univalent cation and anion must be equal. Substitution of Eqs. (122) and (123) into Eq. (160b) yields

$$J_{XS^+}^{1 \to 2(net)} + J_{S^+}^{1 \to 2(net)} = J_{XSA}^{1 \to 2(net)}; \qquad (161a)$$

$$J_{XS^+}^{1 \to 2(net)} = 0 \qquad (161b)$$

Note that the *net* movement of the complex XS^+ must be zero. Nevertheless, XS^+ may contribute a large component to each unidirectional flux of S^+ and could serve as an exchange system for different cations (e.g., for Na^+ and K^+). Since $J_{XS^+}^{1 \to 2(net)} = 0$, Eq. (128) shows that $\Delta\phi'$ (which is related to $\Delta\psi'$, a component of the transmembrane potential difference, $\Delta\psi$) is a function only of $[XS^+]_1'$ and $[XS^+]_2'$ which, in turn, can eventually be related to the ionic composition of the external aqueous phases.

If assumption 1 is modified to allow movement of free (uncomplexed) A^-, then net movement of XS^+ is allowed because it can be matched by the net movement of free A^-.

In biological systems, several types of carriers might exist in the same membrane. For example, in addition to X, which can form a positively charged complex with S^+, there may also be a carrier Y which can form a negatively charged complex with A^-. In this case Eqs. (160b), (161a), and (161b) become

$$J_{S^+}^{1 \to 2(net)} = J_{A^-}^{1 \to 2(net)};$$ (162a)

$$J_{XS^+}^{1 \to 2(net)} + J_{XSA}^{1 \to 2(net)} + J_{YAS}^{1 \to 2(net)} = J_{YA^-}^{1 \to 2(net)} + J_{XSA}^{1 \to 2(net)} + J_{YAS}^{1 \to 2(net)};$$ (162b)

$$J_{XS^+}^{1 \to 2(net)} = J_{YA^-}^{1 \to 2(net)}$$ (162c)

In this case, the charged complexes (XS^+ and YA^-) may contribute to the *net* movement of S^+ and A^- across the membrane.

It is difficult to determine the electrical potential difference across each interface, the distribution of fixed charges in the membrane and the manner in which the intramembrane electric field deviates from constancy. Work in progress on model membrane systems, however, should provide insight into correct ways to account for the effect of the electrical potential.

In view of the complexities inherent in the equations of even the simplified electrolyte carrier model presented above, we have not derived the unidirectional fluxes as functions of *cis*- and *trans*-concentration. The best way to approach this problem involves the use of a computer to solve (via an iterative process) the flux equations for a family of *cis*- and *trans*-concentrations.

III. ACTIVE TRANSPORT

A. A Simplified Model for Active Transport of a Nonelectrolyte

1. *Introduction*

Rosenberg and Wilbrandt[85] showed that a scheme for passive transport could be transformed into a scheme for active transport by the following modifications:

1. *Two forms of free carrier are required.* These two forms, which are interconvertible via a chemical reaction, must have different affinities for the substrate.

2. *On at least one side of the membrane, an off-equilibrium metabolic process must be coupled with the reaction that interconverts the two forms of the free carrier.* The metabolic coupling creates a difference in the amounts of high-affinity and low-affinity form of the carrier on the two sides of the membrane. As a result of this difference, one of the two unidirectional fluxes

is greater, even when the concentrations of substrate are equal on the two sides of the membrane. Consequently, active level flow [see Fig. 1(b)] is observed. In fact, the model can also account for active uphill flow [see Fig. 1(c)] with the result that metabolic energy is partially transduced into osmotic work. Under some conditions, the preceding process is reversed (reversed downhill flow) and osmotic work can be converted into metabolic energy (regenerative flow). A regenerative reversed downhill flow [see Fig. 2(d)] of H^+ across the mitochondrial membrane may serve as the basis for oxidative phosphorylation.[145]*

The scheme shown in Fig. 18 is taken from Rosenberg and Wilbrandt,[88] who based their model on suggestions of Franck and Meyer,[25] Solomon,[146] and Shaw[147] insofar as the affinity of carrier for substrate can assume two different values. In reaction 1, S reacts with X, the high-affinity form of the carrier, to form XS, which is translocated to side 2 (reaction 2). In reaction 3, S is released from XS into aqueous phase 2, thereby re-forming X which is converted in reaction 4 into Z, the low-affinity form, which returns to side 1 (reaction 5). On side 1, reaction 6 converts the low-affinity form into the high-affinity form, thereby completing the cycle. Only a small amount of S returns from side 2 to side 1 by the reverse of reactions 1–6 because reaction 4 maintains X, the high-affinity form, at a low concentration on side 2, thereby repressing the reversal of reaction 3.

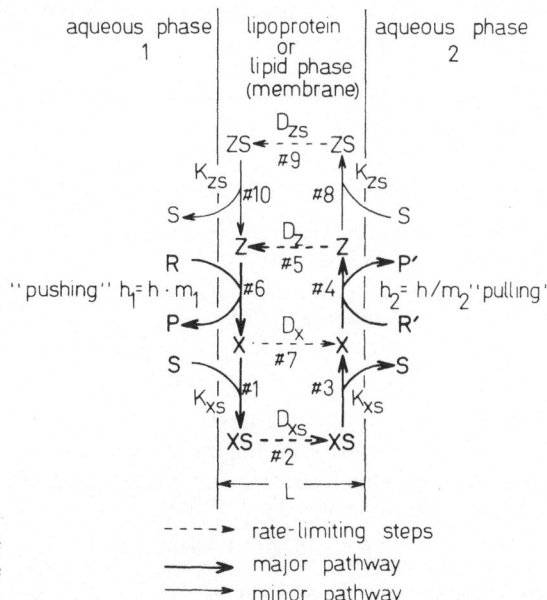

Fig. 18. A model for active transport of a nonelectrolyte across a membrane.

- - - - ▶ rate-limiting steps
⟶ major pathway
⟶ minor pathway

* In Mitchell's chemi-osmotic scheme for oxidative phosphorylation, coupling between transport (flow of H^+) and metabolic flow (oxidative phosphorylation) is mediated by differential permeabilities of the two sides of the membrane to H^+ and OH^- and not by a carrier. His scheme, however, can be modified so that a carrier serves as the mediator of coupling.

In addition to reactions 1–6 which play a major role in the transport process, reactions 7–10 must also be considered. If the high-affinity form, X, is mobile, reaction 7 may contribute to the energetically wasteful cycle 4–7. In this cycle, there is no net movement of S, but there is a continuous degradation of metabolic reactants. Reactions 8–10 serve as the basis for most of the wasteful back movement of S from side 2 to side 1. In reaction 8, the low affinity of Z for S usually allows the formation of only a small amount of ZS, which is translocated to side 1 in reaction 9. In reaction 10, S is readily released by Z into aqueous phase 1. Back movement can be minimized by restricting the mobility of complex ZS so that translocation in reaction 9 is difficult.

Although both reactions 4 and 6 may be coupled to metabolic processes which are not necessarily identical, it is necessary that only one of the reactions be coupled in order to achieve active transport capable of inducing uphill movement. Reaction 4 is called a "pulling" reaction since it is located on side 2 and appears to be "pulling" S to that side from side 1; reaction 6 is called a "pushing" reaction since it is located on side 1 and appears to be "pushing" S from that side to side 2.

A modification of the scheme shown in Fig. 18 is given in Fig. 19. In the modified scheme, Z, the low-affinity form for S becomes the high-affinity form for substrate T; X, the high-affinity form for S, becomes the low-affinity form for T. This transport system, which can actively exchange S and T across a membrane, is similar to that proposed by Shaw[147] to explain exchange of Na^+ and K^+ across erythrocyte membranes and to that proposed by Caldwell[148] to explain a similar exchange across nerve and muscle membranes.*

2. Assumptions and Kinetic Equations

The following assumptions apply to the active transport model of Rosenberg and Wilbrandt[88]:

1. *Nonelectrolytic S is translocated mainly in the form of XS and, to a lesser degree, in the form of ZS.*

2. *Carrier is conserved.* In other words,

$$Q_{X+Z(total)} = Q_X + Q_{XS} + Q_Z + Q_{ZS} = \text{constant} \qquad (163)$$

where Q is the amount of carrier per unit area.

3. *The system is operating in the steady state.* The net amount of carrier in all forms moving from side 1 to 2 must equal the net amount moving from side 2 to 1. In other words,

$$J_{XS}^{1 \to 2(net)} + J_X^{1 \to 2(net)} = J_{ZS}^{2 \to 1(net)} + J_Z^{2 \to 1(net)} \qquad (164)$$

---

* Of course, the active exchange of Na^+ and K^+ may not be so simple as indicated by the scheme in Fig. 19, since the stoichiometry may involve a three-for-two exchange of Na^+ for K^+. In this case, it is necessary to postulate a polyvalent carrier similar to that proposed in Section II, C, 7 to explain *cis*-stimulation. In addition, since Na^+ and K^+ are ionic, some of the considerations of Section II, E must be incorporated into the scheme to account for the effect of the transmembrane electrical potential difference.

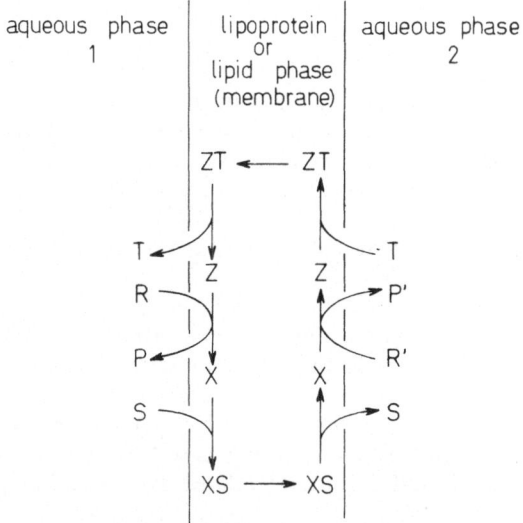

aqueous phase 1 | lipoprotein or lipid phase (membrane) | aqueous phase 2

Fig. 19. A model for active trans-membrane exchange of two sub-strates, S and T.

Only major pathways are shown.

In the steady state, the net flux of S is given by

$$J_S^{1 \to 2(\text{net})} = J_{XS}^{1 \to 2(\text{net})} - J_{ZS}^{2 \to 1(\text{net})} \tag{165}$$

Equation (165) shows that net flux from side 1 to 2 is maximal when the net back flux via the complex with the low-affinity form, Z, is minimized.

4. *The rate-limiting reactions are the transitions of X, XS, Z and ZS across the membrane.* These transitions are described by Fick's equation as follows:

$$J_{XS}^{1 \to 2(\text{net})} = (D_{XS}/L) \times ([XS]_1 - [XS]_2); \tag{166}$$

$$J_X^{1 \to 2(\text{net})} = (D_X/L) \times ([X]_1 - [X]_2); \tag{167}$$

$$J_{ZS}^{2 \to 1(\text{net})} = (D_{ZS}/L) \times ([ZS]_2 - [ZS]_1); \tag{168}$$

and

$$J_Z^{2 \to 1(\text{net})} = (D_Z/L) \times ([Z]_2 - [Z]_1) \tag{169}$$

where J is the flux per unit area of membrane, D is the requisite diffusion constant, and L is the thickness of the membrane.

5. *The formation and dissociation of XS and ZS are so rapid that S, X, Z, XS, and ZS may be considered to be in equilibrium.* The equilibrium or dissociation constants are defined to have units of concentration and to include the partition coefficient of S between aqueous and lipoid phases:

$$K_{XS} = [X]_i \times [S]_i/[XS]_i \tag{170}$$

and

$$K_{ZS} = [Z]_i \times [S]_i/[ZS]_i \tag{171}$$

where i refers to compartment 1 or 2.

The forms X and Z are chosen so that the affinity of Z for S is much less than the affinity of X for S_3, i.e.,

$$\text{affinity of Z for S} \ll \text{affinity of X for S.} \tag{172}$$

Since the dissociation constant is the inverse of the affinity,

$$K_{ZS} \gg K_{XS} \tag{173}$$

In other words, a greater concentration of S is required to saturate the low-affinity form, Z, than to saturate the high-affinity form, X.

6. *An off-equilibrium metabolic process on at least one of the sides creates a difference in the distribution of free carrier between the two forms X and Z.* If the interconversion of X and Z were not coupled by a metabolic reaction, then X and Z could be related by an equilibrium constant, i.e.,

$$[X]_i/[Z]_i = h \tag{174}$$

where the equilibrium constant h is unitless. The greater the value of h, the greater the amount of high-affinity form, X, relative to low-affinity form, Z.

On side 1, the conversion of Z into X may be coupled with a metabolic reaction in which reactants, R, form products, P. Since this step is not rate-limiting, X, Z, P and R may be considered to be in equilibrium, i.e.,

$$[X]_1 \times [P]_1 = p_1 \times [Z]_1 \times [R]_1 \tag{175}$$

where $[R]_1$ and $[P]_1$ represent the product of the activities of all the reactants and all the products, respectively, on side 1, and where the dimension of the equilibrium constant p_1 depends upon the stoichiometry of the metabolic reaction. Equation (175) can be rearranged to resemble the form of Eq. (174) as follows:

$$[X]_1/[Z]_1 = p_1 \times [R]_1/[P]_1 \tag{176a}$$

$$= h_1 \tag{176b}$$

where

$$h_1 = p_1 \times [R]_1/[P]_1 \tag{177}$$

Note that the apparent equilibrium constant, h_1, is not a constant, but depends upon the poise of metabolic reactants and products. Z, X, R, and P may be considered to be in equilibrium, even though R and P are not in equilibrium. In fact, if R and P are in equilibrium, $[X]_1$ and $[Z]_1$ are related by Eq. (174), since coupling to a reaction at equilibrium cannot perturb the natural equilibrium between X and Z. Therefore, at equilibrium, Eq. (177) becomes

$$h = p_1 \times [R]_1^0/[P]_1^0 \tag{178}$$

where the superscript 0 denotes an equilibrium activity. Equation (178) can be solved for p_1, which can be substituted into Eq. (177) to yield

$$h_1 = h \times ([R]_1/[R]_1^0)/([P]_1/[P]_1^0) \tag{179a}$$

$$= h \times m_1 \tag{179b}$$

where the metabolic coupling factor m_1 is given by

$$m_1 = ([R]_1/[R]_1^0)/([P]_1/[P]_1^0) \tag{180}$$

The metabolic coupling factor, m_1, is normally greater than unity since reactants are present at a greater activity, and products at a lesser activity, than that at which they would be if they attained equilibrium. Thus, the "pushing" reaction on side 1 results in an apparent equilibrium constant, h_1, which is greater than the uncoupled equilibrium constant h i.e.,

$$h_1 \geq h \tag{181}$$

where the equality holds only in the absence of coupling or at equilibrium.

On side 2, the symbols for reactants and products are denoted by a prime to indicate that the interconversion of X and Z may be coupled to a different metabolic reaction than on side 1.* Note that on side 2, R′ reacts with X, whereas on side 1, R reacts with Z. Thus, Eq. (182a) is analogous to Eq. (179a) but with reactants and products reversed, i.e.,

$$h_2 = h \times ([P']_2/[P']_2^0)/([R']_2/[R']_2^0) \tag{182a}$$

$$= h/m_2 \tag{182b}$$

where

$$h_2 = [X]_2/[Z]_2 \tag{183}$$

and

$$m_2 = ([R']_2/[R']_2^0)/([P']_2/[P']_2^0) \tag{184}$$

The metabolic coupling factor, m_2, is normally greater than unity. Thus, the "pulling" reaction on side 2 results in an apparent equilibrium constant, h_2, which is less than the uncoupled equilibrium constant, h, i.e.,

$$h_2 \leq h \tag{185}$$

where the equality holds only in the absence of coupling or at equilibrium.

If there is coupling to an off-equilibrium process on at least one of the sides, then Eqs. (181) and (185) may be combined to give

$$h_1 > h_2. \tag{186}$$

* In biological systems, it is common to have two substances such as A and B connected by two different reactions. The first reaction may be involved in the synthesis of B from its precursor, A, and is usually coupled to the hydrolysis of ATP by a particular enzyme. The second reaction may be involved in the degradation of B to A, and usually does not require coupling to a source of metabolic energy. The degradation reaction is mediated by a different enzyme than the synthetic reaction. Similarly, the two forms of the carrier may be coupled to different metabolic reactions by different enzyme systems on the two sides of the membrane.

Equation (186) shows that the high-affinity form X is the preferred form of carrier on side 1 relative to that on side 2, since h_i expresses the ratio of high-affinity to low-affinity form. It is this difference which results in a preferential movement of substrate S from side 1 to side 2.

Although $[X]_i$ and $[Z]_i$ can be related by the apparent equilibrium constant, h_i, they can also be related by expressing $[X]_i$ and $[Z]_i$ as fractions of the total amount of free carrier, i.e.,

$$fX_i = [X]_i/([X]_i + [Z]_i) \tag{187}$$

and

$$fZ_i = [Z]_i/([X]_i + [Z]_i) \tag{188}$$

Note that the sum of the fractions must equal unity, i.e.

$$fX_i + fZ_i = 1 \tag{189}$$

Since

$$[X]_i/[Z]_i = h_i, \tag{190}$$

the fractions may be expressed in terms of the appropriate apparent equilibrium constants as follows:

$$fX_i = h_i/(h_i + 1); \tag{191}$$

$$fZ_i = 1/(h_i + 1) \tag{192}$$

The inequality expressed in terms of h_i by Eq. (186) can be written in terms of the fractions, fX_i and fZ_i, as follows:

$$fX_1 > fX_2; \tag{193}$$

$$fZ_1 < fZ_2 \tag{194}$$

Equations (193) and (194) restate that the high-affinity form X is the favored form on side 1 and the low-affinity form Z is the favored form on side 2.

The fractions fX_i and fZ_i are introduced because the kinetic equations [shown below in Eqs. (196–198)] are more readily expressed in terms of fX_i and fZ_i than in terms of h_i.

The remainder of this section is concerned with a kinetic model based on the six assumptions listed above and on the auxiliary assumption that diffusion constants for all forms of carrier are equal, i.e.,

$$D = D_X = D_{XS} = D_Z = D_{ZS} \tag{195}$$

where the D without a subscript represents all four subscripted constants. This auxiliary assumption is made in order to eliminate *trans*-effects which would only serve to obscure the basic properties of this model for active transport. *Trans*-effects for this model are discussed by Rosenberg and Wilbrandt.[88]

The kinetic model to which Eq. (195) applies leads to Michaelis-Menten kinetics, i.e., to

$$J_S^{cis \to trans} = J_{S,max} \times [S]_{cis}/([S]_{cis} + K_{S,0.5}^{cis \to trans}) \tag{196}$$

where

$$J_{Smax} = Q_{X+Z(total)} \times D/L^2 \tag{197}$$

and where $K_{S,0.5}^{cis \to trans}$, the cis-concentration of S necessary to achieve half-maximal flux, is given by the following expression:

$$K_{S,0.5}^{cis \to trans} = (f X_{cis}/K_{XS} + f Z_{cis}/K_{ZS})^{-1} \tag{198}$$

Since a metabolic reaction on at least one of the sides can create a difference in the fractions ($f X_{cis}$ and $f Z_{cis}$) between the two sides [see Eqs. (193) and (194)],

$$K_{S,0.5}^{2 \to 1} > K_{S,0.5}^{1 \to 2} \tag{199}$$

This difference in apparent dissociation constants can lead to a preferential direction of movement.

Net flux is given by

$$J_S^{1 \to 2(net)} =$$

$$J_{S,max} \times \{[S]_1/([S]_1 + K_{S,0.5}^{1 \to 2}) - [S]_2/([S]_2 + K_{S,0.5}^{2 \to 1})\} \tag{200}$$

If the stipulation is made that the ratio of $[S]_2$ to $[S]_1$ is a constant, the maximal net flux can be determined. $J_S^{1 \to 2(net)}$ from Eq. (200) is differentiated with respect to either $[S]_1$ or $[S]_2$ and $dJ_S^{1 \to 2(net)}/d[S]_i$ is set equal to zero. The equation is then solved for $[S]_i$ which is expressed in the form

$$[S]_1 \times [S]_2 = K_{S,0.5}^{1 \to 2} \times K_{S,0.5}^{2 \to 1} \tag{201}$$

Substitution of Eq. (201) in Eq. (200) yields

$$J_{S,max}^{1 \to 2(net)} = J_{S,max} \times \frac{\sqrt{K_{S,0.5}^{2 \to 1}/K_{S,0.5}^{1 \to 2}} - \sqrt{[S]_2/[S]_1}}{\sqrt{K_{S,0.5}^{2 \to 1}/K_{S,0.5}^{1 \to 2}} + \sqrt{[S]_2/[S]_1}} \tag{202}$$

In summary, Eq. (202) gives the expression for the maximal net flux for any given concentration ratio, $[S]_2/[S]_1$. Equation (201) shows how $[S]_1$ and $[S]_2$ must be related in order to achieve maximal net flux. Equation (202) shows that maximal net flux approaches the theoretical maximum for the system ($J_{S,max}$) only when the concentration ratio is small in comparison with the ratio of the apparent dissociation constants. Before maximum net flux is discussed more completely (see end of Sections III, A, 3, and III, A, 4), active level flow is considered.

3. Active Level Flow

Level flow occurs when there is no gradient of S across the membrane, i.e., when $[S]_1 = [S]_2$. In this case, Eq. (200) reduces to

$$J_S^{1 \to 2(net)} = J_{S,max} \times \{[S]/([S] + K_{S,0.5}^{1 \to 2}) - [S]/([S] + K_{S,0.5}^{2 \to 1})\} \tag{203a}$$

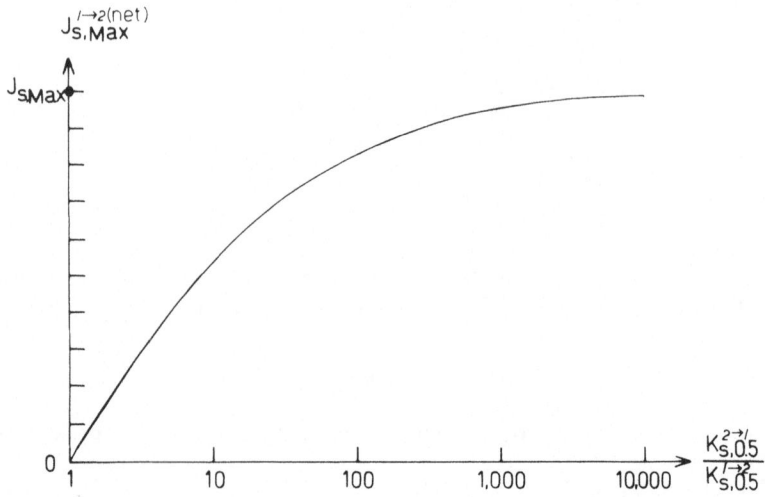

Fig. 20. Maximal net flux (relative to the maximal unidirectional flux) as a function of the ratio of apparent dissociation constants on the two sides of the membrane. This figure applies to the active transport scheme shown in Fig. 18 when all diffusion constants are equal.

$$= J_{S,max} \times \frac{[S] \times (K_{S,0.5}^{2 \to 1} - K_{S,0.5}^{1 \to 2})}{([S] + K_{S,0.5}^{1 \to 2}) \times ([S] + K_{S,0.5}^{2 \to 1})} \tag{203b}$$

where $[S]$ represents both $[S]_1$ and $[S]_2$. Note that there is net movement when $K_{S,0.5}^{2 \to 1} \neq K_{S,0.5}^{1 \to 2}$ and that the direction of this net movement depends upon which of the apparent dissociation constants is greater. For our system, $K_{S,0.5}^{2 \to 1} > K_{S,0.5}^{1 \to 2}$ [Eq. (199)]; therefore Eq. (203b) shows that net movement is from side 1 to 2, i.e.,

$$J_S^{1 \to 2(net)} > 0 \tag{204}$$

Thus, for level flow, the unidirectional flux from side 1 to 2 is larger than the flux from 2 to 1. The larger flux $(J_S^{1 \to 2})$ is termed the forward flux, and the smaller flux $(J_S^{2 \to 1})$ is termed the back flux.

Let us now consider the conditions under which the net flux approaches the theoretical maximal flux. Equation (203b) can be rewritten as follows:

$$J_S^{1 \to 2(net)} = J_{S,max} \times (1 - K_{S,0.5}^{1 \to 2}/K_{S,0.5}^{2 \to 1})/$$
$$\{(1 + K_{S,0.5}^{1 \to 2}/[S]) \times ([S]/K_{S,0.5}^{2 \to 1} + 1)\} \tag{205}$$

If $J_S^{1 \to 2(net)}$ is to approach the theoretical maximum, $J_{S,max}$, all factors to the right of $J_{S,max}$ in Eq. (205) must approach unity. Thus, level flow approaches the theoretical limit when the following three conditions are met:

$$K_{S,0.5}^{2 \to 1} \gg K_{S,0.5}^{1 \to 2} \tag{206}$$

$$[S] \gg K_{S,0.5}^{1 \to 2} \tag{207}$$

$$K_{S,0.5}^{2 \to 1} \gg [S] \tag{208}$$

These three conditions may be rewritten as one condition, namely as

$$K_{S,0.5}^{2 \to 1} \gg [S] \gg K_{S,0.5}^{1 \to 2} \tag{209}$$

Furthermore, it can be shown from Eq. (198) that $K_{S,0.5}^{cis \to trans}$ must lie between K_{ZS} and K_{XS}, Hence,

$$K_{ZS} > K_{S,0.5}^{2 \to 1} \gg [S] \gg K_{S,0.5}^{1 \to 2} > K_{XS} \tag{210}$$

In summary, to achieve appreciable level flow, the two forms of the carrier must have appreciably different dissociation constants. Furthermore, the metabolic processes must poise the system so that $K_{S,0.5}^{2 \to 1} \approx K_{ZS}$ and $K_{S,0.5}^{1 \to 2} \approx K_{XS}$, and the concentration of substrate must lie somewhere between the two apparent dissociation constants.

It is possible to determine the concentration of S at which maximal net flux occurs during level flow. During level flow, $[S] = [S]_1 = [S]_2$. Substitution of [S] into Eq. (201) yields

$$[S]^{\max J} = \sqrt{K_{S,0.5}^{1 \to 2} \times K_{S,0.5}^{2 \to 1}} \tag{211}$$

Equation (211) shows that maximal net flux during level flow occurs when [S] is the geometric mean of the two apparent dissociation constants. Equation (202) reduces to

$$J_{S,max}^{1 \to 2(net)} = J_{S,max} \times (\sqrt{K_{S,0.5}^{2 \to 1}/K_{S,0.5}^{1 \to 2}} - 1)/(\sqrt{K_{S,0.5}^{2 \to 1}/K_{S,0.5}^{1 \to 2}} + 1) \tag{212}$$

Equation (212) shows that maximal net flux approaches the theoretical maximum ($J_{S,max}$) only when

$$K_{S,0.5}^{2 \to 1}/K_{S,0.5}^{1 \to 2} \ggg 1 \tag{213}$$

Figure 20 shows a plot of maximal net flux as a function of $K_{S,0.5}^{2 \to 1}/K_{S,0.5}^{1 \to 2}$. Indeed, in accordance with Eq. (212), maximal net flux approaches within 80% of the theoretical maximum only when the system displays a ratio of $K_{S,0.5}^{2 \to 1}$ to $K_{S,0.5}^{1 \to 2}$ of more than 100 to 1.

4. Active Uphill Flow

It is possible to determine the maximum gradient that can be maintained by this system. To maintain any gradient, it is necessary to maintain net movement of S at zero, i.e.,

$$J_S^{1 \to 2(net)} = 0 \tag{214}$$

Application of Eq. (214) to the equation for net flux [Eq. (200)] yields

$$[S]_1/([S]_1 + K_{S,0.5}^{1 \to 2}) - [S]_2/([S]_2 + K_{S,0.5}^{2 \to 1}) = 0; \tag{215}$$

$$[S]_2/[S]_1 = K_{S,0.5}^{2 \to 1}/K_{S,0.5}^{1 \to 2} \tag{216}$$

Since $K_{S,0.5}^{2 \to 1} > K_{S,0.5}^{1 \to 2}$ [Eq. (199)],

$$[S]_2 > [S]_1, \tag{217}$$

i.e., the mechanism can maintain S at a higher concentration on side 2 than on side 1.

Note that an appreciable concentration ratio is maintained only when $K_{S,0.5}^{2 \to 1}$ is much greater than $K_{S,0.5}^{1 \to 2}$. Since both apparent dissociation constants must lie between K_{XS} and K_{ZS}, the maximum concentration ratio that can be established *regardless of the free energy available from the metabolic processes* is given by

$$[S]_2/[S]_1 = K_{S,0.5}^{2 \to 1}/K_{S,0.5}^{1 \to 2} < K_{ZS}/K_{XS} \tag{218}$$

Under what conditions does net flux approach the theoretical maximal flux? Equation (200) can be rewritten as follows:

$$J_S^{1 \to 2(net)} = J_{S,max} \times \{1 - (K_{S,0.5}^{1 \to 2}/[S]_1) \times ([S]_2/K_{S,0.5}^{2 \to 1})\}/$$
$$\{(1 + K_{S,0.5}^{1 \to 2}/[S]_1) \times (1 + [S]_2/K_{S,0.5}^{2 \to 1})\} \tag{219}$$

If $J_S^{1 \to 2(net)}$ is to approach the theoretical maximum, $J_{S,max}$, the three factors to the right of $J_{S,max}$ in Eq. (219) must approach unity. Thus, net flux approaches the theoretical limit when the following two conditions are met:

$$[S]_1 \gg K_{S,0.5}^{1 \to 2}; \tag{220}$$

$$[S]_2 \ll K_{S,0.5}^{2 \to 1} \tag{221}$$

Thus, to achieve appreciable net forward $(1 \to 2)$ flux, S must saturate the mechanism on side 1 but not on side 2.

We may now return to the discussion of maximal net flux (see end of Sections III, A, 2 and III, A, 3). Figure 21 shows maximal net flux as a function of concentration ratio for four systems with different ratios of $K_{S,0.5}^{2 \to 1}$ to $K_{S,0.5}^{1 \to 2}$ [see Eqs. (201) and (202)]. Figure 21 is divided into four quadrants by a vertical line that indicates level flow $([S]_2/[S]_1 = 1)$ and by a horizontal line that indicates gradient maintenance $(J_S^{1 \to 2(net)} = 0)$. Active downhill movement, manifest when $[S]_2/[S]_1 < 1$, is shown in the upper left quadrant. During active downhill movement, both metabolic and osmotic energy are expended. Active uphill movement, manifest when $1 < [S]_2/[S]_1 < K_{S,0.5}^{2 \to 1}/K_{S,0.5}^{1 \to 2}$, is shown in the upper right quadrant. During active uphill movement, part of the expended metabolic energy is transduced into osmotic work. Note that when the two apparent diffusion constants are equal $(K_{S,0.5}^{2 \to 1}/K_{S,0.5}^{1 \to 2} = 1)$, active uphill movement is not observed. Reversed active downhill movement from side 2 to 1 is observed when the concentration ratio is greater than the maximum that can be maintained, i.e., when $[S]_2/[S]_1 > K_{S,0.5}^{2 \to 1}/K_{S,0.5}^{1 \to 2}$. During reversed flow, shown in the lower right quadrant, osmotic energy is expended and it is possible to store part of the expended osmotic energy in the form of metabolic energy. In our scheme, however, in which transport and metabolism are uncoupled, energy continues to be expended during reversed downhill flow. Further modification of the scheme

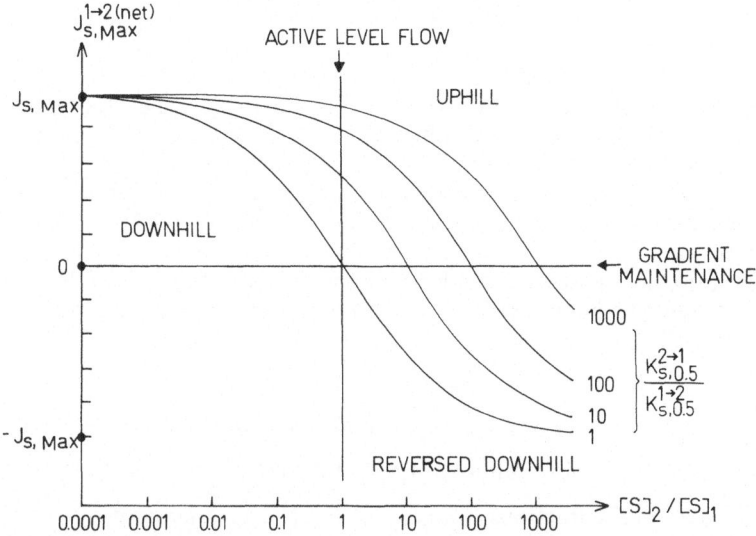

Fig. 21. Maximal net flux as a function of the concentration ratio of substrate for various ratios of apparent dissociation constants. This figure applies to the active transport scheme shown in Fig. 18 when all diffusion constants are equal. Uphill movement is not manifest under any condition when the apparent dissociation constants on the two sides of the membrane are equal ($K_{S,0.5}^{2\to1}/K_{S,0.5}^{1\to2} = 1$).

is necessary to convert it into a model in which part of the expended osmotic energy is stored as metabolic energy. In such a modified scheme, transport and metabolism are said to be tightly coupled. Energetic considerations are discussed further in Section III, A, 7.

5. The Establishment of the Asymmetric Distribution of High- and Low-Affinity Forms of the Carrier

In the previous section (Section III, A, 4) it is shown in Eq. (218) that an appreciable concentration ratio is maintained only when $K_{S,0.5}^{2\to1}$ is much greater than $K_{S,0.5}^{1\to2}$ (also see Fig. 21). Let us examine Eq. (198) for $K_{S,0.5}^{cis\to trans}$ to see how coupling to a metabolic process can make $K_{S,0.5}^{2\to1} > K_{S,0.5}^{1\to2}$. To gain insight into the meaning of Eq. (198), let us consider the equation in terms of affinities. Since the affinity is the inverse of the dissociation constant, Eq. (198) becomes:

apparent affinity of carrier for S on cis-side

$= fX_{cis} \times$ (affinity of high-affinity form, X, for S)

$\quad + fZ_{cis} \times$ (affinity of low-affinity form, Z, for S)   (222)

where the apparent affinity is the inverse of $K_{S,0.5}^{cis\to trans}$. Equation (222) shows that the apparent affinity lies between the low- and the high-affinity values and that each of these forms contributes to the apparent affinity in proportion to its concentration. Thus, the apparent dissociation constant, $K_{S,0.5}^{cis \rightarrow trans}$, must lie between K_{XS} and K_{ZS}.

There is a preferential direction of transport only when the apparent affinities on sides 1 and 2 are different; there is a difference in apparent affinities only when the fractions, fX_i and fZ_i, are different on the two sides. It is shown above in Eqs. (193) and (194) that coupling to an off-equilibrium metabolic process on at least one of the sides establishes a difference in these fractions.

Equation (222) shows that it is necessary to have two forms of the carrier with two different affinities; for if the affinities were equal, Eq. (222) would reduce to

$$\text{apparent affinity on } cis\text{-side}$$

$$= (fX_{cis} + fZ_{cis}) \times (\text{affinity of X and Z for S}) \qquad (223a)$$

$$= 1 \times (\text{affinity of X and Z for S}) \qquad (223b)$$

since the sum of the fractions is unity [see Eq. (189)]. Therefore,

$$\text{apparent affinity on side 1} = \text{apparent affinity on side 2}; \qquad (224)$$

$$K_{S,0.5}^{2 \rightarrow 1} = K_{S,0.5}^{1 \rightarrow 2} \qquad (225)$$

Equations (224) and (225) show that no amount of difference in the fractions can create a difference in the apparent affinity and, hence, in the direction of transport, when the two forms of the carrier display the same affinity for substrate.

Let us assume that two forms of a carrier display two different affinities. On which side must the metabolic process be coupled with the transport process in order to achieve maximal ability to concentrate S? Since concentrating ability is related to the ratio of $K_{S,0.5}^{2 \rightarrow 1}$ to $K_{S,0.5}^{1 \rightarrow 2}$ [see Eq. (218)], maximum concentrating ability is achieved when the ratio is a maximum. From Eq. (198) it is seen that $K_{S,0.5}^{1 \rightarrow 2}$ and $K_{S,0.5}^{2 \rightarrow 1}$ may approach the following minimum and maximum, respectively:

$$K_{S,0.5}^{1 \rightarrow 2} \approx K_{XS} \qquad (226)$$

and

$$K_{S,0.5}^{2 \rightarrow 1} \approx K_{ZS} \qquad (227)$$

The condition expressed by Eq. (226) for side 1 is satisfied when

$$fX_1 \approx 1 \qquad \text{and} \qquad fZ_1 \approx 0 \qquad (228)$$

and the condition expressed by Eq. (227) for side 2 is satisfied when

$$fX_2 \approx 0 \qquad \text{and} \qquad fZ_2 \approx 1 \qquad (229)$$

In order to establish the conditions expressed by Eqs. (228) and (229), is it more advantageous to have a "pushing" metabolic reaction on side 1 or a "pulling" metabolic reaction on side 2? The answer to this question depends upon whether the equilibrium favors the low- or high-affinity form of the carrier when the system is not coupled to metabolic reactions (or when the metabolic reactants and products are in equilibrium).

When the low-affinity form, Z, is highly favored over the high-affinity form X Eq. (174) shows that

$$1/h \gg 1 \tag{230}$$

According to Eqs. (191) and (192) (where $h_i = h$),

$$fX_1 \approx 0 \quad \text{and} \quad fZ_1 \approx 1 \tag{231}$$

and

$$fX_2 \approx 0 \quad \text{and} \quad fZ_2 \approx 1 \tag{232}$$

Equation (232) satisfies the condition expressed by Eq. (229) for side 2, but Eq. (231) does not satisfy the condition expressed by Eq. (228) for side 1 unless h_1 is changed from h (which is very small) to some large value, i.e., the condition for side 1 is satisfied when

$$h_1 \gg 1 \tag{233}$$

Since $h_1 = h \times m_1$ [see Eq. (179b)], the addition of a sufficiently powerful "pushing" reaction (characterized by $m_1 \ggg 1$) can overcome the effect of the small value of h so that Eq. (233) (and, hence, the condition on side 1) is satisfied. The addition of a "pulling" reaction on side 2 does not contribute to the transport process since Eq. (232) already satisfies the necessary condition for maximum concentrating ability. Indeed, a "pulling" reaction would only serve to waste metabolic energy.

When the high-affinity form X is highly favored over the low-affinity form Z, Eq. (174) shows that

$$h \gg 1 \tag{234}$$

According to Eqs. (191) and (192) (where $h_i = h$),

$$fX_1 \approx 1 \quad \text{and} \quad fZ_1 \approx 0 \tag{235}$$

and

$$fX_2 \approx 1 \quad \text{and} \quad fZ_2 \approx 0 \tag{236}$$

Equation (235) satisfies the condition for side 1, but Eq. (236) does not satisfy the condition for side 2 unless h_2 is changed from h (which is very large) to some small value, i.e., the condition for side 2 is satisfied when

$$1/h_2 \gg 1 \tag{237}$$

Since $h_2 = h/m_2$ [see Eq. (182b)], the addition of a sufficiently powerful "pulling" reaction (characterized by $m_2 \ggg 1$) can overcome the effect of the large value of h so that Eq. (237) (and, hence, the condition on side 2) is satisfied. The addition of a "pushing" reaction on side 1 does not contribute to the transport process since Eq. (235) already satisfies the necessary condition for maximum concentrating ability. Indeed, a "pushing" reaction would only serve to waste metabolic energy.

In summary, if one form of the carrier is naturally favored over another form, then a metabolic process must be coupled on the requisite side to reverse the order of preference on that side. As a consequence of this metabolic coupling, one form is favored on one side, and the other form is favored on the other side. During level flow ($[S]_1 = [S]_2$), net transport is from the side where the high-affinity form is favored to the side where the low-affinity form is favored. Only when neither form is favored ($h \approx 1$) is a metabolic process required on both sides in order to achieve maximum concentrating ability.

6. Site of the Metabolic Reaction

Rosenberg and Wilbrandt[88] show how to determine whether the metabolic reaction coupled to a transport system "pushes" or "pulls." For maximal ability to concentrate S on side 2, $K_{S,0.5}^{2 \to 1} \approx K_{ZS}$ and $K_{S,0.5}^{1 \to 2} \approx K_{XS}$ [Eqs. (226) and (227)]. If the metabolic reaction is, however, blocked so that reactants and products come to equilibrium, both $K_{S,0.5}^{1 \to 2}$ and $K_{S,0.5}^{2 \to 1}$ approach the dissociation constant of the favored form of the carrier.

If coupling is to a "pushing" reaction, the low-affinity form Z, is favored, and blockage of the metabolic reaction leads to an increase in $K_{S,0.5}^{1 \to 2}$ from K_{XS} to K_{ZS} while $K_{S,0.5}^{2 \to 1}$ remains essentially unchanged at K_{ZS}. The result of this change in $K_{S,0.5}^{1 \to 2}$ is a decrease in the forward unidirectional flux ($J_S^{1 \to 2}$); during level flow, the forward flux approaches the back flux ($J_S^{2 \to 1}$), which remains essentially unchanged.

If coupling is to a "pulling" reaction, the high-affinity form, X, is favored, and blockage of the metabolic reaction leads to a decrease in $K_{S,0.5}^{1 \to 2}$ from K_{ZS} to K_{XS} while $K_{S,0.5}^{2 \to 1}$ remains essentially unchanged at K_{XS}. The result of this change in $K_{S,0.5}^{2 \to 1}$ during level flow is an increase in the back flux to the level of the forward flux, which remains essentially unchanged.

In summary, blockage of the metabolic reaction coupled to transport leads to a change in the unidirectional flux which originates from the side where the reaction is located, i.e., if blockage results in a change in $J_S^{cis \to trans}$, then the metabolic reaction is located on the cis-side. For example, a decrease in forward flux ($J_S^{1 \to 2}$) following blockage suggests that a "pushing" reaction is located on side 1, and an increase in back flux ($J_S^{2 \to 1}$) following blockage suggests that a "pulling" reaction is located on side 2.

7. Energetics

The six assumptions listed in Section III, A, 2 plus the auxiliary assumption that all diffusion constants are equal [Eq. (195)] serve as the basis for finding the metabolic rate as well as the transport rate. It can be shown that

$$J_M = J_{S,max} \times (h_1 - h_2)/\{(h_1 + 1) \times (h_2 + 1)\} \qquad (238a)$$

$$= J_{S,max} \times h(m_1 m_2 - 1)/\{(hm_1 + 1) \times (h + m_2)\} \qquad (238b)$$

since $h_1 = h_1 \times m$ and $h_2 = h/m$ [see Eqs. (179b) and (182b)]. Note that J_M does not depend upon the actual transport rate, although it does depend upon the theoretical maximum rate, $J_{S,max}$, and upon the metabolic factors, m_1 and m_2, which characterize the off-equilibrium poise of the metabolites.

For either a "pushing" reaction in which

$$m_1 \ggg 1 \quad \text{and} \quad m_2 = 1, \tag{239}$$

or for a "pulling" reaction in which

$$m_1 = 1 \quad \text{and} \quad m_2 \ggg 1, \tag{240}$$

J_M in Eq. (238b) reduces to

$$J_M \approx J_{S,max} \tag{241}$$

In other words, when there is any appreciable degree of metabolic coupling, the metabolic rate is a constant regardless of the net rate of transport. A system in which the metabolic rate is not coupled to the transport rate is said to be completely uncoupled. This system is energetically inefficient because metabolic energy is rapidly expended at the maximal rate even when no osmotic work is performed.

In general, since $m_1 \geq 1$ and $m_2 \geq 1$, the range of J_M is

$$0 \leq J_M \leq J_{S,max} \tag{242}$$

Equation (242) shows that in no case does J_M become negative. Because this system is not capable of exhibiting reversed metabolic flow (regenerative flow) during reversed downhill flow, it cannot serve as a chemi-osmotic mechanism for conversion of osmotic energy into metabolic energy.

In a coupled system, J_M depends upon $J_S^{1 \to 2(net)}$. In general, a system may be partially coupled, and if metabolic rate is a linear function of transport rate, J_M can be represented by:

$$J_M = fU \times J_{S,max} + fC \times J_S^{1 \to 2(net)} \tag{243}$$

where the first term represents the uncoupled moiety and the second term represents the coupled moiety. The fractions fU and fC represent the degree of uncoupling and coupling, respectively, and are chosen in such a way that

$$fU + fC = 1 \tag{244}$$

When $fU \geq fC$, the system is loosely coupled and when $fC > fU$, the system is tightly coupled. Complete uncoupling ($fU = 1$) is a special case of loose coupling and complete coupling ($fC = 1$) is a special case of tight coupling.

The stoichiometry of the metabolic reaction is chosen so that the maximum value of J_M is the theoretical maximum transport rate, $J_{S,max}$. This is evident from Eq. (243) because when $J_S^{1 \to 2(net)} = J_{S,max}$,

$$J_{M,max} = fU \times J_{S,max} + fC \times J_{S,max} \tag{245a}$$

$$= (fU + fC) \times J_{S,max} \tag{245b}$$

$$= J_{S,max} \tag{245c}$$

Note also that the minimal value of J_M is observed when the system is completely coupled and is operating at maximal reversed downhill flow, i.e., when $fC = 1$ and $J_S^{1 \to 2(net)} = -J_{S,max}$:

$$J_{M,min} = 0 + 1 \times (-J_{S,max}) \tag{246a}$$

$$= -J_{S,max} \tag{246b}$$

Let us consider which of the coupled systems are capable of transforming osmotic energy into metabolic energy. Metabolic energy is stored when $J_M < 0$. Assume that the system is operating at maximal reversed downhill flow $(J_S^{1 \to 2(net)} = -J_{S,max})$. From Eq. (243),

$$J_M = fU \times J_{S,max} + fC \times (-J_{S,max}) \tag{247a}$$

$$= -J_{S,max} \times (fC - fU) \tag{247b}$$

J_M is negative when

$$fC > fU \tag{248}$$

Equation (248) defines the condition for a tightly coupled system. Hence, under at least some conditions of reversed downhill flow, all tightly coupled systems are capable of transforming osmotic energy into metabolic energy.

Figure 22 shows the variation of metabolic rate with transport for systems with various degrees of coupling. Note that all lines pass through the point P because of the choice of stoichiometry for the metabolic reaction. In a completely uncoupled system (line A), J_M does not depend upon transport rate. Any experimental attempt to correlate this metabolic rate with transport rate must fail unless the metabolic reaction becomes the rate-limiting step. Line B denotes the boundary between loosely coupled systems (above) and tightly coupled systems (below). Note that only tightly coupled systems have a region of negative metabolic rate, which indicates that osmotic work can be partially converted into metabolic energy. Line C and line D are examples of loosely and tightly coupled systems, respectively. In both examples, J_M varies with the transport rate, and even in the absence of transport $(J_S^{1 \to 2(net)} = 0)$ there is still "wasted" metabolism associated with the transport system. The rate of hydrolysis of ATP has been correlated with the transport of sodium in many tissues. Even in the absence of sodium, however, there is still appreciable hydrolysis of ATP. This moiety may be associated with a process other than sodium transport. On the other hand, this moiety may represent an uncoupled moiety of the metabolic reaction related to transport. In the completely coupled system (line E), no metabolic energy is expended when net transport is zero.

Can the scheme shown in Fig. 18 be regenerative as well as active? In other words, can osmotic energy be transformed into metabolic energy by

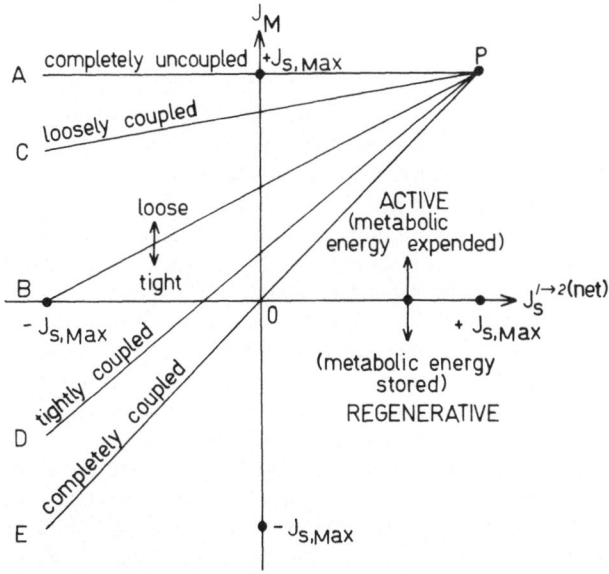

Fig. 22. Metabolic flow as a function of net flux for various modes of metabolic coupling. Line A (the completely uncoupled mode) applies to the active transport scheme shown in Fig. 18 when all diffusion constants are equal. In this case, metabolic flow does not vary with net flux, and osmotic energy cannot be converted into metabolic energy. Only tightly coupled systems are capable of converting osmotic energy into metabolic energy.

this mechanism? We have shown in this section that when all diffusion constants are equal, the system is incapable of regenerating the metabolic reactants from the products. The system is not regenerative because reaction 7 serves as a link in cycle 4–7 which contributes a large positive moiety to J_M. The cycle 1–6 can be driven backward by reversed downhill flow from side 2 to 1. Although this reversed flow contributes a negative moiety to J_M, the magnitude of this negative moiety is not large enough to overcome the positive moiety from cycle 4–7. Therefore, even during reversed downhill flow, metabolic energy is expended. The wasteful contribution of cycle 4–7 to J_M can be decreased by a decrease in the diffusion constant of X (D_X) so that the preferred translocation step becomes reaction 2 rather than reaction 7. Since unequal diffusion constants lead to *trans*-effects, however, the kinetic equations become more complex.

Other energetic aspects of this model are considered by Rosenberg and Wilbrandt.[88]

B. Co-Transport

1. Introduction

The uphill movement of a substrate S can be indirectly coupled with a metabolic process through the combination of a passive co-transport system and an active system. The active transport system creates a concentration gradient of a secondary substrate N which is not a structural analogue of the primary substrate S. The gradient of N drives N through a passive system which co-transports S in the same direction as N. In essence, metabolic energy is transformed into osmotic energy of N via the active transport system; osmotic energy of N is then used to transport S. Thus, the transport of S is not a direct active process, but is indirect inasmuch as it depends upon the gradient of N created by the metabolic process. If the metabolic process is blocked, S cannot move uphill unless a gradient of N is maintained by some other process. From the systemic point of view, the transport of S is active; from the mechanistic point of view, it is passive.

Two configurations of a co-transport system are possible. Figure 23a shows one configuration in which the passive co-transport and active elements are in the same membrane. In this configuration, which resembles a cell, N is ejected from the cell by the active element, thereby creating a gradient of N across the membrane. This gradient then drives S through the passive co-transport element into the cell, where S can be concentrated above its extracellular level. If N is ejected by the active element at the same rate at which it enters by the co-transport element, N can be maintained at a low steady-state level. Since S has no exit path, S must either be used as a metabolite within the cell or accumulate to a sufficiently large concentration to stop net entry ($J_S^{1 \to 2(net)} = 0$). An example of this configuration is the co-transport of amino acids by sodium into erythrocytes[149-152] and into Ehrlich mouse ascites tumor cells.[153-157]

Figure 23b shows the second configuration, in which the passive co-transport and active elements are in different membranes. In this configuration, which resembles an epithelial membrane, N is ejected from the interior of the epithelial cell into the serosal medium by the active element, thereby creating a gradient of N across the membrane on the mucosal side. This gradient then drives S through the passive co-transport element into the cell, where S can be concentrated above its extracellular level. S is then driven downhill through the passive element from the cell into the serosal medium. Thus, S and N are transported through the cell from mucosa to serosa. An example of this configuration is the co-transport of sugars by sodium across intestinal epithelium.[158-168]

Figure 24 shows a scheme based on the mechanism proposed by Crane[167] to explain co-transport of sugars by sodium in intestine. In this scheme, free carrier X has no affinity for S and therefore cannot form the complex XS. The free carrier can, however, form a complex with the secondary substrate N to form XN which does have affinity for S. Thus, S is translocated in the form of complex (XN)S. The sequence of reactions is as follows: In

Fig. 23. Two configurations for a passive co-transport system coupled indirectly to an active transport system.

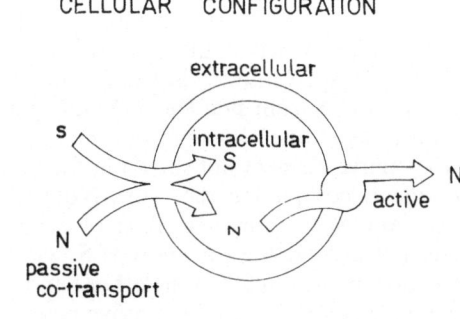

Ⓐ CELLULAR   CONFIGURATION

Ⓑ EPITHELIAL   CONFIGURATION

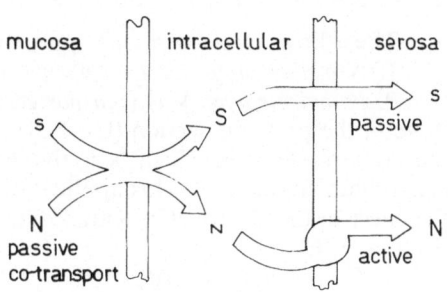

Fig. 24. A model for passive co-transport of a primary non-electrolyte (S) by a secondary nonelectrolyte (N) across a membrane.

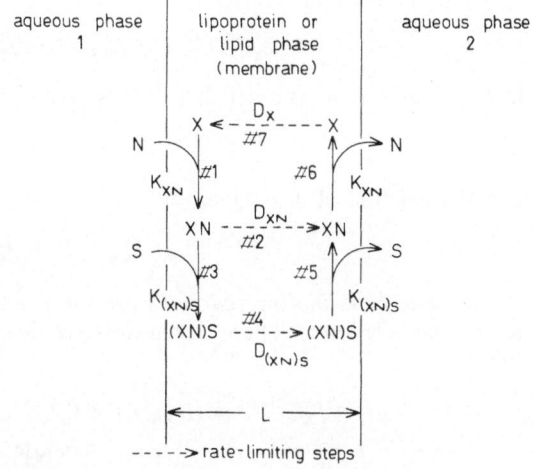

reaction 1, N reacts with X on side 1 to form XN, part of which is translocated to side 2 (reaction 2) and part of which reacts with S (reaction 3) to form (XN)S, which is then translocated to side 2 (reaction 4). On side 2, (XN)S releases S into the aqueous phase (reaction 5), thereby resulting in the transport of S from side 1 to 2. XN, formed by reaction 5, readily releases N to form X because N in compartment 2 is maintained at a low concentration by an active element not shown in the figure. In reaction 7, X is translocated to side 1, completing the cycle. Note that the transport of the secondary substrate, N, may involve either of two cycles: the first cycle (1, 2, 6, and 7) does not involve the transport of S, while the second cycle (1, 3–7) involves the co-transport of S. On the other hand, the transport of S involves only the second cycle (1, 3–7) and net movement occurs only when there is net transport of N.

2. Assumptions and Kinetic Equations

The following assumptions apply to the co-transport model:

1. *Nonelectrolyte S is transported only in the form of $(XN)S$.*

2. *Nonelectrolyte N is transported in the form of either XN or $(XN)S$.* Note in the previous section (III, B, 1) that this model has been applied to biological systems in which N is the electrolyte sodium and S may be an electrolytic amino acid. This application may not be entirely justified since the effect of the electrical potential difference is not taken into account (see Section II, E).

3. *Carrier is conserved.*

$$Q_{X(total)} = Q_X + Q_{XN} + Q_{(XN)S} = \text{constant} \qquad (249)$$

where Q is the amount of carrier per unit area.

4. *The system is operating in the steady state.* The net amount of carrier in all forms moving from side 1 to 2 must equal the net amount moving from side 2 to 1. In other words,

$$J_{XN}^{1 \to 2(net)} + J_{(XN)S}^{1 \to 2(net)} = J_X^{2 \to 1(net)} \qquad (250)$$

In the steady state, the net flux of S is given by

$$J_S^{1 \to 2(net)} = J_{(XN)S}^{1 \to 2(net)} \qquad (251)$$

and the net flux of N is given by

$$J_N^{1 \to 2(net)} = J_{XN}^{1 \to 2(net)} + J_{(XN)S}^{1 \to 2(net)} \qquad (252)$$

5. *The rate-limiting reactions are the transitions of X, XN, and $(XN)S$ across the membrane.* These transitions are described by Fick's equations as follows:

$$J_{XN}^{1 \to 2(net)} = (D_{XN}/L) \times ([XN]_1 - [XN]_2), \qquad (253)$$

$$J_{(XN)S}^{1 \to 2(net)} = (D_{(XN)S}/L) \times \{[(XN)S]_1 - [(XN)S]_2\}, \qquad (254)$$

and

$$J_X^{2 \to 1 \text{(net)}} = (D_X/L) \times ([X]_2 - [X]_1), \tag{255}$$

where J is the flux per unit area of membrane, D is the appropriate diffusion constant, and L is the thickness of the membrane.

6. *The formation and dissociation of XN and (XN)S are so rapid that S, N, XN, and (XN)S may be considered to be in equilibrium.* The equilibrium or dissociation constants are defined to have units of concentration and to include the partition coefficient of S between aqueous and lipoid phases:

$$K_{XN} = [X]_i \times [N]_i/[XN]_i \tag{256}$$

and

$$K_{(XN)S} = [XN]_i \times [S]_i/[(XN)S]_i \tag{257}$$

where i refers to compartment 1 or 2.

The possibility that free X can exhibit a low affinity for S to form the complex XS is discussed by Stein.[169] In this case, dissociation constants must be defined for the complexes XS and (XS)N.

It is also of interest that Vidaver[150–152] found that *two* sodium ions serve as the secondary substrate in the co-transport of glycine in pigeon erythrocytes. Thus, the scheme that adequately describes co-transport in erythrocytes must incorporate features of a divalent carrier scheme (such as was presented in Section II, C, 7) and the scheme presented in this section.

The remainder of this section is concerned with a kinetic model based on the assumptions listed above and on the auxiliary assumption that diffusion constants for all forms of carrier are equal, i.e.,

$$D = D_X = D_{XN} = D_{(XN)S} \tag{258}$$

where D represents all three diffusion constants. This auxiliary assumption is made in order to eliminate *trans*-effects which would only serve to obscure the basic properties of this model for co-transport.

The kinetic model to which Eq. (258) applies leads to the following Michaelis–Menten kinetics:

$$J_S^{cis \to trans} = J_{S,\max} \times [S]_{cis}/([S]_{cis} + K_{S,0.5}^{cis \to trans}) \tag{259}$$

and

$$J_N^{cis \to trans} = J_{N,\max} \times [N]_{cis}/([N]_{cis} + K_{N,0.5}^{cis \to trans}) \tag{260}$$

where

$$J_{S,\max} = J_{N,\max} = Q_{X(\text{total})} \times D/L^2, \tag{261}$$

$$K_{S,0.5}^{cis \to trans} = K_{(XN)S} \times ([N]_{cis} + K_{XN})/[N]_{cis} \tag{262}$$

and

$$K_{N,0.5}^{cis \to trans} = K_{XN} \times K_{(XN)S}/([S]_{cis} + K_{(XN)S}) \tag{263}$$

The equality between $J_{S,max}$ and $J_{N,max}$ depends upon the auxiliary assumption that all diffusion constants are equal. In an actual system, this assumption may not hold, i.e., maximal fluxes of primary and secondary substrates may not be equal.

A more extensive discussion of the kinetics of co-transport is found in Stein.[169]

3. The Influence of the Secondary Substrate, N, upon the Movement of the Primary Substrate, S

The equation for $J_S^{cis \rightarrow trans}$ for this model of co-transport [Eq. (259)] is identical to the equation for the active transport model discussed in Section III, A [Eq. (196)]. Hence, many of the equations and much of the discussion which pertain to the active transport model can be applied to this co-transport model. For example, whenever $K_{S,0.5}^{2 \rightarrow 1} > K_{S,0.5}^{1 \rightarrow 2}$, there is net movement of S from side 1 to 2 during level flow ($[S] = [S]_1 = [S]_2$). In addition, a concentration gradient can be established and maintained so that $[S]_2/[S]_1 = K_{S,0.5}^{2 \rightarrow 1}/K_{S,0.5}^{1 \rightarrow 2}$ at zero net flow. Of course, in the active model, the difference in the two apparent dissociation constants is created by coupling to a metabolic reaction, whereas in the co-transport model, the difference is created by a difference in the concentration of N [see Eq. (262)].

Figure 25a shows the apparent dissociation constant for S ($K_{S,0.5}^{cis \rightarrow trans}$) as a function of $[N]_{cis}$. Note that when N is absent on the cis-side, $K_{S,0.5}^{cis \rightarrow trans}$ is infinite, i.e., the transport system manifests no affinity for S. Also note that the apparent dissociation constant equals twice the actual dissociation constant ($K_{(XN)S}$) when $[N]_{cis} = K_{XN}$ and approaches the actual constant as $[N]_{cis} \rightarrow \infty$. Figure 25b shows that the data of Fig. 25a can be replotted to yield a straight line. K_{XN} and $K_{(XN)S}$ can be evaluated from the intercepts.

Experiments can be performed to measure unidirectional fluxes of S as a function of $[S]_{cis}$ at different levels of $[N]_{cis}$. From these data, $K_{S,0.5}^{cis \rightarrow trans}$ as a function of $[N]_{cis}$ can be determined and plotted as in Fig. 25b. If the data yield a straight line, they are consistent with this model for co-transport.

Let us consider the orientation of the gradient of N required to cause a net flow of S from side 1 to 2 during level flow. According to Eq. (203d), net movement of S is from side 1 to 2 during level flow when

$$K_{S,0.5}^{2 \rightarrow 1} > K_{S,0.5}^{1 \rightarrow 2} \tag{264}$$

Substitution of Eq. (262) into Eq. (264) yields

$$K_{(XN)S} \times ([N]_2 + K_{XN})/[N]_2 > K_{(XN)S} \times ([N]_1 + K_{XN})/[N]_1; \tag{265a}$$

$$1 + K_{XN}/[N]_2 > 1 + K_{XN}/[N]_1; \tag{265b}$$

$$1/[N]_2 > 1/[N]_1; \tag{265c}$$

$$[N]_1 > [N]_2 \tag{265d}$$

Therefore, the maintenance of N at a higher concentration on side 1 results in the net movement of S from side 1 to side 2 during level flow.

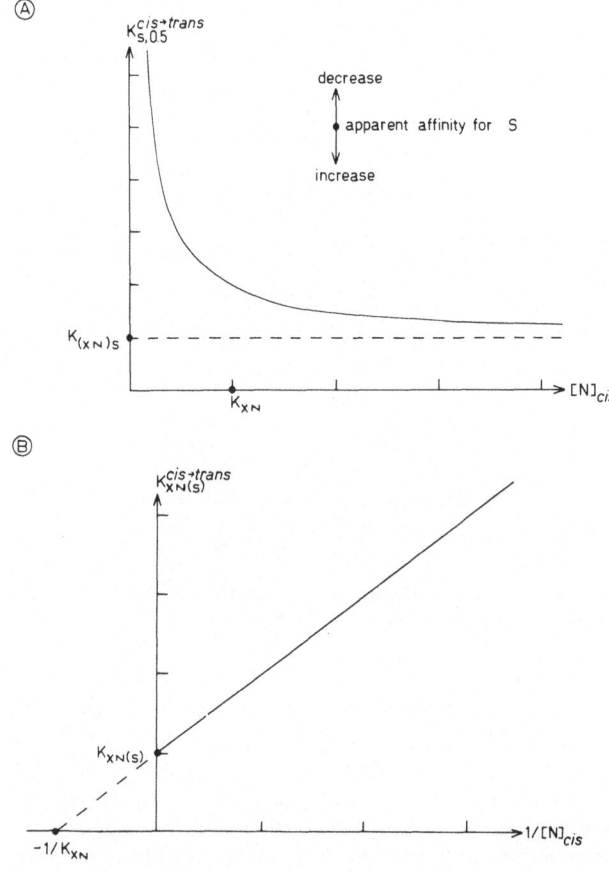

Fig. 25. Effect of the secondary substrate, N, upon the apparent dissociation constant of the carrier for the primary substrate, S. In the absence of N, no affinity is shown toward S. The apparent affinity for S cannot be increased to a value greater than $1/K_{(XN)S}$ even when concentration of N approaches infinity. This figure applies to the co-transport scheme shown in Fig. 24 when all diffusion constants are equal.

Under what conditions does S move uphill? According to Section III, A, 4 and Fig. 21, S moves uphill from side 1 to 2 whenever the concentration ratio does not exceed the ratio of apparent dissociation constants. In other words,

$$J_S^{1 \rightarrow 2(net)} > 0 \tag{266}$$

when

$$[S]_2/[S]_1 < K_{S,0.5}^{2 \rightarrow 1}/K_{S,0.5}^{1 \rightarrow 2} \tag{267}$$

Under what conditions does net flux approach the theoretical maximum flux $(J_{S,max})$? According to Eqs. (220) and (221), appreciable net flux is observed when

$$[S]_1 \gg K_{S,0.5}^{1 \to 2} \tag{268}$$

and

$$[S]_2 \ll K_{S,0.5}^{2 \to 1} \tag{269}$$

From Eq. (262) or Fig. 25, $K_{S,0.5}^{1 \to 2}$ approaches $K_{(XN)S}$ as a minimum when

$$[N]_1 \gg K_{XN} \tag{270}$$

and $K_{S,0.5}^{2 \to 1}$ can be made as large as desired by decreasing $[N]_2$ to a value much less than K_{XN}. Under these conditions, Eqs. (268) and (269) can be rewritten as

$$[S]_1 \gg K_{(XN)S} \tag{271}$$

and

$$[S]_2 \ll K_{(XN)S} \times K_{XN}/[N]_2 ; \tag{272}$$

$$[N]_2 \ll K_{XN} \times K_{(XN)S}/[S]_2 \tag{273}$$

In summary, to achieve appreciable net flux of S from side 1 to 2, the following conditions must be satisfied:
1. N must saturate the carrier on side 1 [Eq. (270)].
2. S must saturate the carrier on side 1 [Eq. (271)].
3. N must be kept low on side 2 to prevent the saturation of the carrier by S on side 2. $[N]_2$ depends upon $[S]_2$ [Eq. (273)]. If S is transported uphill $([S]_2 > [S]_1)$, the factor $K_{(XN)S}/[S]_2$ in Eq. (273) is less than unity, and therefore $[N]_2$ must be maintained at an ultralow level $([N]_2 \lll K_{XN})$.

What is the maximum gradient that can be maintained by the co-transport system? According to Eq. (218),

$$[S]_2/[S]_1 = K_{S,0.5}^{2 \to 1}/K_{S,0.5}^{1 \to 2} \tag{274}$$

when

$$J_S^{1 \to 2(net)} = 0 \tag{275}$$

Substitution of Eq. (262) into Eq. (274) yields

$$[S]_2/[S]_1 = ([N]_1/[N]_2) \times ([N]_2 + K_{XN})/([N]_1 + K_{XN}); \tag{276}$$

$$[N]_1/[N]_2 = ([S]_2/[S]_1) \times ([N]_1 + K_{XN})/([N]_2 + K_{XN}) \tag{277a}$$

$$= ([S]_2/[S]_1) \times \{1 + ([N]_1 - [N]_2)/([N]_2 + K_{XN})\} \tag{277b}$$

If $[N]_1 > [N]_2$, the factor to the right of $([S]_2/[S]_1)$ is greater than unity. Therefore,

$$[N]_1/[N]_2 > [S]_2/[S]_1 > 1 \tag{278}$$

Note that the "maintaining" species, N, cannot maintain a concentration ratio as large as that of the "maintained" species, S. Equation (277b) shows, however, that when $[N] \ll K_{XN}$ on both sides, then

$$[S]_2/[S]_1 \approx [N]_1/[N]_2 \tag{279}$$

In this case, the "maintaining" species can maintain almost the same concentration ratio as the "maintained" species. Note that to maintain S at a higher concentration on a given side, N must be maintained at a higher concentration on the *other* side.

In Section II, C, 5, we discuss the maintenance of a concentration gradient of S by a gradient of an analogue T. In that case, S and T compete for the same site on the carrier. The gradients of S and T are in the same direction, and T must saturate the carrier on both sides if the maximal concentration ratio of S is to be established and maintained. In the case of co-transport, S and N are not structural analogues and, consequently, do not compete for the same site on the carrier. The gradients of S and N are oriented in the opposite direction and N must not saturate the carrier on either side if the maximal concentration ratio of S is to be established and maintained.

4. The Influence of the Primary Substrate, S, upon the Movement of Secondary Substrate, N

In the previous section (Section III, B, 3) we show that the secondary substrate, N, can influence the movement of primary substrate, S. S can, however, also affect the movement of N because the apparent dissociation constant for N ($K_{N.0.5}^{cis \rightarrow trans}$) is a function of $[S]_{cis}$ [Eq. (263)], as is shown in Fig. 26a. Note that when S is absent on the *cis*-side, the apparent dissociation constant is a maximum and is equal to the actual dissociation constant (K_{XN}). The apparent constant is half the actual constant when $[S]_{cis} = K_{(XN)S}$ and approaches zero as $[S]_{cis} \rightarrow \infty$, i.e., the apparent affinity of the carrier for N can be increased to any value by a sufficient increase in the concentration of S. Figure 26b shows that the data of Fig. 26a can be replotted to yield a straight line. K_{XN} and $K_{(XN)S}$ can be evaluated from the intercepts and compared with values obtained from a plot similar to Fig. 25b. The comparison of dissociation constants is a strong test for this co-transport model since the constants are obtained from data on transport of S in Fig. 25b and on transport of N in Fig. 26b.

Arguments similar to those developed in the previous section (Section III, B, 3) can be given to show that a gradient of primary substrate, S, can support level flow of the secondary substrate, N, and can maintain a gradient of N. Since these cases are not pertinent to most biological examples of co-transport, they are not developed in this section.

5. Hetero-cis-stimulation

The kinetic equations for co-transport [Eqs. (259–263)] show that S and N have a hetero-*cis*-stimulatory effect* upon each other. Hetero-*cis*-stimulation of the movement of S by N becomes more apparent when

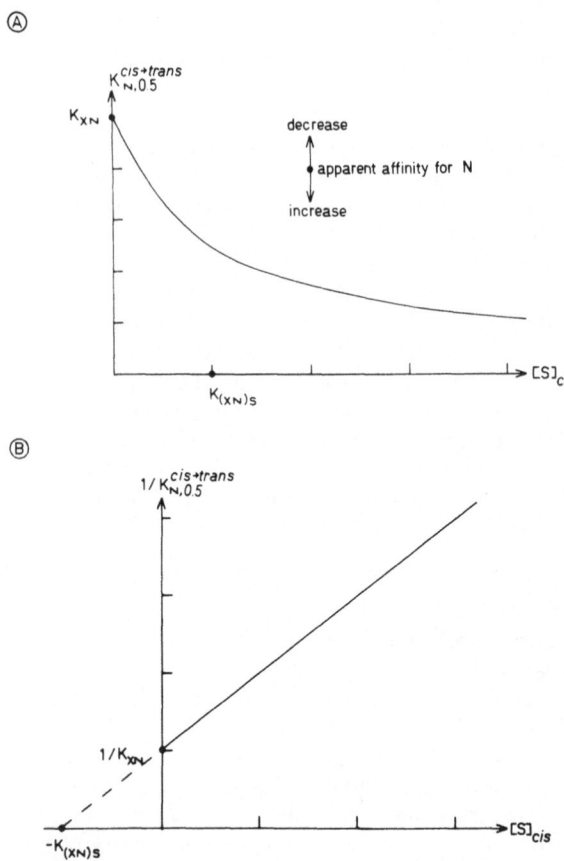

Fig. 26. Effect of the primary substrate, S, upon the apparent dissociation constant of the carrier for the secondary substrate, N. Even in the absence of S, the carrier shows an affinity of $1/K_{XN}$ for N. Apparent affinity for N can be increased to any level desired by a sufficient increase in concentration of S. This figure applies to the co-transport scheme shown in Fig. 24 when all diffusion constants are equal.

$J_S^{cis \to trans}$ is plotted as a function of $[N]_{cis}$ for various values of $[S]_{cis}$, as is shown in Fig. 27a. Note that all increases in $[N]_{cis}$ when $[S]_{cis}$ remains constant result in an increase in $J_S^{cis \to trans}$; this is hetero-*cis*-stimulation. Figure 27b shows that the data of Fig. 27a can be replotted to yield a straight line.

* In Section II,C,1, a modification of clearance rather than flux is used as the criterion for stimulation or inhibition. For hetero effects, modification of the flux is equivalent to modification of the clearance inasmuch as flux and clearance are related by the homo-*cis*-concentration, which is maintained constant. During hetero-effect studies only hetero-*cis*-concentration is varied.

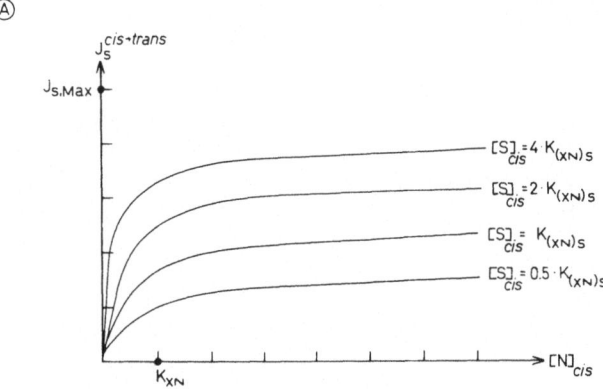

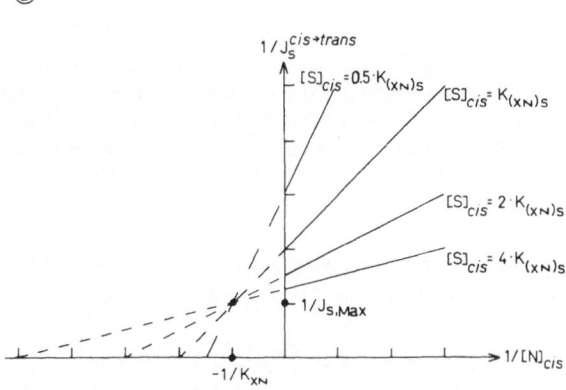

Fig. 27. Unidirectional flux of primary substrate (S) as a function of *cis*-concentration of secondary substrate (N) for various constant *cis*-concentrations of S. This figure applies to the co-transport scheme shown in Fig. 24 when all diffusion constants are equal. There is no unidirectional flux of S when N is absent on the *cis*-side. Hetero-*cis*-stimulation of S movement by N is manifest.

Hetero-*cis*-stimulation of the movement of N by S becomes apparent when $J_N^{cis \to trans}$ is plotted as a function of $[S]_{cis}$ for various values of $[N]_{cis}$. Figure 28 is similar to Fig. 27a except that the roles of S and N are reversed.

6. *Relation Between the Fluxes of S and N*

The ratio of the flux of S to the flux of N can be obtained from Eqs. (259–261) as follows:

$$J_S^{cis \to trans} / J_N^{cis \to trans}$$

$$= [S]_{cis} \times ([N]_{cis} + K_{N,0.5}^{cis \to trans}) / \{[N]_{cis} \times ([S]_{cis} + K_{S,0.5}^{cis \to trans})\} \qquad (280)$$

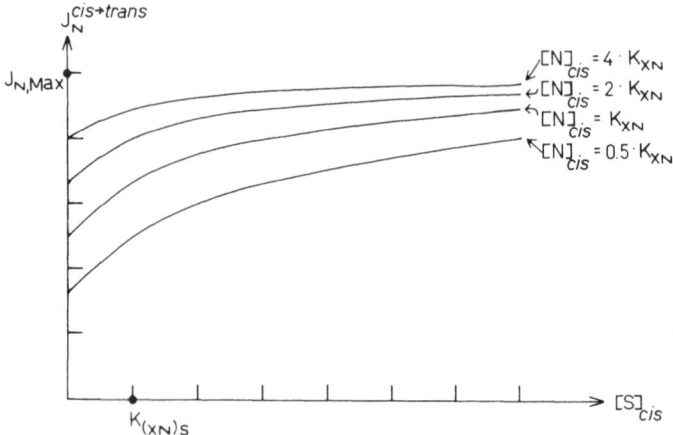

Fig. 28. Unidirectional flux of secondary substrate (N) as a function of *cis*-concentration of primary substrate (S) for various constant *cis*-concentrations of N. This figure applies to the co-transport scheme shown in Fig. 24 when all diffusion constants are equal. There is unidirectional flux of N even when S is absent on the *cis*-side. Hetero-*cis*-stimulation of N movement by S is manifest.

After algebraic manipulation, following substitution of Eqs. (262) and (263) for the apparent dissociation constants, Eq. (280) yields

$$J_S^{cis \to trans}/J_N^{cis \to trans} = [S]_{cis}/([S]_{cis} + K_{(XN)S}) \qquad (281)$$

Equation (281) is plotted in Fig. 29a. Note that the ratio of fluxes assumes the form of a simple rectangular hyperbola which is the same form as the Michaelis–Menten function. Also note that the flux of the primary substrate, S, can approach—but never exceed— the flux of the secondary substrate, N. Furthermore, the ratio depends only upon $[S]_{cis}$ and the dissociation constant between XN and S. Figure 29b shows that the data of Fig. 29a can be replotted to yield a straight line. The intercept can be used to evaluate $K_{(XN)S}$.

C. Transmembrane Gradients Resulting from Metabolic Sources and Sinks

A metabolic reaction can serve as either the *source* of or the *sink* for a substrate. In Fig. 30(a) the reaction on side 1 serves as a source of substrate N and the reaction on side 2 serves as a sink for N. On side 1, reactant R is metabolized to form product P and substrate N, while on side 2, N reacts with reactant R' to form product P'. A transmembrane gradient of N is generated because the metabolic reaction on side 1 creates a high activity

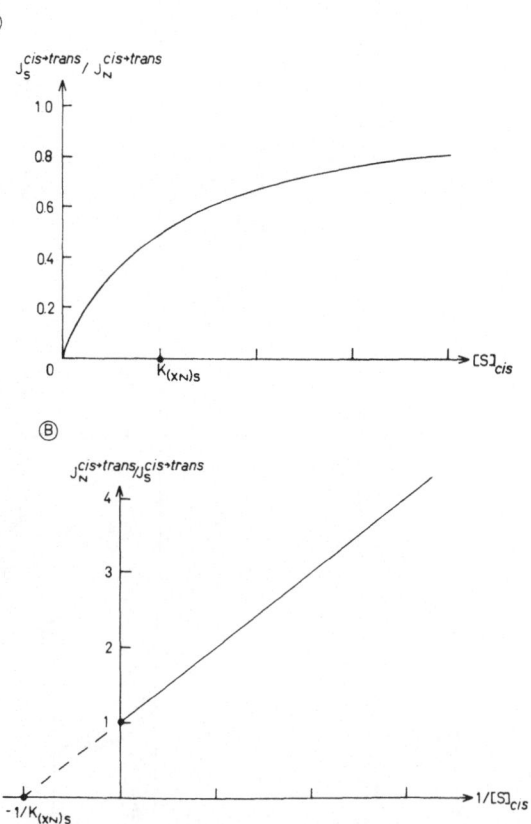

Fig. 29. The ratio of unidirectional flux of primary substrate (S) to that of secondary substrate (N) as a function of *cis*-concentration of S. This figure applies to the co-transport scheme shown in Fig. 24 when all diffusion constants are equal. This ratio is only a function of *cis*-concentration of S. Unidirectional flux of S may approach—but not exceed—that of N.

of N on side 1 and the reaction of side 2 creates a low activity of N on side 2. Only one of the two metabolic reactions, however, is necessary in order to create a transmembrane gradient of N. The gradient of N can result in the downhill movement of N from side 1 to 2. N may drive another substrate S in the same direction (co-transport or symport) or an analogue N′ in the opposite direction (counter-transport or antiport).* The energy for the movement of S or N′ is derived indirectly from the energy expended by the metabolic reactions.

* Mitchell[31,32,66] suggests that co-transport be called symport and counter-transport, antiport.

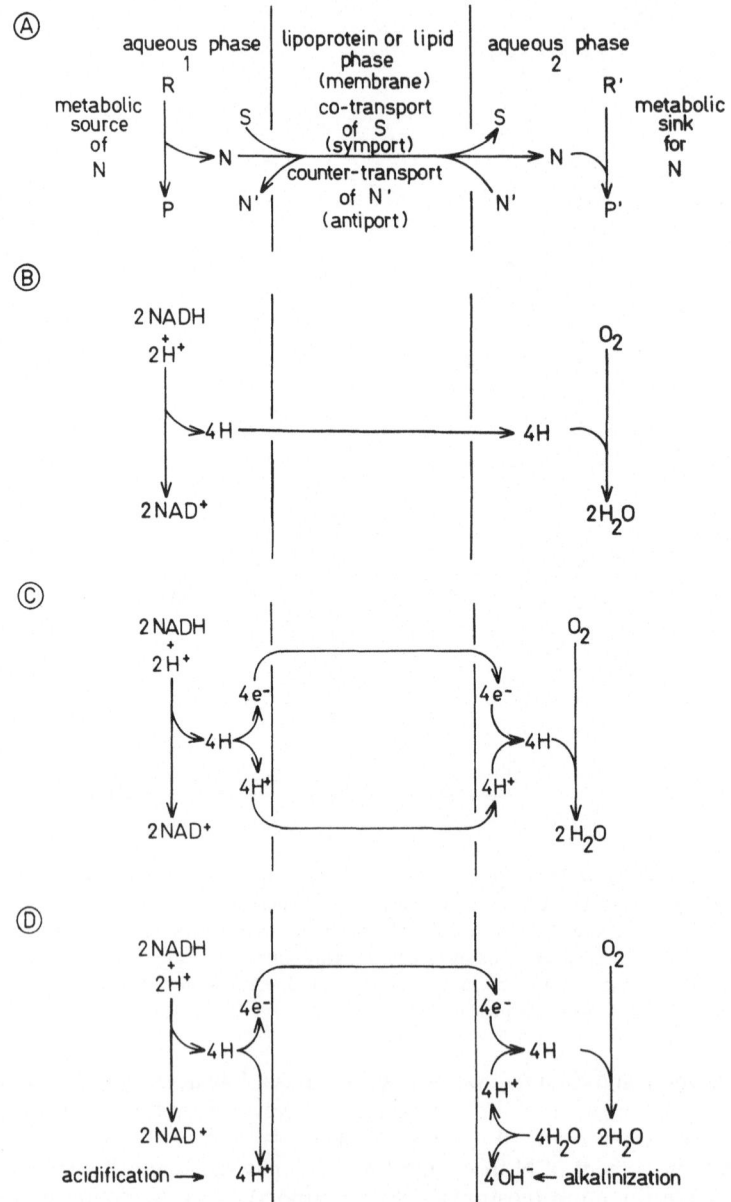

Fig. 30. Transmembrane movement as a result of a metabolic source of or sink for substrate. Panel a shows the general scheme. Panels b–d show redox schemes as special cases of the general scheme. In panel b, the reductant H moves across the membrane. In panel c, the reductant H is transported by the separate movements of H^+ and e^-. In panel d, elimination of the path for movement of H^+ results in acidification on side 1 and in alkalinization on side 2.

An example of metabolic generation of a transmembrane gradient is a redox scheme in which certain metabolites are oxidized on one side of a membrane and other metabolites are reduced on the other side. The reactions on the two sides are coupled by the reductant which is transported from the side of oxidation to the side of reduction. Figure 30(b) shows a redox scheme based on models proposed by E. E. Crane and Davies,[170–172] E. E. Crane, Davies, and Longmuir,[173,174] Davies and Ogston,[175] R. N. Robertson and Wilkins,[176–177] and Conway and Brady[178] and Conway.[179–182] On side 1, NADH is oxidized to form NAD^+ by the removal of atomic hydrogen, H, which is transported across the membrane to side 2. The movement of H may be coupled with the flows of other substrates. On side 2, H reduces O_2 to form H_2O. In Fig. 30(c), atomic hydrogen is transported by the movement of an electron (e^-) in one path and of a proton (H^+) in a separate path. The movements of e^- and H^+ may be coupled with the flows of other substrates. In Fig. 30(d), the transport path for H^+ is eliminated, thereby resulting in a buildup of H^+ on side 1, i.e., aqueous phase 1 is acidified. On side 2, H^+ is required to react with e^- to form reductant, H, which is needed to reduce O_2 to form H_2O. In deriving the required H^+ from H_2O, OH^- is released on side 2, i.e., aqueous phase 2 is alkalinized.

IV. PSEUDO-UPHILL TRANSPORT

A. Introduction

Great care must be exercised in concluding that a given substance moves uphill or downhill across a membrane. Difficulty arises in the following situation: a particular substance is rapidly (and either reversibly or irreversibly) converted to other chemically distinct forms by chemical reactions in the two compartments separated by a membrane; at least one of the forms cannot penetrate the membrane; differences in poise of the chemical reactions in the two compartments result in an asymmetrical distribution of forms on the two sides; and the analytical method used to determine the concentration of the substance in question cannot distinguish between the chemically distinct forms. As the simplest example, consider the case where there are only two chemically distinct, non-ionic substances that can be interconverted: one form, S, readily penetrates the membrane either by simple diffusion or by a passive carrier mechanism and the other form, S′, does not penetrate the membrane. Let the concentration of penetrant, S, be greater on side 1 and the concentration of nonpenetrant, S′, be greater on side 2 (see Fig. 31); this difference in distribution of forms can arise because the poise of the interconversion reaction is different on the two sides. The net movement of S is downhill from side 1 to 2; there is no net movement of S′. If the chemical or isotopic method used to detect flux cannot distinguish between S and S′, a combination, Σ, of S and S′ is measured. It may be difficult to resolve Σ into its penetrating and nonpenetrating components. If the combination of the measurable combination Σ is greater on

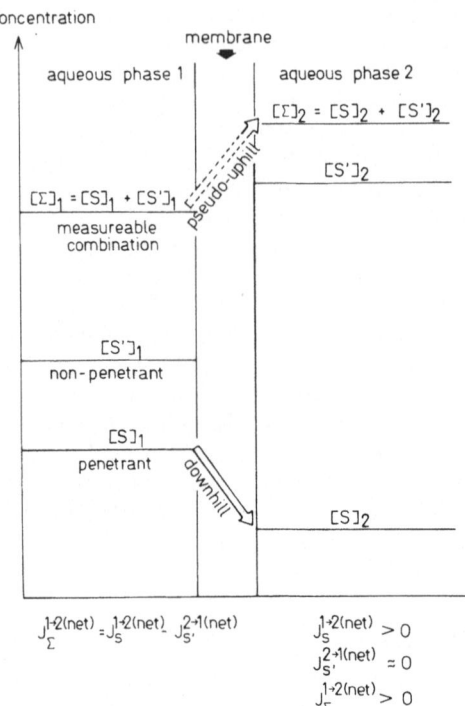

Fig. 31. Pseudo-uphill movement of a substrate. The membrane is permeable only to S which moves downhill from side 1 to 2 and is trapped in the form of S' on side 2. Total substrate ($\Sigma = S + S'$) appears to move uphill from side 1 to 2.

side 2 than on side 1, then Σ appears to move uphill from side 1 to 2 while, in fact, the penetrating form, S, moves downhill from side 1 to 2. (The flux measurement cannot determine the form in which substrate moves.) Since Σ is not a distinct chemical species, an electrochemical potential cannot be defined for it; therefore, the terms uphill and downhill cannot be applied to describe the movement of Σ.

This scheme of apparent uphill movement (pseudo-uphill movement) of Σ is termed "trapping" because Σ is trapped on side 2 in the form of nonpenetrant, S'. "Trapping" schemes were originally suggested by Jacobs[183-185] and later developed by Osterhout,[186] Orloff and Berliner,[187,188] Schwartz et al.,[189] and Schwartz.[190]

B. Modes of Trapping

1. *Group Attachment*

In group attachment, a nonpenetrating group P, such as phosphate, is attached to the penetrant S to form a nonpenetrant, SP. Group attachment may occur by either of two modes: one in which group attachment occurs in one of the aqueous phases and the other in which attachment occurs in the membrane.

In Fig. 32(a), S moves in step 1 from the left side to the right side, where the group donor DP reacts with S in step 2 to form the nonpenetrant SP (see step 3). Translocation of S across the membrane in step 1 may be either by simple diffusion or by carrier mediation.

In Fig. 32(b), translocation must be carrier-mediated because the carrier plays an important role in the group attachment. In this scheme, S enters from the left side in step 1 to form a carrier-substrate complex, XS, which reacts with group donor, DP, in step 2 to form a carrier-substrate-group complex, XSP, which releases the complex, SP, into the aqueous phase on the right in step 3. In this scheme, the two interfaces of the membrane must exhibit different permeabilities to SP. The interface on the right must be permeable to SP, so that it can be released from the carrier into the aqueous phase on the right. On the other hand, the interface on the left must be impermeable to SP. Otherwise, SP could escape into the left side and consequently would not be trapped on the right side.

In bacteria, monosaccharides are trapped within the cell in the form of monosaccharide phosphates which cannot escape.[64] In this case, the attached group is phosphate, the donor group is phosphorylated histidine-containing protein (phospho-HPr), and transfer of the phosphate from the protein to the monosaccharide occurs within the membrane in a manner resembling that shown in Fig. 32(b).

In the examples shown in Figs. 32(a) and 32(b), attachment of the group P to the penetrant S results in the formation of nonpenetrant SP. It is also possible to devise a scheme in which attachment of the group P results in the conversion of a nonpenetrant S into a penetrant SP.

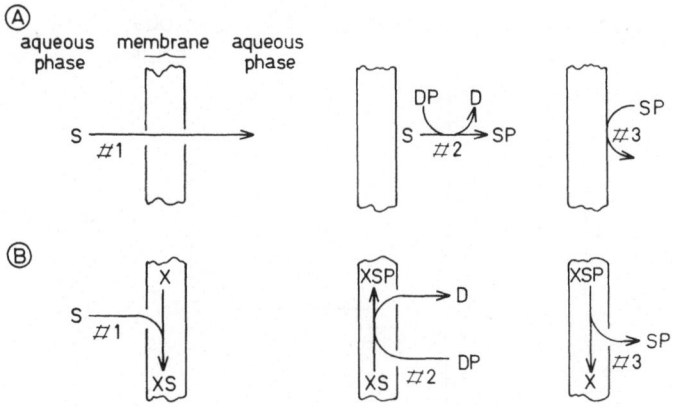

Fig. 32. Modes of group attachment to a substrate. In panel a, carrier is not necessary for group attachment; in panel b, the carrier plays an integral role in the group attachment.

2. Hydrogen Ion Attachment

The schemes shown in Figs. 32(a) and 32(b) can be modified by allowing the group donor, DP, to be identical with the attached group, P. An example of this case is where hydrogen ion serves as both the group donor and attached group. Reaction of substrate with hydrogen ion results in a change in the net charge on the substrate. For example, an anionic substrate may react with hydrogen ion to form a neutral substrate:

$$S^- + H^+ \rightleftharpoons SH \tag{282}$$

or a neutral substrate may react with hydrogen ion to form a cationic substrate:

$$S + H^+ \rightleftharpoons SH^+ \tag{283}$$

Some biological membranes are permeable to the uncharged (neutral or zwitterionic) form of a substrate but not to the charged (anionic or cationic) form and, in such situations, a transmembrane pH gradient leads to the concentration of substrate on one side.

Figure 33 shows a membrane which is permeable to SH but not to S^-. In this case, the attachment of the group H^+ results in the conversion of the nonpenetrant S^- into the penetrant SH. If SH, S^-, and H^+ are in equilibrium in the aqueous phases, their concentrations are related by an equilibrium constant as follows:

$$[S]_i \times [H^+]_i/[SH]_i = K_{SH} \tag{284}$$

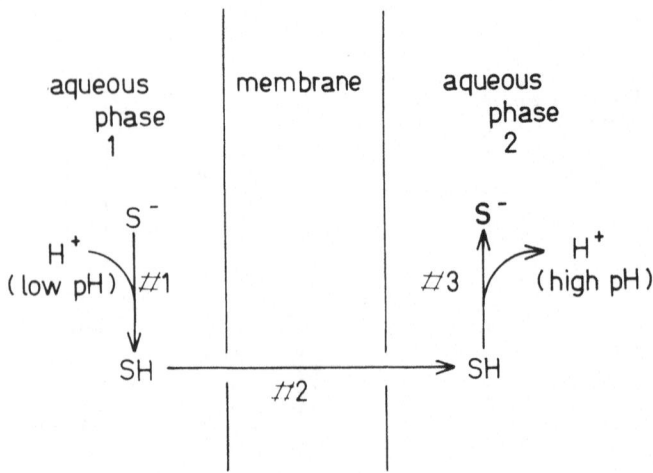

Fig. 33. Trapping of an anionic substrate by a transmembrane difference in hydrogen ion concentration. The membrane is permeable only to the electrically neutral form of substrate (SH).

where i represents either side 1 or 2 and K_{SH} is the dissociation constant for SH.

It is possible to establish and maintain a gradient of substrate across the membrane with a pH gradient. A gradient of substrate is maintained when there is no net flow of substrate across the membrane. Since only the uncharged form, SH, penetrates the membrane, and this penetration is passive, net flux is zero when

$$[SH]_1 = [SH]_2 \tag{285}$$

The measurable concentration is the sum of the concentrations of S^- and SH. i.e.,

$$[\Sigma]_i = [S^-]_i + [SH]_i \tag{286}$$

From Eqs. (284–286), it follows that when net flux is zero,

$$[\Sigma]_2/[\Sigma]_1 = (1 + K_{SH}/[H^+]_2)/(1 + K_{SH}/[H^+]_1) \tag{287}$$

At very low pH's (pH$_1$, pH$_2 \ll pK_{SH}$),

$$[H^+]_1, [H^+]_2 \gg K_{SH} \tag{288}$$

In this case

$$[\Sigma]_2/[\Sigma]_1 \approx 1, \tag{289}$$

i.e., no gradient is established. At very high pH's (pH$_1$, pH$_2 \gg pK_{SH}$),

$$[H^+]_1, [H^+]_2 \ll K_{SH} \tag{290}$$

In this case

$$[\Sigma]_2/[\Sigma]_1 \approx (K_{SH}/[H^+]_2)/(K_{SH}/[H^+]_1) \tag{291a}$$

$$\approx [H^+]_1/[H^+]_2 \tag{291b}$$

Thus, substrate can be maintained at a concentration ratio almost equal to that of hydrogen ion, but with the opposite orientation—i.e., the concentration of substrate is higher on the side where the concentration of hydrogen ion is lower.

Figure 34 shows a membrane which is permeable to S but not to SH$^+$. If SH$^+$, S, and H$^+$ are in equilibrium in the aqueous phases, their concentrations are related by an equilibrium constant as follows:

$$[S]_i \times [H^+]_i/[SH^+]_i = K_{SH^+} \tag{292}$$

Net flow of substrate is zero when

$$[S]_1 = [S]_2 \tag{293}$$

The measurable concentration is given by

$$[\Sigma]_i = [S]_i + [SH^+]_i \tag{294}$$

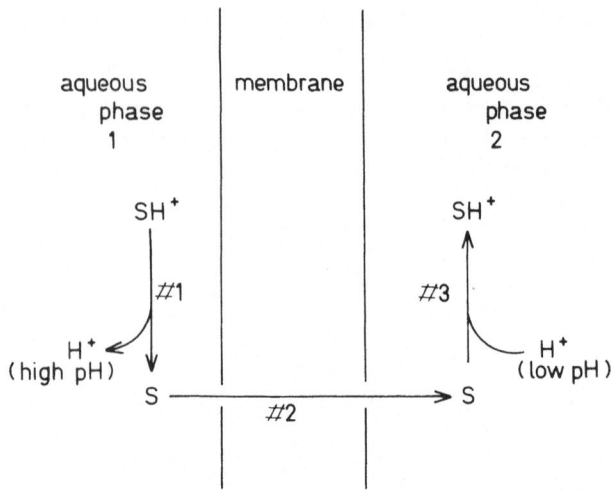

Fig. 34. Trapping of a cationic substrate by a transmembrane difference in hydrogen ion concentration. The membrane is permeable only to the electrically neutral form of substrate (S).

From Eqs. (292–294), it follows that when net flux is zero,

$$[\Sigma]_2/[\Sigma]_1 = (1 + [H^+]_2/K_{SH^+})/(1 + [H^+]_1/K_{SH^+}) \tag{295}$$

At very high pH's (pH_1, $pH_2 \gg pK_{SH^+}$),

$$[H^+]_1, [H^+]_2 \ll K_{SH^+} \tag{296}$$

In this case

$$[\Sigma]_2/[\Sigma]_1 \approx 1 \tag{297}$$

i.e., no gradient is established.

At very low pH's (pH_1, $pH_2 \ll pK_{SH^+}$),

$$[H^+]_1, [H^+]_2 \gg K_{SH^+} \tag{298}$$

In this case

$$[\Sigma]_2/[\Sigma]_1 \approx ([H^+]_2/K_{SH^+})/([H^+]_1/K_{SH^+}) \tag{299a}$$

$$\approx [H^+]_2/[H^+]_1 \tag{299b}$$

Thus, substrate is maintained at a concentration ratio almost equal to that of hydrogen ion, with the same orientation—i.e., the concentrations of substrate and hydrogen are both higher on the same side of the membrane.

Note that in this trapping scheme, a carrier is not required in order to concentrate substrate. The free energy required to concentrate substrate in its nonpenetrating form is derived from the free energy associated with the transmembrane gradient of hydrogen ion.

V. MOLECULAR DESCRIPTION OF CARRIER-MEDIATED TRANSPORT SYSTEMS

A. Introduction

The primary effort of most laboratories studying carrier-mediated transport processes has been directed toward describing the kinetic characteristics of the movement of various substrates through cell membranes as reviewed and further developed in the preceding sections. The kinetic approach has been invaluable for the conceptual formulation of descriptive models of carrier-mediated transport systems. But these models, regardless of their accuracy in fitting and predicting the kinetic data, do not resolve the actual mechanisms of carrier-mediated transport. Ideally, one would like to identify and describe at the molecular level all the interactions responsible for: (a) the identification of substrates by the carrier, (b) substrate penetration of the membrane, (c) the release of substrate following translocation, and (d) the coupling of metabolic processes to supply free energy for active transport. Characterization of this series of events will require at least the isolation and detailed description of the cellular macromolecules responsible for carrier-mediated transport.

As an initial approach to this problem, a number of laboratories have begun isolating cellular proteins that have been indirectly identified as having some role in transport. The isolation of these moieties from the cells or tissues usually presents a troublesome paradox because the techniques used result in the disruption of the cell. This, of course, destroys the unique characteristic by which one might identify a transport system—its capacity to facilitate the movement of substrate across the intact plasma membrane.

Purification of several small molecular weight "soluble" proteins has been achieved and studies of their physicochemical characteristics are proceeding rapidly. The more formidable problem of identifying and purifying those constituents of the plasma membrane that are involved in transport has been less yielding, but considerable progress is being made.

Just as the advances in our understanding of the phenomenological aspects of carrier-mediated transport were dependent upon the development of radioactive and electrophysiological technologies, so future progress in determining the molecular mechanisms of transport seems dependent upon the use of new approaches. These would include techniques of proven utility in mechanistic studies of conventional enzyme systems, as well as the development of new techniques for characterizing the plasma membrane.

In this section we will describe some of the macromolecules that have been implicated in carrier systems up until the present time. No attempt is made to review exhaustively all the recent findings in this area. We propose to outline some of the methodology that is being developed and, recognizing that the full potential of many of these techniques is not yet realized, we will take the liberty of speculating about their future application as this line of investigation develops.

B. Characteristics of Transport Macromolecules

Considering the diversity of cells used in studies of transport, the variety of substrates transported, and the marked substrate specificity of most carrier-mediated systems, it is clear that there will be considerable variation in the detailed structure of different carrier-mediated transport systems. As an initial departure, however, one may make certain general assumptions about the nature of some of the components of carrier-mediated transport systems.

1. The Plasma Membrane is the Major Barrier to Diffusion

Because the cell membrane is the major barrier to dissipative processes, some of the components of the carrier system must be bound to the plasma membrane, either by covalent bonding or by weak interactions such as hydrogen bonding, ionic bonding, or van der Waals forces. Further evidence for the close association of transport systems with cell membranes has been obtained using cells whose major metabolic processes have been irreversibly inhibited or whose intracellular contents have been removed by mechanical means. Experiments by Gardos[191] with erythrocyte ghosts showed that these membranes actively transported potassium when ATP was added to the interior of the cell. Carrier-mediated passive transport of sugars has been extensively studied in erythrocyte ghosts.[192] Keynes[193] found that the cyanide-treated squid axon was capable of transporting sodium against an electrochemical gradient if an appropriate energy source (either ATP or arginine phosphate) were injected into the inside of the axon. Vesiculated fragments of sacroplasmic reticulum actively accumulate calcium ions coincident with the hydrolysis of added ATP.[194]

2. Specificity for Substrate and for Inhibitors

A high degree of specificity of the transport system for its substrate is a characteristic of almost all carrier-mediated systems. Therefore, at least one component of any isolated transport system should have the property of discriminating between the substrate and other closely related chemical species. Systems transporting sugars and amino acids have been used most extensively for studies of carrier specificity because it is possible to obtain a large number of compounds that are structural analogues of the normal substrate. LeFevre[192] has reviewed in detail the substrate requirements for facilitated diffusion in the human erythrocyte. In order for a sugar to be transported by this system it must: (a) have a D-configuration, (b) form a pyranose ring, (c) possess a 2-OH group, (d) not have any large groups added anywhere on the molecule. The affinity of the system for a substrate fulfilling these basic requirements then varies according to the stereochemical configuration of the sugar. For example, the $K_{S,0.5}$'s for the "chair" and "boat" configurations of D-glucose differ by approximately three orders of magnitude (0.006–3 M).

The work of Christensen[195] and others has demonstrated that amino acid transport in animal cells is distributed (with some overlapping) among three transport systems, each with different affinities for the natural amino acids. One system, designated the L-site, primarily transports leucine, valine, and methionine. A second system, the A-site, has a higher specificity for alanine, glycine, methionine, and the unnatural amino acid α-aminoisobutyric acid. A third site has a high affinity for β-alanine and taurine.

Although this characteristic of substrate specificity imposes a stringent requirement on one component of the system, it also gives the researcher a useful device by which subcellular fractions may be identified as being one of the components of a transport system.

3. Molecular Size Considerations

Most of the kinetic models designed to fit carrier-mediated transport data require that one component of the system have a high mobility within the membrane. Whether this presumed mobility is rotational or translational, the assumption speaks for this component having a relatively low molecular weight. The feasibility of a small molecule acting as an intramembrane carrier has been proven by the finding that valinomycin (mol. wt. = 1100) and other macrocyclic antibiotics synthesized by *Streptomyces* accelerate the movement of potassium through natural and artificial membranes.[67] These compounds also have a striking selectivity for potassium in the presence of sodium and other alkali metal ions. The molecular weights of cellular constituents so far directly implicated in transport range from 9500 for the "HPr" protein of *E. Coli*[64] to 28,000 for the calcium-binding protein of vertebrate intestinal mucosa[196] to 670,000 for the sodium potassium dependent adenosine triphosphatase.[197] This wide diversity in size presumably reflects the functional differences between the different components in their respective transport systems.

4. Is the Carrier a Protein?

Cellular plasma membranes are composed almost exclusively of lipid and protein. Phospholipids were implicated in early studies of cellular constituents in ion transport.[198] Solomon, Lionetti, and Curran[199] discovered several phospholipids in blood serum that preferentially bound potassium relative to sodium. Hokin, who has extensively studied the relation of phosphorylation of membrane phospholipids to ion transport, postulated in 1963[200] that a lipoprotein serves as a transducer to convert the free energy released by ATP hydrolysis to the substrate being transported. Although the idea that phosphatides are critical constituents of carrier systems has not recently been as actively pursued as it was several years ago, and the current focus of attention is upon proteins, the role of lipoproteins and phospholipids has not been resolved. The evidence that the Na^+-K^+-dependent adenosine triphosphatase is a lipoprotein is not inconsistent with such a transducer function.[197,201] The more recent studies of transport macro-molecules, however, have produced a great body of evidence indicating that the major constituents of carrier-mediated systems are proteins.

a. Selective Chemical Reagents. The inhibition of carrier-mediated transport by a number of reagents thought to react only with proteins has lent considerable support to the concept of a protein carrier molecule. Dinitrofluorobenzene (DNFB), introduced by Sanger for the identification of N-terminal amino acids, inhibits the transfer of glucose and glycerol in the erythrocyte ghost.[202] Stein[203] found that DNFB reacts with N-terminal histidine residues in the course of inhibiting the glycerol transport system of the erythrocyte. To determine whether the histidine residue was part of the transport active site, Stein used a method of differential covalent labeling similar to that subsequently proposed by Koshland et al. as a general method for the labeling of enzyme and antibody active sites[204] (cf. Section V, C, 3, a). DNFB was also found to inhibit glucose binding by isolated erythrocyte membranes and the kinetics of this inhibition resembled those of the DNFB-inhibition of sugar transport in the intact cell.[205]

N-Ethylmaleimide, a sulfhydryl reagent, has been shown to be a potent inhibitor of glucose transport in the erythrocyte and of β-galactoside transport in E. coli.[206] VanSteveninck et al., using sulfhydryl reagents that only poorly penetrate the erythrocyte membrane (p-chloromercuribenzene-sulfonate and chloromerodrine), found that certain sulfhydryl groups on the external surface of the membrane were essential for the glucose transport system.[207] Stein has tabulated the effects of these and other chemical reagents upon various transport systems.[208]

b. Molecular Architecture to Confer Specificity. Plasma membranes exhibit carrier-mediated transport for a wide range of cations, anions, sugars, amino acids, and, in some cases, proteins. The fact that each transport system is able to discriminate between the transport substrate and its structurally related analogues indicates that some component of each of these systems must possess an active site whose three-dimensional structure permits the proper fit of only the transported species. Solomon, Lionetti, and Curran[199] found that a number of phospholipids preferentially bound potassium relative to sodium. LeFevre et al.[209] studied the binding of erythrocyte membrane fractions and found a phospholipid whose affinity for different sugars corresponded to the affinity of the transport system. But, it is now generally considered that only proteins (or peptides) are able to assume the variety of configurations necessary to satisfy in one membrane all the structural configurations necessary for the recognition of all the transported species. For instance, even the glucose-binding phospholipid isolated from erythrocyte membranes did not distinguish between optical enantiomorphs of glucose.[209] More recent studies of subcellular fractions have discovered a number of small molecular weight proteins whose substrate-binding characteristics closely resemble those of the intact transport systems, even to the extent of discriminating between optical enantiomorphs. The evidence that the proteins are involved in transport is discussed in Section V, D.

c. Transport Processes are Related to Synthesis of Protein.   The active transport of several substrates has been directly related to the synthesis of protein by the transporting cells. Gross and Ring[210] studied the accumulation of α-aminoisobutyric acid (AIB), a nonmetabolizable amino acid, in *Streptomyces*. Treatment of the cells with puromycin, an aminoglycoside inhibitor of protein synthesis, caused a marked reduction in the rate of uptake of AIB by the cells. The same results were obtained when the cells were preloaded with AIB prior to the uptake studies with radioactive AIB. Elsas and Rosenberg[211] studied the effects of puromycin upon the accumulation of AIB by slices of rat kidney cortex. The influx, but not the efflux, of labeled amino acid was diminished after about 120 min of exposure to puromycin. Because there was no alteration in oxygen consumption, ion content, or $K_{S,0.5}$ for the amino acid, the authors postulated that puromycin inhibited the synthesis of a rapidly turning-over transport protein or peptide.

A series of studies from Edelman's laboratory gives evidence that the enhancement by mineralocorticord of sodium transport is dependent upon the synthesis of RNA and protein. Edelman, Bogoroch and Porter[212] used the isolated urinary bladder of the toad to examine the effects of aldosterone upon the active transport of sodium. The bladder was incubated 14 hr to allow for degradation of endogenous aldosterone, and the net sodium transport measured by means of the short-circuit current (SCC). Treatment of the bladder with aldosterone caused, after a 2-hr lag, a sustained increase in sodium transport (SCC) until the rate was three times the pretreatment value. This enhancement of transport could be completely blocked by treating the bladder with either actinomycin D or puromycin prior to aldosterone. Further experiments showed that removal of the added aldosterone after 30 min of exposure did not interrupt the sequence of events leading to the enhancement of sodium transport, and autoradiography of tissue exposed to tritiated aldosterone showed that the hormone was largely localized in the nucleus of the mucosal cells. Because aldosterone and exogenous pyruvate had a synergistic effect upon sodium transport, the authors speculated that the hormone induced the synthesis of enzymes involved in the oxidation of pyruvate thereby enhancing the supply of metabolic energy for transport.

This group also found that aldosterone increased the incorporation of uridine into nuclear RNA, and that this increase in RNA synthesis preceded the enhancement of sodium transport.[213] These data may be interpreted as demonstrating that aldosterone acts upon nuclear DNA to promote the synthesis of a messenger-RNA which in turn promotes the synthesis of a protein related to transport; perhaps a protein coupling metabolism to active transport.

Edelman's group used the kidney of adrenalectomized rats to identify and characterize the nuclear receptor for aldosterone. By fractionating the kidney following injection with tritiated aldosterone, they found that the nuclear binding of the hormone was half-saturated at a plasma concentration (K_m) of 6×10^{-9}.[214] The nuclear aldosterone-binding macromolecules were identified as proteins by chemical analysis, by their susceptibility to proteolytic enzymes, and by their sensitivity to sulfhydryl reagents.[215] The protein competitively bound a series of other steroids in direct relation to their effectiveness *in vivo*. Binding was competitively inhibited by the potent mineralocorticord 9α-fluorocortisol, but the physiologically inactive isomer of aldosterone, 17-isoaldosterone, had no effect. Studies of the effects of aldosterone upon the rat kidney showed that this tissue, like the toad bladder, responded by increasing its synthesis of RNA prior to an increase in sodium reabsorption.[216] It was also possible to demonstrate an increased protein synthetic rate following the enhanced RNA synthesis. These experiments provide good evidence that aldosterone caused the DNA-directed synthesis of an RNA that served as template for synthesis of transport-related protein(s).

Studies relating the synthesis of a specific protein in the intestinal mucosa to the active transport of calcium are presented in Section V,D,2.

Puromycin inhibits the absorption of triglycerides by the rat intestine. Normally, triglycerides accumulate in the mucosal cells of the intestine where they are incorporated into chylomicrons, complex structures with a central core of triglycerides and other lipids bounded by a membrane, probably of β-lipoproteins. Following their formation in the mucosal cell, the chylomicrons are secreted into the intestinal lymphatic circulation. Sabesin and Isselbacher[217] showed that treatment of rats with puromycin, although it apparently did not inhibit the movement of triglycerides across the mucosal membrane, effectively blocked the formation of chylomicrons and the subsequent passage of the triglycerides into the lymphatic system. A similar effect was observed with acetoxycycloheximide. The postulated role of protein in the transport of lipids is supported by the observation that the hereditary deficiency of β-lipoproteins is accompanied by a similar block in triglyceride transport by the intestinal mucosa.[218]

Lubin and Ennis[219] demonstrated in *E. coli* mutants that the rate of protein synthesis is related to the level of intracellular potassium. By using mutants that were unable to transport potassium against an electrochemical gradient, the intracellular potassium concentration was altered at will merely by adjusting the potassium concentration in the culture medium. They demonstrated that the rate of protein synthesis was directly related to the intracellular potassium concentration. To test the same mechanism in mammalian cells where no such mutants were available, Lubin treated *Sarcoma-180* cells with graded doses of amphotericin B, which increased the efflux of potassium from the cells by altering membrane permeability.[220] The intracellular potassium concentration could once again be adjusted to the desired level by the degree of exposure to amphotericin B. Lubin found that the rate of protein synthesis was increased sixfold by raising the concentration of potassium in the medium from 5 to 100 mM. These experiments suggest that the synthesis of protein by the cell and the maintenance by active transport of a potassium-rich internal milieu are linked.

5. Architecture of the Carrier System within the Membrane

a. Mobile Carrier. The model generally accepted for carrier-mediated transport considers that at some point in the transfer of material, substrate is bound to a membrane macromolecule that exhibits mobility. Most models have imparted rotational or translational motion to a membrane carrier macromolecule, but more recent speculations have included allosteric transitions in a membrane enzyme[221,228,124,128] (see Section II, D). The mobile carrier model does not specify the site of initial substrate recognition nor is the mechanism for release of substrate confined to the carrier molecule. While the implicit assumption is often made that the carrier site is alternately exposed to the media on each side of the lipid-protein membrane, another enzyme molecule may be interposed between the carrier site and the media on either side of the membrane (see components E_1 and E_2 in Fig. 3). Moreover, those (active) transport systems requiring an input of free energy (usually ATP) undoubtedly comprise macromolecules not found in passive systems.

In other words, the mobile carrier model for carrier-mediated transport neither specifies nor limits the number of components in the system. Indeed, it seems likely that a series of molecules coordinate the steps leading to (a) recognition of substrate, (b) translocation of substrate, (c) release of substrate, (d) restoration of system, (e) input of free energy (see Fig. 3). This means that in the actual isolation of any one carrier-mediated transport system, one surely would find a number of macromolecules diverse in structure and function rather than a single self-contained unit. This concept is supported by recent studies of the role of the "soluble" transport proteins in membrane translocation processes.

b. Pore Model. The concept of an aqueous channel through the hydrophobic lipid membrane was useful for some time in accounting for the passive movement of small molecules, especially uncharged species, through cell membranes. Accumulating data eventually required that the theoretic pore be modified in order to impart the necessary specificity to the membrane. Patlak[52] suggested that the entrance to the membrane pore is bounded by an enzyme-like protein which recognizes the proper substrate and allows this molecule to pass without further hindrance. Stein and Danielli[223] suggested that the membrane pores were lined with these proteins and they showed that such a pore could give saturation kinetics. But the counterflow of substrate as predicted by Widdas[47] could not be accounted for even by these modifications upon the pore model. The subsequent demonstration of the counterflow of glucose in erythrocytes[91] and of sugars in yeast cells[224,225] seemed to exclude even the most complex pore as a generalized model for carrier-mediated transport. The phenomenon of counterflow has now been reported in a score of transporting systems and the water-filled pore is not generally considered important in carrier-mediated transport.

c. Superficial Enzymes. In studying the hydrolysis and absorption of radioactively labeled sucrose by the hamster intestine, Miller and Crane[226] observed that the moiety of glucose-^{14}C, after hydrolysis of the sucrose, was transported preferentially to free glucose-^{14}C in the medium bathing the tissue. It was also found that the addition of glucose oxidase to the medium had a lesser effect upon the transport of the glucose generated by the hydrolysis of sucrose than upon the transport of free glucose in the medium. Crane[227] suggested that the enzymes hydrolyzing the disaccharides were incorporated into the matrix of the membrane in such a way that their hydrolytic products were presented to the carrier molecule at a kinetic advantage to the same substrate present in the bathing medium. Ugolev et al.[228] obtained similar data which they interpreted as indicating that the disaccharidase was located on the external surface of the cell membrane. In considering the mobility of the carrier molecule within the membrane, Crane[227] suggested that there was local thinning of the lipid membrane accompanying the hydrolysis of sucrose, and that the hydrolytic products were then released through this transient perforation in the lipid layer to the opposing membrane surface. When Marzluf and Metzenberg[229] studied the absorption of sucrose in Neurospora crassa, they were unable to find any evidence that the hydrolytic products of sucrose were transported preferentially with respect to the glucose and fructose in the surrounding medium.

C. Techniques for Identifying Proteins Involved in Transport

1. Substrate Binding by Cell Fractions

The most direct method used so far in the identification of subcellular components of transport systems is the binding of substrate by isolated macromolecules from the cell or tissue. The demonstration of ligand binding generally requires some means by which a solution containing the macromolecule in question is separated from a second solution containing only solvent and the free ligand. At equilibrium the concentration of the ligand in the compartment containing the macromolecule will be enhanced over that in the second compartment in accordance with the concentration of the macromolecule, the concentration of the ligand, the association constant, and the temperature. The primary difference in the various techniques lies in the means by which the two compartments are established.

a. Equilibrium Dialysis. Equilibrium dialysis, the technique most frequently used to study ligand binding, employs a semipermeable membrane to separate the compartment containing the macromolecules from that containing the free ligand.[230,231] The two solutions are stirred until the ligand has reached equilibrium across the membrane, and the concentration of *total* ligand is measured in each compartment. The binding is given by

$$[L]_i = [L]_0 + \bar{v}[P] \tag{300}$$

where $[L]_i$ is ligand concentration in the inner compartment containing the protein, $[L]_0$ the concentration of ligand in the outer compartment, $\bar{v}$ the number of moles of ligand bound per mole protein, and $[P]$ the concentration of protein.

To determine the characteristics of the binding, we consider a protein having only one binding site per molecule, where the reaction is

$$S + P \rightleftharpoons SP \tag{301}$$

and the association constant is given by

$$k_{assoc} = \frac{[SP]}{[S] \times [P]} \tag{302}$$

Considering all the molecules of S and P in solution, we denote the fraction of binding sites occupied by S as $\bar{v}$,

$$\bar{v} = \frac{[SP]}{[SP] + [P]} = \frac{k_{assoc}[S]}{1 + k_{assoc}[S]} \tag{303}$$

If, instead of one binding site per molecule, the protein contains n such binding sites which are independent and equivalent, Eq. (302) may be extended by denoting as SP^j a molecule of P binding a molecule of S at the jth binding site. Then

$$(k_{assoc})_j = \frac{[SP^j]}{[S]\{[P_{total}] - [SP^j]\}} \tag{304}$$

Because we have assumed that the sites are independent and therefore the extent of binding at other sites on P does not affect the binding of S by the jth site, then Eq. (304) describes the binding at site j, averaged over all the sites in the system, and allows one to assume that $\bar{v}_j$, at site j is given by

$$\bar{v}_j = \frac{(k_{assoc})_j[S]}{1 + (k_{assoc})_j[S]} \tag{305}$$

And because all n sites are independent and equal we may drop the j notation. The probability that any of the n sites, chosen at random, is occupied by the ligand, S, is given by

$$\frac{\bar{v}}{n} = \frac{k[S]}{1 + k[S]} \tag{306}$$

where we write k instead of k_{assoc}.

In the treatment of data, this equation is usually rearranged as

$$\bar{v} = \frac{nk[S]}{1 + k[S]} \tag{307}$$

or

$$\frac{1}{\bar{v}} = \frac{1}{n} + \frac{1}{nk[S]} \tag{308}$$

But Eq. (308) heavily weights those experimental points at low concentrations of ligand, and extrapolation of data to higher concentrations may lead to erroneous conclusions. Scatchard[232] rearranged Eq. (307) to

$$\frac{\bar{v}}{n - \bar{v}} = k[S] \tag{309}$$

Then multiplying through by $(n - \bar{v})[S]$, he obtained

$$\frac{\bar{v}}{[S]} = k(n - \bar{v}) \tag{310}$$

Plotting $\bar{v}/[S]$ against $\bar{v}$ gives a straight line if the binding sites are independent and homogeneous. The ordinal intercept is k, the association constant, and the intercept on the abscissa is n, the number of binding sites on each molecule of protein. Heterogeneous binding sites or interaction between binding sites, such as one observes in allosteric enzymes, give deviations from linearity.

 b. *Dynamic Dialysis*.   While the equilibrium dialysis method requires attainment of equilibrium by free ligand, a technique recently introduced by Meyer and Guttman[233,234] measures the binding of ligand in terms of the rate of loss of free ligand from a semipermeable bag containing the protein or other macromolecules.

To characterize the system in use, a measured amount of ligand is introduced into a dialysis bag and the bag immersed in a bath of sufficient size to maintain a sink for the ligand, with provisions for stirring. Sink conditions are maintained by refreshing the bath at frequent intervals and the total amount of escaped ligand in the bath is measured, allowing one to calculate the concentration of ligand within the bag at each period. The loss of ligand from the bag is given by

$$-d[S_t]/dt = k[S_f] \tag{311}$$

where $(-d[S_t]/dt)$ is rate of loss of ligand concentration, $[S_f]$ is the concentration of free ligand in the bag, and $[S_t]$ is the total concentration of ligand in the dialysis bag. The first-order rate constant characteristic of the operational system, k, is calculated from this data.

The dialysis bag is refilled with a solution containing the macromolecule of interest together with ligand and the rate of loss of ligand from the bag again measured. While semilogarithmic plots of the loss of ligand $(-d[S_t]/dt)$ are a linear function of time in the absence of ligand binding, the presence within the dialysis bag of a molecule that binds the ligand causes a deviation from linearity because the relative amount of bound to unbound ligand becomes greater as the concentration of unbound ligand diminishes. The concentration of unbound ligand, $[S_f]$, is calculated from Eq. (311) using the k determined earlier and the rate of loss $(-d[S_t]/dt)$ of ligand from the bag. This may be calculated for any total concentration of ligand $[S_t]$ by estimating $(-d[S_t]/dt)$ graphically. The amount of information obtained in one curve is therefore equivalent to a series of analyses by equilibrium dialyses at several concentrations of ligand.

c. *Ion-Exchange Resins.* Pardee[235,236] introduced a technique for the study of ion binding that allows very rapid equilibrium of ligand and macromolecule. An ion-exchange resin is equilibrated with the appropriate radioactive ion, and the macromolecule added to the mixture. The system is mixed for a few minutes to allow equilibration, the resin beads are allowed to settle, and the radioactivity in the supernatant fluid determined. The amount of ion binding is determined by the ability of the preparation to release ions from the resin into solution. This method was used by Pardee's group to identify and characterize a soluble sulfate-binding protein related to sulfate transport by *Salmonella typhimurium*. Wasserman, Corradino, and Taylor[237] used a similar technique for the assay of calcium-binding protein from intestine and for the determination of the affinity constants of the protein for calcium, strontium, and barium.

d. *Gel Filtration.* Gel filtration media were introduced for the chromatographic separation of molecules on the basis of molecular size.[238,239] This chromatographic technique utilizes the molecular sieving effects of highly cross-linked polymers when exposed to a mixture of molecular species in solution. The molecules in solution partition between the solvent external to the gel matrix and the solvent imbibed by the gel, depending upon the molecular diameter of the solute molecule. This partition may be described by a distribution coefficient,

$$K_d = C_i/C_o \qquad (312)$$

where C_i and C_o are the concentrations of a given solute within and without the gel matrix, respectively. Thus any molecule so large as to be completely excluded by the gel matrix is described by $K_d = 0$, and a small molecule freely penetrating the pores in the gel by $K_d = 1.0$. This consideration does not account for any nonspecific binding of the molecules by the gel, whereupon K_d may be greater than 1.0. Gel filtration chromatography has come into wide use both for the separation of macromolecules according to their molecular weights and for the estimation of the molecular weights of macromolecules.[240]

Hummel and Dreyer,[241] recognizing that the partition of molecules on the basis of molecular size gave another device for separating macromolecules and small molecular weight ligands, extended gel filtration to the study of binding phenomena. The chromatographic column of a material having significantly different K_d's for the ligand and macromolecule is equilibrated with the ligand of interest. The protein, in a solution whose *total* ligand concentration is the same as in the equilibration solution, is loaded on the column and the concentration of ligand measured in the chromatographic effluent. A precise analytical method for measuring ligand, such as spectrophotometry, determination of radioactivity, or a specific chemical reaction, is essential. The binding of ligand by eluted protein is represented by an increase in ligand concentration in the effluent fractions above the equilibration value. Because the larger binding molecule is moving through the relatively stationary ligand and is constantly in reversible equilibrium with the ligand, the peak of the ligand concentration, representing the increment of bound material over that in equilibration in the column, is followed by a trough. The area of the trough also represents the amount of ligand bound by the macromolecule.[242] An essential criterion for the attainment of equilibrium in the system is the return of ligand concentration to the baseline value following the trough.

By choosing gel filtration media that do not exclude all the macromolecules in solution, it is possible to determine whether more than one protein of *differing* molecular weight is binding ligand. By proper calibration of the column, it is also possible to estimate, in a mixture of proteins, the molecular weight of the macromolecule(s) binding the ligand. This technique also offers considerable advantages over equilibrium dialysis for studying competition between ligands for binding sites.[242]

Gel filtration media have also been used in batch and in zonal chromatographic methods for studying binding, though these techniques present a number of technical difficulties.[243]

e. Electron Paramagnetic Resonance. Electron paramagnetic resonance (or electron spin resonance) spectroscopy measures spectral transitions in the unpaired electrons of free radicals. Electrons, being spinning charged bodies, have magnetic moment. Because electrons generally exist in pairs having opposed spins, the net magnetic moment of most molecules is zero. A free radical, however, has an unpaired electron, resulting in a measureable magnetic moment. When these free radicals are placed in a powerful magnetic field, the molecules align themselves with their magnetic axes parallel or antiparallel to the direction of the magnetic field. When irradiated with electromagnetic radiation in the microwave region, some of the free radicals aligned in the parallel, or lower energy state, absorb radiation and are raised to a higher energy, or antiparallel state, their magnetic axes rotating through $180°$ in the process. ESR spectroscopy measures the radiant energy absorbed by these rotating free radicals. The spectral measurements are generally made by varying the field strength, giving one a spectrum of energy absorbed by the sample versus the magnetic field applied.

The ESR spectrum of free electrons would yield only a single invariant value were it not for the fact that the unpaired electron interacts (couples) with those nuclei having a magnetic moment. These nuclei include ^1H, ^{13}C, ^{31}P, ^{15}N, ^{33}S, and ^{17}O. This interaction or coupling occurs only when the electron is in the immediate vicinity of the nucleus, and the degree of coupling varies with each of the isotopes. Therefore, one obtains "hyperfine splitting" with a series of absorption peaks which are more or less characteristic of the environment in which the electron is orbiting. This splitting allows the detection of small shifts in the orbit of the electron.

Until recently, free radicals with their unpaired electrons were considered to exist only as transient intermediates in the course of chemical reactions and the primary usefulness of ESR spectroscopy was the detection of these transient free radicals in the study of reaction mechanisms. Rozantzev and Neiman,[244] however, synthesized a nitroxide free radical that was quite stable and which could be used to form molecules with a variety of reactive groups. A number of workers, primarily McConnell and Griffith and their collaborators, developed a series of these nitroxide free radicals that can be covalently bound to lipid and protein molecules, giving an internal "spin label" to detect conformational changes in the larger molecule.[245,246]

Most studies have used nitroxide molecules with a maleimide, iso-thiocyanate, or iodoacetamide group, which randomly forms covalent bonds with exposed reactive groups of the macromolecules in solution. Landgraf and Inesi[247] used nitroxide-iodoacetamide to label sarcoplasmic reticulum and found evidence of a conformational change in the protein when ATP was added to the membrane vesicles. The inferred conformational changes in these membrane proteins may be related to their ATP-dependent transport of calcium.

Hubbell and McConnell[248] synthesized nitroxide spin-labeled deriva-tives of stearic acid and studied their orientation when they were dissolved in nerve and erythrocyte membranes. It was assumed that the orientation of these molecules within the membranes would reflect local molecular structure and, to some extent, conformational change. The experimental results showed that the preferred orientation of the long hydrocarbon chain was perpen-dicular to the surface of the membrane. Although these techniques have not yet demonstrated conformational changes related to transport or to nerve transmission, this study does indicate that the membranes contain a lipid bilayer structure *in vivo.*

Hsia and Piette[249] synthesized a series of spin-labeled 2,4-dinitro-phenol (DNP) haptens with systematically increasing distances between the nitroxide group and antibody combining site. These nitroxide-haptens were mixed with antibody to DNP and the degree of immobilization of the nitroxide group measured. They estimated that the combining site of the antibody, though perhaps heterogeneous, averaged approximately 10 Å in depth. The sensitivity of this approach for determining active site architec-ture was demonstrated further in studies in which the difference in relative immobilization by the combining site of homologous and cross-reacting antibodies to DNP, o-nitrophenol, p-nitrophenol, and trinitrophenol could be measured.[250]

Although no spin-labeled transport substrates or inhibitors have yet been reported, the recent synthesis[251] of a sulfhydryl analogue of ATP should allow the synthesis of nitroxide-ATP. The use of nitroxide spin labels for probing transport binding sites seems limited only by the synthesis of the appropriate active site-directed nitroxide analogues.

2. Induced Synthesis of Specific Transport Proteins

The genetic regulation of protein synthesis in bacteria is known in considerable detail.[252] It has been shown that the synthesis of enzymes in a given biosynthetic pathway may be repressed by the presence of the end product of the pathway. Conversely, the synthesis of a set of enzymes may be induced by the presence of enzyme substrate. Induction is most common among enzymes catalyzing the catabolic utilization of carbon compounds, but extends also to the synthesis of proteins required for the carrier-mediated transport of substrate. Therefore, by growing bacteria in the appropriate culture medium, the concentration of a given enzyme may be made to vary from vanishingly small amounts to 5% of the total protein of the cell. These techniques are very useful for demonstrating the relationship of particular cellular proteins to the carrier-mediated transport of substrate.[253,254]

Unfortunately for our purposes, the control of protein synthesis is not so readily modulated in vertebrate tissues. The presence of a nuclear membrane, the effects of unrecognized feedback systems, and the more specialized nature of the cells of higher organisms are some of the factors which make the protein synthetic rate in vertebrates less evidently responsive to the environment. Although the synthesis of Na^+-K^+-dependent adenosine-triphosphatase in fish gills is evidently induced by changing the fish from fresh water to sea water, known alterations in vertebrate transport proteins are generally limited to those induced by hormones.[255] Some of these hormonal effects are discussed in Sections V,B,4,c and V,D,2.

3. Covalent Labeling of Transport Active Sites

Techniques have been developed for covalently labeling amino acid residues in the active site of enzyme molecules. After covalently labeling the residues, the protein can be digested by chemical or proteolytic methods and the primary and secondary structure about the active site determined. This approach, which has been so useful in structural studies of enzymes, can also be used to identify and describe the active site in transport proteins.

a. Differential Labeling by Nonspecific Reagents. This labeling technique depends upon the protection by substrate of the reactive amino acid groups in the active site. As developed by Koshland *et al.*[204] the method consists of three steps: (a) the treatment of the protein in the presence of substrate with a specific reagent until all the unprotected groups have reacted; (b) the removal of the substrate and reagent; (c) the treatment of the protein in the absence of substrate with the same reagent into which a radioactive label has been incorporated, thus radioactively labeling the residues in the unprotected active site.

Stein[203] had previously used the same technique (but without radioactive labels) to demonstrate an *N*-terminal histidine residue in the active site of the glycerol transport system in erythrocytes. But perhaps the most elegant and successful use in the transport field of this technique was the identification and labeling of the "M" protein of the lactose permease of *E. coli* by Fox and Kennedy.[206] These workers were studying the lactose permease of *E. coli* regulated by the *y* gene. The transport system, which is very sensitive to inhibition by *N*-ethylmaleimide (NEM) was protected by thiodigalactoside as the cell membranes were treated with NEM. The membranes were washed, and then the unreacted sulfhydryl groups remaining were labeled with radioactive NEM. The membranes were solubilized and the radioactively labeled protein isolated by chromatography. (For further details see Section V,E,1.)

b. Affinity Labeling. Affinity labeling was introduced by Wofsy, Metzger, and Singer[256] as a general method for specifically and irreversibly labeling the active sites of enzymes and of antibodies. With the formation of the covalently bonded label on one of the amino acid residues at the active site, the protein could be digested and the primary and secondary structure in the region of the active site determined. The specificity of the labeling technique allows the labeling of proteins that have not been purified. Moreover the technique is not restricted to catalytic active sites, but may be used for the regulatory active sites of enzymes, for the active sites of antibodies, and inferentially, for active sites of transport proteins.

The technique of affinity labeling takes advantage of the specific, noncovalently bonded reversible complex that is initially formed between an enzyme (protein) and its substrate, which greatly increases the local concentration of the substrate at the active site. For affinity labeling, the substrate is altered chemically so that there is a moiety capable of reacting covalently with one of the amino acid residues in the protein. The requirements for the altered substrate are (a) its configuration must not be altered to the extent that recognition by the enzyme active site is vitiated, and (b) the affinity label reacts covalently with an amino acid side chain under relatively mild conditions. The altered substrate (the affinity label) should then react at a far greater rate with a given amino acid residue in the active site than with the same amino acid in other portions of the protein molecule.

The elementary theory of the method was developed by Wofsy *et al.*[256] Let us consider a protein with a certain amino acid residue (A) in the active site. This residue may react with an appropriate chemical group G. The protein also contains other residues, A', which may be identical with A. S_G is the altered substrate with the reactive group G, AS_G is the specific reversible complex, P the covalently labeled product, and the k's are the appropriate rate constants. The reaction is given by:

$$A + S_G \underset{k_2}{\overset{k_1}{\rightleftharpoons}} AS_G \overset{k_3}{\longrightarrow} P. \qquad (313)$$

The residues A' are competing with A for the affinity label,

$$A' + S_G \overset{k_4}{\longrightarrow} Q \qquad (314)$$

to form an undesired covalently labeled product. When $\dot{P}$, the rate of formation of P, is significantly greater than $\dot{Q}$, the rate of formation of Q, affinity labeling is achieved. The ratio of these two values, $\dot{P}/\dot{Q}$, is a measure of the specificity of the labeling and is given by

$$\dot{P}/\dot{Q} = \frac{k_3 K_{\text{AS}_G}[A]}{k_4[A']} \tag{315}$$

Where [A] is the molar concentration of free A at the active site, [A'] is the concentration of any single unreacted residue outside the active site, and $K_{\text{AS}_G} = k_1/k_2$, the intrinsic association equilibrium constant for the active site and substrate.

Singer's group used this technique to label active sites of antibodies prepared against benzenearsonic acid. The diazonium fluoroborate derivative of the antigen (benzenearsonic acid) was prepared as the affinity label. These workers found good evidence that 0.5 mole of azotyrosine was formed *per* mole of antibody protein, indicating the diazonium group had reacted with a tyrosine residue in the active site. Unmodified antigen protected the specific antibody against the affinity label, and the rate of reaction of the affinity label with normal γ-globulin was rather slow. In similar studies of antibody prepared against 2,4-dinitrophenyl hapten, Metzger et al.[257] also showed that a tyrosine residue reacted covalently with the affinity label.

Schoellman and Shaw[258] independently developed the same general technique to study the active site of chymotrypsin. They synthesized a chloromethylketone derivative of the artificial chymotrypsin substrate N-tosyl-L-phenylalanine. This compound reacted stoichiometrically to form a covalent bond with one histidine residue in the enzyme. Chymotrypsin previously inactivated with diisopropylfluorophosphate did not react with the affinity label and the label did not react with trypsin. Their finding gave strong support to earlier evidence implicating histidine in the active site of chymotrypsin.

More recently Whitney et al.[259] developed a series of chloracetyl sulfonamide derivatives for affinity labeling of carbonic anhydrase. They demonstrated that these compounds reacted with one residue of histidine per mole of carbonic anhydrase with inactivation of the enzyme. These affinity labels may prove useful in studying the sulfonamide-inhibitable chloride transport observed in several tissues. This anion transport has been related to the presence of carbonic anhydrase in the epithelium of these tissues.[260]

The transport of sodium and potassium by epithelial tissues has been related to Na^+-K^+-dependent adenosine triphosphatase which is inhibited by cardiotonic steroids (strophanthidin, ouabain, digitalis). Hokin's group synthesized iodoacetate and bromoacetate derivatives of strophanthidin and demonstrated that these compounds irreversibly inhibited the Na-K-ATPase activity in microsomal membranes prepared from brain. These results encouraged the synthesis of a series of radioactively labeled haloacetate derivatives of strophanthidin and hellebrigenin.[261,262] Hellebrigenin-3-[2-³H]-iodoacetate reacted with partially purified ATPase and yielded a labeled membrane preparation that gave promise of being rather specifically labeled.[201]

It seems clear that the generality of the technique of affinity labeling offers considerable opportunities for the labeling of active sites in transport proteins and that the primary limitation to its use is the synthesis of the altered substrate to serve as the affinity label.

4. Preparation of Antibodies against Transport Proteins

The preparation of a purified protein gives the investigator the opportunity to prepare specific antibodies against that protein. By labeling the antibody with fluorescent or radioactive moieties, the antigen-antibody reaction may be used to establish the presence and subcellular localization of the protein in the tissue. Because the preparation of a specific antibody requires a relatively pure protein antigen, this technique is restricted to those systems where some component has been isolated. Wasserman, Corradino, and Taylor[196] developed an antibody to the calcium-binding protein of chick intestine. This antibody was used in the indirect fluorescent antibody method to demonstrate that the calcium-binding protein was located in PAS-positive goblet cells and over the surface of the villi. Nakane et al.[263] used a similar technique to demonstrate that the leucine-binding protein of E. coli was located in the cell envelope. Whiteside and Salton[264] prepared antibody to the calcium-activated adenosine triphosphatase isolated from membranes of Microcossus lysodeikticus. Antibody inhibition of the phosphatase was complete and was noncompetitive with respect to substrate.

Antibodies to specific transport proteins should also be useful for physiologic studies in isolated membranes such as the intestine and urinary bladder. Inhibition of carrier-mediated transport of a substrate by antibodies to a specific protein would be strong evidence that the protein was intimately involved in the transport of the given substrate.

5. Genetic Control of Transport Proteins

The use of bacteria in studies of transport provides two unique tools for relating transport to a particular protein or "permease". One of these is the facility with which bacteria synthesize vast amounts of inducible proteins under appropriate conditions of growth. The other device is the preparation of bacterial mutants lacking in some specific transport function, which may be related to the lack of synthesis of a specific protein or enzyme. These applications may be seen in the work of Kennedy (M protein), Pardee (sulfate-binding protein), Kaback (HPr protein), and others.

Unfortunately, in adult vertebrate epithelial tissues, the rate of synthesis of any specific protein is an exceedingly small fraction of the rate of synthesis of total protein. Thus it is usually not possible to relate the defect in transport to the deficiency of a specific protein. One exception is the stimulation by vitamin D of the synthesis of calcium-binding protein[196] and calcium-dependent adenosine triphosphatase[265] in the rachitic chick. Although the development of genetic mutants is generally limited to bacteria, studies of a few spontaneous genetic defects in transport systems of vertebrates have yielded valuable information about some renal and intestinal transport systems, especially in regard to amino acids. Studies of amino acid excretion in humans with inherited disorders of transport have generally confirmed data obtained in lower animals indicating the presence of three carrier systems with some overlap in substrate specificities.[266]

The *onset* of synthesis of specific proteins varies considerably during the course of maturation of embryonic vertebrate tissues. This affords an opportunity in some instances to determine whether transport substrates share a common pathway. Rennick et al.[267] used this approach to demonstrate that the development of the renal transport systems for weak acids and weak bases was not coincidental.

Deren et al.[268] observed that accumulation of amino acids in the rabbit yolk sac does not appear until 20–22 days of gestation. The authors showed that the efflux of amino acids did not vary with maturity of the tissue, indicating that the accumulation of amino acid was due to development of a mechanism for the influx of amino acids. The rabbit yolk sac did not develop the facility of accumulating α-methyl glucoside. Butt and Wilson[269] studied the accumulation of proline, glycine, sarcosine, N,N-dimethylglycine, betaine, valine, lysine, and α-methylglucoside by the intestine and yolk sac of the guinea pig fetus. Their data indicate that the intestine of the 40-day fetus actively transports these unsubstituted amino acids and N-methylglycine (sarcosine). At 50 days the tissue develops the capability of accumulating α-methylglucoside, and at 60 days, of dimethyl and trimethylglycine.

A number of studies have demonstrated an inhibitory effect between the transport of sugars and amino acids. The data of Deren et al.[268] and of Butt and Wilson[269] indicate that although there may be indirect interactions between the transport of amino acids and sugars, they do not use the same membrane carrier system.

D. Soluble Proteins

1. *Sulfate-Binding Protein of* Salmonella typhimurium

The enteric pathogen *Salmonella typhimurium* requires sulfate for growth. The sulfate is actively transported into the cell and through a series of enzymatic reductions is converted into cysteine. The transport of sulfate is repressed if cysteine is provided in the culture media. Mutants of *S. typhimurium* have been isolated that are unable to grow on sulfate (or thiosulfate) media but instead require the presence of cysteine. These data are interpreted as evidence that sulfate transport is under the control of an inducible gene, and that the expression of the proper cistron(s) is repressed by the end product of sulfate transport and reduction, i.e., cysteine.[253]

Dreyfuss and Pardee[270] observed that although a mutant strain of *Salmonella* was unable to transport sulfate, the bacterium retained the ability to bind sulfate. This binding capacity was lost when the cell was subjected to osmotic shock, and it was possible to recover from the shock fluid a soluble protein that bound sulfate with kinetics similar to those of the intact cells.[235] At the same time, the shocked cells no longer demonstrated the ability to bind sulfate, indicating that the protein released from the cells by this relatively gentle procedure was responsible for the binding in vivo.[236] Attempts to restore the binding capacity by reshocking the cells in a solution containing the binding protein were unsuccessful. Using the ion-exchange method discussed in Section V,C,I,c Pardee and co-workers found that the capacity for binding was approximately 10^4 molecules per bacterium. The binding showed saturation kinetics with a dissociation constant of 4×10^{-6} M and did not require an energy supply.[236] The binding capacity of the mutant, presumably controlled by the same operon as the transport function in the normal cell, was repressed by growth on cysteine. Purification and crystallization of the soluble protein showed it to be a relatively small (mol. wt. = 32,000) protein with an unremarkable primary amino acid structure and the capacity to bind one sulfate ion per molecule.[271] Subsequent studies of the crystallized protein showed that the binding of sulfate had minimal effects upon the physical properties as measured by a series of optical methods, by nuclear magnetic resonance, and by sedimentation velocity.[272]

All the evidence indicates that the sulfate-binding protein is related to the active transport of sulfate: the genetic mapping, the loss of binding by the bacterium when the protein is released, and the close similarities between the kinetics of the binding by the intact bacterium and by the isolated protein. Unfortunately, these data do not clarify the role of this protein in transport. The characteristics which allow it to be easily isolated, i.e., its release by osmotic shock, and its apparent independence of an energy supply, pose difficulties for assigning its proper role in the active transport of sulfate by the bacterium.

2. Calcium-Binding Protein from Mammalian Tissues

The intestinal epithelium actively transports calcium from the lumen to the plasma and this accumulation of calcium is under the hormonal control of vitamin D.[273,274] Schachter et al.[274] presented evidence that this accumulation occurs in two steps: firstly, the facilitated diffusion of calcium across the mucosal border of the cell, and second, the active transport of intracellular calcium from the cell into the plasma or interstitial fluid. Wasserman's group sought to determine whether the facilitated diffusion process or the active transport step was regulated by vitamin D.[275,276] Using duodenal mucosa, they found that the efflux and influx of calcium was identical both in normal rats and in vitamin D-deficient rats. The flux ratios were unchanged despite a twofold increase in the absolute magnitude of the fluxes in the animals treated with vitamin D. It seemed, therefore, that the hormone was responsible for regulating the facilitated diffusion of calcium across the mucosal surface of the epithelium.

Enhancement of calcium absorption does not occur until several hours following treatment of a deficient animal with vitamin D, and actinomycin D inhibits this effect of vitamin D upon calcium absorption.[277] Wasserman and Taylor[278] found that the supernatant fraction of homogenates of intestinal mucosa from vitamin D-treated rats bound significantly more calcium than those prepared from vitamin D-deficient rats. This increased binding was due to a nondialyzable and heat-labile factor that was inactivated by trypsin and pronase. The proteinaceous nature of this soluble tissue factor was confirmed in subsequent studies. When the protein was isolated and purified it was found to have a molecular weight of 25,000 by equilibrium sedimentation and 28,000 by gel filtration.[237] The affinity constant between the protein and calcium was $2.6 \times 10^5 M^{-1}$.

Antibodies to the purified calcium-binding protein (CaBP) were prepared and these were used to demonstrate that the protein was localized in the surface coat-microvilli region of the mucosal cells.[196] These antibodies were also used to show that the restoration of calcium absorption in the intestine of deficient animals had a close temporal relationship with the synthesis of this specific protein by the intestinal mucosa.[196]

The role of this soluble protein in the transport process is not clarified, but recent experiments by Wasserman[279] show that the properties of the protein may be significantly altered in association with certain lipids. Lysolecithin caused a decrease in the affinity of the CaBP for calcium and a marked change in the electrophoretic mobility of the protein. These alterations in CaBP were reversed by treatment with taurocholate. These experiments suggest that the soluble ion-binding proteins, in association with certain membrane lipids, may act as substrate carriers.

3. Leucine- and Galactose-Binding Proteins

Piperno and Oxender,[280] studying amino acid transport mechanisms in *E. coli*, found that treatment of cells with osmotic shock reduced their ability to actively transport leucine, isoleucine, and valine. These branched amino acids share a common transport system; the transport of other amino acids was not affected. By lyophilizing the fluid in which the cells had been treated, they were able to recover a soluble protein which bound leucine and whose dissociation constant, K_D, was the same as the $K_{S,0.5}$ for leucine transport in the bacteria. The protein was purified and appeared homogeneous by immunodiffusion, ultrafiltration, and acrylamide gel electrophoresis.[281] The binding activity of the purified material was specific for the L-isomers of the branched-chain amino acids. The molecular weight determined by ultracentrifugation (24,000) differed from the value obtained by gel filtration (36,000) and amino acid analysis (34,000).

Anraku[282] independently found a leucine-binding protein that was released from *E. coli* by osmotic shock and his results were in general agreement with those of Piperno and Oxender. Chromatography of the bound leucine recovered from the binding protein showed the protein had not catalyzed any chemical change in the amino acid. The protein bound leucine in a 1:1 molar ratio and the molecular weight, determined by ultracentrifugation, was 36,000.[283] Iodoacetic acid, NEM, zinc ions, and sodium azide (reagents that inhibit transport *in vitro*) did not inactivate the binding activity *in vitro*.

Anraku[284] discovered, however, that the ability of the osmotically shocked cells to transport leucine was partially restored by incubation with the leucine-binding protein. These findings together with kinetic evidence indicating that the number of binding sites on the shocked bacteria, but not their $K_{S,0.5}$, was reduced, gives evidence that the leucine-binding protein serves as a stereospecific binding material necessary for the transport of the branched-chain amino acids. The lack of effect of several transport inhibitors upon the binding activity of the protein suggests that this protein is not involved in the coupling of metabolic energy to the transport processes. Studies in Oxender's laboratory localized this protein in the cell envelope.[263]

Anraku's preparation from the osmotically shocked *E. coli* also contained a galactose-binding protein.[285] The properties of this material are analogous to those of the leucine-binding protein.[282–284] The molecular weight is 35,000, the K_D was equivalent to the transport $K_{S,0.5}$, the binding protein did not catalyze any structural alterations in the sugar, and the binding was selective for either galactose or glucose. Loss of the protein by the cells was associated with a loss in galactose transport and this decrement could be restored by incubating the shocked cells with the protein.

Haškovec and Kotyk[286] found that the membrane fractions prepared from *Saccharomyces cerevesiae* grown on galactose agar bound significant amounts of D-galactose. Attempts to solubilize and purify the binding material from the membranes have not yet been successful.

E. Membrane-Bound Transport Proteins

1. *The Lactose Permease (M Protein)*

Enzymes responsible for the metabolism of the galactoside, lactose, in *E. coli* are synthesized under the regulation of a single operon consisting of three structural genes and a regulator gene.[252] The three structural genes of this operon, designated the *lac* operon, are the *z* gene which controls the synthesis of β-galactosidase, the *a* gene which controls the synthesis of galactoside transacetylase, and the *y* gene which controls the synthesis of the lactose permease or transport protein required for the carrier-mediated accumulation of lactose—both active and passive—by the bacteria.

Extensive analysis of the *lac* operon showed that the *z* and *a* genes each controlled the synthesis of a single enzyme and these enzymes, together with the transport protein(s) controlled by the *y* gene, are inducible. That is, the bacteria synthesize only very small amounts of the proteins controlled by the *lac* operon until the cells are exposed to galactosides, when the bacteria synthesize large amounts of the proteins. Fox and Kennedy[206] designed an experiment to determine whether the transport permease controlled by the *y* gene was a single protein, and to isolate and identify this permease, or transport protein. Because the lactose transport system is extremely sensitive to inhibition by *N*-ethylmaleimide (NEM), these workers used differential

covalent labeling to identify the proteins responsible for galactoside transport (cf. Section V,C,3,a). The sensitivity of the differential labeling was
enhanced by comparing two batches of cells, one induced to synthesize large
amounts of the proteins controlled by the *lac* operon, the other uninduced.
The two batches of cells were treated separately with NEM in the presence
of thiodigalactoside, which has a high affinity for the transport system, to
alkylate the unprotected sulfhydryl groups. The cells, with their lactose
transport systems intact, were washed and treated with radioactive NEM,
the induced cells with ^{14}C-NEM and the uninduced with ^{3}H-NEM. Equivalent amounts of the two batches of cells were mixed, sonified, and fractionated
by differential centrifugation. Because the induced cells contained all the
proteins of the uninduced cells and, in addition, those controlled by the *lac*
operon, the labeled proteins of the *lac* system should have a higher ratio of
^{14}C/^{3}H protein than the other cellular proteins.

The ^{14}C/^{3}H ratio in the particulate fraction was significantly higher than
in the supernatant fraction. Because the other proteins controlled by the *lac*
operon, β-galactosidase and galactoside transacetylase, were in the soluble
fraction, it was evident that the *y* gene regulated the synthesis of a membrane-
bound protein related to lactose transport. This protein was designated as the
M protein.

Similar labeling experiments with sonified *E. coli* demonstrated that the
cell-free membrane fraction retained its affinity for galactoside and its reac-
tivity towards NEM. This property of the membranes permitted the direct
assay cell fractions for the presence of the *M* protein. It was found that the
protein was solubilized by treatment of the membranes with Triton-X.

Subsequently, Kennedy's group demonstrated that the M protein was
absent in membranes prepared from mutant strains of *E. coli* with deletion
mutations in the *y* gene and that significant amounts of the protein were
found in inducible strains only after induction by galactoside.[254] Tempera-
ture-sensitive revertants were isolated whose galactoside transport system
was inhibited by heating to 42°C. It was found that these bacteria synthesized
an M protein that was inactivated by heat and was protected against heat
inactivation by galactosides. These experiments give strong evidence that
the M protein is synthesized by the *y* gene of the *lac* operon. These data also
indicate that the synthesis of membrane-bound transport proteins is not
regulated by a special genetic locus—an operon or operons limited to the
synthesis of membrane proteins—because the proteins synthesized by the
gene on either side of the *y* gene are both soluble enzymes.

2. Phosphotransferase Systems and Sugar Transport

Kundig, Ghosh, and Roseman,[287] studying the synthesis of sialic
acids incorporated into bacterial cell walls, described a novel enzyme
synthesis that catalyzed the transfer of phosphate from phosphoenolpyruvate
(PEP) to a series of sugars. Because previously reported sugar kinase systems
require nucleoside triphosphates as phosphate donors, the PEP requirement

of this reaction made it relatively simple to assay subcellular fractions for the constituent enzymes of the phosphotransferase system. Kundig *et al.*[287] found that in *E. coli* at least three enzymes were necessary for this phosphorylation. These included two soluble proteins, HPr and Enzyme I, and a membrane-bound system, Enzyme II.

The individual reactions catalyzed by these factors are now known to be:

$$\text{PEP} + \text{HPr} \xrightleftharpoons{\text{Enzyme I, Mg}^{2+}} \text{pyruvate} + \text{P-HPr} \tag{316}$$

$$\text{P-HPr} + \text{sugar} \xrightarrow[\text{(factor III)}]{\text{Enzyme II, Mg}^{2+}} \text{sugar-P} + \text{HPr} \tag{317}$$

$$\text{(SUM)} \qquad \text{PEP} + \text{sugar} \xrightarrow[\substack{\text{HPr, Mg}^{2+} \\ \text{Enzyme II}}]{\text{Enzyme I}} \text{sugar-P} + \text{pyruvate} \tag{318}$$

In *Staphylococcus aureus* a fourth protein (factor III) in reaction (317) is also required for the phosphorylation of HPr.

HPr protein has been purified[288] and its molecular weight determined to be approximately 9500. The molecule is heat-stable and contains no carbohydrates, organic phosphates, cysteine, tyrosine, or tryptophan. There are two moles of histidine in the molecule and it is one of these residues that is phosphorylated by PEP. Roseman's group[64] used ^{32}P-PEP to show that the phosphoryl group was linked to N-1 of a histidine imidazole ring. The earlier separation[288] of two HPr proteins was evidently the result of deamination of the native HPr protein during separation.[64]

Enzyme I, also found in the soluble portion of the cell, has been purified approximately 500-fold,[64] and is not heat stable.

Enzyme II activity is found in the membrane fraction of disrupted cells, and this fraction is responsible for the substrate specificity of the entire phosphotransferase system. Kundig and Roseman[289] extracted bacterial membranes with a series of solvents, including urea, butanol, and deoxycholate, and isolated three components from Enzyme II: phosphotidylglycerol and two proteins (II-A and II-B). One of these protein components (II-A) was further fractionated by electrofocusing into three protein bands. The Enzyme II system could be reconstituted by mixing component II-A, component II-B, and phosphotidylglycerol in a specific sequence, and each of the three protein bands of component II-A was specific for catalyzing the phosphorylation of one specific sugar.

Studies of the genetic control of the phosphorylation system have shown that HPr and Enzyme I are constitutive and are not specific for a series of sugars. Tanaka and Lin[290] reported that pleiotropic mutants of *A. aerogenes* and *E. coli* that failed to grow on five carbohydrates lacked Enzyme I or HPr; Simoni *et al.*[291] isolated mutants of *S. typhimurium* that failed to grow on nine sugars and lacked Enzyme I activity. Mutants unable to grow on mannitol alone were shown to lack Enzyme II activity when mannitol was used as substrate[292] and mutants unable to grow on β-glucosides lacked Enzyme II activity for β-glucoside.[293]

Our primary interest in these systems, however, derives from their relation to transport. Kundig et al.[294] found that E. coli subjected to osmotic shock released HPr and Enzyme I into solution and coincidentally lost the capability both of phosphorylating and transporting methyl-β-thio-D-galactopyranoside and α-methylglucoside. Moreover, both the transport and phosphorylating functions of the cells were restored by adding partially purified HPr to the shocked cells. Kaback[295] confirmed these data and showed that the intracellular phosphorylated compound accumulated by the cells was indeed α-methylglucose phosphate. Kaback used an isolated membrane preparation from E. coli to find convincing evidence that the phosphorylation occurred during uptake by the membrane fraction rather than following its entry into the intramembrane sugar pool. Glucose-^{3}H transported from the medium was phosphorylated at a much higher rate than glucose-^{14}C in the intramembrane pool, and there was a stoichiometric loss of PEP and appearance of phosphorylated α-methylglucoside. Thus the transport of glucose and related sugars by bacteria is mediated by the PEP-phosphotransferase system. Although it has been shown that preparations of rat intestinal mucosa catalyze the PEP-dependent phosphorylation of N-acetylglucosamine and fructose, there is no evidence that this activity is related to transport.[296]

3. Sodium-Potassium Dependent Adenosine Triphosphatase

Gardos[191] demonstrated that the free energy required for the active transport of potassium by the erythrocyte was provided by the hydrolysis of ATP. Skou[297] examined the membrane fragments of homogenized crab nerve and found an adenosine triphosphatase activity that required, in addition to Mg^{+2}, both sodium and potassium. Post et al.[298] used erythrocyte ghosts to show that cardiac glycosides had similar effects on both sodium and potassium transport systems and upon the activity of Na^+-K^+-dependent ATPase. Glynn[299] demonstrated that the Na^+-K^+-ATPase of red cell ghosts requires sodium on the internal face of the membrane and potassium on the external. These experiments, and a large amount of subsequent work, have linked the ouabain-inhibitable, Na^+-K^+-dependent ATPase activity of plasma membranes to the active transport of sodium and potassium. Siegel and Albers[300] have recently reviewed the properties of this enzyme in a number of tissues.

This ubiquitous enzyme is the first transport-related enzyme or protein for which a determined attempt at purification has been made. Several laboratories have prepared microsomal membrane fractions with high specific activity and relative homogeneity.[301,302] However, in most cases attempts to solubilize the membrane enzyme have resulted in loss of enzyme activity. Hokin's group, using the detergent Lubrol, was able to solubilize the ATPase from guinea pig brain microsomes.[303] The molecular weight of the active enzyme was estimated by exclusion gel chromatography as approximately 670,000. Further work with enzyme from beef brain showed that the solubilized enzyme could be partially separated from ouabain-insensitive

ATPase by column chromatography.[197] Equilibrium centrifugation of the solubilized enzyme gave a buoyant density of 1.05, suggesting that the enzyme was a lipoprotein; this value may be spuriously low due to the binding of Lubrol by the solubilized enzyme. The active enzyme, whose specific activity was increased ten-fold over that of the microsomal membranes, migrated as a single band on acrylamide gel electrophoresis.[302,304] When this preparation was lipid extracted and dissolved in phenol–acetic acid–urea, there were approximately a dozen protein bands.

Other laboratories, using preparations which appear to be purer than those of Hokin's group, are already attempting to incorporate the ATPase activity into artificial lipid membranes. These have included the use of membrane fragments prepared from rat brain[305] and a solubilized nonparticulate fraction prepared from Streptococcus fecalis.[306] Although the latter ATPase was not Na^+-K^+-dependent, both preparations gave membranes whose conductance was lowered at least two orders of magnitude when ATP was added to the bathing medium. The electrical properties of the microsomal preparation were inhibited by ouabain.

VI. CONCLUSION

In this chapter, we have shown that it is possible to construct kinetic models which can account for the various modes of interaction between substrate molecules and cellular transport systems. Since kinetic data can always be explained by more than one scheme, additional information is needed to determine that particular model which uniquely describes the transport process at the molecular level. Such additional information can be obtained by the following stepwise approach:

1. The individual molecular components of the transport system of interest are isolated, purified, and physiochemically characterized.

2. The transport system is first partially and then totally reconstituted from its component parts.

3. Physiochemical characteristics of the reconstituted system are determined and compared with the original intact transport system. This comparison provides a means for evaluating the nature of the interactions among its component parts. Such data may prove to be related uniquely to a particular molecular model and thereby distinguish this model from others which manifest similar transport kinetics.

Recent work has been aimed primarily at gaining information at the level of the first step, namely, isolation, purification, and physiochemical characterization of molecular components of various transport systems. After more data of this type are adduced, it is not unreasonable to expect to develop insight at the level of the last two steps, namely, reconstitution and determination of the nature of the interaction among the component parts of various transport systems. Thus, we may look forward to the elucidation of the molecular structure of each biological transport system.

ACKNOWLEDGMENTS

This work was supported by U.S. Public Health Service grants AM-10080, AM-13037, and AM-13135 of the National Institute of Arthritis and Metabolic Diseases of the National Institutes of Health, by National Science Foundation grant GB-7764, by the U.S. Atomic Energy Commission, by the Henry C. Kaplan Foundation, and by the Life Sciences Foundation, Inc. The authors are indebted to Norma Linsky, Judith Jacobs, and Rosemyra (Mickey) Dummitt for their conscientious assistance in the editing, proofreading, and typing of this manuscript and to Barbara Panessa and Monroe Yoder for their aid in the preparation of the figures.

VII. REFERENCES

1. W. A. Brodsky, A. E. Shamoo, and I. L. Schwartz, Dissipative transport processes, in *Handbook of Neurochemistry* (A. Lajtha, ed.), Vol. 5, Plenum Press, New York (1971).

2. E. A. Guggenheim, *Thermodynamics, An Advanced Treatment for Chemists and Physicists*, 5th ed. pp. 298–302, North-Holland, Amsterdam (1967).

3. T. Rosenberg, On accumulation and active transport in biological systems I. Thermodynamic considerations, *Acta Chem. Scand.* **2**:14–33 (1948).

4. O. Kedem, Criteria of active transport, in *Membrane Transport and Metabolism* (A. Kleinzeller and A. Kotyk, eds.), pp. 87–93, Academic Press, New York (1961).

5. T. Rosenberg, The concept and definition of active transport, *Symp. Soc. Exp. Biol.* **8**:27–41 (1954).

6. P. J. Garrahan and I. M. Glynn, The incorporation of inorganic phosphate into adenosine triphosphate by reversal of the sodium pump. *J. Physiol. (London)* **192**:237–256 (1967).

7. I. M. Glynn and V. L. Lew, Affinities or apparent affinities of the transport adenosine triphosphatase system, *J. Gen. Physiol.* **54 (pt. 2)**:289s–305s (1969); also in *Membrane Proteins, Proceedings of a Symposium sponsored by the New York Heart Association*, pp. 289–305. Little Brown, Boston (1969).

8. A. Fick, IV. Ueber diffusion, *Poggendorf's Ann. Phys. Chem.* **94**:59–86 (1855).

9. A. Fick, V. On liquid diffusion, *Philosoph. Mag. & J. Science (London, Edinburgh, and Dublin)* **10**:30–39 (1855).

10. P. G. LeFevre and G. F. McGinniss, Tracer exchange *vs* net uptake of glucose through human red cell surface, *J. Gen. Physiol.* **44**:87–103 (1960).

11. T. Rosenberg and W. Wilbrandt, Enzymatic processes in cell membrane penetration, *Internat. Rev. Cytol.* **1**:65–95 (1952).

12. E. Heinz, Kinetic studies on the "influx" of glycine-1-C^{14} into the Ehrlich mouse ascites carcinoma cell, *J. Biol. Chem.* **211**:781–790 (1954).

13. J. F. Danielli, Morphological and molecular aspects of active transport, *Symp. Soc. Exp. Biol.* **8**:502–516 (1954).

14. W. D. Stein, Facilitated diffusion, *Recent Progr. Surface Sci.* **1**:300–337 (1964).

15. W. D. Stein, *The Movement of Molecules across Cell Membranes*, pp. 127–128, Academic Press, New York and London (1967).

16. R. M. Dowben, *General Physiology—A Molecular Approach*, pp. 448–449, Harper & Row, New York (1969).

17. R. Höber, Über Resorption im Dünndarm, *Pflüg. Arch. ges. Physiol.* **74**:246–271 (1899).

18. R. Höber, Correlation between the molecular configuration of organic compounds and their active transfer in living cells, *Cold Spring Harbor Symp. Quant. Biol.* **8**:40–50 (1940).

19. R. Höber, *Physical Chemistry of Cells and Tissues*, 2nd ed., pp. 615–620, Blakiston, Philadelphia and Toronto (1945).

20. F. Verzàr, Probleme und Ergebnisse auf dem Gebiete der Darmresorption, *Ergebn. Physiol.* **32**:391–471 (1931).

21. F. Verzàr, Die Rolle von Diffusion und Schleimhautaktivität bei der Resorption von verschiedenen Zuckern aus dem Darm, *Biochem. Z.* **276**:17–27 (1935).

22. J. A. Shannon and S. Fisher, The renal tubular reabsorption of glucose in the normal dog, *Am. J. Physiol.* **122**:765–774 (1938).

23. J. A. Shannon, The tubular reabsorption of xylose in the normal dog, *Am. J. Physiol.* **122**:775–781 (1938).

24. J. A. Shannon, Renal tubular excretion, *Physiol. Rev.* **19**:63–93 (1939).

25. J. Franck and J. E. Mayer, An osmotic diffusion pump, *Arch. Biochem. Biophys.* **14**:297–313 (1947).

26. J. B. Wittenberg, Oxygen transport—a new function proposed for myoglobin, *Biol. Bull.* (*Woods Hole*) **117** (**abstract**):402–403 (1959).

27. J. B. Wittenberg, The molecular mechanism of hemoglobin-facilitated oxygen diffusion, *J. Biol. Chem.* **241**:104–114 (1966).

28. P. F. Scholander, Oxygen transport through hemoglobin solutions, *Science* **131**:585–590 (1960).

29. F. M. Snell, Facilitated transport of oxygen through solutions of hemoglobin, *J. Theoret. Biol.* **8**:469–479 (1965).

30. J. Wyman, Facilitated diffusion and the possible role of myoglobin as a transport mechanism, *J. Biol. Chem.* **241**:115–121 (1966).

31. P. Mitchell, Active transport and ion accumulation, *in Comprehensive Biochemistry* (M. Florkin and E. H. Stotz, eds.), Vol. 22, pp. 167–197, Elsevier, Amsterdam (1967).

32. P. Mitchell, Translocations through natural membranes, *Adv. Enzymol.* **29**:33–87 (1967).

33. W. J. V. Osterhout and W. M. Stanley, The accumulation of electrolytes. V. Models showing accumulation and a steady state, *J. Gen. Physiol.* **15**:667–689 (1932).

34. W. J. V. Osterhout, Permeability in large plant cells and in models, *Ergebn. Physiol.* **35**:967–1021 (1933).

35. W. J. V. Osterhout, How do electrolytes enter cells? *Proc. Nat. Acad. Sci. U.S.* **21**:125–132 (1935).

36. H. Lundegårdh, Theorie der Ionenaufnahme in lebende Zellen, *Naturwissenschaften* **23**:313–318 (1935).

37. H. Lundegårdh, Investigations as to the absorption and accumulation of inorganic ions, *Ann. Agric. Coll. Sweden* **8**:234–404 (1940).

38. E. Guensberg, Die Glukose Aufnahme in menschliche rote Blut Körperchen. Inaugural-dissertation. Bern, Gerber-Buchdruck, Schwartzenberg (1947).

39. A. L. Hodgkin, The effect of potassium on the surface membrane of an isolated axon, *J. Physiol.* (*London*) **106**: 319–340 (1947).

40. H. H. Ussing, Interpretation of the exchange of radio-sodium in isolated muscle, *Nature* (*London*) **160**:262–263 (1947).

41. H. H. Ussing, Transport of ions across cellular membranes, *Physiol. Rev.* **29**:127–155 (1949).

42. H. H. Ussing, Some aspects of the application of tracers in permeability studies, *Adv. Enzymol.* **13**:21–65 (1952).

43. P. G. LeFevre, Evidence of active transfer of certain non-electrolytes across the human red cell membrane, *J. Gen. Physiol.* **31**:505–527 (1948).

44. P. G. LeFevre, The evidence for active transport of monosaccharides across the red cell membrane, *Symp. Soc. Exp. Biol.* **8**:118–135 (1954).

45. P. G. LeFevre and R. I. Davies, Active transport into the human erythrocyte: evidence from comparative kinetics and competition among monosaccharides, *J. Gen. Physiol.* **34**:515–524 (1951).

46. P. G. LeFevre and M. E. LeFevre, The mechanism of glucose transfer into and out of the human red cell, *J. Gen. Physiol.* **35**:891–906 (1952).

47. W. F. Widdas, Inability of diffusion to account for placental glucose transfer in the sheep and consideration of the kinetics of a possible carrier transfer, *J. Physiol.* (*London*) **118**:23–39 (1952).

48. W. F. Widdas, Comment on Professor Wilbrandt's and Dr. LeFevre's papers, *Symp. Soc. Exp. Biol.* **8**:163–164 (1954).

49. W. F. Widdas, Facilitated transfer of hexoses across the human erythrocyte membrane, *J. Physiol.* (*London*) **125**:163–180 (1954).

50. W. Wilbrandt, Secretion and transport of non-electrolytes, *Symp. Soc. Exp. Biol.* **8**:136–162 (1954).

51. C. S. Patlak, Contributions to the theory of active transport, *Bull. Math. Biophys.* **18**:271–315 (1956).

52. C. S. Patlak, Contributions to the theory of active transport: II. The gate type non-carrier mechanism and generalizations concerning tracer flow, efficiency, and measurement of energy expenditure, *Bull. Math. Biophys.* **19**:209–235 (1957).

53. W. D. Stein, Intra-protein interactions across a fluid membrane as a model for biological transport, *J. Gen. Physiol.* **54 (pt. 2)**:81s–90s (1969); also *in Membrane Proteins, Proceedings of a Symposium sponsored by the New York Heart Association*, pp. 81–90, Little Brown, Boston (1969).

54. J. D. Robertson, New observations on the ultrastructure of the membranes of frog peripheral nerve fibers, *J. Biophys. Biochem. Cytol.* **3**:1043–1047 (1957).

55. J. D. Robertson, Structural alterations in nerve fibers produced by hypotonic and hypertonic solutions, *J. Biophys. Biochem. Cytol.* **4**:349–364 (1958).

56. J. D. Robertson, The ultrastructure of cell membranes and their derivatives, *Biochem. Soc. Symp.* (*Cambridge, England*) **16**:3–43 (1959).

57. J. D. Robertson, The molecular structure and contact relationships of cell membranes, *Progr. Biophys. Biophys. Chem.* **10**:343–418 (1960).

58. J. F. Danielli and H. Davson, A contribution to the theory of permeability of thin films, *J. Cell. Comp. Physiol.* **5**:495–508 (1935).

59. J. F. Danielli, The present position in the field of facilitated diffusion and selective active transport, *in Recent Developments in Cell Physiology, Proceedings of the Seventh Symposium of the Colston Research Society* (J. A. Kitching, ed.), pp. 1–14, Butterworths, London; Academic Press, New York (1954).

60. D. E. Green, An introduction to membrane biochemistry, *Israel J. Med. Sci.* **1**:1187–1200 (1965).

61. D. E. Green and J. F. Perdue, Membranes as expressions of repeating units, *Proc. Nat. Acad. Sci. U.S.* **55**:1295–1302 (1966).

62. G. Vanderkooi and D. E. Green, Biological membrane structure, I. The protein crystal model for membranes, *Proc. Nat. Acad. Sci. U.S.* **66**:615–621 (1970).

63. G. Vanderkooi and M. Sundaralingam, Biological membrane structure, II. A detailed model for the retinal rod outer segment membrane, *Proc. Nat. Acad. Sci. U.S.* **67**:233–238 (1970).

64. S. Roseman, The transport of carbohydrates by a bacterial phosphotransferase system, *J. Gen. Physiol.* **54 (pt. 2)**:138s–184s (1969); also *in Membrane Proteins, Proceedings of a Symposium sponsored by the New York Heart Association*, pp. 138–184, Little Brown, Boston (1969).

65. P. Mitchell and J. Moyle, Group-translocation : a consequence of enzyme-catalyzed group-transfer, *Nature (London)* **182**: 372–373 (1958).
66. P. Mitchell, Molecule, group and electron translocation through natural membranes, *Biochem. Soc. Symp. (Cambridge, England)* **22**:142–169 (1962).
67. B. C. Pressman, Ionophorous antibiotics as models for biological transport, *Fed. Proc.* **27**:1283–1288 (1968).
68. B. C. Pressman, Mechanism of action of transport-mediating antibiotics, *Ann. N.Y. Acad. Sci.* **147**:829–841 (1969).
69. B. C. Pressman, Control of mitochondrial substrate metabolism by regulation of cation transport, *FEBS Symp.* **17**:315–333 (1969).
70. B. C. Pressman and D. H. Haynes, Ionophorous agents as mobile ion carriers, *in The Molecular Basis of Membrane Function* (D. C. Tosteson, ed.), pp. 221–246, Prentice-Hall, Englewood Cliffs, N.J. (1969).
71. S. Ciani, G. Eisenman, and G. Szabo, A theory for the effects of neutral carriers such as the macrotetralide actin antibiotics on the electrical properties of bilayer membranes, *J. Memb. Biol.* **1**:1–36 (1969).
72. G. Eisenman, S. Ciani, and G. Szabo, The effects of the macrotetralide actin antibiotics on the equilibrium extraction of alkali metal salts into organic solvents, *J. Memb. Biol.* **1**:294–345 (1969).
73. G. Szabo, G. Eisenman, and S. Ciani, The effects of the macrotetralide actin antibiotics on the electrical properties of phospholipid bilayer membranes, *J. Memb. Biol.* **1**:346–382 (1969).
74. H. H. Ussing, The distinction by means of tracers between active transport and diffusion. The transfer of iodide across the isolated frog skin, *Acta Physiol. Scand.* **19**:43–56 (1949).
75. A. K. Solomon, The kinetics of biological processes. Special problems connected with the use of tracers, *Adv. Biol. Med. Phys.* **3**:65–97 (1953).
76. C. W. Sheppard, *Basic Principles of the Tracer Method*, John Wiley, New York (1962).
77. T. Rosenberg and W. Wilbrandt, The kinetics of membrane transports involving chemical reactions, *Exp. Cell Res.* **9**:49–67 (1955).
78. W. Wilbrandt, S. Frei and T. Rosenberg, The kinetics of glucose transport through the human red cell membrane, *Exp. Cell. Res.* **11**:59–66 (1956).
79. F. Bowyer, The kinetics of penetration of nonelectrolytes into the mammalian erythrocyte, *Internat. Rev. Cytol.* **6**:469–511 (1957).
80. W. Wilbrandt and T. Rosenberg, The concept of carrier transport and its corollaries in pharmacology, *Pharmacol. Rev.* **13**:109–183 (1961).
81. T. Rosenberg, Membrane transport of sugars. A survey of kinetical and chemical approaches, *Path. Biol.* **9**:795–802 (1961).
82. G. A. Vidaver, Inhibition of parallel flux and augmentation of counter flux shown by transport models not involving a mobile carrier, *J. Theoret. Biol.* **10**:301–306 (1966).
83. H. Lineweaver and D. Burk, The determination of enzyme dissociation constants, *J. Am. Chem. Soc.* **56**:658–666 (1934).
84. C. S. Hanes, Studies on plant amylases. The effect of starch concentration upon the velocity of hydrolysis by the amylase of germinated barley, *Biochem. J. (London)* **26**:1406–1421 (1932).
85. K. Ahmed and P. G. Scholefield, Biochemical studies on 1-aminocyclopentane carboxylic acid, *Canad. J. Biochem. Physiol.* **40**:1101–1110 (1962).
86. M. Dixon, The determination of enzyme inhibitor constants, *Biochem. J. (London)* **55**:170–171 (1953).
87. J. A. Jacquez, The kinetics of carrier-mediated active transport of amino acids, *Proc. Nat. Acad. Sci. U.S.* **47**:153–163 (1961).

88. T. Rosenberg and W. Wilbrandt, Carrier transport uphill. I. General, *J. Theoret. Biol.* **5**:288–305 (1963).

89. D. M. Miller, The kinetics of selective biological transport. I. Determination of transport constants for sugar movements in human erythrocytes, *Biophys. J.* **5**:407–415 (1965).

90. D. M. Miller, The kinetics of selective biological transport. II. Equations for induced uphill transport of sugars in human erythrocytes, *Biophys. J.* **5**:417–423 (1965).

91. T. Rosenberg and W. Wilbrandt, Uphill transport induced by counterflow, *J. Gen. Physiol.* **41**:289–296 (1957).

92. H. G. Britton, Permeability of the human red cell to labelled glucose, *J. Physiol. (London)* **170**:1–20 (1964).

93. D. M. Regen and H. E. Morgan, Studies of the glucose-transport system in the rabbit erythrocyte, *Biochim. Biophys. Acta* **79**:151–166 (1964).

94. M. Levine, D. L. Oxender, and W. D. Stein, The substrate-facilitated transport of the glucose carrier across the human erythrocyte membrane, *Biochim. Biophys. Acta* **109**:151–163 (1965).

95. M. Levine and W. D. Stein, The kinetic parameters of the monosaccharide transfer system of the human erythrocyte, *Biochim. Biophys. Acta* **127**:179–193 (1966).

96. W. D. Stein, *The Movement of Molecules across Cell Membranes*, pp. 152–157 and 162–174, Academic Press, New York and London (1967).

97. H. R. Wyssbrod, Kinetics of a carrier system displaying *trans*-effects, (in preparation).

98. E. Heinz and P. M. Walsh, Exchange diffusion, transport, and intracellular level of amino acids in Ehrlich carcinoma cells, *J. Biol. Chem.* **233**:1488–1493 (1958).

99. R. M. Johnstone and P. G. Scholefield, The influence of amino acids and antimetabolites on glycine retention by Ehrlich ascites carcinoma cells, *Cancer Res.* **19**:1140–1149 (1959).

100. J. A. Jacquez, Transport and exchange diffusion of L-tryptophan in Ehrlich cells, *Am. J. Physiol.* **200**:1063–1068 (1961).

101. A. Lajtha and P. Mela, The brain barrier system—I. The exchange of free amino acids between plasma and brain, *J. Neurochem.* **7**:210–217 (1961).

102. R. M. Johnstone and J. H. Quastel, Effects of lipotropic agents on exchange diffusion in Ehrlich ascites carcinoma cells, *Biochim. Biophys. Acta* **46**:527–532 (1961).

103. R. M. Johnstone and P. G. Scholefield, Factors controlling the uptake and retention of methionine and ethionine by Ehrlich ascites carcinoma cells, *J. Biol. Chem.* **236**:1419–1424 (1961).

104. A. Lajtha and J. Toth, The brain barrier system—V. Stereospecificity of amino acid uptake, exchange and efflux, *J. Neurochem.* **10**:909–920 (1963).

105. D. L. Oxender and H. N. Christensen, Evidence for two types of mediation of neutral amino-acid transport in Ehrlich cells, *Nature (London)* **197**:765–767 (1963).

106. J. A. Jacquez and J. H. Sherman, The effect of metabolic inhibitors on transport and exchange of amino acids in Ehrlich ascites cells, *Biochim. Biophys. Acta* **109**:128–141 (1965).

107. A. Lajtha, Transport as control mechanism of cerebral metabolite levels, *in Brain Barrier Systems (Progress in Brain Research*, Vol. 29), (A. Lajtha and D. H. Ford, eds.), pp. 201–218, Elsevier, Amsterdam (1968).

108. L. Battistin and A. Lajtha, Regional distribution and movement of amino acids in the brain, *J. Neurol. Sci.* **10**:313–322 (1970).

109. R. Blasberg, G. Levi, and A. Lajtha, A comparison of inhibition of steady state, net transport, and exchange fluxes of amino acids in brain slices, *Biochim. Biophys. Acta* **203**:464–483 (1970).

110. D. E. Gentile, A. E. Shamoo, H. R. Wyssbrod, and W. A. Brodsky, Counterflow of sodium across short-circuited acid-killed turtle bladder, *Am. J. Physiol.* **219**:1192–1199 (1970).

111. W. Gross, K. Ring, and E. Heinz, Positive feedback regulation of amino acid transport in *Streptomyces hydrogenans, Arch. Biochem. Biophys.* **137**:253–261 (1970).

112. C. M. Paine and E. Heinz, The structural specificity of the glycine transport system of Ehrlich carcinoma cells, *J. Biol. Chem.* **235**:1080–1085 (1960).

113. K. Ring and E. Heinz, Active amino acid transport in *Streptomyces hydrogenans* I. Kinetics of uptake of α-aminoisobutyric acid, *Biochem. Z.* **344**:446–461 (1966).

114. G. A. Vidaver and S. L. Shepherd, Transport of glycine by hemolyzed and restored pigeon red blood cells, *J. Biol. Chem.* **243**:6140–6150 (1968).

115. M. L. Belkhode and P. G. Scholefield, Interactions between amino acids during transport and exchange diffusion in Novikoff and Ehrlich ascites tumor cells, *Biochim. Biophys. Acta* **173**:290–301 (1969).

116. K. Ring, W. Gross, and E. Heinz, Negative feedback regulation of amino acid transport in *Streptomyces hydrogenans, Arch. Biochem. Biophys.* **137**:243–252 (1970).

117. J. T.-F. Wong, The possible role of polyvalent carriers in cellular transports, *Biochim. Biophys. Acta* **94**:102–113 (1965).

118. W. D. Stein, Spontaneous and enzyme-induced dimer formation and its role in membrane permeability. II. The mechanism of movement of glycerol across the human erythrocyte membrane, *Biochim. Biophys. Acta* **59**:47–65 (1962).

119. W. D. Stein, Spontaneous and enzyme-induced dimer formation and its role in membrane permeability. III. The mechanism of movement of glucose across the human erythrocyte membrane, *Biochim. Biophys. Acta* **59**:66–77 (1962).

120. W. Wilbrandt and A. Kotyk, Transport of sugar mono- and di-complexes in human erythrocytes, *Naunyn-Schmiedebergs Arch. Exp. Path. Pharmak.* **249**:279–287 (1964).

121. H. G. Britton, Fluxes in passive, monovalent and polyvalent carrier systems, *J. Theoret. Biol.* **10**:28–52 (1965).

122. H. R. Wyssbrod, Kinetics of a carrier system displaying cis-stimulation, (in preparation).

123. D. E. Atkinson, J. A. Hathaway, and E. C. Smith, Kinetics of regulatory enzymes. Kinetic order of the yeast diphosphopyridine nucleotide isocitrate dehydrogenase reaction and a model for the reaction, *J. Biol. Chem.* **240**:2682–2690 (1965).

124. W. R. Lieb and W. D. Stein, Quantitative predictions of a noncarrier model for glucose transport across the human red cell membrane, *Biophys. J.* **10**:585–609 (1970).

125. J. Monod, J. Wyman, and J.-P. Changeux, On the nature of allosteric transitions: a plausible model, *J. Mol. Biol.* **12**:88–118 (1965).

126. D. E. Koshland, Jr., G. Némethy, and D. Filmer, Comparison of experimental binding data and theoretical models in proteins containing subunits, *Biochemistry* **5**:365–385 (1966).

127. L. Pauling, The oxygen equilibrium of hemoglobin and its structural interpretation, *Proc. Nat. Acad. Sci. U.S.* **21**:186–191 (1935).

128. J.-P. Changeux, J. Thiéry, Y. Tung, and C. Kittel, On the cooperativity of biological membranes, *Proc. Nat. Acad. Sci. U.S.* **57**:335–341 (1967).

129. J.-P. Changeux and J. Thiéry, On the excitability and cooperativity of biological membranes, *in Regulatory Functions of Biological Membranes* (J. Järnefelt, ed.), BBA Library, Vol. 11, pp. 116–138, Elsevier, Amsterdam (1968).

130. J. A. Jacquez, Carrier-amino acid stoichiometry in amino acid transport in Ehrlich ascites cells, *Biochim. Biophys. Acta* **71**:15–33 (1963).

131. D. L. Oxender and H. N. Christensen, Distinct mediating systems for the transport of neutral amino acids by the Ehrlich cell, *J. Biol. Chem.* **238**:3686–3699 (1963).

132. G. Guroff, G. R. Fanning, and M. A. Chirigos, Stimulation of aromatic amino acid transport by p-fluorophenylalanine in the Sarcoma 37 cell, *J. Cell. Comp. Physiol.* **63**:323–331 (1964).

133. J. A. Jacquez, Competitive stimulation: further evidence for two carriers in the transport of neutral amino acids, *Biochim. Biophys. Acta* **135**:751–755 (1967).

134. J. R. Sachs and L. G. Welt, The concentration dependence of active potassium transport in the human red blood cell, *J. Clin. Invest.* **46**:65–76 (1967).

135. N. Magaña-Schwencke and J. Schwencke, A proline transport system in *Saccharomyces chevalieri, Biochim. Biophys. Acta* **173**:313–323 (1969).

136. B. G. Munck and S. G. Schultz, Interactions between leucine and lysine transport in rabbit ileum, *Biochim. Biophys. Acta* **183**:182–193 (1969).

137. F. Piccoli and A. Lajtha, Some aspects of uptake of non-metabolites in slices of mouse brain, *Biochim. Biophys. Acta* **225**:356–369 (1971).

138. H. R. Wyssbrod, The effect of the electric field upon unidirectional fluxes of ions across membranes, (in preparation).

139. M. Planck, Über die Potentialdifferenz zwischen verdünnten Lösungen binärer Elektrolyte, *Ann. Phys. Chem. N.F.* **40**:561–576 (1890).

140. W. Nernst, Theorie der Reaktionsgeschwindigkeit in heterogenen Systemen, *Z. Physik. Chem.* **47**:52–55 (1904).

141. W. Nernst, Zur Theorie des elektrischen Reizes, *Pflüg. Arch. ges. Physiol.* **122**:275–314 (1908).

142. D. E. Goldman, Potential, impedance, and rectification in membranes, *J. Gen. Physiol.* **27**:37–60 (1943).

143. H. H. Ussing and K. Zerahn, Active transport of sodium as the source of electric current in the short-circuited isolated frog skin, *Acta Physiol. Scand.* **23**:110–127 (1951).

144. F. G. Donnan, Theorie der Membrangleichgewichte und Membranpotentiale bei Vorhandensein von nicht dialysierenden Elektrolyten. Ein Beitrag zur physikalisch-chemischen Physiologie. *Z. Elektrochem.* **17**:572–581 (1911).

145. P. Mitchell, Coupling of phosphorylation to electron and hydrogen transfer by a chemiosmotic type of mechanism, *Nature (London)* **191**:144–148 (1961).

146. A. K. Solomon, The permeability of the human erythrocyte to sodium and potassium, *J. Gen. Physiol* **36**:57–110 (1952).

147. T. I. Shaw, Sodium and potassium movements in red cells, Ph.D. Thesis. Cambridge University, England (1954).

148. P. C. Caldwell, Factors governing movement and distribution of inorganic ions in nerve and muscle, *Physiol. Rev.* **48**:1–64 (1968).

149. H. N. Christensen, T. R. Riggs, and N. E. Ray, Concentrative uptake of amino acids by erythrocytes *in vitro, J. Biol. Chem.* **194**:41–51 (1952).

150. G. A. Vidaver, Transport of glycine by pigeon red cells, *Biochemistry* **3**:662–667 (1964).

151. G. A. Vidaver, Glycine transport by hemolyzed and restored pigeon red cells, *Biochemistry* **3**:795–799 (1964).

152. G. A. Vidaver, Some tests of the hypothesis that the sodium-ion gradient furnishes the energy for glycine-active transport by pigeon red cells, *Biochemistry* **3**:803–808 (1964).

153. H. N. Christensen, T. R. Riggs, H. Fischer, and I. M. Palatine, Amino acid concentration by a free cell neoplasm: Relations among amino acids, *J. Biol. Chem.* **198**:1–15 (1952).

154. H. N. Christensen, T. R. Riggs, H. Fischer, and I. M. Palatine, Intense concentration of α, γ-diaminobutyric acid by cells, *J. Biol. Chem.* **198**:17–22 (1952).

155. T. R. Riggs, L. M. Walker, and H. N. Christensen, Potassium migration and amino acid transport, *J. Biol. Chem.* **233**:1479–1484 (1958).

156. H. Kromphardt, H. Grobecker, K. Ring, and E. Heinz, Über den Einfluss von Alkali-Ionen auf den Glycintransport in Ehrlich-Ascites-Tumorzellen, *Biochim. Biophys. Acta* **74**:549–551 (1963).

157. K. P. Wheeler, Y. Inui, P. F. Hollenberg, E. Eavenson, and H. N. Christensen, Relation of amino acid transport to sodium-ion concentration, *Biochim. Biophys. Acta* **109**:620–622 (1965).

158. T. Z. Czáky and M. Thale, Effect of ionic environment on intestinal sugar transport, *J. Physiol. (London)* **151**:59–65 (1960).

159. T. Z. Czáky and L. Zollicoffer, Ionic effect on intestinal transport of glucose in the rat, *Am. J. Physiol.* **198**:1056–1058 (1960).

160. T. Z. Czáky, H. G. Hartzog III, and G. W. Fernald, Effect of digitalis on active intestinal sugar transport, *Am. J. Physiol.* **200**:459–460 (1961).

161. T. Z. Czáky, Significance of sodium ions in active intestinal transport of nonelectrolytes, *Am. J. Physiol.* **201**:999–1001 (1961).

162. R. K. Crane, D. Miller, and I. Bihler, The restrictions on possible mechanisms of intestinal active transport of sugars, *in Membrane Transport and Metabolism* (A. Kleinzeller and A. Kotyk, eds.), pp. 439–449, Academic Press, New York (1961).

163. I. Bihler and R. K. Crane, Studies on the mechanism of intestinal absorption of sugars. V. The influence of several cations and anions on the active transport of sugars, *in vitro*, by various preparations of hamster small intestine, *Biochim. Biophys. Acta* **59**:78–93 (1962).

164. I. Bihler, K. A. Hawkins, and R. K. Crane, Studies on the mechanism of intestinal absorption of sugars. VI. The specificity and other properties of Na⁺-dependent entrance of sugars into intestinal tissue under anaerobic conditions, *in vitro*, *Biochim. Biophys. Acta* **59**:94–102 (1962).

165. S. G. Schultz and R. Zalusky, Ion transport in isolated rabbit ileum. II. The interaction between active sodium and active sugar transport, *J. Gen. Physiol.* **47**:1043–1059 (1964).

166. S. G. Schultz and R. Zalusky, Ion transport in isolated rabbit ileum. I. Short-circuit current and Na fluxes, *J. Gen. Physiol.* **47**:567–584 (1964).

167. R. K. Crane, Na⁺-dependent transport in the intestine and other animal tissues, *Fed. Proc.* **24**:1000–1006 (1965).

168. R. K. Crane, G. Forstner, and A. Eichholz, Studies on the mechanism of the intestinal absorption of sugars. X. An effect of Na⁺ concentration on the apparent Michaelis constants for intestinal sugar transport, *in vitro*, *Biochim. Biophys. Acta* **109**:467–477 (1965).

169. W. D. Stein, *The Movement of Molecules across Cell Membranes*, pp. 192–206, Academic Press, New York and London (1967).

170. E. E. Crane and R. E. Davies, Chemical energy relations in gastric mucosa, *Biochem. J. (London)* **43**:xlii (1948).

171. E. E. Crane and R. E. Davies, Electric energy relations in gastric mucosa, *Biochem. J. (London)* **43**:xlii–xliii (1948).

172. E. E. Crane and R. E. Davies, Chemical and electrical energy relations for the stomach, *Biochem. J. (London)* **49**:169–175 (1951).

173. E. E. Crane, R. E. Davies, and N. M. Longmuir, Relations between hydrochloric acid secretion and electrical phenomena in frog gastric mucosa, *Biochem. J. (London)* **43**:321–336 (1948).

174. E. E. Crane, R. E. Davies, and N. M. Longmuir, The effect of electrical current on HCl secretion by isolated frog gastric mucosa, *Biochem. J. (London)* **43**:336–342 (1948).

175. R. E. Davies and A. G. Ogston, On the mechanism of secretion of ions by gastric mucosa and by other tissues, *Biochem. J. (London)* **46**:324–333 (1950).

176. R. N. Robertson and M. J. Wilkins, Studies in the metabolism of plant cells, VII. The quantitative relation between salt accumulation and salt respiration, *Austr. J. Sci. Res. Ser. B.* **1**:17–37 (1948).

177. R. N. Robertson and M. Wilkins, Quantitative relation between salt accumulation and salt respiration in plant cells, *Nature (London)* **161**:101 (1948).

178. E. J. Conway and J. G. Brady, Source of hydrogen ions in gastric juice, *Nature (London)* **162**:456–457 (1948).

179. E. J. Conway, The biological performance of osmotic work. A redox pump, *Science* **113**:270–273 (1951).

180. E. J. Conway, *The Biochemistry of Gastric Acid Secretion*, C. C. Thomas, Springfield, Ill. (1952).

181. E. J. Conway, A redox pump for the biological performance of osmotic work, and its relation to the kinetics of free ion diffusion across membranes, *Internat. Rev. Cytol.* **2**:419–445 (1953).

182. E. J. Conway, Some aspects of ion transport through membranes, *Symp. Soc. Exp. Biol.* **8**:297–324 (1954).

183. M. H. Jacobs, The influence of ammonium salts on cell reaction, *J. Gen. Physiol.* **5**:181–188 (1922).

184. M. H. Jacobs, The exchange of material between the erythrocyte and its surroundings, *Harvey Lectures* **22**:146–164 (1927).

185. M. H. Jacobs, Some aspects of cell permeability to weak electrolytes, *Cold Spring Harbor Symp. Quant. Biol.* **8**:30–39 (1940).

186. W. J. V. Osterhout, Is living protoplasm permeable to ions? *J. Gen. Physiol.* **8**:131–146 (1925).

187. J. Orloff and R. W. Berliner, Relationship between urine pH and weak electrolyte excretion in the dog, *Fed. Proc.* **13 (abstract)**:107 (1954).

188. J. Orloff and R. W. Berliner, The mechanism of the excretion of ammonia in the dog, *J. Clin. Invest*, **35**:223–235 (1956).

189. I. L. Schwartz, N. A. Thorn, J. H. Thaysen, and A. R. Feinstein, pH and p-aminohippurate in human sweat, *Fed. Proc.* **14 (abstract)**:135 (1955).

190. I. L. Schwartz, Extrarenal regulation with special reference to the sweat glands, *in Mineral Metabolism, An Advanced Treatise* (C. L. Comar and F. Bronner, eds.), Vol. I, Part A, Chap. 10, pp. 337–386, Academic Press, New York (1960).

191. G. Gardos, Accumulation of K$^+$ ions in human blood cells, *Acta Physiol. Acad. Sci. Hung.* **6**:191–199 (1954).

192. P. G. LeFevre, Sugar transport in the red blood cell: Structure-activity relationships in substrates and antagonists, *Pharmacol. Rev.* **13**:39–70 (1961).

193. R. D. Keynes, The energy source for active transport in nerve and muscle, *in Membrane Transport and Metabolism* (A. Kleinzeller and A. Kotyk, eds.), pp. 131–139, Academic Press, New York (1961).

194. A. Martonosi and R. Feretos, Sarcoplasmic reticulum. II. Correlation between adenosine triphosphatase activity and Ca^{++} uptake, *J. Biol. Chem.* **239**:659–668 (1964).

195. H. N. Christensen, Methods for distinguishing amino acid transport systems of a given cell or tissue, *Fed. Proc.* **25**:850–853 (1966).

196. R. H. Wasserman, R. A. Corradino, and A. N. Taylor, Binding proteins from animals with possible transport function, *J. Gen. Physiol.* **54 (pt. 2)**:114s–134s (1969); also *in Membrane Proteins, Proceedings of a Symposium sponsored by the New York Heart Association*, pp. 114–134, Little, Brown, Boston (1969).

197. S. Uesugi, A. Kahlenberg, F. Medzihradsky, and L. E. Hokin, Studies on the characterization of the sodium-potassium transport adenosine triphosphatase. IV. Properties of a Lubrol-solubilized beef brain microsomal enzyme, *Arch. Biochem. Biophys.* **130**:156–163 (1969).

198. H. N. Christensen and A. B. Hastings, Phosphatides and inorganic salts, *J. Biol. Chem.* **136**:387–398 (1940).

199. A. K. Solomon, F. Lionetti, and P. F. Curran, Possible cation-carrier substances in blood, *Nature (London)* **178**:582–583 (1956).

200. L. E. Hokin and M. R. Hokin, Phosphatidic acid metabolism and active transport of sodium, *Fed. Proc.* **22**:8–18 (1963).

201. L. E. Hokin, On the molecular characterization of the sodium-potassium transport adenosine triphosphatase, *J. Gen. Physiol.* **54 (pt. 2)**:327s–342s (1969); also *in Membrane Protein, Proceedings of a Symposium sponsored by the New York Heart Association*, pp. 327–342 Little, Brown, Boston (1969).

202. F. Bowyer, and W. F. Widdas, The action of inhibitors on the facilitated hexose transfer system in erythrocytes, *J. Physiol. (London)* **141**:219–232 (1958).

203. W. D. Stein, N-terminal histidine at the active centre of a permeability mechanism, *Nature (London)* **181**:1662–1663 (1958).

204. M. E. Koshland, F. Englberger, and D. E. Koshland, Jr., A general method for the labeling of the active site of antibodies and enzymes, *Proc. Nat. Acad. Sci. U.S.* **45**:1470–1475 (1959).

205. H. Bobinski and W. D. Stein, Isolation of a glucose-binding component from human erythrocyte membranes, *Nature (London)* **211**:1366–1368 (1966).

206. C. F. Fox and E. P. Kennedy, Specific labeling and partial purification of the M protein, a component of the β-galactoside transport system of *Escherichia coli, Proc. Nat. Acad. Sci. U.S.* **54**:891–899 (1965).

207. J. VanSteveninck, R. I. Weed, and A. Rothstein, Localization of erythrocyte membrane sulfhydryl groups essential for glucose transport, *J. Gen. Physiol.* **48**:617–632 (1965).

208. W. D. Stein, *The Movement of Molecules across Cell Membranes,* pp. 289–295, Academic Press, New York and London (1967).

209. P. G. LeFevre, K. I. Habich, H. S. Hess, and M. R. Hudson, Phospholipid-sugar complexes in relation to cell membrane monosaccharide transport, *Science* **143**:955–957 (1964).

210. W. Gross and K. Ring, Effect of chloramphenicol on active amino acid transport, *FEBS Letters* **4**:319–322 (1969).

211. L. J. Elsas and L. E. Rosenberg, Inhibition of amino acid transport in rat kidney cortex by puromycin, *Proc. Nat. Acad. Sci. U.S.* **57**:371–378 (1967).

212. I. S. Edelman, R. Bogoroch, and G. A. Porter, On the mechanism of action of aldosterone on sodium transport: The role of protein synthesis, *Proc. Nat. Acad. Sci. U.S.* **50**:1169–1177 (1963).

213. G. A. Porter, R. Bogoroch, and I. S. Edelman, On the mechanism of action of aldosterone on sodium transport: The role of RNA synthesis, *Proc. Nat. Acad. Sci. U.S.* **52**:1326–1333 (1964).

214. D. D. Fanestil and I. S. Edelman, Characteristics of the renal nuclear receptors for aldosterone, *Proc. Nat. Acad. Sci. U.S.* **56**:872–879 (1966).

215. T. S. Herman, G. M. Fimognari, and I. S. Edelman, Studies on renal aldosterone-binding proteins, *J. Biol. Chem.* **243**:3849–3856 (1968).

216. G. M. Fimognari, D. D. Fanestil, and I. S. Edelman, Induction of RNA and protein synthesis in the action of aldosterone in the rat, *Am. J. Physiol.* **213**:954–962 (1967).

217. S. M. Sabesin and K. J. Isselbacher, Protein synthesis inhibition: Mechanism for the production of impaired fat absorption, *Science* **147**:1149–1151 (1965).

218. K. J. Isselbacher, Biochemical aspects of lipid malabsorption, *Fed. Proc.* **26**:1420–1425 (1967).

219. M. Lubin and H. L. Ennis, On the role of intracellular potassium in protein synthesis, *Biochim. Biophys. Acta* **80**:614–631 (1964).

220. M. Lubin, Intracellular potassium and macromolecular synthesis in mammalian cells, *Nature (London)* **213**:451–453 (1967).

221. O. Jardetzky, Simple allosteric model for membrane pumps, *Nature (London)* **211**:969–970 (1966).

222. T. L. Hill, A proposed common allosteric mechanism for active transport, muscle contraction, and ribosomal translocation, *Proc. Nat. Acad. Sci. U.S.* **64**:267–274 (1969).

223. W. D. Stein and J. F. Danielli, Structure and function in red cell permeability, *Disc. Faraday Soc.* **21**:238–251 (1956).

224. M. Burger, L. Hejmová, and A. Kleinzeller, Transport of some mono- and di-saccharides into yeast cells, *Biochem. J. (London)* **71**:233–242 (1959).

225. V. P. Cirillo, The mechanism of sugar transport into the yeast cell, *Trans. N.Y. Acad. Sci.* **23**:725–734 (1961).

226. D. Miller and R. K. Crane, The digestive function of the epithelium of the small intestine. I. An intracellular locus of disaccharide and sugar phosphate ester hydrolysis, *Biochim. Biophys. Acta* **52**:281–293 (1961).

227. R. K. Crane, in Structural and functional organization of an epithelial cell brush border, *in Intracellular Transport* (K. B. Warren, ed.), pp. 97–99, Academic Press, New York (1966).

228. A. M. Ugolev, N. N. Jesuitova, and P. deLaey, Localization of invertase activity in small intestinal cells, *Nature (London)* **203**:879–880 (1964).

229. G. A. Marzluf and R. L. Metzenberg, Studies on the functional significance of the trans-membrane location of invertase in *Neurospora crassa*, *Arch. Biochem. Biophys.* **120**:487–496 (1967).

230. J. T. Edsall and J. Wyman, *Biophysical Chemistry*, Vol. I, pp. 594–595, Academic Press, New York (1958).

231. U. Westphal, Assay and properties of corticosteroid-binding globulin and other steroid-binding serum proteins, *in Methods in Enzymology* (R. B. Clayton, ed.) Vol. 15, pp. 761–796 (1969).

232. C. Scatchard, The attraction of proteins for small molecules and ions, *Ann. N.Y. Acad. Sci.* **51**:660–672 (1949).

233. M. C. Meyer and D. E. Guttman, Novel method for studying protein binding, *J. Pharmaceut. Sci.* **57**:1627–1629 (1968).

234. M. C. Meyer and D. E. Guttman, Dynamic dialysis as a method for studying protein binding II: Evaluation of the method with a number of binding systems, *J. Pharmaceut. Sci.* **59**:39–48 (1970).

235. A. B. Pardee and L. S. Prestidge, Cell-free activity of a sulfate binding site involved in active transport, *Proc. Nat. Acad. Sci. U.S.* **55**:189–191(1966).

236. A. B. Pardee, L. S. Prestidge, M. B. Whipple, and J. Dreyfuss, A binding site for sulfate and its relation to sulfate transport into *Salmonella typhimurium*, *J. Biol. Chem.* **241**:3962–3969 (1966).

237. R. H. Wasserman, R. A. Corradino, and A. N. Taylor, Vitamin D-dependent calcium-binding protein, *J. Biol. Chem.* **243**:3978–3986 (1968).

238. G. H. Lathe and C. R. J. Ruthven, The separation of substances and estimation of their relative molecular sizes by the use of columns of starch in water, *Biochem. J. (London)* **62**:665–674 (1956).

239. J. Porath and P. Flodin, Gel filtration: A method for desalting and group separation, *Nature (London)* **183**:1657–1659 (1959).

240. P. Andrews, Estimation of the molecular weights of proteins by Sephadex gel-filtration, *Biochem. J.* **91**:222–233 (1964).

241. J. P. Hummel and W. J. Dreyer, Measurement of protein-binding phenomena by gel filtration, *Biochim. Biophys. Acta* **63**:530–532 (1962).

242. G. F. Fairclough, Jr., and J.S. Fruton, Peptide-protein interaction as studied by gel filtration, *Biochemistry* **5**:673–683 (1966).

243. G. C. Wood and P. F. Cooper, The application of gel filtration to the study of protein binding of small molecules, *Chromatog. Rev.* **12**:88–107 (1970).

244. E. G. Rozantzev and M. B. Neiman, Organic radical reactions involving no free valence, *Tetrahedron* **20**:131–137 (1964).

245. C. L. Hamilton and H. M. McConnell, Spin labels, *in Structural Chemistry and Molecular Biology* (A. Rich and N. Davidson, eds.), pp. 115–149, W. H. Freeman and Co., San Francisco and London (1968).

246. O. H. Griffith and A. S. Waggoner, Nitroxide free radicals: spin labels for probing bio-molecular structure, *Accts. Chem. Res.* **2**:17–24 (1969).

247. W. C. Landgraf and G. Inesi, ATP dependent conformational change in "spin labelled" sarcoplasmic reticulum, *Arch. Biochem. Biophys.* **130**:111–118 (1969).

248. W. L. Hubbell and H. M. McConnell, Orientation and motion of amphiphilic spin labels in membranes, *Proc. Nat. Acad. Sci. U.S.* **64**:20–27 (1969).

249. J. C. Hsia and L. H. Piette, Spin-labeling as a general method in studying antibody active site, *Arch. Biochem. Biophys.* **129**:296–307 (1969).

250. J. C. Hsia and L. H. Piette, Spin-labeled hapten studies of structure heterogeneity and cross-reactivity of the antibody active site, *Arch. Biochem. Biophys.* **132**:466–469 (1969).

251. A. J. Murphy, J. A. Duke, and L. Stowring, Synthesis of 6-mercapto-9-β-D-ribofuranosyl-purine 5'-triphosphate, a sulfhydryl analog of ATP, *Arch. Biochem. Biophys.* **137**:297–298 (1970).

252. F. Jacob, Genetics of the bacterial cell, *Science* **152**:1470–1478 (1966).

253. J. Dreyfuss, Characterization of a sulfate- and thiosulfate-transporting system in *Salmonella typhimurium*, *J. Biol. Chem.* **239**:2292–2297 (1964).

254. C. F. Fox, J. R. Carter, and E. P. Kennedy, Genetic control of the membrane protein component of the lactose transport system of *Eschericha coli*, *Proc. Nat. Acad. Sci. U.S.* **57**:698–705 (1967).

255. F. H. Epstein, A. I. Katz, and G. E. Pickford, Sodium- and potassium-activated adenosine triphosphatase of gills: role in adaptation of teleosts to salt water, *Science* **156**:1245–1247 (1967).

256. L. Wofsy, H. Metzger, and S. J. Singer, Affinity labeling—a general method for labeling the active sites of antibody and enzyme molecules, *Biochemistry* **1**:1031–1039 (1962).

257. H. Metzger, L. Wofsy, and S. J. Singer, Affinity labeling of the active sites of antibodies to the 2,4-dinitrophenyl hapten, *Biochemistry* **2**:979–988 (1963).

258. G. Schoellmann and E. Shaw, Direct evidence for the presence of histidine in the active center of chymotrypsin, *Biochemistry* **2**:252–255 (1963).

259. P. L. Whitney, G. Fölsch, P. O. Nyman, and B. G. Malmström, Inhibition of human erythrocyte carbonic anhydrase B by chloroacetyl sulfonamides with labeling of the active site, *J. Biol. Chem.* **242**:4206–4211 (1967).

260. W. N. Scott, Y. E. Shamoo, and W. A. Brodsky, Carbonic anhydrase content of turtle urinary bladder mucosal cells, *Biochim. Biophys. Acta* **219**:248–250 (1970).

261. A. E. Ruoho, L. E. Hokin, R. J. Hemingway, and S. M. Kupchan, Hellebrigenin 3-halo-acetates: potent site-directed alkylators of transport adenosine-triphosphatase, *Science* **159**:1354–1355 (1968).

262. A. E. Ruoho, P. A. Meitner, and L. E. Hokin, Studies on characterization of the sodium-potassium transport adenosine-triphosphatase III. Synthesis of strophanthidin 3-[1-14C]-bromoacetate for affinity labeling of the cardiotonic steroid site, *Anal. Biochem.* **28**:119–129 (1969).

263. P. K. Nakane, G. E. Nichoalds, and D. L. Oxender, Cellular localization of leucine-binding protein from *Escherichia coli*, *Science* **161**:182–183 (1968).

264. T. L. Whiteside and M. R. J. Salton, Antibody to adenosine triphosphatase from membranes of *Micrococcus lysodeikticus*, *Biochemistry* **9**:3034–3040 (1970).

265. M. J. Melancon, Jr., and H. F. DeLuca, Vitamin D stimulation of calcium-dependent adenosine triphosphatase in chick intestinal brush borders, *Biochemistry* **9**:1658–1664 (1970).

266. H. Harris, *The Principles of Human Biochemical Genetics*, pp. 177–183, North-Holland, Amsterdam and London (1970).

267. B. Rennick, B. Hamilton, and R. Evans, Development of renal tubular transports of TEA and PAH in the puppy and piglet, *Am. J. Physiol.* **201**:743–746 (1961).

268. J. J. Deren, H. A. Padykula, and T. H. Wilson, Development of structure and function in the mammalian yolk sac. III. The development of amino acid transport by rabbit yolk sac, *Develop. Biol.* **13**:370–384 (1966).

269. J. H. Butt, II, and T. H. Wilson, Development of sugar and amino acid transport by intestine and yolk sac of the guinea pig, *Am. J. Physiol.* **215**:1468–1477 (1968).

270. J. Dreyfuss and A. B. Pardee, Evidence for a sulfate-binding site external to the cell membrane of *Salmonella typhimurium, Biochim. Biophys. Acta* **104**:308–310 (1965).

271. A. B. Pardee, Regulation of active transport, National Cancer Inst. Monograph No. 27: 249–257 (1967).

272. R. Langridge, H. Shinagawa, and A. B. Pardee, Sulfate-binding protein from *Salmonella typhimurium*: physical properties, *Science* **169**:59–61 (1970).

273. D. Schachter and S. M. Rosen, Active transport of Ca^{45} by the small intestine and its dependence on vitamin D, *Am. J. Physiol.* **196**:357–362 (1959).

274. D. Schachter, S. Kowarski, J. D. Finkelstein, and R. W. Ma. Tissue concentration differences during active transport of calcium by intestine, *Am. J. Physiol.* **211**:1131–1136 (1966).

275. R. H. Wasserman and F. A. Kallfelz, Vitamin D_3 and unidirectional calcium fluxes across the rachitic chick duodenum, *Am. J. Physiol.* **203**:221–224 (1962).

276. R. H. Wasserman, A. N. Taylor, and F. A. Kallfelz, Vitamin D and transfer of plasma calcium to intestinal lumen in chicks and rats, *Am. J. Physiol.* **211**:419–423 (1966).

277. R. Eisenstein and M. Passavoy, Actinomycin D inhibits parathyroid hormone and vitamin D activity, *Proc. Soc. Exp. Biol. Med.* **117**:77–79 (1964).

278. R. H. Wasserman and A. N. Taylor, Vitamin D_3-induced calcium-binding protein in chick intestinal mucosa, *Science* **152**:791–793 (1966).

279. R. H. Wasserman, Interaction of vitamin D-dependent calcium binding protein with lysolecithin: Possible relevance to calcium transport, *Biochim. Biophys. Acta* **203**:176–179 (1970).

280. J. R. Piperno and D. L. Oxender, Amino acid-binding protein released from *Escherichia coli* by osmotic shock, *J. Biol. Chem.* **241**:5732–5734 (1966).

281. W. R. Penrose, G. E. Nichoalds, J. R. Piperno, and D. L. Oxender, Purification and properties of a leucine-binding protein from *Escherichia coli, J. Biol. Chem.* **243**:5921–5928 (1968).

282. Y. Anraku, Transport of sugars and amino acids in bacteria. I. Purification and specificity of the galactose- and leucine-binding proteins. *J. Biol. Chem.* **243**:3116–3122 (1968).

283. Y. Anraku, Transport of sugars and amino acids in bacteria. II. Properties of galactose– and leucine-binding proteins, *J. Biol. Chem.* **243**:3123–3127 (1968).

284. Y. Anraku, Transport of sugars and amino acids in bacteria. III. Studies on the restoration of active transport, *J. Biol. Chem.* **243**:3128–3135 (1968).

285. Y. Anraku, The reduction and restoration of galactose transport in osmotically shocked cells of *Escherichia coli, J. Biol. Chem.* **242**:793–800 (1967).

286. C. Haškovec and A. Kotyk, Attempts at purifying the galactose carrier from galactose-induced baker's yeast, *Eur. J. Biochem.* **9**:343–347 (1969).

287. W. Kundig, S. Ghosh, and S. Roseman, Phosphate bound to histidine in a protein as an intermediate in a novel phospho-transferase system, *Proc. Nat. Acad. Sci. U.S.* **52**:1067–1074 (1964).

288. B. Anderson, W. Kundig, R. Simoni, and S. Roseman, Further studies of carbohydrate permeases, *Fed. Proc.* **27 (abstract)**:643 (1968).

289. W. Kundig and S. Roseman, Further studies on bacterial permeases, *Fed. Proc.* **28 (abstract)**:463 (1969).

290. S. Tanaka and E. C. C. Lin, Two classes of pleiotropic mutants of *Aerobacter aerogenes* lacking components of a phosphoenolpyruvate-dependent phosphotransferase system, *Proc. Nat. Acad. Sci. U.S.* **57**:913–919 (1967).

291. R. D. Simoni, M. Levinthal, F. D. Kundig, W. Kundig, B. Anderson, P. E. Hartman, and S. Roseman, Genetic evidence for the role of a bacterial phosphotransferase system in sugar transport, *Proc. Nat. Acad. Sci. U.S.* **58**:1963–1970 (1967).

292. S. Tanaka, S. A. Lerner, and E. C. C. Lin, Replacement of a phosphoenolpyruvate-dependent phosphotransferase by a nicotinamide adenine dinucleotide-linked dehydrogenase for the utilization of mannitol, *J. Bacteriol.* **93**:642–648 (1967).

293. C. F. Fox and G. Wilson, The role of a phosphoenolpyruvate-dependent kinase system in β-glucoside catabolism in *Escherichia coli*, *Proc. Nat. Acad. Sci. U.S.* **59**:988–995 (1968).

294. W. Kundig, F. D. Kundig, B. Anderson, and S. Roseman, Restoration of active transport of glycosides in *Escherichia coli* by a component of a phosphotransferase system, *J. Biol. Chem.* **241**:3243–3246 (1966).

295. H. R. Kaback, The role of the phosphoenolpyruvate-phosphotransferase system in the transport of sugars by isolated membrane preparations of *Escherichia coli*, *J. Biol. Chem.* **243**:3711–3724 (1968).

296. M. M. Weiser and K. Isselbacher, Phosphoenolpyruvate-activated phosphorylation of sugars by intestinal mucosa, *Biochim. Biophys. Acta* **208**:349–359 (1970).

297. J. C. Skou, The influence of some cations on an adenosine triphosphatase from peripheral nerves, *Biochim. Biophys. Acta* **23**:394–401 (1957).

298. R. L. Post, C. R. Merritt, C. R. Kinsolving, and C. D. Albright, Membrane adenosine triphosphatase as a participant in the active transport of sodium and potassium in the human erythrocyte, *J. Biol. Chem.* **235**:1796–1802 (1960).

299. I. M. Glynn, Activation of adenosinetriphosphatase activity in a cell membrane by external potassium and internal sodium, *J. Physiol. (London)* **160**:18P–19P (1961).

300. G. J. Siegel and R. W. Albers, Nucleoside triphosphate phosphohydrolases, *in Handbook of Neurochemistry* (A. Lajtha, ed.), Vol. 4, pp. 13–44, Plenum, New York (1971).

301. W. Schoner, C. von Ilberg, R. Kramer, and W. Seubert, On the mechanism of Na^+- and K^+-stimulated hydrolysis of adenosine triphosphate. I. Purification and properties of a Na^+- and K^+-activated ATPase from ox brain, *Eur. J. Biochem.* **1**:334–343 (1967).

302. D. W. Towle and J. H. Copenhaver, Jr., Partial purification of a soluble $(Na^+ + K^+)$-dependent ATPase from rabbit kidney, *Biochim. Biophys. Acta* **203**:124–132 (1970).

303. F. Medzihradsky, M. H. Kline, and L. E. Hokin, Studies on the characterization of the sodium-potassium transport adenosinetriphosphatase. I. Solubilization, stabilization, and estimation of apparent molecular weight, *Arch. Biochem. Biophys.* **121**:311–316 (1967).

304. A. Kahlenberg, N. C. Dulak, J. F. Dixon, P. R. Galsworthy, and L. E. Hokin, Studies on the characterization of the sodium-potassium transport adenosine-triphosphatase. V. Partial purification of the Lubrol-solubilized beef brain enzyme, *Arch. Biochem. Biophys.* **131**:253–262 (1969).

305. M. K. Jain, A. Strickholm, and E. H. Cordes, Reconstitution of an ATP-mediated active transport system across black lipid membranes, *Nature (London)* **222**:871–872 (1969).

306. W. R. Redwood, H. Müldner, and T. E. Thompson, Interaction of a bacterial adenosine triphosphatase with phospholipid bilayers, *Proc. Nat. Acad. Sci. U.S.* **64**:989–996 (1969).

SUBJECT INDEX

This is a joint index for Volume V, Parts A and B. Pages 1-438 will be found in Part A and pages 439-837 in Part B.